Long Zhang, Changsheng Gong, Bin Dai (Eds.)
Green Chemistry and Technologies

Also of Interest

Ecosystem Science and Applications
Sustainable Biofuels
An Ecological Assessment of the Future Energy
Bhardwaj, Zenone, Chen (Eds.), 2017
ISBN 978-3-11-027584-1, e-ISBN 978-3-11-027589-6
ISSN 2196-6737

Green Chemical Processing
Volume 1
Sustainable Green Chemistry
Benvenuto (Ed.), 2017
ISBN 978-3-11-044189-5, e-ISBN 978-3-11-043585-6
ISSN 2366-2115

Green Chemical Processing
Volume 2
Green Chemical Processes
Developments in Research and Education
Benvenuto (Ed.), 2017
ISBN 978-3-11-044487-2, e-ISBN 978-3-11-044592-3
ISSN 2366-2115

GREEN - Alternative Energy Resources
Volume 1
Pyrolysis of Biomass
Wang, Luo, 2016
ISBN 978-3-11-037457-5, e-ISBN 978-3-11-036966-3
ISSN 2509-7237

Green Processing and Synthesis
Hessel, Volker (Editor-in-Chief)
ISSN 2191-9542, e-ISSN 2191-9550

Green Chemistry and Technologies

Edited by
Long Zhang, Changsheng Gong, Bin Dai

DE GRUYTER

华中科技大学出版社
http://www.hustp.com

Editors

Prof. Long Zhang
Changchun University of Technology
Changchun
China
Zhanglongzhl@163.com

Prof. Changsheng Gong
Wuhan Institute of Technology
Wuhan
China
gcs4412@163.com

Prof. Bin Dai
College of Chemistry and Chemical Engineering
Shihezi University
ShiheziChina
dbinly@126.com

ISBN 978-3-11-047861-7
e-ISBN (PDF) 978-3-11-047931-7
e-ISBN (EPUB) 978-3-11-047978-2

Library of Congress Cataloging-in-Publication Data
Names: Zhang, Long (PhD of engineering), editor.
Title: Green chemistry and technologies / edited by Long Zhang, Changsheng
  Gong, Dai Bin.
Description: Berlin ; Boston : De Gruyter, [2018] | Series: De Gruyter
  textbook
Identifiers: LCCN 2018007860| ISBN 9783110478617 (softcover) | ISBN
  9783110479317 (pdf) | ISBN 9783110479782 (epub)
Subjects: LCSH: Green chemistry.
Classification: LCC TP155.2.E58 G73 2018 | DDC 660.028/6--dc23 LC record available at https://lccn.
loc.gov/2018007860

Bibliographic information published by the Deutsche Nationalbibliothek
The Deutsche Nationalbibliothek lists this publication in the Deutsche Nationalbibliografie;
detailed bibliographic data are available on the Internet at http://dnb.dnb.de.

© 2018 Huazhong University of Science & Technology Press, Wuhan and Walter de Gruyter GmbH,
Berlin/Boston
Typesetting: Integra Software Services Pvt. Ltd.
Printing and binding: CPI books GmbH, Leck
Cover image: v alex / iStock / Getty Images Plus

www.degruyter.com

About chief editors

Zhang Long is second-grade professor and PhD advisor in Changchun University of Technology, dean of School of Chemical Engineering, and director of Jilin Provincial Engineering Laboratory for the Complex Utilization of Petroresources and Biomass. Jilin provincial outstanding achievements expert. His research interests include synthesis of advanced energy materials and novel catalysts with green chemical technologies, and the design and development of green transformation processes of biomass and petrochemical products. Up to now, he has more than 40 science citation index and engineering indexed publications, 27 issued patents and 6 books and 6 textbooks, and 5 scientific and technologic awards granted by China or Jilin Province.

Gong Changsheng graduated from the Department of Chemistry of Wuhan University in July 1969. Since February 1972, he has served as a teaching and research work at Wuhan Institute of Technology. The key areas of scientific research are phosphorus chemistry, chemical engineering, and green chemical engineering. He has published more than 100 scientific papers, 10 academic monographs, and 4 teaching materials. Professor Gong Changsheng was appointed member of the expert committee of the China Phosphate Industry Association, member of the expert committee of China Inorganic Salt Industry Association, and member of the National Committee of experts on phosphorus chemical industry.

Bin Dai obtained his PhD in applied chemistry from Dalian University of Technology (DUT) in 2002. He is director of Key Laboratory for Green Processing of Chemical Engineering of Xinjiang Bingtuan and a professor at Shihezi University. Professor Dai has over 20 years of experience in advanced functional materials, working for government, industry, and university. His research interests include porous micro-nanostructured materials for powerful energy storage, plasma catalysis, catalyst for acetylene hydrochlorination, and heterocyclic carbene catalysis.

https://doi.org/10.1515/9783110479317-201

Contributing Authors

Gongxiao Zhang
School of Chemistry Pharmaceutical Engineering
Taishan Medical University
Changcheng Road 619
Taian Shandong, China
e-mail: gxzhang@tsmc.edu.cn
Chapter 13

Jianxin Yang
College of Materials and Chemical Engineering
Hainan University
No. 58, Renmin Avenue, Haikou, Hainan
Province, 570228
E-mail: yangjxmail@ sohu.com
Chapter 2, 6

Zhong-ming Li
Green Chemical Technology, Chemical
Engineering and Technology
School of Chemical and Environmental
Engineering
Jianghan University
8 Sanjiaohu Road, Wuhan city, Hubei
Province
P. R. China
e-mail: lizhongm@jhun.edu.cn
Chapter 7

Zaifeng Li
College of Chemistry and Molecular
Engineering,
State Key Laboratory Base of Eco-chemical
Engineering
Qingdao University of Science and Technology
53 Zhengzhou Road
Qingdao266042, China
e-mail: lizfengphd@126.com
Chapter 5

Xu Jun
School of Chemical Engineering and Energy
University of Zhengzhou
Zhengzhou, P. R. China
e-mail: jxuzzu@aliyun.com
Chapter 2

Guang-ming Du
College of Chemical Engineering
Xinjiang Agricultural University
Urumqi, China
e-mail: 375900653@qq.com.cn
Chapter 9

Xin Liu
Resource Recycling Science and Engineering
College of Chemical Engineering
Changchun University of Technology
No. 2055 Yanan Avenue, Changchun City, Jilin
Province
P. R. China
e-mail: liuxin83@ccut.edu.cn
Chapter 9

Xueqiong Yin
Hainan Provincial Fine Chemical Engineering
Research Center
College of Materials and Chemical Engineering
58th Renmin Avenue, Hainan University
Haikou, Hainan
China
e-mail: yxq@hainu.edu.cn
Chapter 10

Mei Jin
Chemical Engineering and Technology, Catalysis
School of Chemical and Environmental
Engineering
Jianghan University
8 Sanjiaohu Road, Wuhan city, Hubei Province
P. R. China
e-mail: jinmei@jhun.edu.cn
Chapter 7

Xiaowei Ma
School of Chemistry and Chemical Engineering
of Shihezi University
Shihezi China
e-mail:mxw_tea@shzu.edu.cn
Chapter 11, 12

https://doi.org/10.1515/9783110479317-202

Preface

Green chemistry and technologies is the forefront of international chemistry and chemical engineering development. It combines the latest achievements of modern chemistry, chemical engineering, physics, biology, material science, environmental science and information technology, and a new interdiscipline with defined scientific destination and essential social demands.

In a recent decade, green chemistry and technology has gained more and more attention by governments, academics, and industries. The direct participation of governments, close cooperation of industries with research institutions, and international technological exchange and cooperation boost the development of green chemistry and technology. As the research on green chemistry is comprehensive, its key role in the sustainable development of the world is clearly revealed.

Therefore, elucidation and summary of the latest progress and applications of green chemical technologies is an essential task, especially, employing these new achievements to solve the problems of the efficient, clean, and high-quality utilization of resources, and the clean production of chemicals and materials has been a new task undertaken by green chemistry and technology in view of sustainable development.

This book comprises 14 chapters. Chapter 1 describes the origins, background of development, and the nature of green chemistry and technology. Chapter 2 elucidates the 12 principles of green chemistry and green chemical process. Chapters 3 and 4 summarize concisely the principles and advanced technologies in green inorganic synthesis and green organic synthesis, respectively. Chapter 5 describes the green synthetic methods and technologies of polymers. Chapter 6 describes mainly the green synthetic processes in medical industry, agricultural drugs, and industrial assistants. Chapter 7 describes the environment-friendly chemical and the concept of green engineering. Chapter 8 summarizes the green processes for the complex utilization of carbon dioxide. Chapter 9 describes the green chemistry and chemical processes for the biomass transformations. Chapter 10 describes the green chemistry and chemical processes for the ocean resources transformations. Chapter 11 elucidates mainly the evaluation principles and index used in the evaluation of a green chemical process and application procedures. Chapter 12 describes the clean coal combustion technologies and renewable energies. Chapter 13 summarizes the chemical process intensification technologies and their utilizations in clean chemical processes. Chapter 14 introduces the essential concept and models of recycling economy and ecological industrial parks. In a word, this book is focused on the extended utilizations of the principles of green chemistry and chemical technologies in new fields of clean synthesis and resources utilization. It characterizes the combination of theory with practice and creative development.

https://doi.org/10.1515/9783110479317-203

This book's chief editors are Zhang Long, Dai Bin, and Gong Changsheng and co-chief editors are Zhang Gongxiao, Yang jiangxin, Li Zhongming, and Li Zaifeng. The division of work is as follows: Chapter 1 (Gong Changsheng, Wuhan University of Engineering; Yang jiangxin, Hainan University), Chapter 2 (XuJun, Zhengzhou University), Chapters 3 and 4 (Zhang Long, Changchun University of Technology), Chapter 5 (Li Zaifeng, Qingdao University of Science and Technology), Chapter 6 (Yang jiangxin, Hainan University, Chapter 7 (Li Zhongming, Mei Jin, Jianghan University, Chapter 8 (Zhang Long, Changchun University of Technology, Chapter 9 (Liu Xin, Changchun University of Technology and Du Guangming, Xinjiang agricultural University), Chapter 10 (Yin Xueqiong, Hainan University, Chapters 11, 12, and14 (Dai Bin, Ma Xiaowei Shihezi University, and Chapter 13 (Zhang Gongxiao, Taishan Medical College. Zhang Long made the whole revision and modification of the manuscript.

We sincerely thank the editors from Huazhong Science and Technology University Press and Gemany De Gruyter Press for their excellent work to make this book publishable.

As green chemistry and technologies is a newly launched and multidiscipline research area with various sciences and technology understandings and the limits of the authors academic level, there should be some mistakes in the book. We sincerely welcome all suggestions and comments from the readers! And much appreciation is given for all the authors from China and all over the world, whose publications have been cited in this book!

Compliers
AUG.28, 2017
CHANGCHUN

Contents

1 Introduction

Green chemistry, also called sustainable chemistry, is an area of chemistry and chemical engineering that focuses on the designing of products and processes that minimize the use and generation of hazardous substances [1]. Environmental chemistry focuses on the effects of polluting chemicals on nature, whereas green chemistry focuses on technological approaches to preventing pollution and reducing consumption of nonrenewable resources [2–4]. Green chemistry overlaps with all subdisciplines of chemistry but with a particular focus on chemical synthesis, chemical process, and chemical engineering in industrial applications. To a lesser extent, the principles of green chemistry also affect laboratory practices. The overarching goals of green chemistry – namely, more resource-efficient and inherently safer design of molecules, materials, products, and processes – can be pursued in a wide range of contexts.

Green chemistry originated in the 1990s, targeting explicit social needs and scientific goals, and has now become the international frontier of chemistry research [1]. It is also one of the most significant research areas of chemistry and chemical engineering in the twenty-first century. The core of green chemistry is to apply chemical theories and novel chemical technologies, guided by the principle of atom economy, to reduce or eliminate pollution from the source, to meet the needs of sustainable development to a maximum extent in terms of rational utilization of resources, ecological balance, environmental protection, and so on, and moreover to realize the coordination and harmony between human being and nature. Therefore, green chemistry and its corresponding applicable technologies have become key research focuses and are pursued by numerous governments, academia, and industries.

1.1 Rise and Development of Green Chemistry

1.1.1 The Crisis of the Ecological Environment Calling for Green Chemistry

The ecological environment of human being is deteriorating due to the rapid growth of world population, acceleration of industrialization processes, excessive consumption of nonrenewable resources and energy, emissions of industrial and agricultural pollutants and domestic wastes, and so on. The main manifestation includes atmosphere pollution, acid rain, global warming, destruction of ozone layer, scarcity and pollution of freshwater resources, marine pollution, degradation and desertification of land resources, forest decline, reduction of biodiversity, and pollution of solid waste.

The essence of contemporary global issues is the crisis of human survival, since the natural environment of humans is being destroyed. In other words, the

https://doi.org/10.1515/9783110479317-001

contradiction between humans and nature is intensifying. Green stands for life and is also regarded as a symbol of harmony between humans and nature. Therefore, green chemistry is an inevitable choice for human survival and the development of social sustainability.

1.1.2 Promotion of Green Chemistry by Propagation of Environment Protection and Regulations

We have only one Earth. To protect our homeland, to strengthen pollution management, and to protect the ecological environment have become the common aspirations and the concerned focuses of people all over the world. The promulgation of laws and regulations pertains to environmental protection and also promotes the rise and development of green chemistry.

In 1962, R. Carson, a woman marine biologist, published a book *Silent Spring*, which describes in detail the negative effects of dichlorodiphenyltrichloroethane (DDT) and other pesticides to various birds. The pesticides, for example DDT, not only reduce the population of bald eagles sharply via food chain but also endanger other bird species. Therefore, the previous beautiful and rigorous spring turns into "silence." In addition, these chemicals enter human bodies through skin contact and enter digestive tract thereby poisoning the victims. With the circulation of earth's atmosphere, they can be brought to any corner of the world. Even Arctic seals and Antarctic penguins were detected to have DDT in their bodies. This book greatly arouses the attention for protecting the ecological environment and is considered as a masterpiece of warning.

In 1972, the United Nations Conference on Human Environment was held, and then the declaration of the environment was issued.

In 1987, the United Nations Committee of Environment and Development released a long report, titled as "Our Common Future."

In 1990, the US Congress passed the act of pollution prevention, and set it as a national policy aiming to prevent the pollution at the origin.

In 1991, the American Chemical Society (ACS) and the Environmental Protection Agency (EPA) initiated the program of green chemistry. The purpose is to promote and develop novel or improved chemical products and technological processes that are benign to human health and ecological environment.

In June 1992, the United Nations Conference on Environment and Development was held in Rio de Janeiro, Brazil. Rio declaration and Agenda 21 were signed.

In 1994, the Chinese government published a white paper of "China's agenda 21," relying on the strategies of prospering the nation with science and education and sustainable development, and declared to follow the harmonious development of economy and society, and meanwhile place the clean production as a priority in key areas.

1.1.3 The Development of Chemical Industry Declaring Green Chemistry

Chemistry, as a creative subject, has acquired brilliant achievements from its birth to date. Chemistry and chemical industry have provided abundant chemical products for daily life, and so far more than 6 million compounds have been synthesized, including more than 50,000 industrially produced products. Recently, the global annual output of chemical products exceeds 1.5 trillion dollars. China manufactures over 40,000 chemicals, and the output of petroleum and chemicals in 2001 is 1.099 trillion yuan RMB, accounting for 9.8% of the national industrial output. These chemical products create enormous material wealth and enrich the material life for human beings, and moreover promote the social civilization and progresses. Therefore, the chemical industry is playing an extremely important role in national economy, and is becoming the fundamental and pillar of national economy.

However, harmful substances originated from the production and usage of a huge amount of chemicals result in serious pollution toward the ecological environment, and the contemporary challenges of global ecological environment issues are directly or indirectly related to chemical pollution. Table 1.1 lists eight typically global pollution incidents since the 1930s.

It is necessary to point out that the industrial development experience of western developed countries should be introspected and taken into consideration. The extensive model of "treatment after pollution" causes the waste of natural resources and energy, leads to excessive investment without resolving virtual issues, and might even trigger the risk of secondary pollution. Therefore, the development of traditional chemical industry continuously seeks the solutions of reduction and elimination of chemical wastes to the environment. Green chemistry and its applicable technologies are just the effective ways to solve the environmental pollution by chemical industry. The core and goal of green chemistry is to prevent the pollution from its origin and realize "zero emission" by implementation of clean production technologies [5, 6].

1.1.4 Sustainable Development Promoting the Green Chemistry

Sustainable development is a new scientific concept of development which has been developed since the late 1980s. With the development of science and technology and the great improvement of social productivity, humans have created unprecedented material wealth and accelerated the progress of social civilization. Meanwhile, the world's rapid growing population, excessive consumption of the exhausted resources

Table 1.1: The world's eight major pollution incidents in the twentieth century.

Events	Contaminants	Time and location	Causes and symptoms	Causes
Mas vale smog	SO_2, Smoke, and dust	Dec 1930 Maas River Valley, Belgium	$SO_2 \rightarrow SO_3 \rightarrow$ chest pain, pharyngalgia, difficulty breathing, cough, tears	Pollutants accumulated in industrial areas under foggy weather.
Donora smog		October 1948, Donora, USA	SO_2+ smoke$\rightarrow$ sulfuric acid $\rightarrow$ eye pain, cough, chest tightness, sore throat, vomiting	
London smog		December 1952. London, UK		
Los Angeles photochemical smog	Photochemical smog	May to December 1955 Los Angeles, USA	The petroleum industry, automobile exhaustion/ultraviolet rays, eye and throat inflammation	Oxynitride, hydrocarbon $\rightarrow$ atmosphere
Minamata event	Methylmercury	From 1953 to 1979, Kyushu, Japan	Fish feeds on methylmercury, people feed on fish $\rightarrow$ disorder	Mercury-based catalysts from chemical companies
Yokkaichi	SO_2, coal dust ...	1955–1972 Yokkaichi, Japan	Heavy metal particles, $SO_2 \rightarrow$ lungs, bronchial asthma	Co/Mn/Ti dust, SO_2
Rice bran oil event	PCBs	1968, 23 counties including Aichi-ken, Kyushu, Japan	Consumption of rice bran oil containing PCBs, body roseola, Vomiting, nausea, muscle pain	Production of rice bran oil polluted by PCBs
Toyama bone pain disease	Cadmium	From 1955 to 1965 Toyama, Japan	Feed on rice and water contaminated by cadmium; joint, nerve, and whole body bone pain, skeletal atrophy	Cadmium-containing wastewater from zinc refined plant

PCBs, polychlorinated biphenyls.

and energy, deterioration of ecological environment by industrial, agricultural, and domestic wastes not only seriously hinder the development of national economy and the improvement of people's quality of life, but also threaten human survival and development.

Facing these harsh challenges, humans have to reexamine their own social and economic behavior as well as the development of industrialization, to realize the fact that the sacrifice of vast amounts of resources and energy to pursue the "high input, high consumption, and high pollution" traditional economic growth mode cannot meet the requirements of current and future developments. Alternatively, seeking a new and mutually coordinated "resources-economy-environment-society" mode can satisfy the needs of modern people without affecting the benefits of future generations. The sustainable development theory was established under such historical background.

In April 1987, the United Nations Environment Programme published a long report "Our Common Future" after four years of scientific research, in which the sustainable development was set as a basic guideline and was defined as a new development mode by both meeting the needs of current and future human generations.

In June 1992, the United Nations Conference on Environment and Development was held in Rio de Janeiro, Brazil. More than 180 national and regional representatives, 102 government leaders, and more than 60 international organizations attended the conference and cosigned five documents, including Rio Declaration and 21st Century Agenda. A new strategy of global development, namely sustainable development strategy, was formally founded based on the consensus of reasonable resources usage, protection of ecological environment, and sustainable development of social economy.

Sustainable development strategy consists of economic sustainability, ecological sustainability, and social sustainability. Economic sustainable development should be emphasized on both quantity and quality of economic growth, conversion of traditional production, and consumption modes into the modes of clearer production and civilized consumption, in which every basic unit should strictly comply with the principles of sustainable development, from the utilization of resources and energy, products design, and production technologies to consumption patterns. Ecological sustainable development requires economic and social developments should be based on natural resources, coordinated with ecological environment and be within the bearing capacity of natural resources and ecological environment. The real sustainable development of economic society can be realized by considering the human's self-development under the premise of respecting the natural development law and safeguarding the integrated interests of humans and nature. In the domain of social sustainable development, it is regarded that the stages and goals of development can be different according to different countries, but the essence of development should include improvement of human life quality and health level, development of high-tech and superior education, guarantee of human equality and globally coordinated development, fair treatment of next generation without endangering their future needs, and ensurement of intergenerational equity. In short, economic sustainable development is the foundation, ecological sustainable development is the premise, and social sustainable development is the final destination. It is regarded that the sustainable development strategy is guided by global significance, adaptive to the contemporary theme of "peace and development," based on the premise of harmonious development of humans and nature, and is a development view of social comprehensive progress and coordinated economic development. And green chemistry is the center of science in the twenty-first century, which prevents and eliminates the pollution from the origin, meets the sustainable development of humans to a maximized extent in terms of rational utilization of resources, environmental protection, ecological balance, and others. It is the scientific and technological support to realize the sustainable developments of economy, ecology, and society, and also an important part of sustainable development theory.

In September 2002, the global sustainable development summit was held in Johannesburg, South Africa. This meeting further discussed and evaluated the progress of green chemistry, established the action plans of green chemistry and sustainable development, strengthened the pledge of participating countries, and vigorously promoted the implementation of the strategy of sustainable development [7, 8].

1.1.5 Green Chemistry and Technology is Becoming the Hotspot of Various Governments and Academia

In recent years, green chemistry and technology have become one of the governments' most focusing issues and are the significant research areas interested by a large variety of industries and academia. The vigorous development of green chemistry has been promoted by direct involvement of the governments and the mode of "Production–Study–Research." One of the most striking events was that the US President Bill Clinton announced the foundation of the "Presidential Green Chemistry Challenge Award" on March 16, 1995. This is the only award issued by the government. The categories of the awards included award of changing synthesis routes, award of changing the solvent/reaction conditions, award of designing safer chemicals, award of small business, and award of academy for those awarded chemists, companies, or enterprises whose achievements are fundamental and creative, reflecting the basic principles of green chemistry during the design, manufacture, and use of chemical products, eliminating the chemical pollutants at source, and radically reducing the environmental pollution. It has been awarded for 12 times since 2007.

It is obvious to find out the enormous social and economic benefits brought by green chemistry from all previous Presidential Green Chemistry Challenge Awards.

As reported, green chemistry reduced the enterprise emission of organic solvents up to 2.5 billion liters per year, saved industrial process water 143.8 billion liters per year, saved energy for 90 trillion British thermal units (1 btu = 1,055.06 J), and reduced CO_2 emission by 430,000 tons. Meanwhile, the residual amount of harmful chemicals in organisms was decreased by 800,000 tons. Although these numbers are still far less than the ideal emission reduction figures that the whole chemical industry should have, the prospect of improvement is very promising.

The British "Green Chemistry Award," sponsored by the Royal Society of Chemistry (RSC), Salter Inc., Jerwood Charitable Foundation, the Ministry of Industry and Commerce, and the Ministry of Environment, began in 2000 with the purpose of encouraging more people participated in the green chemistry research and promoting the industry's latest scientific research achievements.

Japan set up a "Green and Sustainable Development Award in Chemistry," which was initiated by the Japanese Green and Sustainable Development of Chemical Network (GSCN). This committee consists of the representatives from Japanese chemical and chemical-related industries. The award was first conferred in 2002 and then once every year [9].

Clean development mechanism (CDM) is one of the three flexible mechanisms under the framework of the Kyoto protocol. The main content of CDM is that the developed countries cooperate with the developing countries for projects by providing funding and technology. The developed countries fulfilled their commitment on reduction of greenhouse gas emission under article 3 of the protocol by carrying out the projects met by "certified emission reductions (CER)." CDM is considered to be a "win–win" mechanism: On the one hand, the developing countries receive capital and technology by cooperation and realize the sustainable development of them; on the other hand, the cost of emission reduction abroad can be much lower for the developed countries comparing the cost if the emission reduction is done in their own countries. This type of cooperation project is known as CDM project. The additional and veritable reduction of CO_2 emission is termed as certified emission reductions, which is owned by Chinese project enterprises and can be sold. Each CDM project has a project implementation period. This period is regulated by the CDM executive board (CDM EB), which is an official organization of the United Nations Framework Convention on Climate Change (UNFCCC) and the detailed flowchart is shown in Figure 1.1.

1.2 The Contents and Characteristics of Green Chemistry

1.2.1 The Meaning of Green Chemistry

Green chemistry is the frontier area of modern chemical science research, which is incorporated with the latest theoretical achievements and technologies of chemistry,

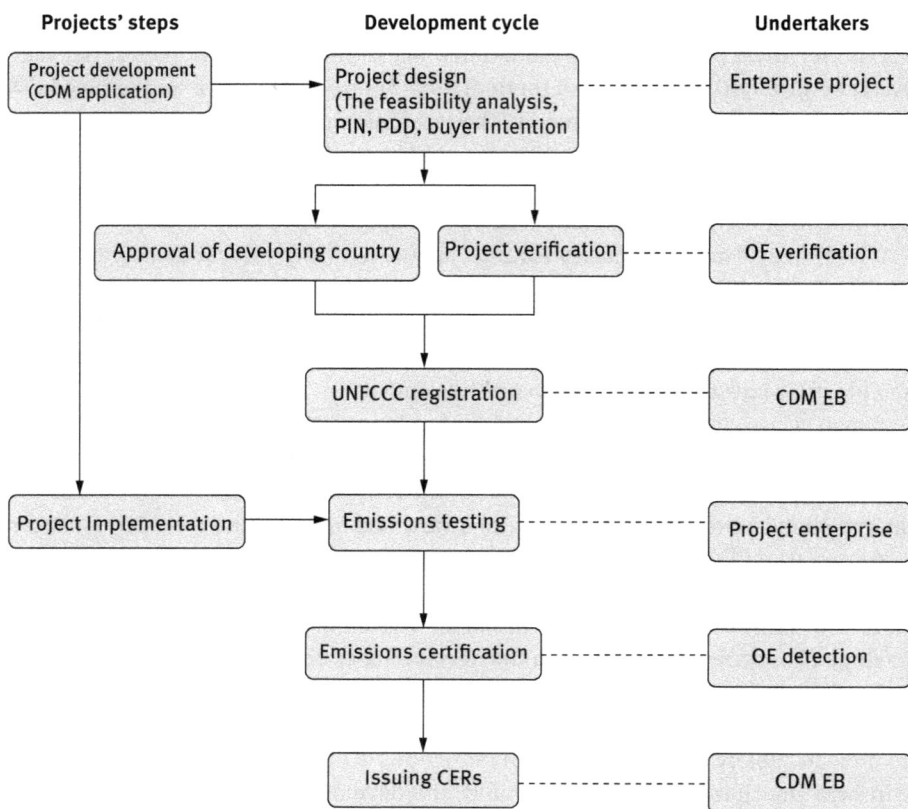

Figure 1.1: Flowchart of CDM project implementation.

chemical engineering, physics, biology, materials, environment, information, and others. It is also a newly emerging interdisciplinary with clear social needs and scientific goals, and has already become a center of science in the twenty-first century [10].

Green chemistry, also known as environmental-friendly chemistry or sustainable chemistry, is aimed at reducing and eliminating the use and formation of harmful substances during the design, production, and application of chemicals and making the developed chemicals and technological processes much safer and more environmental friendly.

The technology developed based on green chemistry is called green technology or clean production technology. The ideal green technology applies the chemical reactions with certain conversion rate and high selectivity for the production of destination products. It is featured to produce less or no byproducts; achieve or get close to "zero discharge" of the wastes, use harmless raw materials, solvents, and catalysts, and produce environmental-friendly products [11].

1.2.2 Research Contents of Green Chemistry

Green chemistry is the study of developing environmental-friendly chemicals and its technological processes, which avoids pollution from the origin. Therefore, the research contents of green chemistry are listed as follows:

(1) Clean synthetic process and technology with reduced waste emissions and its goal is "zero emissions"
(2) Convert the existing processes into clean production
(3) Design and development of safe chemicals and green new materials
(4) Improve the utilization rate of raw materials and energy and extensively increasing the use of renewable resources
(5) Utilization of biotechnology and biomass
(6) New separation technologies
(7) Evaluation of green technology and technological process
(8) Education of green chemistry for changing social life and coordinating development of social economy and environment

The core of the green chemistry based on "atom economy" is to investigate the novel reaction systems by means of chemical principles and new chemical technologies. These reaction systems include new synthetic methods and technology, discovery of new chemical raw materials (e.g., biomass), exploration of new reaction conditions (e.g., environmental-friendly reaction mediums), and design and development of safe, environmental–friendly, and healthy green products.

1.2.3 The Characteristics of Green Chemistry

Green chemistry is the frontier subject of the international chemical society and is also the important direction of chemical science in the twenty-first century. Its remarkable features are as follows:

The difference of green chemistry and traditional chemistry is that the former is more focused on social sustainable development and the promotion of the coordinated relationship between human and nature. Green chemistry is a higher level of chemistry defined as new knowledge, new thoughts, and new science, which is recognized from the environmental crisis.

Green chemistry is also different from environmental chemistry because green chemistry is the environment-friendly chemical reaction and technology, especially referring to the new catalytic technology, new biological engineering technology and new clean synthetic technology. However, the environmental chemistry is the study of chemistry-related issues that affect the environment. The difference between green chemistry and environmental chemistry is obvious, since the former prevents pollutants from its origin, namely pollution prevention, while the latter is the posttreatment of the existing polluted environment known as end treatment. As revealed from the practice, this end treatment serves as an extensive management, which is only targeting on the purification and treatment of pollutants, neglecting the prevention of the pollutants formation and emission from its source. Therefore, this type of treatment is considered as a temporary treatment while not the permanent one. Its low treatment efficiency is also reflected from its high operation cost and the waste of resources and energy [12].

In a word, from a scientific point of view, green chemistry is not only the innovation of the fundamentals of chemistry and chemical engineering but also the integration and expansion of these two subjects; from an environmental point of view, it is the new science and technology for the protection of environment and resolution of the gradually worsening ecological environment; from an economic point of view, it is in line with the requirements of economic sustainable development, since it is following the guidelines of rational utilization of resources and energy and reduction of production cost. Because of the above fact, scientists believe that green chemistry is one of the most important areas of scientific developments in the twenty-first century [13].

1.3 The Developments of Green Chemistry in Domestic and at Abroad

Green chemistry is an important direction of the developments of chemistry and chemical engineering in the twenty-first century, and is the inevitable choice for social and economic sustainable developments. Therefore, the research of green chemistry attracts tremendous attention in both domestic and abroad. Numerous achievements have been acquired in terms of theories, knowledge, and technological innovation for further development of green chemistry [14–17].

1.3.1 The Developments of Green Chemistry at Abroad

Due to the frontier status of green chemistry in today's international chemical science research, it is therefore highly active for the theoretical research and technological developments of green chemistry. For instance, establish professional journals of green chemistry, publish green chemistry monographs, and hold various academic seminars of green chemistry. All of these are beneficial for further promotion of the international communication and development of green chemistry [18–20].

1.3.1.1 Main Organizations and Research Institutions
The US EPA is responsible for Green Chemistry Program, the management of green chemistry project, release of the related information (such as Presidential Green Chemistry Challenge Award), as well as the related activities.

In 1997, the Green Chemistry Institute was set up in the United States based on national laboratories, universities, and companies. Its main purpose is to promote the academic, educational, and research cooperation among governments, enterprises, universities, and national laboratories. The main activities consisted of research, education, resources, conference, publication, and international collaboration of green chemistry.

The UK RSC created the Green Chemistry Network with the main purposes to popularize and promote the propagation, education, training, and practice of green chemistry among industry, academia, and schools.

Britain, Italy, Australia, and other countries also established the research centers of green chemistry (or research centers of clean technologies). For example, a green chemistry research center was set up in York University, led by Prof. J. Clark with the main research direction of catalysis and clean synthesis. The research team of Prof. M. Poliakoff at the University of Nottingham is mainly working on the research and development of supercritical fluid and education. The green design institute in Carnegie Mellon University is focusing on the green design and development of product process and manufacture.

In 2000, Japan set up the GSCN, which aimed at promoting the research of green chemistry that is environmentally friendly and beneficial to human health and social security. Its main activities include green and sustainable development of chemical research, education, reward, information exchange, and international cooperation.

1.3.1.2 American Presidential Green Chemistry Challenge Awards

The first Presidential Green Chemistry Challenge Award was conferred on National Academy of Sciences in July 1996 at Washington, the United States. In total, 67 projects had been nominated, and four companies and one professor of chemical engineering were finally awarded. Monsanto Company developed a new production process of amino sodium diacetate, by selecting nontoxic diethanolamine as raw material and catalytic dehydrogenation as the major approach. This new production process changed the previous two-step synthetic route which is applying ammonia, formaldehyde, and hydrocyanic acid as the raw materials, and thus obtained Presidential Green Chemistry Challenge Award in 1996 for changing the synthetic route. The award of changing the solvent/reaction conditions was granted to Dow Chemical Company for application of carbon dioxide as a forming reagent instead of environmentally harmful chlorofluorocarbons for the production of polystyrene foam plastic. Rohm and Hass Company won the American Presidential Green Chemistry Challenge Award for design of safer chemicals because of the development of an environmental-friendly marine biological antifouling reagent. Small business award was conferred on Donlar Inc., for its development of Thermal Polyaspartic Acid (TPA), which is a biodegradable product for replacement of polyacrylic acid. Professor M. Holtzapple from the University of Texas A&M won the academy award in 1996 because of his technologies of converting waste biomass into animal feeds, industrial chemicals, and fuels. Up to 2017, 22 awards had been granted as shown in Table 1.2.

1.3.1.3 Establishment of Professional Journals

Green Chemistry, established in 1999, by the RSC, is an international professional journal of green chemistry and its contents include the research achievements, reviews, news reports, and other information of green chemistry [21].

Clean Products and Processes, founded in 1998, is another professional journal in the field of green chemistry, with the theme of clean-related technologies, for example, product development, industrial design, technological process, and experimental model.

Journal of Cleaner Production, initiated by Dutch Elsevier in 1996, mainly reports the research papers and comments on clean production technologies.

In addition, some other journals such as *Industrial & Engineering Chemistry Research*, *Pure and Applied Chemistry*, *Catalysis Today*, and *Journal of Industrial Ecology* also set up the column of green chemistry, and publish the green chemistry-related papers regularly or irregularly.

1.3.1.4 Publishment of Green Chemistry Monographs

Since the occurrence of green chemistry, many countries published a large number of monographs to introduce green chemistry, and some of the most representative ones are as follows:

In 1998, P. T. Anastas and J. C. Warner published *Green Chemistry: Theory and Practice*, in which the definition, principles, evaluation methods, and development trend of green chemistry were systematically discussed. It is a masterpiece of green chemistry.

In 2000, P. Tunds and P. T. Anastas published *Green Chemistry: Challenging Perspectives*, also published by Oxford University Press. This book further describes the production, opportunities, and challenges of green chemistry as well as its prospect of development.

1.3.1.5 Organization of Various International Academic Conferences

In August 1994, the 208th ACS National Meeting hosted a seminar of "Design for the Environment: New Paradigms of the 21st Century," which was focused on the environment-harmless chemistry, environmental-friendly technologies, green technologies, and others.

Gordon Conference, themed as green chemistry, has been held alternately in the United States and the Europe since 1996. Gordon Conference in 1996 selected

Table 1.2: The Presidential green chemistry challenge awards of the United States.

Name Year	Academic Award	Small Business Award	Greener Synthetic Pathways Award	Greener Reaction Conditions Award	Designing Greener Chemicals Award
2001	Quasi-Nature Catalysis: Developing Transition Metal Catalysis in Air and Water Prof. Chao-Jun Li, Tulane University	Messenger®: A Green Chemistry Revolution in Plant Production and Food Safety EDEN Bioscience Corporation	Baypure™ CX (Sodium Iminodisuccinate): An Environmentally Friendly and Readily Biodegradable Chelating Agent Bayer Corporation and Bayer AG	BioPreparation™ of Cotton Textiles: A Cost-Effective, Environmentally Compatible Preparation Process Novozymes North America, Inc.	Yttrium as a Lead Substitute in Cationic Electrodeposition Coatings PPG Industries, Inc.
2002	Design of Non-Fluorous, Highly CO₂-Soluble Materials Prof. Eric J. Beckman, University of Pittsburgh	SCORR – Supercritical CO₂ Resist Remover SC Fluids, Inc.	Green Chemistry in the Redesign of the Sertraline Process Pfizer, Inc.	NatureWorks™ PLA Process Cargill Dow LLC	ACQ Preserve®: The Environmentally Advanced Wood Preservative Chemical Specialties, Inc.
2003	New Options for Mild and Selective Polymerizations Using Lipases Prof. R. A. Gross, Rensselaer Polytechnic Institute	Serenade®: An Effective, Environmentally Friendly Biofungicide AgraQuest, Inc.	A Wastewater-Free Process for Synthesis of Solid Oxide Catalysts Sud-Chemie, Inc.	Microbial Production of 1,3-Propanediol (PDO) Dupont	EcoWorx™ Carpet Tile: A Cradle-to-Cradle Product Shaw Industries, Inc.
2004	Benign Tunable Solvents Coupling Reaction and Separation Processes Prof. C. A. Eckert and C. L. Liotta, Georgia Institute of Technology	Rhamnolipid Biosurfactant: A Natural, Low-Toxicity Alternative to Synthetic Surfactants Jeneil Biosurfactant Co.	Development of a Green Synthesis for Taxol® Manufacture via Plant Cell Fermentation and Extraction Bristol-Myers Squibb Company	Optimyze: A New Enzyme Technology to Improve Paper Recycling Buckman Laboratories International, Inc.	Engelhard Rightfit™ Organic Pigments: Environmental Impact, Performance, and Value Engelhard Corporation (now BASF Corporation)

(continued)

Table 1.2: (continued)

Year	Academic Award	Small Business Award	Greener Synthetic Pathways Award	Greener Reaction Conditions Award	Designing Greener Chemicals Award
2005	A Platform Strategy Using Ionic Liquids to Dissolve and Process Cellulose for Advanced New Materials Prof. R. D. Rogers, Alabama University	Producing Nature's Plastics PHA Using Biotechnology Metabolix, Inc.	① NovaLipid™: Low Trans-Fats and Oils Produced by Enzymatic Interesterification of Vegetable Oils Using Lipozyme ADM Novozymers Co. ② A Redesigned, Efficient Synthesis of Aprepitant, the Active Ingredient in Emend: A New Therapy for Chemotherapy-Induced Emesis Merck & Co., Inc.	A UV-Curable, One-Component, Low-VOC Refinish Primer: Driving Eco-Efficiency Improvements BASF	Archer RC™: A Nonvolatile, Reactive Coalescent for the Reduction of VOCs in Latex Paints ADM
2006	Synthesis of propanediol and polyhydric alcohol from natural propanetriol Prof. Galen J. Suppes, University of Missouri – Columbia	Environmentally Safe Solvents and Reclamation in the Flexographic Printing Industry Arkon and NuPro Technology Companies	Novel Green Synthesis for β-Amino Acids Produces the Active Ingredient in Januvia™ Merck & Co., Inc.	Directed Evolution of Three Biocatalysts to Produce the Key Chiral Building Block for Atorvastatin, the Active Ingredient in Lipitor Codexis Inc.	Greenlist™ Process to Reformulate Consumer Products S. C. Johnson & Son Companies
2007	Hydrogen-Mediated Carbon–Carbon Bond Formation Prof. Krische M. J., University of Texas at Austin	Environmentally Benign Medical Sterilization Using Supercritical Carbon Dioxide NovaSterilis, Inc.	Development and Commercial Application of Environmentally Friendly Adhesives for Wood Composites Prof. Li K. C., Oregon State University, and Colombia Wood Inc., and Hercules Group Company	Direct Synthesis of Hydrogen Peroxide by Selective Nanocatalyst Technology Headwaters Technology Inc.	Developed Biobased BiOH™ Polyols for Several Polyurethane Applications Cargill Inc.

Year					
2008	Green Chemistry for Preparing Boronic Esters Prof. R. E. Maleczka, Jr. and M. R. Smith, Michigan State University	New Stabilized Alkali Metals for Safer, Sustainable Syntheses SiGNa Chemistry, Inc.	Development and Commercialization of Biobased Toners Battelle	3D TRASAR® Technology for Water Cooling System Monitoring Nalco Company	Spinetoram: Enhancing a Natural Product for Insect Control Dow AgroSciences LLC
2009	Atom Transfer Radical Polymerization: Low-Impact Polymerization Using a Copper Catalyst and Environmentally Friendly Reducing Agents Prof. Kof Matyjaszewski, Carnegie Mellon University	BioForming® Process: Catalytic Conversion of Plant Sugars into Liquid Hydrocarbon Fuels Virent Energy Systems, Inc.	A Solvent-Free Biocatalytic Process for Cosmetic and Personal Care Ingredients Eastman Chemical Company	Innovative Analyzer Tags Proteins for Fast, Accurate Results without Hazardous Chemicals or High Temperatures CEM Corporation	Chempol® MPS Resins and Sefose® Sucrose Esters Enable High-Performance Low-VOC Alkyd Paints and Coatings Cook Composites and Polymers Company
2010	Recycling Carbon Dioxide to Biosynthesize Higher Alcohols James C. Liao, University of California, Los Angeles, Easel Biotechnologies, LLC	Microbial Production of Renewable Petroleum™ Fuels and Chemicals LS9, Inc. (Acquired by REG LifeSciences LLC)	Innovative, Environmentally Benign Production of Propylene Oxide via Hydrogen Peroxide The Dow Chemical Company and BASF Corporation	Greener Manufacturing of Sitagliptin Enabled by an Evolved Transaminase Merck & Co., Inc. and Codexis, Inc.	Larvicide: Adapting Spinosad for Next-Generation Mosquito Control Clarke, Natular™
2011	Towards Ending Our Dependence on Organic Solvents Prof. Bruce H. Lipshutz, University of California, Santa Barbara	Integrated Production and Downstream Applications of Biobased Succinic Acid BioAmber, Inc.	Production of Basic Chemicals from Renewable Feedstocks at Lower Cost for Greener Reaction Conditions Genomatica	NEXAR™ Polymer Membrane Technology Kraton, Performance Polymers, Inc.	Water-Based Acrylic Alkyd Technology The Sherwin-Williams Company

(continued)

Table 1.2: (continued)

Name / Year	Academic Award	Small Business Award	Greener Synthetic Pathways Award	Greener Reaction Conditions Award	Designing Greener Chemicals Award
2012	① Organic Catalysis: A Broadly Useful Strategy for Green Polymer Chemistry Prof. Robert M. W., Stanford University; Dr. James L. Hedrick, IBM Almaden Research Center ② Synthesizing Biodegradable Polymers from CO_2 and CO Prof. Geoffrey W. C., Cornell University	Using Metathesis Catalysis to Produce High-Performing, Green Specialty Chemicals at Advantageous Costs Elevance Renewable Sciences, Inc.	An Efficient Biocatalytic Process to Manufacture Simvastatin Codexis, Inc., Prof. Yi Tang, University of California, Los Angeles	MAX HT® Bayer Sodalite Scale Inhibitor Cytec Industries Inc.	Enzymes Reduce the Energy and Wood Fiber Required to Manufacture High-Quality Paper and Paperboard Buckman International, Inc.
2013	Sustainable Polymers and Composites: Optimal Design Prof. Richard P. Wool of the University of Delaware	Functional Chrome Coatings Electrodeposited from a Trivalent Chromium Plating Electrolyte Faraday Technology, Inc.	Safe, Sustainable Chemistries for the Manufacturing of PCR Reagents Life Technologies Corporation	EVOQUE™ Pre-composite Polymer Technology The Dow Chemical Company	Vegetable Oil Dielectric Insulating Fluid for High-Voltage Transformers Cargill, Inc.
2014	Aerobic Oxidation Methods for Pharmaceutical Synthesis Prof. Shannon S. Stahl, University of Wisconsin-Madison	Farnesane: A Breakthrough Renewable Hydrocarbon for Use as Diesel and Jet Fuel Amyris	Tailored Oils Produced from Microalgal Fermentation Solazyme, Inc.	Greener Quantum Dot Synthesis for Energy Efficient Display and Lighting Products QD Vision, Inc.	RE-HEALING™ Foam Concentrates–Effective Halogen-Free Firefighting The Solberg Company

Year					
2015	Greener Condensation Reactions for Renewable Chemicals, Liquid Fuels, and Biodegradable Polymers Prof. Eugene Y.-X. Chen, Colorado State University	The Plantrose® Process: Supercritical Water as the Economic Enabler of Biobased Industry Renmatix E.	LanzaTech Gas Fermentation Process LanzaTech Inc.	A Novel High Efficiency Process for the Manufacture of Highly Reactive Polyisobutylene Using a Fixed Bed Solid State Catalyst Reactor System Synthetic Oils and Lubricants of Texas Inc.	Hybrid Non-isocyanate Polyurethane/Green Polyurethane™ Hybrid Coating Technologies, Nanotech Industries
2016	Catalysis with Earth Abundant Transition Metals Prof. Paul J. Chirik, Princeton University	Renewable Nylon Through Commercialization of BIOLON™ DDDA Verdezyne	An Inherently Safer Technology for the Production of Gasoline Alkylate CB&I. Albemarle. AlkyClean® Technology	Instinct® Technology – Making Nitrogen Fertilizers Work More Effectively for Farmers and the Planet (Summary) Dow AgroSciences LLC.	AirCarbon: Greenhouse Gas Transformed into High-Performance Thermoplastic Newlight Technologies
2017	Simple and Efficient Recycling of Rare Earth Elements from Consumer Materials Using Tailored Metal Complexes Prof. Eric J. Schelter, University of Pennsylvania	The UniSystem™: An Advanced Vanadium Redox Flow Battery for Grid-Scale Energy Storage UniEnergy Technologies LLC.	Letermovir: A Case Study in State-of-the-Art Approaches to Sustainable Commercial Manufacturing Processes in the Pharmaceutical Merck & Co., Inc.	Green Process for Commercial Manufacture of Etelcalcetide Enabled by Improved Technology for Solid Phase Peptide Synthesis Amgen Inc., Bachem	Breakthrough Sustainable Imaging Technology for Thermal Paper The Dow Chemical Company, Papierfabrik August Koehler SE

environment-friendly organic synthesis as the theme, and discussed the issues such as atom economy and environmental-friendly solvents. This is the first time of world's high academic-level discussion of green chemistry. The meeting together with the American "Presidential Green Chemistry Challenge Award" was commented as "two significant first time" in 1996 by "Brealow–Green Chemistry." The Gordon Conference in 1997 was held in Oxford University, with the topics as follows:
(1) Catalysis (including homogeneous, heterogeneous, and biological catalysis)
(2) New reaction medium
(3) Clean synthesis and process
(4) New reactor technology
(5) Environmental-friendly materials

In 1997, the first Green Chemistry and Engineering Conference was held by the US National Academy of Sciences and exhibited the important research achievements of green chemistry from 64 papers, including biological catalysis, reactions in supercritical fluid, process and reactor design, and technology outlook for 2020. The following year held the second Green Chemistry and Engineering Conference with the theme as "Green chemistry: A global outlook." The first summer workshop of green chemistry, sponsored by the EU parliament, was held in Venice, Italy, in the same year.

In June 2001, the 14th Chemrawn meeting was held by IUPAC in Boulder, CO, the United States, with the theme of "Green chemistry – Harmless processes and products towards the environment." The conference proceeding was compiled and published in Issue 4, Volume 73 of *Pure & Applied Chemistry* in 2001.

On March 3–6, 2001, the Royal Society-sponsored "Green Chemistry – Sustainable Products and Processes" meeting was held in the Wals Swansea University. This meeting covered the frontier domains of chemistry and chemical engineering including many subjects such as green chemistry, clean process, and pollution minimization.

In March 2003, the first Green and Sustainable Development of Chemistry International Conference (GSC Tokyo, 2003) was held in Tokyo, Japan, focusing on the innovation of chemical technology for the reduction of resources consumption. Moreover, advocate the reduction of waste emission during the entire production process so as to guarantee the human health and security and protect the ecological environment. GSC Tokyo Statement was issued during the meeting.

In May 2003, the Engineering Conference of Green Chemistry was held in Sandestin, Florida, the United States, during which nine additional engineering principles of green chemistry was determined and the development concept of green project was also put forwarded. And this will expand the green chemistry and engineering to the entire engineering field.

1.3.2 Green Chemistry Research Receiving Much Attention in China

China plays a great role in the research and development of green chemistry, actively follows the globally latest research achievements and development trends, advocates clean technology, and implements sustainable development strategy.

In 1995, the Chinese Academy of Sciences – Chemistry Department set up an academician-consulting subject of "Green chemistry and technology to promote the sustainable development of chemical production."

In 1997, Xiangshan Science Meeting was held, themed as "Challenge of sustainable development towards science – green chemistry." Academician Qingshi Zhu from the University of Science and Technology of China gave a special report on "The challenge of sustainable development strategy to science and technology." Academician Enze Min from China Petrochemical Corporation made a talk, entitled as "Green chemistry and technology in the production of essential organic chemical materials." Academician Jiayong Chen from the Chinese Academy of Sciences – Institute of Chemical Metallurgy gave a speech on "Green chemistry and technology – Challenges and opportunities of metallurgy and inorganic chemistry." All the talks mentioned above together with the reports from other specialists greatly promoted the development of the green chemistry research in China.

In 1997, the National Natural Science Foundation and China Petroleum and Chemical Corporation jointly funded the important basic research project (known as "ninth five-year" plan) of "Environment-friendly petroleum catalysis chemistry and chemical engineering." In 1999, the National Natural Science Foundation set up a key project of "Investigation of basic problems with metal organic chemistry." In 2000, the green chemistry was listed as a preferentially subsidized area in the "tenth five-year" plan.

The 16th "Core Sciences in the 21st Century – The Green Chemistry Basic Scientific Questions" seminar was held in JiuHua Hills, Beijing, on December 21–23, 1999. Up to 40 experts from chemistry, life science, materials science, and so on attended this meeting and proposed the recent research emphases: ① Green synthesis technology, methodology, and process study; ② the fundamentally scientific issues during the utilization and conversion of renewable resources; and ③ key scientific issues of green chemistry in utilization of mineral resources.

Successive advanced symposiums on green chemistry have been organized successively in China, since the first one was held in the University of Science and Technology of China, Hefei, in 1998. The 7th International Green Chemistry Seminar was held in Zhuhai, Guangdong, May 24–26, 2005, and the main subjects include green chemical reactions (mechanisms and processes), design, process, and utilization of environment-friendly chemicals, efficient utilization of biomass resources, and the computer-aided design and simulation of green chemistry On May 21–24, 2007, the 8th International Seminar of Green Chemistry was held in Jiuhua Shanzhuang, Beijing, focusing on green chemistry and sustainable development. The detailed contents include the utilization and development of sustainable materials, research of green synthesis routes, process, technology, and integration of green chemical engineering, and new opportunities of green chemistry [22].

In July 12, 2006, the Chinese Chemical Society – Green Chemistry Professional Committee was formally founded with the aim to promote the research and development of green chemistry, and strengthen the academic communication and cooperation.

1.4 Green Chemistry is the Only Way for the Sustainable Development of Chemical Industry in China

1.4.1 Industrial Revolution Originated from Green Chemistry

Green chemistry is a multidisciplinary subject emerging from the 1990s, and has become the frontier of today's international chemical industry development. The industrial revolution caused by green chemistry is rising rapidly in the world, influencing various walks of life, not only bringing fundamental changes for the traditional chemical industry, but also promoting the establishment and development of ecological material industry, green manufacturing industry, green energy industry, green ecological agriculture, and so on [23].

1. **Greening the petrochemical industry**

   The needs of petroleum-based chemicals have increased rapidly by the fast development of modern industries. The structural changes of petrochemical raw materials and serious challenges of protection of biological environment require the petrochemical technologies of twenty-first century overcome the restrictions of raw materials, processes, and equipment of the first generation of petrochemical industry. It is necessary to apply the principles of green chemistry and gradually realize the greening of petrochemical industry, which is an inevitable trend of sustainable development. For example, using zeolite catalyst instead of aluminum trichloride for the production of ethylbenzene and isopropylbenzene; selecting ion exchange resin as catalyst instead of sulfuric acid for the production of sec-butyl alcohol; phosgene-free production of polycarbonate, toluene diisocyanate, and diphenyl methane diisocyanate; carbonylation of acetic acid with methanol under low pressure; catalytic oxidation of acetic acid by ethylene; direct oxidation production of acetic acid by methane; two-step production route of ethylene from methane by silicon aluminophosphate molecular sieves; and oxidation production of phenol by benzene and oxidation production of p-dihydroxybenzene and o-dihydroxybenzene by phenol in TS-1 and H_2O_2 clean catalytic oxidation system. These new petrochemical technologies will play a leading role in the petrochemical industry of twenty-first century.

2. **Greening chemical pharmaceutical industry**

The chemical pharmaceutical industry is featured by many characteristics, such as numerous varieties, fast updating, multistep reactions, high consumption of the raw and auxiliary materials, low overall yield, high emission of "Three Wastes," complicated composition, and easy cause of environmental pollution. Therefore, greening chemical pharmaceutical industry not only has important economic benefits but also has profound social benefits and environmental benefits [24].

For instance, sertraline, known as the active component of antidepressant medicine –Zoloft, is a highly selective serum reuptake inhibitors and is applied for the therapies of most melancholia, trauma psychic, and psychological suppression. It has been listed as the designated antidepressant since its approval from 1990. Its market sales were \$2.4 billion and \$2.7 billion in 2001 and 2002, respectively. During the previous synthetic method for sertraline, the carbonyl group of dichlorophenyl tetralone is converted to imino group using $TiCl_2$ as dehydrant. The isolated imine compound is further hydrogenated (Pd/C as catalyst) to amine isomer mixtures (cis/trans 6:1). And then the isolated cis-isomer is reacted with D-mandelic acid for the production of (S, S)-cis-isomer. The mandelate is finally converted into its corresponding hydrochloride product. The above synthesis route is too long, multiple steps are involved, and at the same time produces a large amount of wastes.

Pfizer Company improved and innovated the synthetic process of sertraline. First, select ethanol, an environment-friendly reagent, as the solvent instead of the previous tetrahydrofuran and toluene. Since the imine compound has low solubility in ethanol, it is readily precipitated out, avoiding the dehydration with $TiCl_4$ and separation of intermediates. More importantly, the selectivity of cis-amine isomer has been greatly improved, and the cis/trans ratio is up to 18:1. The unreacted methylamine can be recycled by distillation. This new process simplifies the reaction operation, saves raw materials, improves the yield, and reduces the emission of wastes at the same time. According to the new process of Pfizer, every ton of sertraline production demands 22,710 dm^3 solvents, which is sharply reduced from the former 227,100 dm^3/ton product. On the other hand, raw materials such as methylamine, dichlorophenyl tetralone, and D-mandelic acid have been saved for 60%, 45%, and 20%, respectively. Additionally, annual emission reductions of titanium dioxide-methylamine hydrochloride, HCl, and 50% NaOH waste residual are 440, 150, and 100 tons, respectively. Therefore, Pfizer Company is the winner of the American Presidential Green Chemistry Challenge Awards in 2002 for changing the synthetic route.

3. **Greening the pesticide industry**

The toxicity of chemical pesticide and its impacts on the environment have always been the concerned issues. Especially when the environmental regulations are becoming more and more standardized and the genetically modified crops are rapidly promoted, the traditional pesticide is facing severe challenges. Gradual realization of green chemical pesticides and establishment of sustainable development of the pesticide industry are the new patterns and trends of the development of world's pesticide industry.

Green pesticides refer to the pesticides that can efficiently eliminate pests and pathogens, while are safe to human, livestock, and crops, easy to decompose in the natural environment, and revealed with low or no residue in crops. Currently, the research and developments in this area are as follows:

Bioregulator: Using different active substances to adjust, change, and restrain harmful organisms in different stages of growth, development, and reproduction so as to reach the goal of prevention and control of diseases and pests. For example, Rohm & Hass Company developed CONFIRM – a series of pesticides (e.g., tebufenozide, halofenozide, and methoxyfenozide), which belong to the slough-induced pesticides and can prompt the molting of pests in advance with a result of dehydration, stopped feeding, and quick death.

Nerve paralysis agent: Dow agricultural science company developed a highly selective insecticide – Spinosad – which belongs to this category. The effects of Spinosad on pests are different from the known like products. The Spinosad-treated pests exhibited neurological syndromes, such as lack of coordination, fatigue, tremors and muscle twitching, paralysis, and death.

Insect pheromones: Disparlure synthesized by Baker Company is a type of insect sex pheromone, which can interfere with the breeding of insects and therefore control diseases and insect pests effectively.

Photoactivated pesticide: The key component in photoactivated pesticide is a photosensitizer that kills pests by singlet oxygen ($^1O_2$) catalytically produced by photosensitizer under the conditions of light and oxygen.

4. **Green new materials**

Green materials, also known as ecomaterials or environmentally conscious materials, refer to a big family of materials that have good performance or functions, consume less resource and energy, cause less pollution to the ecological environment, are beneficial to human health, are biodegradable for recycling, have high utilization rate of regeneration, and are well coordinated with the environment during the stages of preparation, use, discard, and recycling.

In recent years, the "white pollution" caused by unbiodegradable plastic products has affected water, soil, and urban environment, and the situation is getting worse and worse. Therefore, the development and application of biodegradable polymeric materials are becoming hot spots of social concern. The desired materials include biodegradable polymeric materials, photodegradable polymeric materials, and photobiodegradable polymeric materials.

At present, many biodegradable polymers have been investigated and developed. For example, polyhydroxybutyric acid ester, polymer hyaluronic acid, chitosan and its derivatives, polylactic acid, polyanhydride, polyparadioxanone, polyvinyl alcohol, starch-grafted copolymer, and phosphazene polymer. Many of these products can be served as biomedical materials.

The function of photodegradable polymeric materials is realized by addition of appropriate amount of photosensitizer during the synthetic process of the polymers.

5. **Greening the energy industry**

Energy is the important material basis for the survival and development of human society. At present, our common energy refers to petroleum, natural gas, and coal. From the beginning of the current century, the consumption of fossil energy has been growing exponentially, leading to the gradually exhausted status of the reservation of fossil resource. It was predicted in the 13th Global Petroleum Conference that the global oil reservation is about 300 billion tons. If selecting the global oil output in 1992, namely 3 billion barrels, as a basis, the world's current oil reservation can only last for 46 years and the natural gas can last for 66 years. Although the world's coal reservation is up to 1.0391 trillion tons, it still can only last 232 years according to the current production level. Furthermore, the production and consumption of these fossil energy also bring about the severe environmental issues, such as air pollution, acid rain, greenhouse effect, and ozone depletion. Therefore, the development of new energy and green energy is becoming an extremely significant issue for every government and scientific academia.

According to our national conditions, three key aspects need to be resolved during the development of green energy industry [25].

(1) Clean combustion technology of coal

China is the world's largest coal producer and consumer, and this type of coal-centered energy structure is predicted to remain for a long period in twenty-first century. It is urgent to promote the clean coal combustion technology, since China has already turned into a typical coal–smoke-polluted country. For example, improvement of combustion methods, adoption of catalytic combustion, development of coal gasification technology, promotion of water–coal slurry technology, and enhancement of comprehensive treatment of soot gas.

(2) Conversion technology of biomass

Biomass contributes 15% of the world's total energy consumption, next to oil, coal, and natural gas. The development and utilization of biomass is converting itself into industrial chemicals and fuels, such as ethanol gasoline, and biodiesel, so as to alleviate the pressure on resources and the environment.

(3) Development of green energy

Solar energy, hydroenergy, oceanic energy, wind energy, and biomass energy all belong to clean energy. The reasonable development and utilization of these types of energy can both replace a considerable portion of the fossil energy and reduce environmental pollution.

1.4.2 Green Chemistry is the Preferential Mode for the Sustainable Development of Chinese Chemical Industry

Under the guidance of sustainable development strategy in China, the clean production and environmental protection have been highly concerned by all levels of government departments. In 1994, the standing committee of the state council passed the "China's Agenda 21," and put it as a white paper on China's population, environment, and development in the twenty-first century. It was clearly presented in the third section of "sustainable development" that the implementation of cleaner production technology is an important task for chemical industry system. It was also emphasized to strengthen the "three wastes" treatment and comprehensive utilization of wastes, save resources, and protect the environment, by relying on the scientific and technological progresses. The former ministry of chemical engineering invested 3.25 billion yuan and 5.27 billion yuan during the periods of "seventh 5-year" and "eighth 5-year," respectively. Meanwhile, 16,912 environmental protection projects have been arranged. All of the above-mentioned efforts have been returned by eminent achievements of environmental protection. The million yuan of chemical engineering industrial output value of comprehensive energy consumption was reduced from 7.41 tons of standard coal in 1990 to 4.8 tons of standard coal in 1996. The decreasing amplitude is 35%. The comprehensive utilization value of chemical wastes increased from 1.257 billion yuan in 1990 to 4.977 billion yuan in 1996 and meanwhile the annual reductions of emission of wastewater, exhaust gas, and waste residue are 1.3 billion tons, more than 3,000 m^3, and nearly 10 million tons, respectively. However, due to the huge amounts of discharged industrial and domestic wastes originated from the large population base and the fast progress of industrialization, the Chinese people are facing the increasingly serious issues of sparse resources and polluted ecological environment. The Industrial pollution in China takes up 70% of the national environmental pollution load. Coal is the primary energy source in China and each year the emission of SO_2 from exhaust gas is up to 1.6×10^7 tons, which enlarges the acid rain areas to 22 provinces and cities in China. The disaster cultivated area is 2.67×10^4 km^2 and exhibits the spreading trend from southwest and south to northeast, central, and northeast parts of China. The annual emission of wastewater in China is 36.6 billion tons, in which industrial wastewater takes up 23.3 billion tons. Water quality of 86% of urban rivers is unqualified. The situations of heavy metal pollution and eutrophication in rivers and lakes are increasingly serious and the seven water systems have all been polluted. Moreover, China is a country that is extremely in shortage of water resource. Although the total water reservation is 2.8 trillion m^3, the amount per capita is 2,300 m^3 that is only a quarter of world per capita, ranking 88th in the world. There are more than 300 cities in short of water in China and the number of cities severely in deficiency of water is more than 100. The situation of water deficiency is even worse in northern part of the country, which has become one of the key restricting factors of social and economic development.

Chemical engineering industry due to its own characteristics (such as large varieties, many synthetic steps, long process flow, and the current dominating status of small- and medium-scaled chemical companies) has adopted the extensive production mode with high consumption and low efficiency. Therefore, the environment has been severely polluted with the development of chemical industry, and China is becoming a major client of "three wastes" emission. The annual emissions of industrial wastewater, exhaust gas, and waste residue in Chinese chemical industry are 50×10^8, 8.5×10^{11}, and 4.6×10^7 tons, respectively. Their percentages of national industrial "three wastes" emission are 22.5%, 7.82%, and 5.93%, respectively. When compared with other industrial departments, the emissions of mercury, chromium, phenol, arsenic, fluorine, cyanogen, and ammonia-nitrogen pollutants from chemical industry rank first. The dye industry, for instance, the emission of industrial wastewater is 1.57×10^8 tons per year, and its wastewater is high in chemical oxygen demand concentration, deep in color, and hard to biodegradation. For chromic salts industry, the emission of chromic slag is about 13×10^4 tons every year, and total stocked amount over the past years is up to 200×10^4 tons. In addition, there are more than 1,000 tons of Cr^{6+} ions, lost into the environment, which brings about severe damages to groundwater quality and human health.

To sum up, the ecological environment and resources in China have been seriously polluted and destructed due to the traditional mode of "vast consumption of resources and extensive management," unreasonable industrial structure, and undeveloped scientific techniques and administration levels. Therefore, it is required to renew the outlook; to establish the new mode of "raw materials-industrial production-product use-waste recycling-the secondary resource"; to adopt clean technology of "waste source prevention and full monitoring of production process" instead of the past "end-treatment" strategy; to develop green chemical industry based on the progresses of science and technology; to follow the road of "resources-environment-economy-harmonious development of society." These are the only ways for the development of chemical industry and even the entire industrial modernization.

1.4.3 Corresponding Solutions

Green chemistry is the central science of the twenty-first century and is regarded as the new science and technology to fulfill the sustainable development of society and economy, which has become the hot spot by the governments all over the world, scientific academia, and industries. Green chemistry itself and its corresponding industrial revolution are rapidly emerging around the world, not only presenting challenges to chemical – and chemical engineering – researchers but also bringing great opportunities for development.

1. **Strengthen the propagation and education of green chemistry**
   Over the past 10 years, green chemistry and its applicable techniques have developed rapidly in European countries and the United States. Green chemistry, considered as government behavior, has always been organized and implemented in many countries. On March 16, 1995, for example, the US government set up the "Presidential Green Chemistry Challenge Award," aiming to reward the achievements by application of chemical principles to reduce or eliminate chemical pollution fundamentally. This award has been granted every year since 1996, and it is the first time in the world to set up an incentive policy for development of green chemistry by government. In Britain, the British Green Chemistry Award, sponsored by RSC, Salters Company, Jerwood Charitable Foundation, DTI, and DETR, has also been granted since 2000. Other European countries, such as Sweden, Netherlands, Italy, Germany, and Denmark, also actively promote clean production technologies and implement the method of minimum waste assessment, which receive remarkable success. The governments at all levels in China should fully recognize the positive influence of green chemistry and its industrial revolution on future development of human society and economy, and adjust the industrial structure in time for fast development of green technology and green industry. Green chemistry and its related industries not only accommodate China's current economic development mode but also suit the national features of the country. Green chemical industry selects protection and saving of resources as its goals, promotes the harmony and coordination between humans and nature, pursues sustainable development, and can be almost involved in all industries.

   In order to comprehensively promote the development of green chemistry and its industry, it is necessary to strengthen the propagation of green chemistry and technology, and set up certain awards and supporting policies.

2. **Select key areas for the research and development of green chemical technologies**
   The aim of green chemistry is to investigate and develop new environment-friendly reactions, new technologies and new products by using the principles of chemistry and new chemical technologies. From the point of view of sustainable development, realize the coordinated and harmonious developments of environmental resource and social economy.

   (1) Pollution-prevented clean coal technology
       Clean coal technology includes the purification techniques prior, during, and after coal combustion as well as the coal conversion technique. China is the world's largest coal producer and consumer. The research and development of clean coal technology is beneficial to save energy, improve air quality, and reduce environmental pollution. Therefore, it is the priority to realize the strategy of green industrial revolution.

(2) Green biological chemical technology

The conversion of cheap biomass resources into useful chemical industrial products and fuels is the strategic goal for development of green chemistry in China. Development of green biological chemical technologies includes genetic engineering technology, cell engineering technology, microbial fermentation technology, and enzyme engineering technology. Plant resource is the most abundant renewable resource on earth, and regenerates with a rate of 160 billion tons per year, equal to the energy released from 80 billion tons of oil. The annual output of agricultural straws in China is more than 1 billion tons, but the utilization rate is less than 5% (mainly used for papermaking). If these residues are well converted to organic chemical raw materials using biological chemical technology, at least 2×10^5 tons of ethanol, 8×10^7 tons of furfural, and 3×10^5 tons of lignin can be manufactured, which are worthy of billions of dollars of value. Therefore, green chemistry and its corresponding techniques are considerably promising for conversion and utilization of biomass resources.

(3) Green technology for efficient utilization of mineral resources

China is a country with a huge population and relative shortage of resources. Development of green technology for efficient utilization of mineral resources and recycling of low-grade mineral resources is thus an important goal of green chemistry research. At present, the biological catalytic technology, microwave chemical technology, ultrasonic chemical technology, membrane separation technology, and so on have attracted tremendous attention; some of which have already been served for industrial applications and exhibit broad prospects.

(4) Green synthesis of fine chemicals

Fine chemicals are the foundation for the development of high-techs, pertaining to the national economy and people's livelihood. However, the overall yields of many fine chemicals are relatively low, due to their multiple steps for preparation. On the other hand, large amounts of raw and auxiliary materials have been consumed. Thus, it is fairly crucial for the preparation of fine chemicals to explore and investigate those green synthetic techniques of high selectivity and high atomic economy. For example, using asymmetric catalytic synthesis technology for the preparation of fine chemicals has already become the hot topics in the domain of green chemistry research. Combinatorial synthesis has become an effective shortcut for the realization of molecular diversity in green chemistry [26].

(5) Green technology for ecological chemical engineering

Ecological chemical engineering takes a compound system cross-coupled by ecological system and chemical system as its research subject, and is a modern chemical engineering mode linked by material circulation, energy flow, information transmission, and value increment. Guided by the principles of industrial ecology based on circular economy concept and by means of green synthesis and conversion, biological chemical engineering technology promotes the balance and good circulation of ecosystem as well as ensuring the sustainable development of global social economy during the production of environment-friendly materials. Green chemistry and ecology chemical engineering technology is representing the development direction of modern chemical industry. It is highly pursued by numerous governments, companies, and academia, and has become a core issue of chemistry and chemical industry in the twenty-first century.

3. **Vigorous implementation of clean production processes**
   Another significant research project is to evaluate the currently existing production processes by taking the reference of the latest advanced scientific techniques as well as the principles and techniques of green chemistry, strengthen the technological renovation, and implement clean production processes. It is both practical and highly profitable from plenty of successful experiences in domestic and abroad [27].

4. **Enhancement of the strength for scientific renovation**
   Innovation is the soul of a nation and the eternal power for continuous progresses of scientific technologies. Throughout the past winning projects of American "Presidential Green Chemistry Challenge Award," all of them are featured as concepts of innovation, varieties of innovation, and technological innovation. With the target of fast development of green chemical industry in China, it is required to keep up with the times; innovate independently; strengthen the exploration of new ideas, new theories, new methods, and new processes; break through the key techniques; promote the combination of production–study–research; and accelerate the conversion and application of scientific achievements.

   The main body of innovation is individual and in order to form the backbone elements for the greening of chemical industry in China, it is highly demanded to poster and train a large variety of high-level technical personnel engaged in the research and development of green synthetic technology and the management of clean production.

5. **Strengthen the international communication and cooperation**
   Green chemistry is the center of science in the twenty-first century. Green chemistry and its application technology have developed rapidly in Europe and the United States. As for China, we should actively pursue the international trends of research and industry development, strengthen the international academic communication and cooperation, absorb the advanced process and technology from abroad, and finally promote the continuous, healthy, and fast developments of Chinese chemical industry.

Questions

1. What is green chemistry? Please discuss the background for the origin of green chemistry briefly.
2. Why is green chemistry considered as a newly emerging interdiscipline with clear social demands and science goals?
3. What are the main research subjects of green chemistry?
4. Why is green chemistry considered as one of the most significant areas of scientific developments in the twenty-first century?
5. What is the essential difference between green chemistry and environmental treatment?
6. Which important inspirations can we learn from the winning projects of the American Presidential Green Chemistry Challenge Award?
7. Why is green chemistry regarded as the only way for the sustainable development of chemical industry in China?
8. How to accelerate the development and progress of green chemistry in China?

References

[1] Ritter, S. K. Green chemistry. Chemical and Engineering News. United States Environmental Protection Agency. 2006-06-28. 2001, 79(29): 27–34.
[2] Sheldon, R. A., Arends, I. W. C. E., Hanefeld, U. Green Chemistry and Catalysis. 2007, ISBN 9783527611003
[3] Poliakoff, M., Licence, P. Sustainable technology: Green chemistry. Nature, 2007, 450 (7171): 810–812.
[4] Clark, J. H., Luque, R., Matharu, A. S. Green chemistry, biofuels, and biorefinery. Annual Review of Chemical and Biomolecular Engineering, 2012, 3: 183–207.
[5] Anastas, P. T., Bartlett, L. B., Kirchoff, M. M., et al. The role of catalysis in the design, development, and implementation of green chemistry. Catalysis Today, 2000, 55: 11–22.
[6] Anastas, P. T., Warner, J. C. Green Chemistry: Theory and Practice. London: Oxford University Press, 1998.
[7] Rouhi, A. M. Green chemistry for pharma. C & EN, 2002, 80(16): 30–33.
[8] Matlack, A. Some recent trends and problems in green chemistry. Green Chemistry, 2003, 5(1): G7–G12.
[9] Li Chao-jun. Developing metal-mediated and catalyzed reactions in air and water. Green Chemistry, 2002, 4(1): 1–4.
[10] Min Enze and Fu Jun. The recent development of green chemistry. Chemistsry, 1998, 20(1): 10–15.
[11] Zhu Qingshi. Green chemistry and sustainable development. Bulletin of Chinese Academy of Sciences, 1997 (6): 415–420.
[12] Liang Wenping, Tang Jin. Green chemistry – one of the important frontiers in chemistry. Progress in Chemistry, 2000, 12(2): 228–230.
[13] Hjeresen, D. L., Schuff, D. L., Boese, J. M. Green chemistry and education. Journal of Chemical Education, 2000, 77(12): 1543–1547.

[14] Min Enze, Wu Wei. Green Chemistry and Chemical Engineering. Beijing: Chemical Industry Press, 2000.

[15] Xu Hansheng. The Introduction of Green Chemistry. Wuhan: Wuhan University Press, 2002.

[16] Hu Changwei, Li Xianjun. Principles and Applications of Green Chemistry. Beijing: Sinopec Press, 2002.

[17] Gong Changsheng, Zhang Keli. Practical Techniques of Green Chemistry and Chemical Engineering. Beijing: Chemical Industry Press, 2002.

[18] Tundo, P., Anastas, P. T., Black, D. S., et al. Synthetic pathways and processes in green chemistry, introductory overview. Pure Applied Chemistry, 2000,72(7): 1207–1228.

[19] Clark, J. H., Lancaster, M. Green chemistry: The path to a sustainable, competitive chemical industry. Nature. 2000, 22(1): 1–6.

[20] Graedel, T. E. Green chemistry as systems science. Pure Applied Chemistry, 2001, 73(8): 1243–1246.

[21] Min Enze. Green chemistry and chemical engineering – 2003. Journal of Chemical Industry and Engineering, 2004, 55(12): 1933–1937.

[22] Wang Jingkang, Bao Ying. Advances in development of green chemistry and engineering and eco-industrial-park construction. Modern Chemical Industry, 2007, 27(1): 2–6.

[23] Xiao Wen de. Green chemistry – science of 21th century. Chemistry World, 2000 (3): 27–28.

[24] Gong Changsheng. Accelerate the development of green fine chemicals in China. Modern Chemical Industry, 2003, 23(12): 5–9.

[25] Zhang Kunmin, Zhu Da, Cheng Yawei. China's environmental protection and sustainable development: action, policy and target. Environmental Protection, 1998 (1): 3–6.

[26] Blaser, H. U. Heterogeneous catalysis for fine chemicals production. Catalysis Today, 2000, 60: 161–165.

[27] Swindall, W. J. Environmental policy and clean technology in Europe. Clean Technologies and Environmental Policy, 2002, 4: 1–2.

2 Basic principles of green chemistry

Green chemistry is a field of chemistry and chemical engineering that focuses on designing products and processes that minimize the use and generation of hazardous substances. Whereas environmental chemistry focuses on the effects of chemical pollution on nature, green chemistry focuses on technological approaches that prevent pollution and reduce consumption of nonrenewable resources. In 1998, Paul Anastas and John C. Warner formulated the following 12 principles of green chemistry [1]:

(1)   It is better to prevent waste formation than to treat it after it is formed.
(2)   Design synthetic methods to maximize incorporation of all materials used in the process into the final product.
(3)   Wherever practicable, synthetic methods should be designed to use and generate substances that possess little or no toxicity to human health and the environment.
(4)   Chemical product design should aim to preserve efficacy while reducing toxicity.
(5)   Auxiliary materials (solvents, extractants, etc.) should be avoided if possible or otherwise made innocuous.
(6)   Energy requirements should be minimized: syntheses should be conducted at ambient temperature or pressure.
(7)   Where feasible and practicable, any raw material should be renewable.
(8)   Unnecessary derivatization (such as protection or deprotection) should be avoided, where possible.
(9)   Selectively catalyzed processes are superior to stoichiometric processes.
(10) Chemical products should be designed to be degradable to innocuous products when disposed of and not be environmentally persistent.
(11) Process monitoring should be used to avoid excursions leading to the formation of hazardous materials.
(12) Materials used in a chemical process should be chosen to minimize hazard and risk.
      These principles address a range of ways to reduce environmental and health impacts of chemical production; moreover, they indicate research priorities for the development of green chemistry technologies.

To assess the potential for a process ,its relative "greeness" when compared with other processes, Winterton proposed 12 more green principles [2]:

(1)   Identify and quantify by-products.
(2)   Report conversions, selectivities, and productivities.
(3)   Establish full mass balance for the process.
(4)   Measure catalyst and solvent losses in air and aqueous effluent.
(5)   Investigate basic thermochemistry.

https://doi.org/10.1515/9783110479317-002

(6)   Anticipate heat and mass transfer limitations.
(7)   Consult a chemical or process engineer.
(8)   Consider the effect of the overall process on choice of chemistry.
(9)   Help develop and apply sustainability measures.
(10) Quantify and minimize the use of utilities.
(11) Recognize where safety and waste minimization are incompatible.
(12) Monitor, report, and minimize the laboratory waste emitted.

2.1 Waste Prevention Instead of Remediation

The development of modern chemistry has enabled the transformation of our natural resources as well as creation of new matter from existing ones to benefit society. These have greatly enriched our modern living and increased the quality of life. Traditionally, people focused on maximization of selectivity and yield of a product in a chemical process for economic reasons without considering the extent of wastage that occurred in the process. However, massive production and widespread use of chemicals also led to global environmental problems, including global climate changes, sustainable energy production, food production, resource depletion, and the dissipation of toxic and hazardous materials in the environment. In fact, the traditional chemical industry was a hazardous and polluting one. It generated stoichiometric amounts of waste, causing extensive air, water, and soil pollution. In order to reduce the effect of environmental pollution by the chemical industry, end-of-the-pipe treatment was selected in the early 1980s. End-of-the-pipe treatment was the remediation of the generated pollution by physical, chemical, and biological methods. These traditional technologies focus on mitigating a hazardous material. This approach has failed to meet the demands of sustainable development.

 Green chemistry targets pollution prevention at the source, during the design stage of a chemical product or process, and thus prevents pollution before it begins. A waste is not a waste if a valuable use can be found for it. However, it is better to find an improved process that eliminates the waste. Therefore, green chemistry need not be expensive. Pollution prevention is often cheaper than end-of-the-pipe treatments. Green chemistry avoids pollution by utilizing processes that are "benign by design." Ideally, these processes use nontoxic chemicals and produce no waste, while saving energy and helping our society achieve a transition to a sustainable economy. Rarely do chemical processes give the targeted product with perfect specificity even after much effort. Improved process design can minimize waste.

 The challenge is to reduce the incidence and severity of accidents, waste, the toxicity of chemicals, and the amount of energy used, while still providing the goods that society needs.

2.2 Atom Economy

Traditionally, chemists have judged the efficiency of a reaction by percent yield of product formed, minimizing the steps or synthesizing a completely unique chemical. This view ignores the quantity and nature of by-products generated by the reaction. The concept of atom economy was developed by Barry Trost in 1991 [3]. The goal of atom economy is to create syntheses in which most of the atoms of the reactants become incorporated into the desired final product leading to fewer waste by-products. Atom economy can be written as

Atom economy = molecular mass of desired product/molecular mass of all reactants × 100%

In an ideal chemical process, the amount of starting materials or reactants equals the amount of all products generated and no atom is wasted. The goal is to achieve a product with high yield and low waste instead of a high yield product with high waste. Atom economy of some common organic reactions (rearrangement, addition, substitution, and elimination) is calculated as shown in eqs (2.1), (2.2), (2.3), and (2.4).

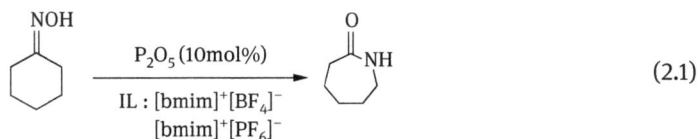

$$(2.1)$$

Percentage of atom economy = (113.16/113.16) × 100 = 100%

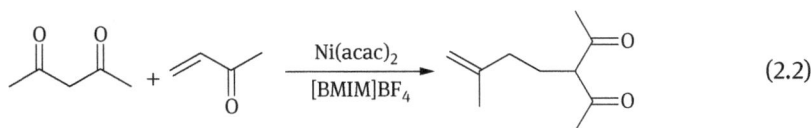

$$(2.2)$$

Percentage of atom economy = [170.20/(100.11 + 70.09)] × 100 = 100%

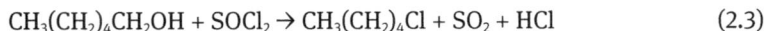

$$CH_3(CH_2)_4CH_2OH + SOCl_2 \rightarrow CH_3(CH_2)_4Cl + SO_2 + HCl \qquad (2.3)$$

Percentage of atom economy = [120.63/(102.18+118.97)] × 100 = 54.55%

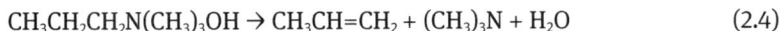

$$CH_3CH_2CH_2N(CH_3)_3OH \rightarrow CH_3CH=CH_2 + (CH_3)_3N + H_2O \qquad (2.4)$$

Percentage of atom economy = (42.08/119.21) × 100 = 35.30%

The rearrangement and addition reactions is a 100% atom economic reaction, since all the reactants are incorporated into the product. This is followed by the substitution reaction (54.55%) and the least economical is the elimination reaction (35.30%) (equations 2.1–2.4). Poor atom economy is common in fine chemicals or pharmaceuticals synthesis. Synthetic methods should be designed to maximize the incorporation of all materials used in the process into the final product.

2.3 Less Hazardous Chemical Synthesis

Many industrial chemical processes make use of starting materials (reactants) that can be harmful to human health and the environment. If a chemical process uses hazardous substances, there is the danger that accidents can occur, exposing workers and others to these chemicals or releasing those chemicals into the environment. For example, chemicals such as carbonyl chloride, hydrocyanic acid, and dimethyl sulfate are highly reactive and toxic, but these dangerous chemicals are used in industry. Sometimes it may not be possible to replace them completely. This leads to a hierarchy of approaches from least change to the most change.

Phosgene is quite toxic. The largest industrial use of phosgene is in the preparation of isocyanates and polycarbonates. Dimethyl carbonate (DMC) is a versatile compound that represents an attractive eco-friendly alternative to both methyl halides (or dimethyl sulfate) and phosgene for methylation and carbonylation processes, respectively. DMC, produced nowadays by a clean process, possesses properties of no toxicity and biodegradability, which makes it a true green reagent to be used in syntheses that prevents pollution at the source.

DMC can also be used in preparation of isocyanates (TDI), instead of toxic phosgene (eqs (2.5) and (2.6)) [4].

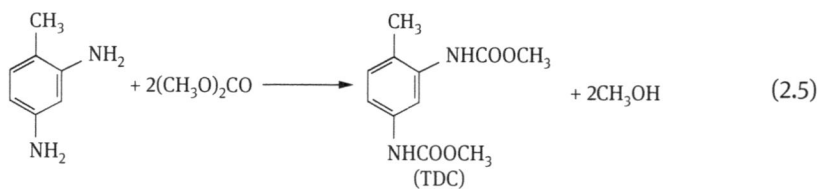

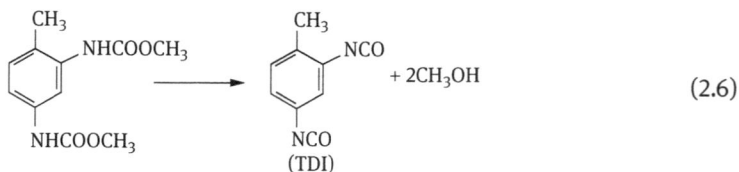

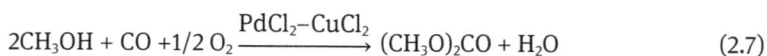

$$ \text{(2.5)} $$

$$ \text{(2.6)} $$

$$ 2CH_3OH + CO + 1/2\,O_2 \xrightarrow{\;PdCl_2\text{--}CuCl_2\;} (CH_3O)_2CO + H_2O \qquad (2.7) $$

DMC is obtained again by the by-product methanol (eq. (2.7)). The entire process constitutes an ideal zero-emission green synthetic route. Therefore, wherever practicable, synthetic methods should be designed to use and generate substances that possess little or no toxicity to human health and the environment.

2.4 Designing Safer Chemicals

An increased understanding of reaction mechanisms and toxicology allows chemists to better predict which compounds/functional groups may pose an environmental hazard. Design new products to be safer and more benign impacts over their life cycle. This information assists in the design of safer chemicals while maintaining the desired function of the product.

The principles of mechanistic toxicology and understanding the physicochemical properties allow for a molecular design for reduced toxicity. Environmentally friendly metal-complexed textile dye eliminates the source of wastewater containing toxic heavy metal.

Pesticides have been designed to be more selective and less persistent than many traditional organic pesticides. Polymers and surfactants have been developed to degrade in the environment at the end of their useful lifetime. γ-poly(glutamicacid) can been synthesized by microbial fermentation from renewable feedstocks. This water-soluble polymer is a biodegradable polyanion that is a suitable replacement for polyacrylic acid. Another example is replacement of insecticides by pheromones; pheromones are non-toxic and environmental benign, which will be hastened through the design of more affordable synthetic methods.

2.5 Safer Solvents and Auxiliaries

Organic solvents such as benzene, carbon tetrachloride, and chloroform pose a particular concern to the chemical industry in synthesis, processing, and separations. Many kind of organic solvents are volatile, flammable, toxic, and carcinogenic. The replacement of toxic and noxious solvents by environmentally benign solvents and solventless systems is important.

Water is not normally a good solvent for organic molecules. Water promotes many reactions at high temperatures, including condensation, decarboxylations, and hydration of alkenes. Water at 250–300 °C can remove organic compounds from soil, catalysts, and some sludges just as effectively as organic solvents do.

Carbon dioxide is nontoxic, nonflammable, relatively inert, abundant, and inexpensive. As a solvent, it is used most commonly as a supercritical fluid (i.e., above its critical temperature of 31.1 °C and critical pressure of 7.38 MPa). It is often used with foods, for which it eliminates the possibility of leaving toxic residues of solvents. The use of supercritical carbon dioxide is further expanded as environmentally benign solvents for extraction and fractionation, chemical reactions, polymeric synthesis, material science, supercritical chromatography, painting, dyeing and cleaning, and emulsions.

Ionic liquids are salts that melt below 100 °C. They have very low vapor pressures and can dissolve many organic and inorganic compounds. Their density, viscosity, solubility, miscibility, stability to water, thermal stability, and other properties can be varied by the choice of cation and anion. Ionic liquids are being considered as green solvents for various separation processes. Many syntheses, including aldol reactions, cyclotrimerizations of alkynes, conversion of epoxides to carbonates, hydroformylation, hydrogenation ketal formation, oxidation, and ring-closing metathesis, have been carried out using ionic liquids. Ionic liquids have also been used successfully in the industry. PetroChina has a 65,000 tons/year plant for the dimerization of isobutene to isooctene with an aluminum chloride–based ionic liquid followed by hydrogenation to isooctane.

Auxiliary substances are used to promote a reaction, but incorporated into the final product. Sometimes, they become part of the waste stream and may be pose an environmental hazard. The use of auxiliary substances (e.g., solvents and separation agents) should be made unnecessary wherever possible and innocuous when used.

2.6 Design for Energy Efficiency

Each step of the chemical reaction or chemical process involves the transformation and transmission of energy. Energy usage comes in many forms: heating, cooling, sonication, high pressure, vacuum, and energy needed for separation and purification of products. The use of electricity, light, ultrasound, and microwaves possess advantages for many chemical reactions. Many chemical reactions must be heated to obtain adequate rates. Catalysis provides an excellent tool for lowering energy requirements in a particular reaction. Selectivity may be improved. The time required for the reaction can occasionally be shortened dramatically, which would permit a greater throughput.

Electricity is used on a large scale in the production of inorganic compounds such as chlorine and sodium hydroxide. Light is environmentally benign. It can catalyze some reactions that are difficult or impossible to run in other ways. Ultrasound can help when there are two immiscible liquids. In this case, mechanical stirring for macromixing may be combined with ultrasound for micromixing. The use of microwave energy, instead of conventional heating, often results in good yields in very short times. In the synthesis of zeolite MCM-41, heating for 80 min at 150 °C is replaced by the use of microwaves for 1 min at 160 °C.

Energy requirements of chemical processes should be recognized for their environmental and economic impacts and should be minimized. If possible, synthetic methods should be conducted at ambient temperature and pressure. The catalytic synthesis of polyaspartate, which is non-toxic, environmentally safe, and biodegradable, can be carried at a lower temperature, thus decreasing energy demands.

2.7 Use of Renewable Feedstocks

Most organic chemicals in use today are derived from nonrenewable petroleum, natural gas, and coal. The utilization of benign, renewable feedstocks is a needed component of addressing global depletion of resources. For a sustainable future, these must be based on renewable resources from fields and forests. Renewable feedstocks are often agricultural products, or they are waste from other processes. Biocatalysis with renewable raw materials has become more popular with the recent escalation of prices for oil and gas. Carcinogens and very toxic chemicals will be avoided where possible. Energy-intensive processes will be avoided.

It may be possible to prepare valuable new materials from renewable raw materials. It should be possible to make the plastics that we need from cellulose, hemicellulose, starch, chitin, lignin, proteins, or other natural products. The majority of these polymers and their derivatives will be biodegradable and compostable. Biocatalysis and biosynthesis of chemicals have the potential to allow synthesis under environmentally friendly conditions while achieving high specificity and yield. Biocatalysis has proven effective in converting biomass to ethanol. Synthesis gas is usually made from natural gas or petroleum. It can also be made from biomass and from municipal solid waste. A recent method used flash evaporation over a rhodium/cerium catalyst at about 800 °C with a contact time of <50 ms. This mixture of carbon monoxide and hydrogen can then be converted to a variety of materials by standard petrochemical methods using a variety of catalysts. The 1,3-propanediol can be produced from glucose or glycerol by biocatalysis. It is used in making poly(trimethylene terephthalate), which is made from renewable raw materials. A new route to tetrahydrofuran starts with furfural obtained by the acidic dehydration of pentose sugars found in oat hulls, maize cobs, and other vegetable waste. Catalytic decarbonylation gives furan, which is reduced to tetrahydrofuran. Tetrahydrofuran can be polymerized to glycols used in making polyurethanes by reaction with diisocyanates.

2.8 Reduce Derivatives

Derivatization, including use of blocking groups, protection/deprotection, and temporary modification of physical/chemical processes, is a common method in the synthesis of fine chemicals, pharmaceuticals, pesticides, and so on. If an organic molecule contains two reactive groups (e.g., hydroxyl, amine, or other reactive group), and one wants to use only one of these groups, then the other group has to be protected, the desired reaction should be completed, and the protecting group should be removed. The protecting groups that are needed to solve a chemoselectivity problem should be added to the reaction in stoichiometric amounts only and must be removed after the reaction is complete. Because these protecting groups are not incorporated into the final product, their use makes a reaction less atom-economical. Therefore, one must avoid the use of blocking or protecting groups, or any temporary modifications; such steps require additional reagents and can generate waste.

2.9 Catalysis

If all the reactants are incorporated into the product in the case of no other reagents and additional products, these stoichiometric reactions are environmentally benign. However, for most stoichiometric reactions, it is more often the case that [5–6]

(1) One of the starting materials is a limiting reagent, and consequently, unreacted starting material will be left over.
(2) One or both of the starting materials are only partially needed for the end product, and therefore the rest of the molecules go to the waste stream.
(3) Additional reagents are needed to carry out or facilitate the reaction, and when the reaction is completed, those reagents must be discarded in the waste stream.

One of the primary causes of waste generation is associated with the use of stoichiometric inorganic reagents. Manufacture of fine chemicals is rampant with antiquated classical "stoichiometric" technologies, for example, stoichiometric reductions with metals (Na, Mg, Fe, Zn) and metal hydrides (LiAlH$_4$, NaBH$_4$) and oxidations with permanganate or chromium(VI) reagents.

Catalytic reactions are preferred in environmentally friendly green chemistry because of the reduced amount of waste generated, as opposed to stoichiometric reactions in which all reactants are consumed and more side products are formed. Catalyzed reactions have low activation energy (rate-limiting free energy of activation) than the corresponding uncatalyzed reaction, resulting in a higher reaction rate at the same temperature and for the same reactant concentrations. Catalysts are not consumed by the reaction itself. Catalytic reagents (as selective as possible) are superior to stoichiometric reagents.

A catalyst works by providing an alternative reaction pathway to the reaction product.

The catalyst opens a selective route to the desired product. There are various kinds of product selectivity. Chemical selectivity, or chemoselectivity, denotes a situation where two different chemical reactions can occur, giving two different products. Similarly, regioselectivity occurs when the same chemical reaction in different regions of the molecule leads to different products. When a reaction gives two (or more) diastereomers, the selectivity to each of these is called diastereoselectivity. In a special case when two products are mirror-image diastereomers, or enantiomers, we talk about enantioselectivity.

2.10 Design for Degradation

Chemical products should be designed so that at the end of their function they break down into innocuous degradation products and do not persist in the environment. The polymers and insecticides are problem of products not being biodegradable.

The durability of synthetic polymers, include polyvinyl chloride and polystyrene, can present problems for wildlife and long-time disposal. More than 90% of the plastic material in municipal waste are resistant to biodegradation. Aliphatic polyesters are generally biodegradable. Aliphatic polyesters are not only naturally biodegradable but also thermoplastic.

It is well known that farmers use different types of insecticides to protect crops from insects. The ideal pesticide might affect only one pest species and nothing else. The more widely used insecticides are organochlorides, organophosphates, and carbamates. Organochlorides (e.g., aldrin, dieldrin, and DDT) are persistent for a long time in the environment, which tend to bioaccumulate in many plant and animal species, and incorporate into the food chain. Some of the insecticides cause the population decline and genetic aberration of beneficial insects and animals. Therefore, in addition to the high efficiency of ideal synthesized insecticides, it is important that during degradation, the products themselves should not posses any toxic effect on human health.

2.11 Real-time Analysis for Pollution Prevention

Continuous process monitoring assists in optimizing the use of feedstocks and reagents while minimizing the formation of hazardous substances and unwanted by-products. Reaction condition can control the generation of unwanted by-products. New analytical tools are needed for real-time industrial process monitoring and for preventing the formation of toxic materials. Analytical techniques should be so designed that they require minimum usage of chemicals. Placement of accurate sensors to monitor the generation of hazardous by-products during chemical reaction is also advantageous.

2.12 Inherently Safer Chemistry for Accident Prevention

The challenge for the chemical industry is to reduce the incidence and severity of accidents, waste, toxicity of chemicals, and the amount of energy used, while still providing the goods that society needs. A number of accidents have been found to occur in industrial units. The biggest accident happened in the chemical industry when 40 tons of toxic methyl isocyanate escaped from a pesticide plant into a densely populated area of Bhopal, India, on December 3, 1984. This resulted in 5,000 deaths and 200,000 injuries. At the time of the accident, a refrigeration system and on-line continuous monitoring of reactions were not functioning. As accidents cannot be completely prevented, substances and the form of a substance used in a chemical process should be chosen to minimize the potential occurrence of chemical accidents, including releases, explosions, and fires. Furthermore, placement of accurate sensors to monitor the generation of hazardous by-products during a chemical reaction is also advantageous.

References

[1]   Anastas, P. T., Warner, J. C. Green Chemistry: Theory and Practice. Oxford University Press, 1998, 29–56.
[2]   Winterton, N. Twelve more green chemistry principles. Green Chemistry, 2001, 3(6): G73–G75.
[3]   Trost, B. M. The atom economy – a search for synthetic efficiency. Science, 1991, 254: 1471–1477.
[4]   Matlack, A. S. Introduction to Green Chemistry. CRC Press, 2010, 28–33.
[5]   Sheldon, R. A., Arends, I., Hanefeld, U. Green Chemistry and Catalysis. Weinheim, Germany, Wiley-VCH Verlag GmbH & Co.KGaA, 2007, 5–6.
[6]   Ibanez, J. G., Hernandez-Esparza, M., Doria-Serrano, C., Fregoso-Infante, A., Singh, M. M. Environmental Chemistry: Fundamentals. Springer, 2007, 302.

3 Green technologies in inorganic synthesis

3.1 Hydrothermal Synthesis

3.1.1 Introduction

Hydrothermal synthesis, also termed as "hydrothermal method," includes the various techniques of crystallizing substances from high-temperature aqueous solutions at high vapor pressures.

Geochemists and mineralogists have studied hydrothermal phase equilibria since the beginning of the twentieth century. Hydrothermal synthesis can be defined as low temperature (<100 °C), medium temperature (100–300 °C) and high temperature (>300 °C). Hydrothermal synthesis has been successfully applied in the preparation of various zeolites and other porous materials [1].

Advantages of the hydrothermal method over other types of crystal growth methods include the ability to create crystalline phases that are not stable at its melting point. Also, materials with a high vapor pressure near their melting points can be grown by the method. It is also particularly suitable for the growth of large good-quality crystals while maintaining control over their composition. Disadvantages of the method include the need of expensive autoclaves and the safety concern during operation.

3.1.2 Principle

Hydrothermal synthesis usually takes metal salts, oxide or hydroxide aqueous solution as the precursors, and makes the precursor solution to nucleate at supersaturation, grow and form the material under higher temperature above 100 °C and atmospheric pressure. The factors affecting hydrothermal synthesis include reaction temperature, heating rate, stirring speed and reaction time. The flowchart of hydrothermal synthesis process is shown in Figure 3.1.

3.1.3 Application Examples of Hydrothermal Synthesis

1. **Diamond Synthesis**
   Diamond has the highest hardness and thermal conductivity of any bulk material, and its Mohs' hardness could reach to 10 and elastic modulus to 460–542 GPa. Those properties determine the major industrial application of diamond in cutting and polishing tools and the scientific applications in diamond knives and diamond anvil cells. While the amount of diamond in nature is limited, synthetic

https://doi.org/10.1515/9783110479317-003

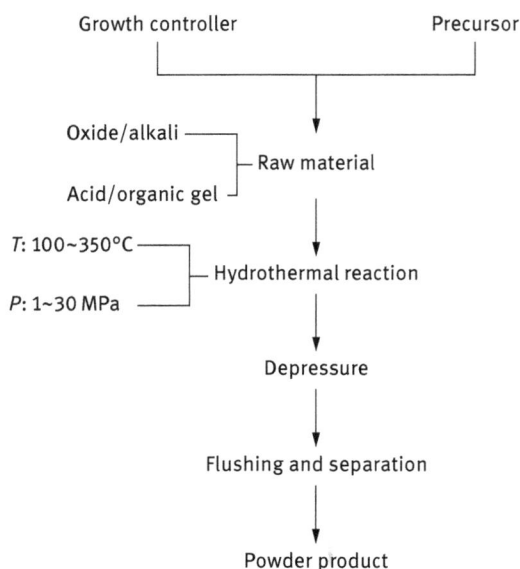

Figure 3.1: Process flow diagram of the hydrothermal synthesis.

diamonds are necessary. The typical process for synthetic diamonds includes 800 °C, 1.4×10^8 Pa, 3% nickel powder (particle diameter: 3 μm, purity: 99.7%), 95% glassy carbon (particle diameter:3 μm) and 2% diamond (particle diameter: 0.25 μm) mixed with water, then react for 50–100 h under hydrothermal condition. The key issue is that Ni is used as the catalyst.

2. **Preparation of titania nanocrystals**

   Nanotitanium dioxide is widely used in chemical industry, environmental treatment industry, medicine and electronics industry because of its brightness, very high refractive index and excellent light catalytic activity [1]. Hydrothermal synthesis of titanium dioxide has the advantages of simple equipment and technology, high product purity and so on. The typical process for preparation of titania nanocrystals is as follows:

   In a vessel containing dodecamine surfactant and deionized water, stir to dissolute the surfactant, add acid or base to adjust the pH of the solution, then rapidly add isopropyl titanium of 0.25 mol/L, and then stir for 30 min to get gel-like precipitate, which endures a hydrothermal treatment in a Teflon-lined autoclave at 140 °C for 20 h upon constant magnetic stirring. After the hydrothermal reaction, the titania nanotubes were filtered in vacuum, washed several times with deionized water and dried at 80 °C for 6 h.

3. **Molecular sieve preparation**

   There are various reports on the hydrothermal synthesis of molecular sieve.

3.2 Sol–gel Method

3.2.1 Introduction

The process involves conversion of monomers into a colloidal solution (sol) that acts as the precursor for an integrated network (or gel) of either discrete particles or network polymers. There are several steps in process procedure, such as hydrolysis, polymerization, gelling, dehydration and sintering [2]. Typical precursors are metal alkyloxides or metal chlorides, which undergo hydrolysis and polycondensation reaction to form a colloid. The sol–gel process is a wet chemical technique and can be used for the fabrication of both glassy and ceramic materials.

Advantages: (1) The basic structure or morphology of the solid phase can range anywhere from discrete colloidal particles to continuous chain-like polymer networks. (2) It can reduce the temperature of material preparation, allowing the synthesis of ceramic, glass and other functional materials in a mild condition. (3) Rheology facilitates the preparation of materials with various morphology and size, such as membranes, fibers or particles.

Disadvantages: (1) Raw materials are expensive, and some are toxic. (2) The whole process spends longer time, i.e., for several days to weeks. (3) There are a lot of micropores in the gel. During the drying process, the gas and organic volatile will emit, resulting in size change and even cause structural rupture of the material.

3.2.2 Principle

The main reaction steps of gelling is that the precursors dissolve in a homogeneous solution, then hydrolysis or alcoholysis occurs to form particles that are around 1 nm. These particles make sol, and then gel is obtained after drying.

1. **Solvation**

   Metal cation M^{Z+} in the precursor could dissociate the absorbed water to form solvent units of $M(H_2O)_n^{Z+}$ and releases H^+:

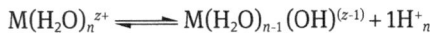

   $$M(H_2O)_n^{z+} \rightleftharpoons M(H_2O)_{n-1}(OH)^{(z-1)} + 1H^+{}_n$$

2. **Hydrolysis**

   Nondissociated precursor, such as metal alcohol salt $M(OR)_n$ (R represents alkyl group) reacts with water. The reaction could continuously proceed until $M(OH)_n$ is produced:

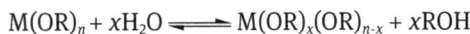

   $$M(OR)_n + xH_2O \rightleftharpoons M(OR)_x(OR)_{n-x} + xROH$$

3. **Condensation**

   Condensation includes dehydration and polycondensation.

Dehydration:

$$-M-OH + HO-M \longrightarrow M-O-M- + H_2O$$

Polycondensation:

$$-M-OR + HO-M \longrightarrow M-O-M- + ROH$$

3.2.3 Application of Sol–gel Method

The traditional nuclear fission fuel ThO_2 microspheres often induce radioactive pollution, and sol–gel method makes up for this problem [3]. The ThO_2 sol could be obtained through following hydrolysis:

$$Th(NO_3)_4 + 4NH_3 \cdot H_2O \longrightarrow ThO_2 + 4NH_4NO_3 + 2H_2O$$

Nitric acid solution is often added to the sol prepared by dispersion, being every 1 mol Th at least adsorbs 0.1 mol nitric acid to make stable sol, then the electrolyte was removed by dialysis. The process of ThO_2 sol by sol–gel method is as follows: proper amount of ammonia was added into the thorium nitrate solution which was vigorously stirred in order to achieve completely hydrolysis at 70 °C, the pH of the final sol was 3.0, the viscosity was about 3 mPa.s, the particle size was 3–4 nm and the ionic surface was positively charged. Then, excess ammonia was added further to make pH and viscosity increase sharply. The positive charges on the particle surface decrease and the particle condensation energy barrier decreases to form a gel.

The microsol droplets were prepared by the stable sol through the nozzle. After ammonia treatment, they were collected in the gel column, NH_4NO_3 on the gel surface was washed out, and calcinated in air at 1,100 °C for 1 min in the air. The theoretical density can reach to 99.0 %, which can be obtained at 1,700 °C for several hours calcination by traditional methods.

3.3 Local Chemical Reaction Method

The local chemical reaction is referred to a process of solid materials preparation by locochemical and local regular reaction. It featured that reactivity is controlled by the structure of reactants, and the main structure is basically kept unchanged. This provides new path for the synthesis of materials under mild condition [4, 9]. The structure of the product is firmly related to the raw materials. This method can prepare some solid materials which can be made or hard to make by other processes. It includes dehydration, intercalation, ion exchange reaction, decomposition and oxidation–reduction reaction.

3.3.1 Dehydration

A dehydration reaction is a chemical reaction between two compounds where one of the products is water. For example: the preparation of unique crystal structure ($Mo_{1-x}W_xO_3$) is finished by the dehydration reaction. Solid chemists have discovered that WO_3 with ReO_3 structure can accommodate the layered MoO_3 and form a special kind of coplanar crystalline state. By means of the isomorphism of the hydrates $MoO_3 \cdot H_2O$ and $WO_3 \cdot H_2O$, MoO_3 and WO_3 was first dissolved in concentrated acid, and the mixed solution was allowed to crystallize to give $Mo_{1-x}W_xO_3 \cdot H_2O$ hydrate solid solution crystal. It was modified at 500 K by dehydration to form $Mo_{1-x}W_xO_3$ crystal with a modulated structure.

3.3.2 Intercalation

The intercalation is a reaction that directs some foreign ions or molecules into the solid matrix lattice without inducing a significant change in the structure of vigilance. The intercalation usually occurs in layered compounds, sometimes accompanied by redox reactions. The interlayer interactions between layers are very weak, and the chemical bonds in the layers are very strong. As a result, foreign ions or molecules are easier to form new compounds from intercalation. The inverse process of the intercalation is deintercalation. The intercalation has become an effective method for the construction of new materials, and has been applied in the synthesis of superconductor, electrolyte and membrane catalytic materials.

The main body of the intercalation reaction exists in the form of solid, and the external ions exist in the form of liquid and gas. There are four ways to process intercalation: (1) Direct reaction of the same intercalation agent; (2) electrochemical intercalation using cathodic reduction; (3) solvation of three-member compounds (A_xMCh_2); and (4) exchange reactions between cations and solvent.

1. **Preparation of cathode materials for lithium ion batteries**

   Lithium ion batteries have been developed rapidly in recent years due to their unique properties such as high voltage, small size, light weight and no memory effect. With the $Li(OAc) \cdot 2H_2O$ and $Mn(NO_3)_2 \cdot 6H_2O$ as the starting material, first heat them at 100 °C to melt the mixture evenly, then the Li-Mn-O precursor was obtained at 250 °C with an oxygen flow, and calcinated at a certain temperature to obtain $Li_{1-x}Mn_{2-x}O_4$, with particle size ranging from 0.1 to 2.0 μm. When $x = 0.125$, $Li/ Li_{1-x}Mn_{2-x}O_4$ demonstrates good electrochemical properties at 4.0 V.

2. **Synthesis of new microporous materials**

   Clay and some hydrogen phosphates (such as zirconium phosphate) are layered compounds, Their porous materials can be prepared by embedding inorganic compounds (such as $[Al_{13}O_4(OH)_{24}(H_2O)_{12}]$, silane and colloidal particles (such as Cr_2O_3, ZrO_2). This is a new way of pore forming, which opens up a new way for the synthesis of new catalytic materials. The basic process is to mix the embedded material solution with clay or hydrogen phosphate, and proceed the embedded reaction at a given pH and temperature condition, and then the embedded products are heat treated for making the embedded material to cross-link with layered clay or hydrogen phosphate. As the embedding amount of substance is significantly affected by the upper layer or between, this determines the embedding amount of substance is limited. At the same time as the embedded material has a certain size, after cross-linking, two layers were enlarged by the intercalate like a pillar to form a new channel. A new type of microporous material with different pore size and distribution can be obtained by choosing various embedded molecules or atom groups.

3.3.3 Ion Exchange Reaction

Ion exchange reaction refers to the topochemical reaction in which a material with exchangeable ion is modified by ion exchange process. This approach provides a unique way to prepare solid materials that many other methods can't succeed to prepare such as clay, molecular sieves and some oxide materials.

1. **The synthesis of new oxide materials**
   $LiNbO_3 + H^+ \rightarrow HNbO_3 + Li^+$
   $LiTaO_3 + H^+ \rightarrow HTaO_3 + Li^+$
   $LiNbWO_6 + H^+ \rightarrow HNbWO_6 + Li^+$
   $LiTaWO_6 + H^+ \rightarrow HTaWO_6 + Li^+$

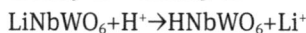

   $HTiNbO_5$, $H_2Ti_3O_7$, $H_2Ti_4O_9$ and $HCa_2Nb_3O_{10}$ and similar oxides can also be prepared through hydrogen ion exchange. These oxides have enough acid and can be used for embedding.

2. **The preparation of solid acid catalysts or carriers**
   Layered hydroxide has strong ion exchange capacity; its exchange process is as follows:

$$LDH-A+X \rightleftharpoons LDH-X+A$$

In recent years, studies found that anion exchange layered hydroxide was a fairly effective catalyst. When tungsten phosphorus compound containing transition metal, Li and Co is introduced into zinc and aluminum hydroxide, the newly formed compounds show the characteristic of a strong acid. In addition, the catalytic activity of Zn_2Al-heteropoly acid compound is higher than that of HY molecular sieve.

3.3.4 Isomorphous Substitution

Isomorphous replacement is referred to a reaction that a new compound is formed by ion exchange reaction while the matrix structure remains unchanged. It is similar to ion exchange reaction to some extent, but has some differences: in the porous material with exchangeable ion, ion exchange reaction occurred in foreign ions with the ions in the cavity, while the isomorphous replacement takes place between foreign ions and the skeleton elements.

Isomorphous replacement reaction can proceed commonly by a gas–solid or liquid–solid reaction. Usually a higher temperature is required for gas–solid reaction, as the foreign ions are required in gas state. Liquid–solid reaction temperature can be lower as the foreign ions are in a liquid. This method plays an important role in the modification of some catalytic materials, such as molecular sieves for the development of new catalytic material.

Aluminum phosphate (APO) molecular sieve is a new type of molecular sieve catalyst. It is composed of Al^{3+} and P^{5+} through the oxygen connection with channel 3d frame, and it is only a weak acid. The introduction of heteroatoms to phosphate molecular structure will improve its acidity. Valence of the heteroatoms, ways of introducing and aluminum phosphate amount will have effect on the properties of modified aluminum phosphate molecularsieve. The SAPO–25 molecular sieve, MAPO–25 molecular sieve and MSAPO–25 molecular sieve were synthesized by this method.

3.3.5 Decomposition

Decomposition was characterized by the decomposition of reactant to give product. Decomposition can be in accordance with the local chemical reaction or nonlocal reaction. The materials used as precursor in the reaction are usually the compounds that are easy to decompose, such as carbonate, nitrate, organic metal complexes and cyanide. So many preparations have been done through decomposition reaction of solid materials.

1. **Precursor thermal decomposition to prepare metal oxide nanorods**
   First, prepare a mixture of surfactant and oxalic acid metal salt precursor, and then the oxalate precursor is decomposed to oxide nanorods at appropriate calcination temperature.
2. **Organic template removal**
   In the process of zeolite synthesis, the organic template used is removed by decomposition, and the real channels were formed in a molecular sieve without the disruption of molecular sieve frame structure.

3.3.6 Redox Reaction

Redox reaction is referred to as the redox reaction of transition metal elements for the synthesis of solid materials. Its essence is to change coordination unit of the transition metal ions through the electronic gain loss, resulting in a new structure, forming different stationary phase interface. This effect is achieved by controlling the oxidation and reduction atmosphere. Mesostabilized metal oxide (such as La_2O_3, NiO and CoO) cannot be obtained by ceramics preparation process, but it can be made conveniently through a redox reaction.

For example:

$$2LaNiO_3 + H_2 \xrightarrow{\boxed{350–400\,°C}} La_2Ni_2O_5 + H_2O$$

$$2LaCoO_3 + H_2 \xrightarrow{\boxed{350–400\,°C}} La_2Co_2O_5 + H_2O$$

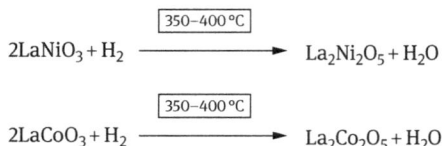

3.4 Low-temperature Solid-phase Reaction

3.4.1 Introduction

The solid-state reaction is referred to a reaction of the solid participation, including mainly solid–solid reactions [5–9]. Low-thermal solid-phase reaction is considered as a solid chemical reaction to proceed at room temperature or near-room temperature. According to the number of reactants, solid-phase reaction can be divided into one-component solid-phase reaction and multicomponent one. In industrial applications, the solid-phase reaction has a short production cycle without solvent employment, high selectivity, high purity of the product, and easy separation and purification of the product.

3.4.2 Mechanism of Solid-phase Reaction

1. **Reaction mechanism**

   Like liquid-phase reaction, solid-phase reaction is initiated from the diffusion contact of two reactant molecules and then proceeds a chemical reaction to create product molecules. Now the product molecules are dispersed in the matrix of reactants. It can appear crystal nucleus of product when the molecules gathered to a certain size, thus completing the process of nucleation. It appears independent crystallographic orientation when the crystal nucleus grown up to a certain size. Solid-phase reaction is composed of diffusion, reaction, nucleation and growth. But different stages have different rates in different systems or the same reaction system under different reaction conditions. Thus we can't clearly distinguished every stages. Overall reaction characteristics only show the response characteristic of the rate control steps. In general, the rate control step of high-temperature solid-phase reaction are diffusion and nucleation growth, while the rate control step of low-thermal solid-phase reaction may be chemical reactions.

2. **Law of the chemical reaction**
   1) **The incubation period**

      Multicomponent solid-phase reaction starts at the contact part of the two phases. When the reaction product layer is formed, reactant mass transfer in the form of diffusion through the product is to keep the reaction continuing. And this diffusion in most solids is slower. At the same time, product can only be nucleated when it is gathered to a certain size. Solid diffusion between the reactants and the products nucleation process make up the special incubation period of the solid-phase reaction. The two processes are affected by temperature significantly. The higher the temperature, the faster the diffusion, the faster the product nucleation, and the shorter the incubation period is. On the contrary, the incubation period is longer.

   2) **Nonchemical equilibrium limitation**

      According to the knowledge of thermodynamics, if the reaction changes a little, the system Gibbs function will change correspondingly. Even if the reaction is proceeded under the isothermal and constant pressure, its molar Gibbs function will also change. We suppose to participate in the reaction of N kinds of substances with n kinds of gases, and the rest is pure condensed phase (pure solid or pure liquid). And the gas pressure is not high, so it can be seen as an ideal gas. Obviously, when the reaction of gaseous substances are involved in the reaction, it does have an effect on reaction Gibbs function. When gases are products, the gases escape and partial pressure becomes small. Thus once the reaction begins, the Gibbs function of reaction system is less than zero and can maintain until all the reactants are consumed. When the gases are reactants, the Gibbs function can be less than zero as long as they maintained certain pressure after the start of the reaction, making all the reactants convert into products. When some of these gases are reactants and some are products, we can also make the reaction to end by keeping a constant pressure of gas reactants to have certain pressure and the gaseous product components escape from the system. Therefore, solid-phase reaction can be proceed completed with no chemical equilibrium limitation.

   3) **Topological chemical control principle**

      In solution, the reactant molecules are surrounded by solvents and an even molecular collision opportunity is achieved; hence, the reaction is mainly decided by the molecular structure of the reactants. But in the solid-phase reaction, the crystal lattice of solid reactant is high-ordered array, lattice molecules moving is more difficult. It can provide the appropriate reaction centers when appropriate orientation of the molecules on the crystal is enough. This is a unique topological chemical control principle to solid instead. For example, it tends to be highly symmetrical in the solution of planar-type atomic cluster compounds when MoS_4^{2-} and Cu^+ react. But in solid reaction, first it often generates class cubane structure of atomic cluster compounds. This may be related to MoS_4^{2-} on the crystal surface, and an S atom is buried deeply in the lower level of lattice.

4) **Distribution of reactions**

In solution, complexes need to be balanced step by step. All kinds of coordination compounds can coexist and be balanced. For example, metalions M and ligand L have the following balance:

$$M+L \rightleftharpoons ML+L \rightleftharpoons ML_2+L \rightleftharpoons ML_3+L \rightleftharpoons ML_4+L \rightleftharpoons \ldots$$

Concentration of various intermediates is related to the pH of ligand solution. As there is often no chemical equilibrium in solid-state reaction, we can achieve reaction step by step to obtain the target compound by controlling the reactant ratio accurately.

5) **Intercalation reaction**

The intercalation can occur in solids (such as MoS_2, TiS_2 and graphite) that have lamellar or sandwich structure, producing intercalation compound as there is enough distance between the layers of the material to be embedded. For example, $Mn(OAc)_2$ and oxalic acid first embeds in the reaction to form intermediate intercalation compounds, and then produces the final product.

3.4.3 Applications of Low-temperature Solid-phase Reaction

1. **The manufacture of printed circuit board**

The basic process of the traditional printed circuit board manufacturing includes sequential processing in aqueous solution: ① sensitization in $SnCl_2$ aqueous solution and surface activation of palladium sediment particles; ② copper chemical plating, deposition of copper on the surface of an insulating plate with trailing particles in the presence of formaldehyde; and ③ electroplating copper. During these stages, water is used to wash repeatedly, and the heavy metal ions (Cu^+, Sn^+, Pd^{2+} etc.) were brought into the wastewater and induced serious pollution. Although copper in the third stage in waste liquor can be recovered, the copper in the waste liquid of the second stage is in the form of complex, and the concentration is very low, so it cannot be recovered. It is provided that producing 1.0 m^2 circuit boards will make more than 0.1 g Pd loss, greatly increasing product expense.

Solid-phase synthesis technology provides a new process for the manufacture of printed circuit boards. Its key step is the solid decomposition of copper hypophosphite, this makes the lively copper deposit on the insulation board to facilitate the following copper plating. This process eliminated the pretreatment of $SnCl_2$ solution, the surface activation and washing of Pd particles, and the process of copper electroplating. It doesn't require precious metal Pd; hence, it is more economical and environmental friendly.

2. **Preparation of industrial catalysts**

The principle of topological chemical control is one of the characteristics of solid-phase reaction. The molecular design of the final product can be achieved by selecting different structure precursors. This has been demonstrated in the preparation of some important industrial catalysts. For example, we compose the oxide catalyst (amorphous V_2O_5) with unique structure and properties by using complexes as precursors. Amorphous V_2O_5 is commonly used as a catalyst for the oxidation of SO_2 to SO_3 in industry. In the traditional process, V_2O_5 is prepared by the decomposition of NH_4VO_3. In NH_4VO_3, VO_4 tetrahedral forms long chains, so the thermal decomposition product V_2O_5 also retains the long-chain and crystalline structure. Therefore, other methods are needed to convert V_2O_5 from crystalline to amorphous. And the VO_4 tetrahedral in amorphous V_2O_5 is spaced apart and has no long-chain structure. Selection of precursor complexes in accordance with the structural characteristics — $(NH_3CH_2CH_2CH_2NH_3)_2V_2O_7 \cdot 3H_2O$. The highly reactive amorphous V_2O_5 with an average particle diameter of about 100 nm was prepared by thermal decomposition. The anion $V_2O_7^{4-}$ is separated by a large cation $(NH_3CH_2CH_2CH_2NH_3)^{2+}$ in the complex precursor. This feature is retained in the product structure during the thermal decomposition process.

3.5 Rheological Phase Reaction

3.5.1 Introduction

In the seventeenth century, British scientists Hooke and Newton's study of elastic solid and fluid flow established material viscoelastic theory [4]. In 1929, the American chemist Bingharm E. C. put forward the concept of "rheology." It studied the relationship between the stress, deformation and time of a substance. People usually identify the solid or liquid based on low stress response of gravity and the observed time interval is usually a few seconds to a few minutes.

If a very wide range of stress is applied in a very wide range of time or frequency spectrum, using rheological instrument, we will observe liquid nature in a solid or the solid nature in a liquid. Therefore, sometimes it is very difficult to make sure a material is a kind of solid or liquid, and rheological system can be disguised as a homogeneous system solid–liquid coexistence.

Ju-tang Sun and others combined closely rheology with chemical reaction. They first proposed the concept of "rheological phase reaction" and introduced rheological technology in synthetic chemistry. Rheological phase reaction refers to a reaction system in which a rheological phase exists. It is a soft chemical synthesis method with the combination of rheology and synthetic chemistry.

3.5.2 The Principle of Rheological Phase Reaction

Rheological phase reaction is a kind of process for preparing new compounds by rheological mixing process. Adding water or other solvent to the solid reactants to prepare a viscous mixture of solid particles and liquid material with even distribution, and then reaction takes place under appropriate conditions to obtain the required product. Rheological system generally has complex chemical composition or structure, and shows the dynamic properties of both solids and liquids. In the physical composition, contains both the solid particles and liquid, and can be considered as a macrouniform complex system of flow or slow flow characteristic.

The uniform mixture of solid and liquid as a rheological body has many advantages, such as effective utilization of solid particle surface area, and the tight and uniform contact with solid, good heat exchange without local overheating and easy to adjust the temperature. In this state, many substances will demonstrate some super-concentration phenomena and new reaction characteristics. Rheological phase reaction is a kind of energy-saving, high-efficiency and pollution-reducing green synthetic route.

3.5.3 Applications of Rheological Phase Reaction

1. **Preparation of magnesium phthalate**
    The given amount of magnesium oxide and benzoic acid is mixed evenly, then add appropriate amount of distilled water to adjust the reactant into a rheological phase and put into a reactor. The reactor was controlled at 100°C to react for 14 h, then the product was washed three times with anhydrous ethanol and dried at 120°C for 3 h, and magnesium benzoate was obtained.
2. **Synthesis of rare earth spinel SnY_2O_4**
    Accurately weigh yttrium oxide, stannous oxide and oxalic acid in the mass ratio of 1:1:4.1. The mixture was added in a mortar and finely grounded, and then moved to the reactor, and finally a small amount of distilled water was added to form a rheological phase, followed by the reaction at 100 °C for 10 h. The sample was removed, washed with distilled water to remove excess oxalic acid, dried, grounded and kept in a desiccator. The thermogravimetric analysis of the precursor was conducted at a temperature rate of 20/min in a Shimadzu DT40 thermal analysis apparatus (nitrogen flow rate is 50 mL/min) to determine temperature of the oxide formation by precursor decomposition and the intermediate production. The precursor is placed in a reaction vessel of porcelain and is heated in a tube furnace of up to 550 °C and a temperature of 700 °C for 10 h. A powdered product is obtained. The valence of tin in the final product was determined by X-ray photoelectron spectroscopy.

3.6 The Precursor Method

3.6.1 Summary

The precursor method is to synthesize a precursor with a desired composition, structure, and chemical properties by accurate molecular design, and then treat the precursor under mild conditions to obtain the desired material. The key lies in the molecular design and preparation of precursors.

In this method, people usually choose some inorganic compounds (such as nitrate, oxalate, carbonate, hydroxide and cyanide complexes), and organic compounds (such as citric acid) to prepare precursor with metal cations required. In a precursor, the reactants are present in the stoichiometric relationship that is required to overcome the problem of uniform mixing of reactants in the process of ceramics preparation. The process of making pottery is usually to react directly with solid materials at a high temperature, while in the precursor method, first a precursor is made, and then the product is obtained by calcination.

Complex metal complexes are a class of important precursors that are usually synthesized in solution, and their compositions and structures can be well controlled. These compounds are generally decomposed at 400 °C to form the corresponding oxides. This provides a way to prepare high-quality composite oxide materials. Another precursor is metal carbonate, which can be used to prepare highly homogeneous oxide solid melt. A lot of metal carbonates are isomorphic, such as calcium, magnesium, manganese, iron, platinum and zinc, with same calcite structure. It can be used to prepare a certain group of metal carbonates by recrystallization, followed by heat treatment at low temperature, and finally get the uniform composition of metal oxide solid melt. The cathode material $LiCoO_2$, $LiCo_{1-x}Ni_xO_2$ of lithium battery is prepared by carbonate precursor.

The precursor method is characterized by a high degree of uniformity of mixing, an accurate ratio of the amount of the cationic substance and a lower reaction temperature. In principle, the precursor method can be used in a variety of solid-state reactions, but for some reactions, it is difficult to find a suitable precursor. The method does not apply to the following situations: 1. The solubility of the two reactants in water is very different; 2. the reactants do not crystallize at the same rate; and 3. formed oversaturated solution.

3.6.2 Application of Precursor Method

1. **Synthesis of spinel $ZnFe_2O_4$**
   The mixed solution of $n_{Fe}:n_{Zn} = 2:1$ of zinc nitrate and iron nitrate was prepared and to react with oxalic acid to form iron and zinc oxalate coprecipitate. The resulting coprecipitate is a solid melt, which contains cations that have been mixed

at atomic scale. The oxalate precursor was calcinated to obtain $ZnFe_2O_4$. Due to the high degree of homogenization of the mixture, the reaction temperature can be reduced a lot, such as the reaction temperature for $ZnFe_2O_4$ is about 1,000 °C. The reaction is as follows:

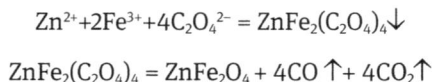

$$Zn^{2+} + 2Fe^{3+} + 4C_2O_4^{2-} = ZnFe_2(C_2O_4)_4 \downarrow$$

$$ZnFe_2(C_2O_4)_4 = ZnFe_2O_4 + 4CO \uparrow + 4CO_2 \uparrow$$

3.7 Melting Method

Melting method refers to material synthesis at high temperatures (200–600 °C) in the melting system. For example, the preparation of a low-dimensional structure of metal chalcogenide. Chalcogenide elements (sulfur, selenium and tellurium) usually have a variety of structures, such as atomic clusters, atomic chain or a layered compound. The structure with metal ions can construct a variety of fiber materials having peculiar photoelectric properties. Because the material is easy to decompose, solid reaction method or vapor deposition method can't be used. In addition, a simple solution reaction can only obtain the smaller powder solid. In recent years, those materials have synthesized by melting method. For example, using polysulfide potassium, molten salt reacts with copper; some low-dimensional sulfide can be obtained; CuS can be obtained above 350 °C; KCu_2S_3 can be obtained at 250–350 °C; and α-$KCuS_4$ and β-$KCuS_4$ can be obtained at 250–350 °C, respectively.

3.8 Chemical Vapor Deposition

3.8.1 Introduction

Chemical vapor deposition (CVD) is conducted by introducing one or more gases composed of the target material into the reaction chamber to form the desired product [4, 9, 10]. CVD for material synthesis has the following features: ① to prepare a material in a temperature far lower than its melting point; ② accurately adjust the composition of the material when two or more reactants were used; ③ to control the crystal structure of materials; ④ to control the morphology of materials (powder, fibrous, dental, tubular, block, etc.); ⑤ don't need sintering additives to prepare high purity and high density material; ⑥ structure control can be generally from micron grade to submicron, and under certain conditions can reach nanoscale; ⑦ complex-shaped products can be made; ⑧ can carry out the coating to the substrate in the form of complex; ⑨ the gradient coating and multilayer coatings can be achieved; ⑩ ability of metastable substances and new material synthesis, usually the CVD technology is a kind of thermal CVD, with deposition temperature of 900–2,000 °C, depending on the characteristics of thin-film materials. This technology has been widely used in composite material synthesis, machinery manufacturing, metallurgy and other fields.

3.8.2 The Principle of CVD

There are two main types of chemical reaction of CVD: one is through a variety of reactions between the initial gases to generate the deposit and another is by gas-phase reaction between a component and the substrate surface to deposit. CVD sediment formation involved all kinds of chemical equilibrium and kinetics, the chemical process design, CVD reactor process parameters (temperature, pressure, gas mixing ratio, gas flow rate and gas concentration), gas properties, basic performance and the function of matrix factors. Considering all the factors used to describe the complete CVD process model is almost impossible; therefore, some simplification and hypothesis must be made. Among them, the most typical one is the concentration boundary-layer model (see Figure 3.2). It is simple to illustrate the main phenomenon in CVD process – the process of nucleation and growth.

The driving force of thermal CVD is heat. Figure 3.3 shows the basic principles of thermo-CVD. A represents solid raw materials, B represents matrix material, C represents carrier gas; T_1 represents the temperature of the reaction of A and C; T_2 is the decomposition temperature of both A and C. A generated on B when temperature is T_2.

Because of the high temperature needed in the thermal CVD, and special requirements to the substrate material, these limit the application of CVD technology. Therefore, CVD technology is to develop in the direction of low temperature and high vacuum, in combination with plasma technology and laser technology.

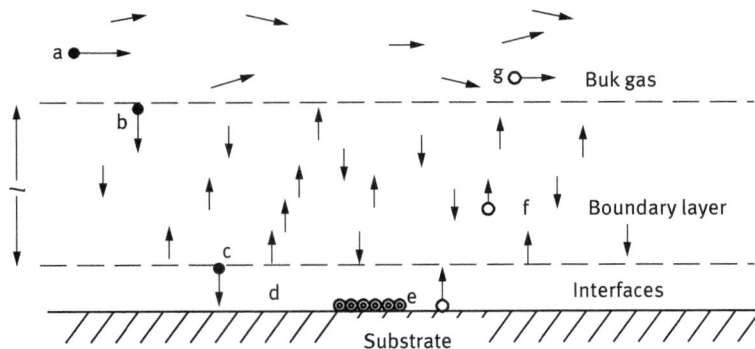

Figure 3.2: Diagram of concentration boundary-layer model.

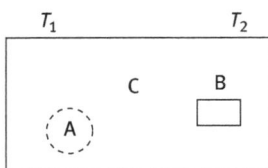

Figure 3.3: Basic principles of thermochemical vapor deposition.

3.8.3 The Application of CVD Method

1. **Preparation of TiO$_2$ thin film on glass substrate**

   With four isopropyl alcohol titanium as the precursor and N$_2$ as carrier gas, after dehydration by molecular sieve, reactants vaporized at inner evaporator were spread into the substrate of 5 mm × 5 mm glass to deposit and form a film in a reactor with a nozzle [11]. During the deposition process, the flow rate of carrier gas was kept at 1.0 L/min; the substrate and nozzle distance is 25 mm. If the substrate temperature is 350 °C, reactant vaporization temperature change has no effect on phase of TiO$_2$ thin film. With the increase of reactant vaporization temperature, crystallization degree of TiO$_2$ film gradually gets better. When the evaporation temperature is 200 °C, sediment grain size becomes small to pile. When vaporizing chamber temperature is 140 °C, N$_2$ flow rate is 1.0 L/min, and deposition temperature has an effect on the phase and microstructure of TiO$_2$ film. When the deposition temperature is not higher than 300 °C, amorphous thin film is formed; when the deposition temperature is 300–400 °C, thin film is of an anatase structure. When the deposition temperature is above 450 °C, a small amount of rutile structure appeared in the film [12–17].

3.9 Polymer Template Method

3.9.1 Introduction

The template method is to use the material with nanopores and cheap and easily obtained as the template,to import the target materials or precursor to the nanopore, and react to obtain the material with the same morphology and size of the template. And finally obtain novel materials with micro- and macrostructures controlled. This method is called the template method [18–21].

Commonly used templates include porous glass, molecular sieve, macroporous ion exchange resin, polymer and surfactant. Depending on the type of the micropores used in the template, people can prepare various shapes of materials, such as granular, tubular, linear and layered structure. Polymer template has the characteristics of various types and is easy to remove by changing the type, concentration or proportion. Many kinds of nanometer materials can be built (such as nanoparticles, nanowires and nano-mesoporous materials).

The main characteristics of polymer template method include: ① most templates can be easily synthesized and available, and their nature can be precisely controlled in a wide range; ② synthetic process is relatively simple and mass production is practical; ③ the stability of size and dispersion problems can be overcome simultaneously; ④ especially suitable for one-dimensional nanomaterial preparation, such as nanowires and nanotubes, polymer template method has become the most important nanomaterial synthetic method. Polymers used as templates include natural polymers and synthetic polymer resins.

3.9.2 Applications of the Template Method

1. **The synthesis of hydroxyapatite (HAP)**
   HAP has good biocompatibility, strong ability of adsorption and exchange function, effects of temperature and humidity sensitivity. Columnar, whisker and flake types of HAP can be used to prepare biomedical functional materials and polymer matrix composites. Porous HAP can be used as catalyst, as catalytic carrier, for the separation of proteins or enzymes and as green environmental protection material. Granular, columnar and flake HAP can be used as smart sensitive material.

   The preparation of HAP with template method is described as follows: the pre-mixed and accurately weighed diammonium phosphate and calcium nitrate hybrid sol is mixed with urea, poured in soluble organic templates (starch, collagen and dextrin) at room temperature, then the mixture is put into an autoclave and slowly heated to 110 °C and reacted for 8 h at the temperature. After reaction ends, the reaction solution is filtered and the residue was dried at 100 °C for 5 h, to obtain a white, fluffy HAP powder.

2. **Nanosilver synthesis with natural polymer as the template**
   Because polyhydroxy polymer can form supermolecules by hydrogen bonding, it can be used as a template to guide the growth of nanocrystals. Starch is a kind of natural polyhydroxy biopolymer. More importantly, by simple processing, starch can be dissolved in water. It can avoid using organic solvent, and thus it is used in the pollution-free, nontoxic green synthesis of nanoparticles. For example, mix up a given concentration of $AgNO_3$ solution and soluble starch solution, then add a certain amount of glucose and react for 20 h at 40 °C with the protection of Ar, and finally silver particles of around 4 nm diameter were obtained. This indicated that starch solution plays a key role in controlling the nanometer metal particle size.

Questions

1. Explain the principles of hydrothermal synthesis methods and its application areas.
2. Explain the main steps in sol–gel process and their functions.
3. What kinds of reactions are included in local chemical reaction and their characteristics?
4. Explain the principle of low-temperature solid-phase reaction.
5. To elucidate the principle and procedure for the preparation of amorphous V_2O_5 catalyst.
6. To elucidate the rheological phase preparation procedure of copper diphthalate.
7. To elucidate the factors influencing the CVD.
8. What properties should the polymer template possess?
9. To elucidate the principle and procedure for the preparation of nano-silver with starch as the template.
10. To elucidate the principle of ionic exchange method.

References

[1]   Ma Zhiguo, Meng Chaohui, Li liping. TiO$_2$ nanocrystal by hydrothermal preparation. Fine and Specified Chemicals, 2006, 14(14): 20–25.
[2]   Zeng Qingbing, Li Xiaodong, Lu Yi. Fundamental principles of sol-gel method and applications in ceramatic material preparation. Polymer Science and Engineering, 1998, 14(2): 138–143.
[3]   Gabriela, Z., Emmanuelle, S., Jan, K. Characterization of CrAlPO$_5$ materials in test reactions of conversion of 2-methyl-3-butynol and isopropanol. Journal of Catalysis. 2002, 208(2): 270–275.
[4]   Japan Chemistry Society. Inorganic solid phase reactions. Translated by Dong Wantang, Dong Shaojun, Beijing: Science Press, 1985.
[5]   Zhou, Y. M., Xin, X. Q. Synthetic chemistry for solid state reaction at low-heating temperatures. Chinese Journal of Inorganic Chemistry, 1999, 15(3): 273–292.
[6]   Hou, H. W., Xin, X. Q., Song, L. Q. Synthesis and third-order nonlinear optical absorptive properties of two novel cluster compounds. Acta Chimica Sinica, 2000, 58(3): 283–286.
[7]   Long, D. L., Liang, B., Xin, X. Q. Solid state reaction at low heating and room temperatures: application in synthetic chemistry. Chinese Journal of Applied Chemistry, 1996, 13(6): 1–6.
[8]   Yang, Y., Jia, D. Z., Xin, X. Q. Synthesis of inorganic nano-materials by solid state reaction at low heating temperatures. Chinese Journal of Inorganic Chemistry, 2004, 20(8): 881–888.
[9]   Sun Wanchang. Preparation and Synthesis of Advanced materials. Chemical Industry Press, 2016.
[10] Zhou Yiming, Yue Xiangrong, Xi Xingquan. XRD diagram analysis of solid-solid phase reaction by diffusion control. Journal of Chemistry in Universities, 1999, 20(3): 361–363.
[11] Li Lina, GuJinghua, ZhanG Yue, et al. Preparation of thin TiO$_2$ film on glass substrate by chemical vapor deposition of metal organic compounds. Journal of Synthetic Crystal, 2005, 34(5): 902–906.
[12] Lin, C. H., Chien, S. H., Chao, J. C., et al. The synthesis of sulfated titanium oxide nano-tubes. Catalysis Letters, 2002, 80: 153–159.
[13] Zhang, X. Y., Wen, G. H., Chan, Y. F., et al. Fabrication and magnetic properties of ultrathin Fe nanowire arrays. Applied Physics Letters, 2003, 83: 3341–3343.

[14] Stacy, A. J., Elaine, S. B., Patricia, J. O., et al. Effect of micropore topology on the structure and properties of zeolite polymer replicas. Chemistry of Materials, 1997, 9: 2448–2458.

[15] Darmstadt, H. C., Ryoo, R. Surface and pore structures of ZSM-5 ordered mesoporous carbons by adsorption and surface spectroscopy. Chemistry of Materials, 2003, 15: 3300–3307.

[16] Huang Zhiliang, Zhang Lianmeng, Liu Yu, et al. The controlled growth of HAP crystal morphology by organic template with homogeneous precipitation. Journal of Synthetic Crystal, 2006, 35 (2): 261–264.

[17] Liu Xuening, Yang Zhizhong, Tang Kangtai, et al. Preparation of peculiar structure of nano ZnO material. Chemistry Communications, 2000 (11), 46–48.

[18] Huang Liyan, Zhang Yan, Liu Zhengping, et al. Gellous porous microsphere by template method. Journal of Beijing Normal University, 2006, 42(2): 177–179.

[19] Raveendran, P., Fu J., Wallen, S. L. Completely "green" synthesis and stabilization of metal nanoparticles. Journal of the American Chemical Society, 2004, 125: 13940–13941.

[20] Ji Xuelai, Tao Jie, Deng Jie, et al. The green synthesis of nanometals and nano inorganic material. Electronic Appliances and Materials, 2005, 11(24): 66–69.

[21] Huang Zhiliang, Zhang Lianmeng Liu Yu, et al. The controlled growth of hydroxyapatite crystal by organic template induction and homogenous precipitation. Journal of Synthetic Crystal, 2006, 35(2): 261–264.

4 Green organic synthesis

Organic synthesis is the key to the preparation of new organics with specific functions. These substances with specific functions can be used as pharmaceuticals, fuels, various fine chemicals and organic chemical raw materials. As the organic reaction conversion is generally not high, the organic reaction processes bring more pollution. The study of the green technology of organic synthesis is of importance for maximizing the efficiency of the synthesis process and minimizing the process pollution. This chapter focuses on the basic content, principles and applications of the green technologies commonly used in various organic reactions.

4.1 Efficient Chemical Catalytic Organic Synthesis

The organic reaction rate is generally slow and the product yield is low, so most of the organic synthetic reaction process (85%) employed a catalyst to improve the efficiency of a reaction process. Therefore, the choice of excellent catalyst for the synthetic process has a key role in promoting the overall efficiency of a process.

4.1.1 Organic Synthesis with Solid Acid Catalysts

1. **Solid superacid catalyst**
   1) **Introduction**
      Solid superacids are solid acids whose acidity is higher than the acid strength of 100% sulfuric acid [1–3]. The acid strength is commonly expressed as an acidity function H_0 of the Hammet indicator, $H_0 = pK_a$. At present, 100% sulfuric acid has been measured as $H_0 = -11.93$. Thus, a solid acid of $H_0 < -11.93$ can be regarded as a solid superacid. The smaller the H_0 value is, the greater the acid strength becomes. Solid superacids can be divided into three types.
      (1) Metal oxide-supported solid superacids (SO_4^{2-}/M_xO_y), such as SO_4^{2-}/TiO_2, SO_4^{2-}/Al_2O_3 and the type of composite oxide carrier, such as SO_4^{2-}/TiO_2-Al_2O_3. These catalysts possessed a high catalytic activity, noncorrosive, high temperature stability and reusable characteristics, and they are suitable for all reactions that require strong acid catalysis.
      (2) Strong Lewis acid-supported solid superacid mainly refers to BF_3, $AlCl_3$ and SbF_5 and other components loaded on porous oxide, graphite and polymer carrier, such as $AlCl_3$/ion exchange resin and BF_3/SiO_2. The Lewis acid and the carrier are mainly bounded by physical and chemical adsorption. This kind of catalyst has the problem of dissolution of the active components, and halogen atoms have a strong corrosion effect on the equipment and are not suitable for working at higher temperature.

https://doi.org/10.1515/9783110479317-004

(3)   Other types of solid superacid such as hetero-poly acid, molecular sieve and acidic polymer resin.

The formation of SO_4^{2-}/M_xO_y solid superacid center is mainly due to the coordinative adsorption of SO_4^{2-} on the surface of the solid oxide, resulting in the electron cloud in the M–O bond to strongly shift to produce a strong Lewis acid center which adsorbs water molecule. The center has a strong attraction for the electrons in water molecule, so that it dissociates and produces proton acid center.

Solid superacids undergo catalytic reactions following a positive carbon ion mechanism. In general, factors that affect the activity of the catalyst (expressed as conversion or selectivity) include reaction temperature, pressure, reactant ratio, space velocity, catalyst, particle size and reactor types. In addition, the physicochemical properties (such as specific surface area, pore volume, pore size and active component distribution) of the catalyst itself are a prerequisite for determining the activity of the catalyst.

The main factors influencing the activity of SO_4^{2-}/M_xO_y-type catalysts are metal compounds, precipitates, crystal forms of metal hydroxides, solvents, SO_4^{2-} introduction and their concentration.

2) SO_4^{2-}/M_xO_y preparation method

(1)   Precipitation impregnation method. Precipitation impregnation method is currently the most widely used method for the preparation of SO_4^{2-}/M_xO_y. The method has the advantages of simple process, easy operation and low raw material price. It includes direct precipitation, homogeneous precipitation and co-precipitation. The precipitation method is to react with an appropriate metal salt solution with the alkali to form a hydroxide precipitate, then filtered, washed, dried and calcinated to obtain the target metal oxide, and then impregnated it with H_2SO_4 solution, filtered and calcinated to obtain solid superacid. The precipitant used is alkali or hydroxylamine, pH is 9–11. For example, the preparation of SO_4^{2-}/TiO_2 catalyst is the reaction of $TiCl_4$ with dilute aqueous ammonia to produce white $Ti(OH)_4$ precipitate, which is filtered, washed, dried and pulverized to obtain TiO_2 and impregnated with 0.5 mol/L H_2SO_4 solution for 12 h, activated at 500 °C for 3 h, SO_4^{2-}/TiO_2 solid superacid catalyst was obtained.

(2)   Sol–gel method. Sol–gel method is to make an organic metal salt (or inorganic salt) dispersed in the solution, its hydrolysis to produce active monomer, and then the active monomer to polymerize into a sol, aging it to produce a certain structure of the gel. Finally, the gel was dried and calcinated to form metal oxide carrier, and then it is impregnated with H_2SO_4 solution, filtered and dried and calcinated to get the solid superacid. The method is characterized by large specific surface area and the uniformity of the catalyst particles. However, the preparation period is

longer, generally takes a few days or tens of days to complete an operation run, and the preparation cost is high. For example, the preparation procedure of SO_4^{2-}/ZrO_2 catalyst is as follows: isopropyl alcohol is added dropwise to the propanol solution of tetrapropyl titanate and a small amount of nitric acid as a catalyst. The sol was hydrolyzed at the temperature of 150 °C for a certain time to obtain the gel. The gel was dried at 420 °C, pulverized, immersed in H_2SO_4 solution and calcinated at 550 °C to obtain SO_4^{2-}/ZrO_2 solid superacid. The specific surface area is 188 m^2/g, which is 50% larger than that of the sample prepared by precipitation impregnation method.

In a sol–gel process, the obtained catalyst will have better performance if combined with supercritical fluid drying techniques.

(3) Solid-phase synthesis. The solid-phase synthesis is referred to as a solid superacid as it is prepared by mixing a metal salt and a metal hydroxide in a predetermined ratio and then firing to obtain a corresponding metal oxide. The method has the advantages of simple equipment and process, easy control of reaction conditions, high yield and low cost, but the product size is not uniform and easy to agglomerate, thus affecting the acidity distribution and activity of the catalyst. Hino conducted the solid-state synthesis of Pt-SO_4^{2-}/ZrO_2 solid superacid. The preparation process was as follows: The dry $Zr(OH)_4$ powder was immersed in H_2PtCl_6 solution, dried at 200 °C, mixed with $(NH_4)_2SO_4$ powder and calcinated at 600 °C for 3 h to obtain Pt-SO_4^{2-}/ZrO_2. The experimental results show that the stability of the catalyst can be improved greatly by loading Pt.

3) Applications of solid superacid catalyst

(1) Alkylation. Alkylation is an important process for the preparation of surfactants, alkyl benzene, antioxidants and other aromatic hydrocarbons [4–5]. These reactions are all strongly acid-catalyzed processes. Traditionally, H_2SO_4, HF and $AlCl_3$ have been used as catalysts; at present solid superacid has been used in the alkylation process. In the alkylation of isobutane and butene with SO_4^{2-}/ZrO_2 as catalyst, the conversion of butene can reach 100%, and the content of C_8 in the product is more than 80%.

(2) Esterification. Solid superacid has been successfully used in the esterification process to synthesize various esters, plasticizers, surfactants, preservatives and fragrances. For example, when the ratio of stearic acid to polyethylene glycol (PEG400) is 1:2 in the PEG mono stearate (PEGMS) nonionic surfactant synthetic process with SO_4^{2-}/ZrO_2 as catalyst, the amount of catalyst used is about 15 g (1 mol/mol of acid), the reaction temperature is 125 °C, the reaction time is 6 h, the obtained esterified crude product is 44.8% and the diester is 1.0%. After purification, monoester content is 97% and stearic acid conversion is above 90%.

(3) Isomerization. It is a typical superacid-catalyzed reaction, mainly used for the preparation of isoparaffins from straight-chain alkanes, followed by dehydrogenation to produce isomeric olefins for many other purposes. The isomerization of n-butane was carried out at 20–50 °C with SO_4^{2-}/ZrO_2 as catalyst. The selectivity to isobutane was 97.9%. The Pt-containing SO_4^{2-}/ZrO_2 catalyst exhibits sound activity stability during this process and there is no deactivation after 1,000 h run.

The catalyst is also used for etherification, F-T synthesis, hydration (or dehydration) and cyclization.

2. **Molecular sieve catalyst**
1) **Introduction**

Molecular sieves are polymers of porous inorganic aluminosilicates capable of sieving substances at a molecular level, including natural zeolites and synthetic molecular sieves [5]. There are more than 50 kinds of natural zeolites, representing mordenite and montmorillonite. There are more than 170 types of synthetic molecular sieves. MCM-2, ZSM-5, SAPO, Ferrierite, RH and SAPO-34 have gained applications in industries. Those are being developed, including phosphorous zeolite, vanadium aluminum zeolite, nanomolecular sieve, and mesoporous and macroporous molecular sieve. Most of these molecular sieves can be used as a solid catalyst for oil refining, fine chemical preparation, gas purification, adsorption separation and preparation of special functional materials. Synthetic molecular sieves have become the most widely used catalyst in modern chemical industry.

2) **The classification and structural characteristics of zeolites**

According to the international classification, zeolite is identified into 10 groups: cubicite, sodium zeolite, heulandite, phillipsite, mordenite, chabasite, faujasite, laumonite and Pentasil (ZSM-5 and ZSM-11) and clathrate (ZSM-39). Chinese scholar divides the molecular sieves into five groups based on the characteristics of the zeolite skeleton and the crystal system. The first group is a zeolite skeleton with a quaternary ring and a six-member ring with a cubic configuration; the second group has a hexagonal or triple; the third group is its skeleton composed of five elements, usually orthogonal or monoclinic crystal; the fourth family is its skeleton composed of four-member or eight-member ring; the fifth group is those that do not have the above four characteristics of the zeolites.

The simplest structural unit of the zeolite framework is a tetrahedron composed of the central atom T (T is Si, Al, Ti, Fe, V, B, Ga, Be, Ge, etc.). Common central atoms are Si, Al and Ti. The central atom and the surrounding four oxygen atoms bond together with sp^3 hybrid orbital. The cubic structure and planar structure of the tetrahedral are shown in Figure 4.1.

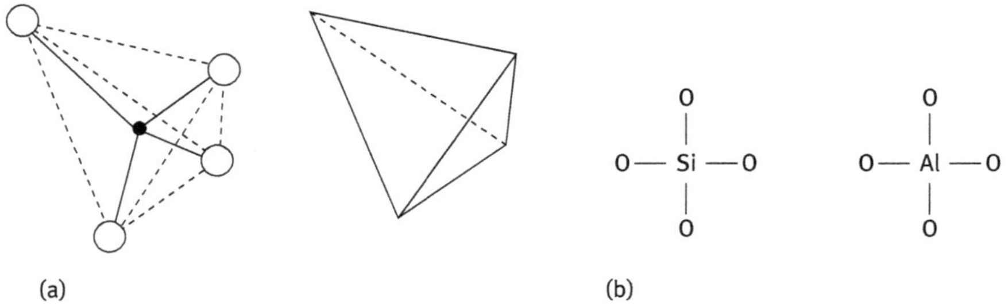

Figure 4.1: The basic structure of zeolite skeleton: (a) the three-dimensional structure of the basic structural unit and (b) the planar structure of the basic structural unit.

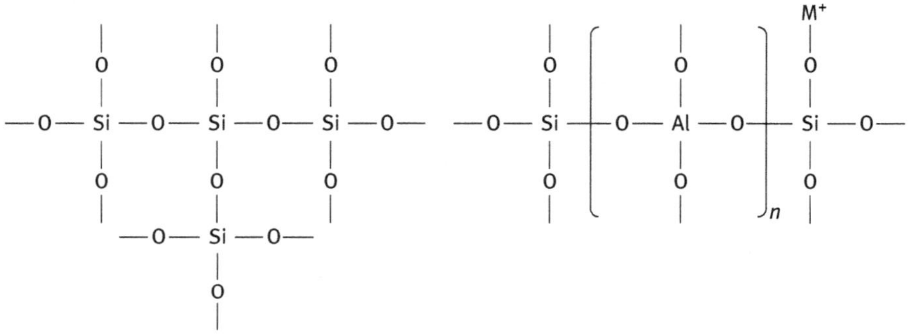

Figure 4.2: The connection between tetrahedra in zeolite.

The zeolite molecular sieve is composed of a plurality of TO_4 tetrahedral, and the tether is a tetrahedral vertex oxygen atom (also called oxygen bridge). The connection ways are shown in Figure 4.2.

During the connection process, two aluminum–oxygen tetrahedral can't be adjacent. The tetrahedral bodies interconnected by oxygen bridges form secondary structural units with different ring structures. One or more secondary structural units constitute a skeletal structure of complex molecular sieves. The secondary structure unit further forms the cage structure through the oxygen bridge. Cage structure is the main structural unit of various zeolites. The secondary structure unit is surrounded by a new larger hole in the process of assembling. The hole cage is connected with other holes through the multi-ring window, so that the inner pore cages formed between many channels, called the channel. The pore size of the zeolite refers to the maximum multimodal window size of the zeolite main pore cage. The pore structure data of different zeolite molecular sieves are shown in Table 4.1. The geometry of the different zeolites is not the same. According to the size of the aperture, the zeolite can be divided into micropores, mesopores, macropores and super-macropores.

Table 4.1: The maximum diameter of the multivariate ring.

Ring	Four-member ring	Five-member ring	Six-member ring	Eight-member ring	Ten-member ring	Twelve-member ring
The maximum diameter (nm)	1.15	1.6	2.8	4.5	6.3	8.0

In addition to the differences in the shape and size of the channels, the pore distribution (dimension) of the different zeolites is different. In the coordinate axes, the channels are called n-dimensional, including three-dimensional (zeolite A, Faujasite, ZSM-5), two-dimensional (Mordenite, Ferrierite) and one dimensional (ZSM-48).

3) The characteristics of zeolite catalyst

(1) The characteristic pore size of the zeolite allows it to be selectively adsorbed on the reactants. In general, molecules with a kinetic diameter greater than 0.1 nm or less than the pore size of the zeolite pore can enter the pores of the zeolite. Thus zeolites have excellent selectivity as catalysts.

(2) The specific pore structure of the zeolite directly affects the diffusion behavior in the reaction system. The pore and macroporous zeolite molecular sieve are easily inactivated due to the limitation of the product diffusion, and the mesoporous zeolite is not easy to inactivate by coking.

(3) Some zeolites are acidic. It has been proven that zeolites have Lewis acid or Brönsted acid sites that can be used for acid-catalyzed reactions.

(4) Zeolites possess high thermal stability and hydrothermal stability.

(5) The zeolite catalyst is nontoxic, environmentally friendly and no corrosion on the equipment. Therefore, it is a kind of solid catalyst that meets the requirements of green chemistry.

4) Formation of zeolite acidic sites and its catalytic action principles

The formation of zeolite acid sites has the following four ways:

(1) The zeolite leaves the cation to form an acid center.

(2) Exchange of zeolite with polyvalent cation to form acid centers.

(3) The protons, cations and skeletal oxygen in a zeolite can be moved, and the surface diffusion of oxygen ions and cations on the zeolite surface lead to the dissociation of O–H, which enhances the acidity of the zeolite.

(4) The Brönsted acid sites were produced by cation exchange of zeolite, or by dehydration of the Lewis acid center and can be applied to the reactants to form carbon-positive ions, and it follows carbon cationic mechanism in catalytic transformation.

5) Typical preparation methods of zeolite catalyst

(1) Sol–gel method

It is the most primitive synthetic method of zeolites. It includes the soluble aluminum salt and soluble silicon salt to form an active gel under strong basic condition, and the gel under heat aging can form zeolite structure. A further improvement is the use of quaternary ammonium salt-based modulating agents to regulate the pore structure of the zeolite. The reaction is as follows:

$$NaAlO_2 + Na_2SiO_3 + NaOH + R_4N^+OH^- \rightarrow Silica\ gel \rightarrow Zeolite$$

A-type zeolite preparation: The chemical composition of A-type zeolite formula is $Na_2O \cdot Al_2O_3 \cdot 2SiO_2 \cdot 5H_2O$. Its raw materials are sodium silicate, sodium aluminate, sodium hydroxide and water. The synthetic process is as follows: First of all, the mixture is made up of Na_2O, Al_2O_3, SiO_2 and H_2O, and the mass ratio of the substances is 3:1:2:185. The corresponding component concentrations were Na_2O 0.9 mol/L, Al_2O_3 0.3 mol/L and SiO_2 0.6 mol/L, respectively. Every solution was measured by metering tanks, and then putting the sodium aluminate solution, sodium hydroxide and water into a gel-mixing tank and pre-heated to 30 °C under stirring, then the water glass is quickly thrown into the tank, stirred around for 30 min and then formed an uniform gel. Heat and stir the mixture for 20–40 min at 98–102 °C, and conduct static crystallization for 5 h at the temperature. The product is fully precipitated at lower part of the reaction tank; crystal product from the sampling port is checked with a microscope, and appearance of total clear square crystal may be the end of the synthetic process. The mothor liquor in the upper part of the tank was transferred into a liquor storage tank and the lower part liquor was transfered into storage tank to recover the left mother liquor by precipitation of the molecular sieve product. Put an agitator into the product storage tank and separate the product through a plate filter, and then the product was washed with water till the pH of the filtrate is 9–10. Products will be discharged from the filter and dried at 110 °C for 5–7 h to obtain crystal powder of zeolite A. The mother liquid is dilute sodium hydroxide solution and can be recycled. The synthetic process flowchart is shown in Figure 4.3.

(2) Hydrothermal synthesis

Hydrothermal synthesis is the most widely used method for synthesizing zeolite. Hydrothermal synthesis of zeolite consists of two basic processes: the formation of aluminosilicate hydrated gels and the crystallization of the hydrated gels. The crystallization process generally includes the following four steps: polysilicate and aluminate

Table 4.2: The raw materials used for synthesis of zeolite.

Material type	Specific material name
Aluminum compound	Inorganic salt-activated alumina, aluminum hydroxide, sodium aluminate and other aluminum
Silicon resources	Silica gel, silica sol, sodium silicate, silicate, powdered SiO_2 and quartz glass
Alkali	Sodium hydroxide, potassium hydroxide, fluoride
Template	Quaternary ammonium salts, amines, diamines, alkanolamines, alcohols, glycols, quaternary phosphonium bases
Water	Deionized water

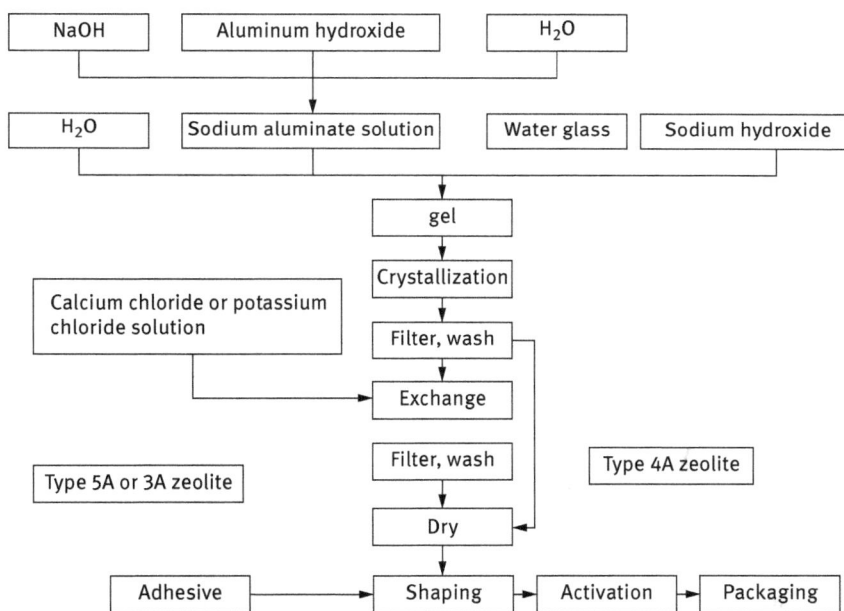

Figure 4.3: Flowchart of A-type zeolite synthesis.

repolymerization, zeolite nucleation, nuclear growth, and zeolite crystal growth and secondary nucleation.

There are many factors influencing the synthetic process of zeolite. At present, the effect of reaction conditions on the synthetic process is mainly studied, such as composition of reactants, type and nature of reactants, aging conditions, crystallization temperature and time and pH value.

The main raw materials for the synthesis of zeolite are aluminum-containing compounds, silicon-containing compounds, alkali, water and template agent. The composition is usually expressed as $xM_2O \cdot Al_2O_3 \cdot ySiO_2 \cdot zH_2O$. The commonly used raw materials for preparing zeolite are shown in Table 4.2.

The general process of synthesizing zeolite by hydrothermal method is

Reactant → Gel → Zeolite crystal → Zeolite crystal powder

Adding the reactants arranged in advance according to a certain proportion to a lined plastic or stainless steel reactor and performing the crystallization reaction at 100–450 °C for a period of time. After the crystallization is completed, the precipitate is subjected to filtration, washing and drying, and zeolite powder having a size of 1–10 μm can be obtained. When a high silica zeolite is prepared, the crystallization temperature should be more than 150 °C, and the general zeolite crystallization temperature is 100 °C; washing pH is 9–10; drying temperature is 110 °C. Zeolite with low silica alumina ratio can be obtained with inorganic base, and zeolite with high silica alumina ratio can be obtained with organic base.

Aluminophosphate zeolite: It is a kind of heteroatom molecular sieve, which uses element phosphor to replace the silicon in the zeolite skeleton. It is composed of tetrahedral and phosphorus oxygen tetrahedron. Generally expressed as $(AIPO_4)_n$; composition of $xRAl_2O_3 \cdot (1 \pm 0.2)P_2O_3 \cdot yH_2O$ (R on behalf of a template), pore size 0.5–1 nm. More than 60 kinds of aluminophosphate zeolites have been synthesized.

Aluminophosphate zeolite was synthesized by hydrothermal crystallization. The equal amount of active hydrated Al_2O_3 and phosphate in water reacted to form aluminophosphate gel, adding organic amine or quaternary ammonium salt as template, stir the mixture evenly and put into an autoclave lined of polytetrafluoroethylene for static crystallization under 125–200 °C to obtain zeolite crystals. The crystal obtained was calcinated at 400–600 °C to form the aluminophosphate zeolite.

In addition, silicon-aluminum-phosphate molecular sieve and titanium-aluminum-phosphorus molecular sieve can be obtained by hydrothermal crystallization method.

Mesoporous molecular sieve with larger pore size and adjustable pore size, specific surface area (up to 1,000 $m^2 \cdot g^{-1}$), high porosity, surface rich in unsaturated groups, high thermal stability and hydrothermal stability has become an important new catalytic material. Mesoporous molecular sieves are divided into six classes according to their structure: the MCM-41 of the hexagonal phase, the MCM-48 of the cubic phase, the MCM-50 of the layered, the SBA-1 of the six phase, the SBA-2 of the cubic structure and the MSU-n of disordered arrangement of the hexagonal phase structure.

Mesoporous molecular sieves were also synthesized by hydrothermal method or solvothermal method. First, surfactant, acid or alkali is added to water to form a mixed solution, then the silicon source or other materials are added to the solution, and the system endures hydrothermal treatment or room temperature aging. The product was washed, filtered and finally by calcinating or chemical treatment the organic matter is removed to obtain mesoporous molecular sieve.

Nanomolecular sieve: Nano-molecular sieve is a molecular sieve with particle size less than 100 nm. It has a larger surface area, more crystal cell exposure, to increase reactivity and contacting area; short regular pore structure is more favorable to the diffusion and catalytic reaction; with more accessible active sites; more uniform distribution of active sites; easy structure adjustment through ion exchange, skeleton scheduling and surface modification to demonstrate better catalytic performance.

6) **Applications of molecular sieve catalyst**
   (1) Acid-catalyzed reaction in petroleum processing. Acid-catalyzed reaction is the most important application field of molecular sieve catalyst, which has been widely used in the cracking, isomerization, alkylation, disproportionation, hydration, dehydration, oxygenation and dehydrogenation.
   (2) Oxidation: Oxidation reaction is an important reaction for the preparation of a variety of fine chemicals; at present, oxygen or hydrogen peroxide as oxidant is preferred. When using hydrogen peroxide as oxidant with TS-1 molecular sieve or β-zeolite as catalyst, a series of fine chemicals can be prepared. The TS-1 molecular sieve as catalyst for direct hydroxylation of phenol to hydroquinone and catechol has been industrialized in Italy; selectivity to diphenol is calculated as 90%.
   (3) N-alkylation and O-alkylation: Many important fine chemical products can be prepared by N-alkylation and O-alkylation. At present, the main alkylating agents are dimethyl sulfate and halohydrocarbons. These two kinds of alkylating agents will produce serious environmental problems in process. With molecular sieve as catalyst, and methanol or dimethyl carbonate as alkylating agents, a series of environmental-friendly production processes can be designed. With hydrogen-type zeolite as catalyst and methanol as alkylating agent, the green synthesis process of the drug intermediate, dimethylimidazole, is developed as follows:

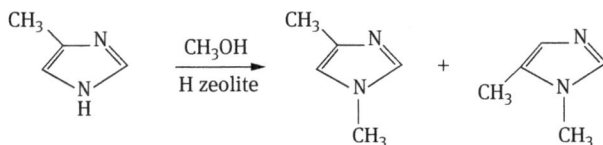

Rhone-Poulenc Company developed a new process for the preparation of vanillin from catechol, with mordenite as a catalyst, and the reaction process is as follows:

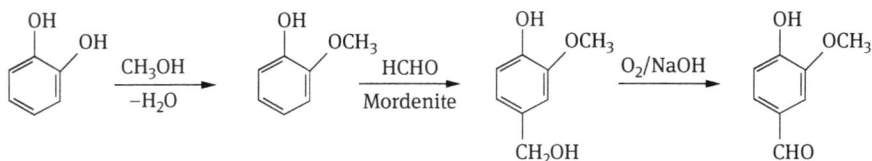

(4) Aromatic nitration: The traditional nitrification reaction uses HNO_3 and H_2SO_4 mixed acid as nitrating agent. This process causes severe environmental pollution. Now a new process for gas-phase catalytic nitration with benzene and 65% nitric acid, dealuminated mordenite as catalyst have been developed. This process not only realizes the environmental-friendly production but also has a controllable product distribution.

(5) Amination: The traditional method for the production of aniline is catalytic hydrogenation of nitrobenzene. Recently a new process of direct catalytic amination of phenol to aniline was developed. Aminophenol is produced with resorcinol as raw material and molecular sieve as catalysts by direct amination, an environmental-friendly production process.

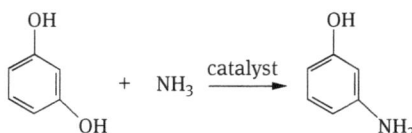

(6) Nitrogen heterocyclic reaction: With the rapid development of medical and related industries, the demand for pyridine and its alkyl substituted derivatives is increasing. Scientists successfully developed a new method to prepare pyridine compounds with aldehydes and molecular sieve as catalyst by amination. For example, when acetaldehyde (or formaldehyde) is used as raw material and HZSM-5 as catalyst, pyridine and 3-picoline are successfully prepared. The reaction equation is as follows:

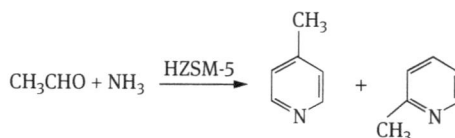

3. Heteropoly compound catalyst

1) Introduction

Heteropoly compound (HPC) is a general designation of heteropoly acid and its salts. Heteropoly acid is the polyacid of two or more inorganic oxygen acids. In HPCs, its heteroatoms (P, Si, Fe, Co, etc.) and coordination atoms

(Mo, W, V, Nb, Ta, etc.) construct a certain structure through the oxygen atom coordination oxygen to form polyacid [6]. Polyoxometalate is the product when some or all hydrogen of the heteropoly acid is substituted by metal ions or organic amine. Solid HPC is composed of heteropoly anions, cations and crystal water. Among them, the heteropoly anion is a polynuclear coordination structure which connected with central atom and coordination atoms through an oxygen bridge.

At present, HPCs can be divided into five categories: ① Keggin structure, such as $H_3PW_{12}O_{40}$; ② Anderson structure, such as $[TeMo_6O_{24}]^{6-}$; ③ Silverton structure, such as $[CeMo_{12}O_{42}]^{8-}$; ④ Wangh structure, such as $(NH_4)_6MnMo_9O_{32}$; and ⑤ Dawson structure, such as $K_6P_2W_{18}O_{62}$. HPCs used as catalysts mainly refer to substances with Keggin structure. The Keggin structure is a cage-like molecule formed by the connection of $12\,MO_6$ (M = Mo, W) octahedron around a PO_4 tetrahedron, which is similar to the cage-like structure of zeolites.

2) **Types and essential properties of HPC catalysts**

HPC catalysts include heteropoly acid, polyoxometalate, their supported types and their structure-modified ones. They generally have high molecular weight, and heteropoly acid and salts of small metal ions dissolve easily in water and in other polar solvents. As for large cationic salts (Cs^+, Ag^+, NH_4^+), it is insoluble or slightly soluble. The compounds are stable in aqueous solutions of low pH and in organic solvents, but are readily dissociated in solutions of high pH, and their free acids are highly acidic.

HPCs generally have good thermal stability below 350 °C. Due to the existence of metal element with various valences, they possessed redox properties and are called a bifunctional catalytic material and can be used as the acidic and redox catalyst.

The main properties of HPCs as catalysts are as follows.

(1) They have strong Brönsted acidity. Its acidity is stronger than that of its constituent elements (such as $H_3PW_{12}O_{40}$, $H_0 = -13.2$), but the corrosion is much weaker than that of common inorganic acids.

(2) They can easily solute in water and common organic solvents, which makes heteropoly acid and reaction mixture to easily form a homogeneous system, which was beneficial for the reaction.

(3) Their acidity can be designed and controlled by changing the composition of anions, the formation of salt and the loading and so on; therefore, the catalyst can be designed according to the characteristics and requirements of the reaction. Large groups of modified HPCs have been designed and prepared for catalytic purposes.

(4) They have good chemical and thermal stability. In a common acid and alkali media, the HPCs can maintain their structure with the thermally stable temperature of 350 °C, so they can be applied to most of the reactions.

(5) They are suitable for homogeneous or heterogeneous reaction systems. For nonpolar molecules, reaction takes place only in surface, and for polar molecules, they can diffuse into the lattice in the bulk to react (called pseudo-liquid-phase behavior). This behavior makes the reaction to occur both in the solid surface and inner solid phase, exhibiting very high catalytic activity.

To overcome its small specific surface area, high cost and product separation and recovery difficulties and to improve the utilization efficiency of the catalyst active component, researchers have proposed the preparation of supported HPC catalyst. The so-called supported HPC refers to an HPC loaded on a carrier with inert pore structure (SiO_2, C and zeolite). This can greatly reduce the amount of catalyst used in a reaction process and is favorable to the separation of catalyst from the liquid phase, making the HPCs to be suitable for gas–solid catalytic system. However, when the HPCs are loaded, they will change in acidity; there are still problems in the dissolution of the active components of the catalyst.

3) Preparation of HPC catalysts

(1) Heteropoly acid catalyst

As early as 1826, Berzelius J.J. obtained the phosphomolybdic acid by acidifying the mixed solution of molybdate and phosphate. With the improvement of the preparation methods, some more advanced preparation methods have been invented.

① Acidification. The heteroatom oxygen acid and polyatomic oxygen-containing acid or polyatomic oxide are well mixed in a certain ratio and the mixed solution is acidified. After heating to reflux for 1–12 h, the heteropoly acid catalyst is obtained by extraction with ether or crystallization. The synthetic procedure of $H_4PMO_{11}VO_{40}$ is as follows: 3.58 g of $Na_2HPO_4 \cdot 12H_2O$ is dissolved in 50 mL of distilled water; meanwhile, 26.65 g of $Na_2HPO_4 \cdot 12H_2O$ is dissolved in 60 mL of distilled water. The two solutions were mixed and heated to boiling for 30 min. Then 0.91 g of V_2O_5 was dissolved in 10 mL of Na_2CO_3 solution, and this solution is added to the above mixed solution under stirring, after 30 min of reaction at 90 °C; a certain amount of 1:1 H_2SO_4 was added while stirring, and the solution was divided into three layers. The middle part of the bright red oil is heteropoly acid ether complex, remove the ether from the complex, add a small amount of distilled water and then place it in a vacuum dryer, until the crystal is completely crystallized, recrystallized and dried to obtain the product.

② Ion exchange: The heteropoly acid salt is used as the raw material, and the aqueous solution of the heteropoly acid salt is exchanged through a strongly acidic cationic exchange resin, so that the metal ions in the salt are exchanged with the hydrogen ions. The effluent solution is a heteropoly acid solution. and then the crystalline or powdered pure heteropoly acid is obtained by extracting with ether or evaporating crystallization.

③ Degradation: By controlling the pH value of the heteropoly acid solution, the heteropoly anions are partially degraded, thereby the heteropoly acids containing less atoms are obtained.

④ Electro-osmotic method: The electro-osmotic method is one of the newly developed methods to prepare heteropoly acids. H_3PO_4 and Na_2WO_4 as raw materials are circulated through the anode box and the alkali solution circulated in the cathode boxes were separated by the cation exchange resin film. When the current passes, the cation enters the cathodic electrolyte through the semipermeable membrane, and the anodic electrolyte is acidified. The following reaction occurs in the electro-osmotic device:

$$Na_2WO_4 + H_3PO_4 \rightarrow H_3PW_{12}O_{40}$$

After the removal of the cation, the cationic electrolyte is evaporated to crystallize to give the heteropoly acid. In this process, the yield of heteropoly acid is above 99% and the current efficiency is 15–30%.

(2) Heteropoly acid salt catalyst.

There are two kinds of methods to prepare heteropoly acid salt.

① Heteropoly acid partial neutralization method: A saturated solution of an alkali metal or alkaline earth metal ion is added to the saturated solution of the heteropoly acid to obtain a heteropoly acid salt directly.

② Solid-phase reaction: The synthesis of nanometer heteropoly acid salt can be carried out by utilizing the advantage of solid-phase reaction without using solvent. The synthesis of the ammonium phosphomolybdate is given as an example here. The equation of the reaction process is as follows:

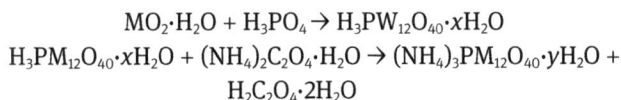

$$MO_2 \cdot H_2O + H_3PO_4 \rightarrow H_3PW_{12}O_{40} \cdot xH_2O$$
$$H_3PM_{12}O_{40} \cdot xH_2O + (NH_4)_2C_2O_4 \cdot H_2O \rightarrow (NH_4)_3PM_{12}O_{40} \cdot yH_2O + H_2C_2O_4 \cdot 2H_2O$$

The specific preparation process is as follows: First, weigh 48.6 g (0.3 mol) of molybdic acid, 7.0 g (0.06 mol) of phosphoric acid, respectively, and place them into the gate mortar in full grinding, then add ammonium oxalate 5.1 g (0.036 mol) to continue to full grinding. At the beginning it was sticky, then gradually dried up. The resulting mixture powder was washed with absolute ethanol and centrifuged after grinding for 40 min, then repeated the runs for four to five times and then vacuum dried at 50–60 °C for 24 h to give ammonium phosphomolybdate powder of about 48.0 g.

(3) Preparation of supported HPC catalyst.

The preparation methods of the supported HPC catalyst are mainly impregnation method, sol–gel method and hydrothermal dispersion method. The commonly used method is the impregnation. The carriers used in the loading process are activated carbon, SiO_2, MCM-41, ion exchange resin and carbonized resin.

① Impregnation: A certain amount of carrier is immersed in a known concentration of heteropoly acid solution, heated to reflux for a certain period of time, filtered, washed and dried, and then activated at a given temperature to obtain the supported catalyst. By changing the concentration of the heteropoly acid solution and the reflux time, the catalyst with a different loading amount can be obtained.

② Sol–gel method. Orthosilicate is hydrolyzed under the acidic condition and in the presence of the heteropoly acid to form a silica sol. The silica sol is gelatinized and dried to form the supported heteropoly acid catalyst. The preparation process is as follows: Orthosilicate, n-butanol, 12-tungstophosphoric acid and deionized water are well mixed at a certain ratio, stirred and heated to reflux for 2 h to hydrolyze the tetra-ethoxysilane to form a transparent sol. The sol was transferred into a plastic mold, placed it in a water bath at 80 °C for 2 h to form a transparent gel, then dried at 100 °C, to get the corresponding catalyst after grinding. The heteropoly acid catalyst prepared by this method has a relatively lower leaching of active component, but its acidity is lower than the catalyst obtained by the impregnation method.

③ Hydrothermal dispersion method: This method refers to the carrier and the known concentration of heteropoly acid solution is mixed in a certain ratio, added in a stainless steel autoclave, reacted at 90–110 °C for 24 h, then rapidly removed the moisture in the solid, finally dried at 110–120 °C, grounded and calcinated at a given temperature for 4 h to obtain the catalyst.

4) **The applications of HPC catalyst**

(1) Alkylation: Straight-chain dodecylbenzene is an important raw material for the production of anionic detergents. At present, mesoporous molecular sieve MCM-41-loaded silicotungstic acid catalyst is used in the synthesis of dodecylbenzene. In the reaction of isobutene and butene alkylation to prepare the clean gasoline, $Cs_{2.5}H_{0.5}PW_{12}O_{40}$ was used as the catalyst. The product yield and selectivity achieve 79.4%

and 73.3%, respectively. When 40% of HPW_{12}/SiO_2 was used as catalyst, the conversion of butene was 98.8%, C_8 alkane accounted for 59.5% of the liquid product and no loss of the active component was found during the reaction.

(2) Condensation: Xylene–formaldehyde resin is an important raw material for the production of paint and new polyester. Sulfuric acid was used as the catalyst in the traditional preparation process. There are many problems, such as complex process, environmental pollution and unstable product quality. When the unsaturated silicon-molybdenum-tungsten mixed HPC is used as the catalyst, the catalyst dosage is 0.8–10.0% and the reaction temperature is 160 °C, the yield achieves 95%. In addition, the catalyst can be reused, and the whole process is of no pollution to the environment and the process has been greatly simplified.

(3) Nitration: Heteropoly acid and its salts show good catalytic activity in the gas-phase nitration of benzene. $H_3PW_{12}O_{40}/SiO_2$-Al_2O_3 can catalyze the gas-phase nitration of benzene with NO_2 at 270 °C and no dinitrobenzene was produced in the product. The rate increases with the increase of the loading of $H_3PW_{12}O_{40}$, and the yield of nitrobenzene achieves 56% when the loading of phosphotungstic acid is 30%.

(4) Hydration: HPCs are highly efficient catalysts for hydration reactions. Silicotungstic acid, phosphotungstic acid, borotungstic acid, phosphomolybdic acid and silicolybdic acid can be used as catalysts for the synthesis of ethanol from ethylene, propylene and butene. At the reaction temperature of 170–350 °C, the pressure of 10.0 MPa, and the catalyst concentration of 10^{-5}–10^{-3} mol/L, the selectivity can reach 95–99%.

(5) Polymerization: The tetrahydrofuran homopolymer (PTMEG) obtained by cationic ring-opening polymerization of tetrahydrofuran is an important material to produce PU elastomer. Studies have shown that phosphotungstic acid and molybdenum-tungsten heteropoly acid are highly efficient catalysts for this polymerization and have been successfully used in the 10,000 ton industrial installation.

4. Polymer acid catalyst

1) Overview

Polymer acid catalyst refers to the ion exchange resin which is a cross-linked polymer-bearing sulfonic acid groups, simply expressed as RSO_3H (R for the polymer matrix), such as strong acid cationic exchange resin and perfluoric acid resin [7]. Compared with the traditional acidic catalyst, the catalyst has the following characteristics: ① the complete separation of the catalyst from the system can be achieved by simple filtration, greatly simplifying the process; ② high selectivity for the target reaction; ③ easy to achieve continuous production; ④ corrosion free; and ⑤ less than "three wastes" generated during the reaction.

2) The main properties of polymer acid catalyst

(1) The exchange volume

The exchange volume of a strong acid cationic exchange resin is an important index to characterize the acidity of resin. It refers to the number of all sulfonic acid groups in unit mass or volume cationic exchange resin, expressed in mmol/g or mmol/mL. The measuring method is as follows.

When the hydrogen type cationic exchange resin is immersed in a calcium chloride solution, only the strong acid group (sulfonic acid group) can react, and the exchange capacity of the strong acid group is calculated by the titration of the replaced hydrogen ion (H^+). The reaction is

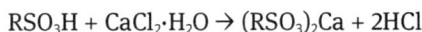

$$RSO_3H + CaCl_2 \cdot H_2O \rightarrow (RSO_3)_2Ca + 2HCl$$

The exchange capacity of the resin is calculated as follows:

$$Q'_s = \frac{4(V-V_0)C_{NaOH}}{m}$$

$Q's$ is the cationic exchange resin wet-base strong acid group exchange capacity, in mmol/g; C_{NaOH} is the concentration of NaOH standard solution in mol/L; V is the volume of NaOH standard solution consumed by soaking solution in mL; V_0 is the volume of NaOH standard solution consumed by blank solution in mL; m is the mass of the resin sample in g.

(2) Particle size: The ion exchange resin is generally spherical particles having a particle size of 0.03–1.20 mm. The particle size of the resin is often expressed as a standard mesh number, and the number of American standard sieve and millimeters can be converted using the following empirical formula:

Particle size (mm) = 16/mesh number

(3) Pore structures: Pores in an ion exchange resin are divided into two categories; one is the gel pore, it refers to the distance between the resin macromolecule chains, rather than a authentic pore, and it changes with the change of external conditions, so it is difficult to determine; the other is the macroporous, which is the real pore, can be measured by low-temperature nitrogen adsorption method.

(4) Stability: The stability of the ion exchange resin refers to the circumstances changed in external force, heat and chemical effects, mainly including mechanical strength, heat resistance and chemical stability. These directly relate to the service life span of the resin. Mechanical strength refers to the ability of the resin to resist the deformation of

all kinds of machinery. Heat resistance refers to the temperature range in which the resin does not undergo thermal decomposition during use. The results show that the heat resistance of ion exchange resin is closely related to its structure. The maximum working temperature of ordinary gel-type strong acid resin is 100 °C, and the maximum working temperature of macroporous resin can reach 130 °C. Chemical stability refers to its ability to withstand the attacks of chemicals and oxidants.

When the ion exchange resin is used as catalyst, the composition of the resin, the pore structure, the degree of cross-linking, the nature of the exchange group and the nature of the counter-ion all affect the reaction.

3) Preparation of polymeric acidic catalysts

(1) Preparation of the strongly acidic cationic exchange resin: The preparation of strong acid cationic exchange resin mainly includes suspension polymerization and sulfonation.

① The synthesis of the styrene–divinylbenzene copolymer microspheres by the suspension polymerization. With the styrene as the monomer, the divinylbenzene as the cross-linking agent, the polymer is obtained by suspension copolymerization after stirring and heating in the presence of the initiator in the aqueous medium containing the dispersant and the pore-opener. Commonly used initiator is benzoylperoxide or azo-bis-isobutyronitrile with a proportion of monomer mass of 0.5–1.0%. The dispersant is generally a polyvinyl alcohol solution of 0.1–0.5%, a polyvinyl alcohol solution having the degree of hydrolysis of about 88% or an aqueous solution of photographic gelatin in sodium chloride of 0.5–1.0%. The mass ratio of water to monomer is (2–4):1. Magnesium phosphate, magnesium carbonate and calcium phosphate are used as dispersants. The polymerization reaction is as follows:

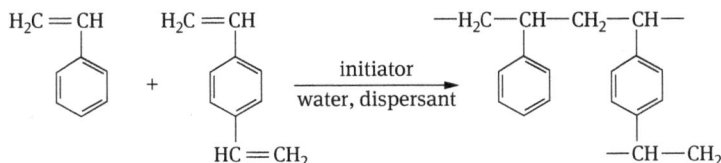

② Sulfonation of copolymer microspheres: The polystyrene cross-linked by divinylbenzene is stable, and the reaction can be carried out by the reactivity of the hydrogen atom on the benzene ring to proceed sulfonation, then the strong acidic cation resin is obtained. The sulfonation reaction is as follows:

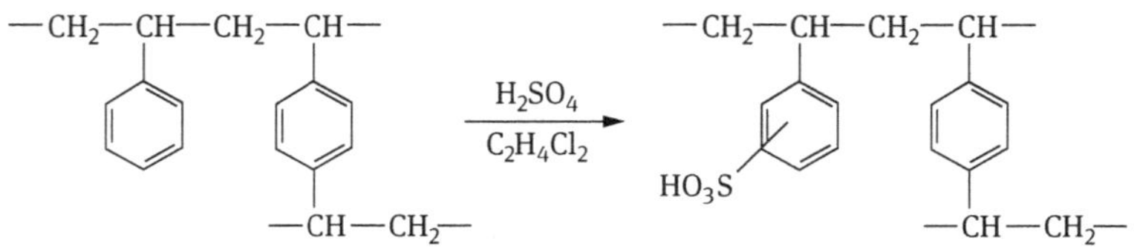

(2) Preparation of perfluorosulfonic acid resin (Nafion-H): Nafion-H is the strongest solid acid as known. Its chemical structures and main preparation methods are as follows:

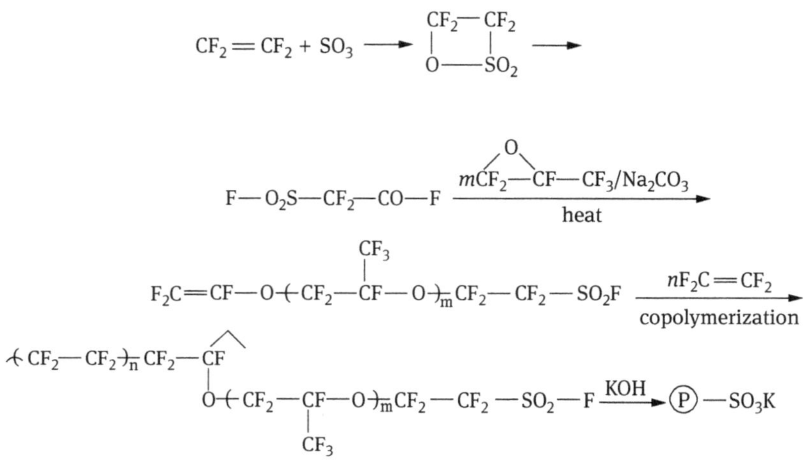

As Nafion-H preparation process is more complex, the Nafion-H was prepared by ion exchange of Nafion-K resin in the laboratory. For example, 50 g of Nafion-K resin was boiled with 150 mL of deionized water for 2 h, filtered, then with the concentration of 20–25% HNO_3 solution 200 mL, stirring at room temperature for 4–5 h, filtering, treating with 20–25% HNO_3 solution, repeat three to four times. Eventually washed with water to neutral, filtered, vacuum dried for 24 h above 105 °C.

4. **The applications of polymer acid catalyst**

   1) **Alkylation**

   Nafion can catalyze alkylation and the reaction can be carried out in the gas phase or liquid phase. The relative catalytic activity of the other ion exchange resins was compared using the alkylation reaction of benzene with propylene: the reaction rates of Amberlyst 15, Nafion-H and Nafion-H/SiO_2 are, respectively, 0.6 (10.7%), 2.0 (2.2%), 87.5 (16.2%) at 100 °C, respectively.

   2) **Acylation**

   Acylation reaction generally requires a high reaction temperature and a higher acid strength of the catalyst. Macroporous polystyrene sulfonic acid resin can only catalyze the acylation of highly active aromatic rings. Various ion exchange resin catalysts have been tested on the reaction of anisole and acetic anhydride; the selectivity to the product was 100%; and the catalytic activity of Amberlyst-36 shows the highest activity among them.

3) **Condensation**

Methyl isobutyl ketone is an important solvent, using strong acidic cationic exchange resin Amberlyst IR-120 as catalyst, acetone is condensed, dehydrated and hydrogenated to form methyl isobutyl ketone, and this process has been industrialized.

4.1.2 Solid Base-catalyzed Organic Synthesis

1. Introduction

Solid base catalysts (heterogeneous base) become more and more attractive, and have been widely used in the oxidation, amination, hydrogenation, reduction, addition and other typical organic reactions [8]. With the development of green chemistry, researchers are paying more and more attention to the environment-friendly new catalytic process. Due to the high activity, high selectivity, mild reaction conditions, easy separation of products and recyclable, solid base plays a more and more important role in the synthesis of fine chemicals and is expected to be a new generation of environmental-friendly catalytic materials.

Solid base refers to the solid that causes an acid indicator to change color or chemically adsorb acid. According to the definition of Brönsted and Lewis, a solid base refers to a solid substance that has the ability to accept protons or give electronic pairing.

At present, solid base mainly includes three categories: organic solid base, organic/inorganic composite solid base and inorganic solid base. The organic solid base mainly refers to the alkaline resins having a terminal group as a tertiary amine or a tertiary phosphine group.

Organic/inorganic composite solid base mainly refers to molecular sieves loaded with organic amines or quaternary ammonium. The alkali active site of a molecular sieve loaded with organic amines is mainly the nitrogen atom which can provide the lone pair electrons; however, the base active site of the loaded quaternary ammonium base molecular sieves is mainly hydroxide ions. Since the active sites of this kind of solid base are organic bases, which are connected by chemical bonds with molecular sieves, the active components in the reaction process are kept, and the alkalinity is uniform, but it is not suitable for high-temperature reactions.

Inorganic solid base has the advantages of simple preparation, wide-base distribution and good thermal stability, and has become the main species of solid base catalyst. These include metal oxides, metal hydroxides, hydrotalcites and supported solid bases.

1) **Metal oxide inorganic solid base**

The metal oxide-type inorganic solid base mainly refers to the alkali metal and the alkaline earth metal oxide. The alkali sites of metal oxide are mainly derived from the hydroxyl groups produced by the surface adsorption of water and the negatively charged lattice oxygen. As MgO, the alkali active sites on the surface of MgO at low temperature are mainly weakly basic hydroxyl groups; however, at high temperature, the surface appears; line, point defects in the surface made the original six coordinated Mg into five coordination, four coordination or three coordination, and increase the charge density of oxygen atom, so that different degrees of alkali-active sites exist in the surface of MgO. Generally speaking, the basic strength of alkali metal and alkaline earth metal oxide catalyst increased with the increase of its atomic number, ordered as $Cs_2O > Rb_2O > K_2O > Na_2O$, $BaO > SrO > CaO > MgO$. In addition, the calcination temperature and the types of precursors also significantly affect the alkali strength of alkali and alkaline earth metal oxides; generally, a high calcination temperature is favorable for the formation of strong alkali-active sites. The alkaline strength of the alkaline earth oxide obtained by calcination of different precursors is carbonate > hydroxide > acetate. All in all, by changing the preparation conditions or selecting different precursors, it is possible to prepare the alkali metal and alkaline earth metal oxide which has the strongly basic active sites or even the super-alkaline active sites, but these solid alkali are powder with low mechanical strength, hard to separate and has small surface area (except MgO, all of them were less than 70 m^2/g); its specific surface area reduced with the significant increase in alkali-active site.

2) **Hydrotalcite-based inorganic solid base**

The hydrotalcite-like material is a layered double hydroxide (LDH), and its structure is $[M_1^{2+}M_2^{3+}(OH)_{2(x+1)}](A_{1/m}^{m-}) \cdot nH_2O$, among them, M_1 is Mg, Zn or Ni and M_2 is Al, Cl or Fe, A^{m-} can be Cl^-, CO_3^{2-}. The plates with positive charge were made up of M_1^{2+} and M_2^{3+} centric octahedrons by edge shared. A^{m-} and H_2O are various anions and water molecules located between the layers, A^{m-} plays a role in balancing laminar positive charge. When M_1 is Mg, M_2 is Al, the surface of this hydrotalcite catalyst has both acid- and alkali-active sites, appropriately changing the ratio of magnesium to aluminum and neutralizing anions can change the charge density of the laminar oxygen atoms, thereby adjusting the ratio of the acid–base active sites on the surface of such catalysts. Its preparation methods include co-precipitation, hydrothermal, ion exchange, calcination recovery and crystallization isolation.

3) **Supported inorganic solid base**

At present, the carriers commonly used to prepare supported inorganic solid bases include alumina, molecular sieves or activated carbon, calcium oxide, zirconium dioxide and titanium dioxide. The precursors are mainly alkali metal, alkali metal hydroxides, carbonates, fluorides, nitrates, acetic acid salt, amide or azide. Because alkaline earth metal hydroxides and carbonates are insoluble matters, but its nitrate decomposition temperature is generally higher, there are also a few reports with alkaline earth metal acetate as the precursor. The active sites of the supported inorganic solid base are mainly alkali metal or alkaline earth metal oxides, hydroxides and carbonates and alkali metals, there are also active sites that are produced by reaction with the carrier after calcination at high temperatures.

2. **Methods for preparing solid bases**

There are several methods to prepare solid base catalysts: impregnation method, co-precipitation method, hydrothermal method, ion exchange method and others. In this chapter, we take the preparation of hydrotalcite as an example to illustrate the preparation process of typical solid bases.

1) **Hydrotalcite-like inorganic solid base**

LDH, also called the layered structure of double hydroxyl anionic clay, is a kind of novel inorganic functional material that has developed rapidly in recent years. In order to prepare pure hydrotalcite-like compounds, we should rationally select the ratio of cationic to anionic first. It is required that the anions in LDHs must be present at a higher concentration in the solvent and have a strong affinity with the LDH layer. Meanwhile, we should also avoid the occurrence of metal salt anions into the interlayer surface. In the preparation of noncarbonate anion LDHs, carbon dioxide in the air is very easy to enter the reaction system, so it is always prepared by ion exchange or under the protection of nitrogen.

LDHs are usually synthesized by co-precipitation. The advantages of this method are as follows: (1) almost all of the M_1^{2+} and M_2^{3+} can form the corresponding LDHs; (2) by adjusting the ratio of M_1^{2+} to M_2^{3+}, series of LDHs with different M^{2+}/M^{3+} can be prepared; and (3) different anions could exist between the layer boards.

The specific preparation process is the precipitate is added into the solvent containing metal salt, and the insoluble metal hydrate hydroxide or gel is formed by the metathesis reaction, and its crystallization is carried out under the condition of precipitation, filtrated, washed, dried and LDHs are obtained. Analysis of the specific surface area, acidity and basicity and thermal stability (TG) is necessary.

2) **Supported inorganic solid base**

The methods to prepare a supported inorganic solid base mainly include co-precipitation and hybrid method. Co-precipitation refers to that the co-precipitate is produced by two or more components of the precursor composed of the catalyst in the solution, then catalyst is obtained after treatment. It is characterized by uniform mixing of components in molecular level, and heat treatment (calcination) can accelerate the solid-state reaction between components. Hybrid method is to make the two kinds of oxide mechanically mixing, and then the catalyst was obtained after heat treatment. Its properties are greatly influenced by the oxide particle size and the grinding time. The method has the advantages of simple equipment and convenient operation, and it can be used to prepare the high content of the multicomponent catalyst, especially the mixed oxide catalyst. But the dispersion by this method is poor. The preparation method of supported inorganic solid catalyst is introduced in detail as follows.

(1) Preparation of Mg(Al)O solid base: Weigh a certain amount of solid Na_2CO_3 and NaOH, dissolve in water and pour into a three-necked flasks, then drop the solution of $Mg(NO_3)_2$, $Al(NO)_3$ into the flasks with vigorous stirring, heat the solution to (60 ± 5) °C for 1 h after the reaction is finished. The product is cooled to room temperature, filtered out the precipitate. It is washed with deionized water three to five times, and dried under 100 °C for 16 h, and then calcinated at 450 °C for 12 h, ground and the 12–30 mesh particles are selected as the Mg(Al)O solid base.

(2) Preparation of MgO/Al_2O_3 solid base catalyst: Accurately weigh a certain amount of powder MgO and Al_2O_3, mix in a mortar and grind evenly, squeeze into strips after adding proper amount of distilled water, then dry at 100 °C for 8 h, at 200 °C for 8 h, ground, and select 12–30 mesh particles as the target solid base.

(3) Preparation of supported CoPcS: A certain amount of CoPcS was dissolved in anhydrous methanol, after impregnation of Mg–Al composite oxide solid base carrier for 24 h, methanol was removed under vacuum and CoPcS /Mg(Al)O or MgO/Al_2O_3 solid base catalyst was obtained.

3. **The applications of solid bases**

1) **Alkylation**

Mixtures containing MO (M is Zn, Cu, Cs, Ba, Mn, Co) and FeO are excellent catalysts for the methylation of phenol to produce methyl phenol and 2,6-dimethylphenol.

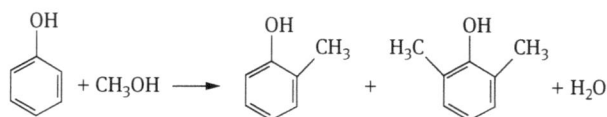

Side-chain alkylation of toluene and methanol over an alkali metal exchange of zeolite MX(Na^+, K^+, Rb^+, Cs^+ exchanged X-type zeolite) can be performed.

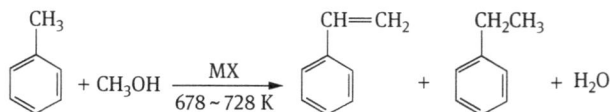

The side-chain alkylation of toluene with methanol can also be carried out on activated carbon-containing alkali metal oxides.

2) **Dehydrogenation**

The styrene is prepared by the ethylbenzene dehydrogenation in the industry.

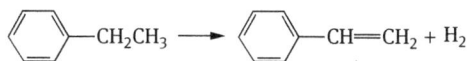

$$\text{\textlangle}\bigcirc\text{\textrangle}-CH_2CH_3 \longrightarrow \text{\textlangle}\bigcirc\text{\textrangle}-CH{=}CH_2 + H_2$$

In the presence of water vapor, the catalytic promotion effect of alkali was reduced by Cs, K, Na, Li in Fe oxide catalyst (Fe-Cr-K, Fe-Ce-Mo, Fe-Mg-K oxide). Function of K is to form $K_2Fe_2O_3$ alkaline active phase. In addition, K can also reduce the carbon deposit on catalyst surface and accelerate product desorption. Ce is effective in promoting dehydrogenation reaction. The role of Mo is to regulate the activity to inhibit the production of benzene and toluene. Mg increased the number of active centers, and formed a solid melt in Fe_2O_3 to improve the thermal stability of the catalyst.

3) **Aldol condensation**

Aldol condensation refers to the self-condensation of aldehydes and ketones (dimer) or cross-condensation to produce β-hydroxyl aldehyde or β-hydroxyl ketone. The reaction is given below:

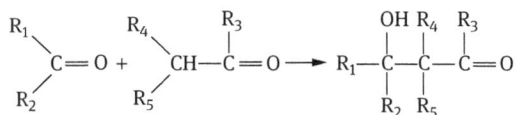

$$\begin{array}{c} R_1 \\ \searrow \\ C{=}O \ + \\ \nearrow \\ R_2 \end{array} \begin{array}{c} R_4 \\ \searrow \\ CH{-}C{=}O \\ \nearrow \quad | \\ R_5 \quad R_3 \end{array} \longrightarrow \begin{array}{c} OH \ R_4 \ R_3 \\ | \quad\ | \quad\ | \\ R_1{-}C{-}C{-}C{=}O \\ | \quad\ | \\ R_2 \ R_5 \end{array}$$

The most commonly used alkaline catalyst is $Ba(OH)_2$, and alkali metal and alkaline earth metal compounds. Strong alkaline ion exchange resin is also active in the self-condensation reaction of aldehydes and ketones. Hydrotalcite $Mg_6Al_2(OH)_{16}CO_3\cdot4H_2O$ has a high activity for the cross-condensation of formaldehyde and acetone to produce methyl vinyl ketone.

4.1.3 Ionic Liquid Catalyst

1. **Summary**

Ionic liquid is a kind of salt, which is composed of organic cations and inorganic anions and is a liquid at room temperature or near-room temperature [9]. It has the following characteristics: ① colorless, odorless, nonvolatile and low vapor pressure; ② wider range of temperature stability; ③ good chemical stability and wide potential range; ④ good solubility in organic, inorganic compound and polymer; ⑤ it can be acidic or alkaline. Therefore, the ionic liquid can be used as both a clean medium and a catalyst for the reaction.

Ionic liquids mainly include four types: ① the cationic alkyl quaternary ammonium ion $[NR_xH_{4-x}]^+$; ② the alkyl phosphonium ion $[PR_xH_{4-x}]^+$; ③ 1,3-dialkyl-substituted imidazole ion or N, N′-dialkyl-substituted imidazole ions of $[R'R_3im]^+$

or $[R_0^1R^2R^3im]^+$; ④ N-alkyl-substituted pyridine ions $[RPy]^+$. The main anion is halide salt, such as $AlCl_3$ and $AlBr_3$, whose acidity and basicity of ionic liquids composed of these anions are changed as the composition change of the anions, for $[C_4mim](AlCl_3)$, when $x = 0.5$, the ionic liquid is neutral; when $x < 0.5$ is acidic; when $x > 0.5$ is alkaline. This kind of ionic liquid is sensitive to water and is easy to be decomposed, so it must be used in an anhydrous environment. Ionic liquids composed of BF_4^-, PF_6^- with substituted imidazolium ion $[R^1R^3im]$ is stable to water and air. There are CF_3^-, SO_3^{2-}, $C_3F_7COO^-$, CF^3COO^-, SbF_6^-, AsF_6^- and other anions can be used to synthesize various ionic liquids.

2. **Method for preparing ionic liquids**
    1) **One-step synthesis**

        One-step synthetic method includes acid–base neutralization and quaternary ammoniumization method. They have the advantages of simple operation, easy purification and no by-products. Acid–base neutralization is the method of neutralization of alkali and acid and then to obtain product after purification, such as the preparation of nitroethylamine ionic liquids, first, the ethylamine solution and nitric acid are neutralized, the water in the system is removed by vacuum dehydration and then the ionic liquid product is dissolved in acetonitrile or tetrahydrofuran. Next is decolorization with activated carbon, removing the organic solvent by vacuum to obtain the target ionic liquid product. The quaternary ammonium reaction refers to the reaction of halogenated hydrocarbon with methyl imidazole, such as the preparation of $[C_4mim]Cl$, the purified chlorobutane reacts with methyl imidazole under reflux condition for 10 h, to get the crude $[C_4mim]Cl$, then the crude product was washed with ethyl acetate two to three times, finally the ethyl acetate was removed by vacuum rotary evaporation to obtain $[C_4mim]Cl$ ionic liquid.

    2) **Two-step synthesis**

        If the ionic liquid can't be synthesized by one-step synthesis, two-step synthetic method can be used. First, the quaternary ammonium was used to prepare the halide salt ([cation]X) containing the target cation, and then the target ion Y^- was replaced with X^- or Lewis acid MXy to obtain the target ionic liquid. Following are the reactions:

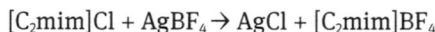

        $$C_2Cl + mim \rightarrow [C_2mim]Cl$$

        $$[C_2mim]Cl + AgBF_4 \rightarrow AgCl + [C_2mim]BF_4$$

        or

        $$[C_2mim]Cl + NH_4BF_4 \rightarrow NH_4Cl + [C_2mim]BF_4$$

        $$[C_2mim]Cl + HBF_4(aq) \rightarrow [C_2mim]BF_4 + HCl$$

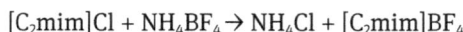

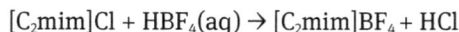

        The chiral ionic liquids can also be synthesized by the method.

3. **Characterization of ionic liquids**

Ionic liquids, as an important solvent and catalyst, play a more and more important role in organic synthesis. Therefore, it is the key to determine the composition and catalytic properties of ionic liquids.

1) **Structure identification of anions**

The anion of the ionic liquid was determined by an Fast Atom Bombardment mass spectrometry (FAB), and the FAB spectrum of the ionic liquid formed by $AlCl_3$ and [C_4mim]Cl was shown in Figure 4.4.

2) **Determination of cations**

Nuclear magnetic resonance instrument (^{1}H NMR) is used to determine the structure of ionic liquids by D_2O and DMSO-δ_6 as solvent. [C_4mim]BF_4 NMR data are shown in Table 4.3.

3) **Ultraviolet (UV) spectrometry analysis**

Three kinds of ionic liquids of 2.5×10^{-4} mol in methanol, and their UV absorption spectra at 20 °C were shown in Figure 4.5.

4) **Determination of ionic liquids by IR spectroscopy**

The main functional groups were determined by the characteristic peaks of infrared spectra of ionic liquids. The featured absorption peaks of the infrared spectra of ionic liquids are shown in Table 4.4 and Figure 4.6.

Table 4.3: ^{1}H-NMR data of [C_4mim]BF_4.

Carbon number in imidazolium cation	[C_4mim]Cl δ	[C_4mim]BF_4 δ
H at No. 2 carbon	8.69	8.68
H at No. 4 carbon	7.46	7.46
H at No. 5 carbon	7.41	7.42
H at No. 6 carbon	4.18	4.19
H at No. 10 carbon	3.88	3.88
H at No. 7 carbon	1.84	1.84
H at No. 8 carbon	1.31	1.31
H at No. 9 carbon	0.91	0.92

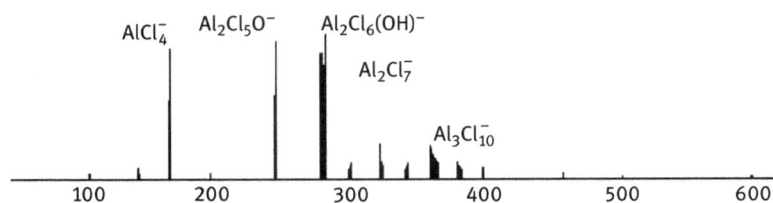

Figure 4.4: FAB spectrum of the ionic liquid formed by $AlCl_3$ and [C_4mim]Cl.

Table 4.4: Infrared spectral data of typical ionic liquids.

v_{max} (cm^{-1})	Band attribution	v_{max} (cm^{-1})	Band attribution
3,168, 3,120	Aromatic C–H stretching	1,170	In-plane deformation vibration of aromatic ring C–H
2,966, 2,912, 2,878	Aliphatic C–H stretching	1,059	B–F vibration of BF$_{4-}$
1,577, 1,456	Aromatic ring skeleton	838	P–F vibration of PF$_{6-}$
1,467, 1,385	Me C–H deformation vibration		

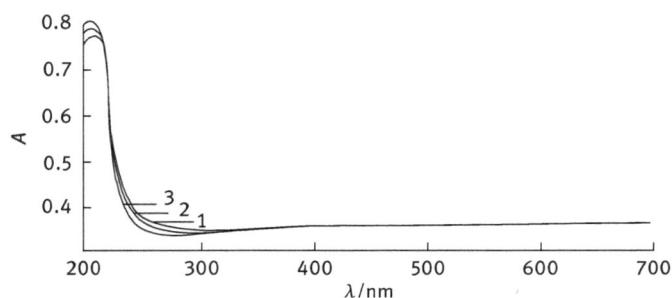

Figure 4.5: UV absorption spectra of ionic liquids (1-[C$_4$mim]Cl, 2-[C$_4$mim]BF$_4$, and 3-[C$_4$mim]PF$_6$).

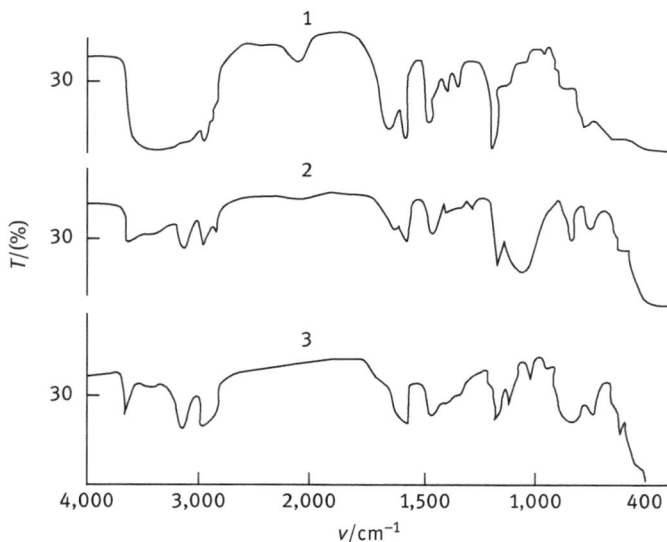

Figure 4.6: Infrared spectra of ionic liquids (1-[C$_4$mim]Cl, 2-[C$_4$mim]BF$_4$, and 3-[C$_4$mim]PF$_6$).

5) **Acidity measurement and characterization of ionic liquids**

There are two methods to determine the acidity of ionic liquids: Hammett indicator method and infrared spectrum probe method. The IR probe method is used to make the ionic liquid react with alkaline substances (pyridine and acetonitrile), then analyze the product characteristic peaks in infrared spectrum to understand the acidic type and strength of an ionic liquid. When pyridine is used as the probe molecule, the peaks of 1,450 cm^{-1} and about 1,540 cm^{-1}, respectively, represent the Lewis and Brönsted acidity of ionic liquids. When acetonitrile was used as probe molecule, the peak near 2,330 cm^{-1} was the characteristic absorption peak of Lewis acid, and the peak shifted to higher value when the acidity increases. The IR spectra of acetonitrile can also be used to determine the acid strength of an ionic liquid.

4. **Applications of ionic liquid in organic synthesis**

1) **Friedel–Crafts reaction**

Alkylation of isobutane with butene is used for the synthesis of high octane to protect gasoline. The above reaction was carried out by using Cu^{2+} modified $AlCl_3$ type ionic liquid as catalyst, and the component of C_8 in the product was more than 75%, which was close to or reached the level of industrial sulfuric acid alkylation, and the catalyst could be reused for eight times without activity loss.

2) **Esterification**

Because the separation of the products in the traditional esterification is difficult, ionic liquid is used as the catalyst and solvent. Since the ionic liquids and esters are immiscible, they can be separated from each other automatically, and the ionic liquid can be reused after dehydration at a higher temperature. Experimental results show that the ionic liquid has a good catalytic performance in the esterification.

3) **Carboxyl synthesis**

Carboxyl synthesis refers to the process by which olefins react with syngas (CO and H_2) to form aldehydes, which is the key method for the industrial production of aldehydes. C_2–C_5 carboxylation is generally carried out in water/organic two-phase systems; higher carbon atoms of olefin with a low solubility in water can't work in water/organic two-phase reaction system, and the ionic liquid as the reaction media can easily solve the above problems.

4.2 Biocatalysis in Organic Synthesis

4.2.1 Introduction

Since the 1980s of the twentieth century, the application of biocatalysts in organic synthetic chemistry has attracted more and more attention, and the research on organic synthesis has been extremely well in recent decades [10, 11]. A lot of organic chemists do their best to research on the synthesis of special functions and complex natural products. In recent years, synthesis program required the highly selective scientific application and biological reactions. Asymmetric synthesis has become the forefront of organic synthesis, and has an increasingly higher demand for asymmetric synthesis. Therefore, the developing direction of organic synthesis is to select highly region-selective and stereo-specific reactions, and biocatalysis is conform to this trend. Among them, the enzyme-catalyzed organic synthesis has the following advantages:

(1) Enzymes are very efficient catalysts. The amount of enzymes is very less in the enzyme-catalyzed reactions, but the reaction rate of enzyme catalyzed is above 106 times faster than nonenzymatic reactions.

(2) Enzymes are environmentally friendly. Enzymes can completely degrade into environmental nonpolluting substances.

(3) Enzymes play a role in very mild conditions. Enzymes can work under mild conditions; its reaction of pH values is in the range of 5–8, generally about 7.0, and the reaction temperature is 20–40 °C.

(4) Enzymes can be compatible with each other. The conditions of different enzyme-catalyzed reactions are often the same or similar, so several enzymatic reactions can be carried out in the same reactor.

(5) The substrates and environments of the enzyme-catalyzed function are wider. The enzymes show great tolerance; they can catalyze the reactions of nonnatural materials, and they do not require the water environment.

(6) Wide applications of Enzyme-catalyzed organic reactions, including oxidation and dehydrogenation, reduction, deamination and hydroxylation and methylation, ring oxidation and esterification, amidation, phosphorylation, open-loop, isomerization, condensation and halogenated.

(7) Enzymes are highly selective. Enzyme catalyst exhibits regioselectivity, stereoselectivity and enantioselectivity, which is the key difference of an enzyme-catalyzed reaction than that of the general chemical catalysts.

Enzyme-catalyzed reaction also has the following disadvantages.

(1) The parameters of enzyme action required are very narrow, the enzyme reaction parameters (such as temperature, pH and ion strength) must be controlled accurately; otherwise, the enzymes will lose the activity.

(2) Natural enzymes can form only one kind of enantiomer.

(3) Enzymes can only exhibit high catalytic activity in water.

(4) Enzyme-catalyzed reaction may cause an allergic reaction.

(5) Enzyme activity was often inhibited by high substrate concentrations or product, even loss in the activity.

(6) Some enzymes subject to its natural cofactor.

Existing research results show that many enzymes can catalyze organic reactions in anhydrous organic solvents or organic solvent containing a small amount of water. Compared with water, the stability of enzyme in nonaqueous media can be greatly improved, and selecting a proper solvent can control the specificity for the substrate and even control the stereoselectivity.

4.2.2 Basic Principle of Enzyme Catalysis

1. **Lock-key theory**
   The mechanism of enzyme catalysis is first proposed by Fischer E. He proposed the "lock key theory" in 1894; some researchers call this as "template theory." By the theory's view, there is a strict complementary between the enzyme and the substrate in structure, and the interaction between enzyme and its substrate is like a lock-and-key mechanically. If the substrate molecule has minor changes in the structure, it cannot be embedded in the molecule, and cannot be worked by enzyme, as shown in Figure 4.7. Although this theory was quite successful at that time, but it's believed that the structure is completely rigid does not meet the actual state.
2. **Inducing fit theory**
   Koshland proposed the inducing fit theory in 1958; this retained the concept of complementary between substrate and enzyme. He thought that the enzyme itself is not totally rigid, but has a very fine and soft structure. The combination between enzyme and substrate is a dynamic fit, when the enzyme closed to a specific substrate, the substrate will induce changes in the structure of enzyme active site to make the catalytic groups accurately be located and the site are located properly in the substrate-sensitive bond. The two fit well, forming an enzyme–substrate complex, as shown in Figure 4.8. This theory can explain the experimental facts that the formal theory cannot explain, especially the structure modification information of lysozyme and elastase substrate binding by X-ray diffraction. It is fairly consistent with the expectations of this theory.
3. **The catalytic mechanism of enzymes**
   When the enzyme is binding with a substrate, it will release the part of binding energy; thus, free energy of the transition state in enzyme-catalyzed reactions

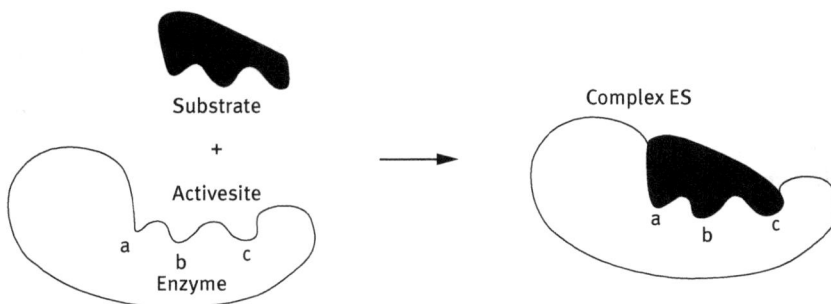

Figure 4.7: Diagram of "lock" and "key" of substrate and enzyme action.

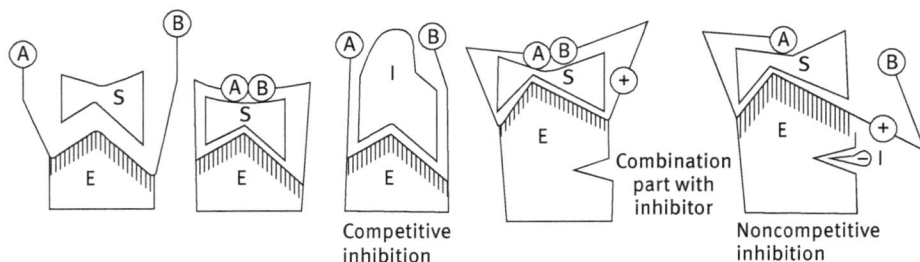

Figure 4.8: The schematic diagram of inducing fit theory of the interaction between substrate and enzyme. A, acid base; B, alkaline base; S, substrate; E, enzyme; I, inhibitors.

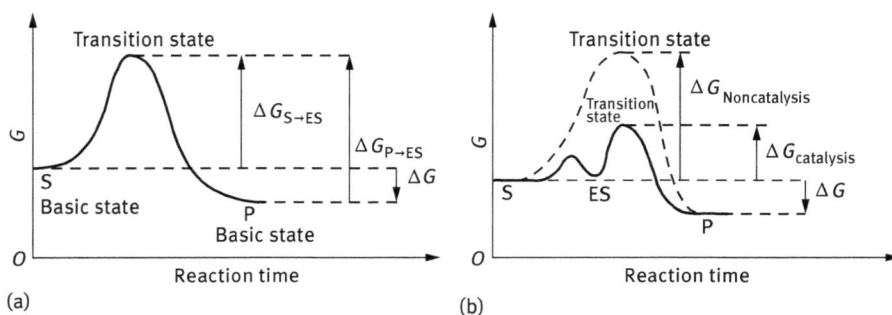

Figure 4.9: Free energy change during the reaction process.

was lower than that of a nonenzyme-catalyzed reactions. The activation energy of enzyme-catalyzed reactions dropped; therefore, its reaction rate changes to be fast, and makes the reaction equilibrium to achieve shortly.

For a simple single-substrate enzyme-catalyzed reaction, it can be presented as the following expression:

$$E + S = ES \longrightarrow P + E$$

where E, S, P and ES represent enzyme, substrate, products and intermediate complex of substrate, respectively. Figure 4.9 represents the free energy change in the enzyme-catalyzed reaction, As shown in Figure 4.9, the activation energy in the presence of enzyme is lower than that of the activation energy in a nonenzyme reaction.

Enzyme-catalyzed reactions contain two processes of enzyme and substrate combination and acceleration of a reaction, which is achieved by the following mechanisms.

1) **Proximity and orientation effect**

When a substrate molecule and a group of enzyme react, they must stay close to each other. Maintain proper angles with each other to form the secondary bond (hydrogen bonds, van der Waals forces and so on). The molecular orbitals of reactive groups need overlapping. It seems to fix the substrates in the enzyme's active site, and existing as a certain conformation, maintain the correct orientation to function effectively. When the distance and direction of the substrate molecules achieve the best, the catalytic efficiency is the highest.

2) **Acid–base catalysis**

Acid–base interaction exists in most of the enzyme-catalyzed reactions, and certain groups on the active site can be used as proton donors or acceptors for substrates to proceed acid–base catalysis. Enzyme molecule contains several generalized acid–base functional groups, such as amino, carboxyl, mercapto, hydroxyl groups in arginine phenol and imidazole groups in histidine and so on.

3) **Covalent catalysis**

The catalyst with the substrate can form covalent intermediates of higher reactivity to reduce reaction activity energy, so as to increase the reaction rate. This is called covalent catalysis. And it referred to the nucleophilic or electrophilic catalysis, and during the process, nucleophilic or electrophilic catalyst acts on the substrate by electron-deficient centers or electron-rich center, resulting in the rapid formation of instable covalent compound intermediate, so the activation energy greatly reduced, and the substrate can go across a lower barrier to form the products.

4) **Substrate deformation**

Many active sites do not fit with the substrates, but for combining with substrates, the enzyme's active sites have to deform (induced fit) to fit with the substrates. Once combining with substrates, enzyme can induce the substrate deformation, making sensitive bonds easy to be broken, and promoting the formation of a new bond.

When an enzyme combined with a substrate, a number of weak interactions can be formed. They did not really reach matching at first, while the enzyme can induce the substrate distortion and force the substrates to convert to transitional state. Only when the substrate reached the transition state, the weak interaction between substrate and enzyme can achieve the so-called fitting. That only in the transition state, enzyme and substrate molecules have the greatest interaction. As shown in Figure 4.10, enzyme and substrate combined, making the substrate to deform and generate products.

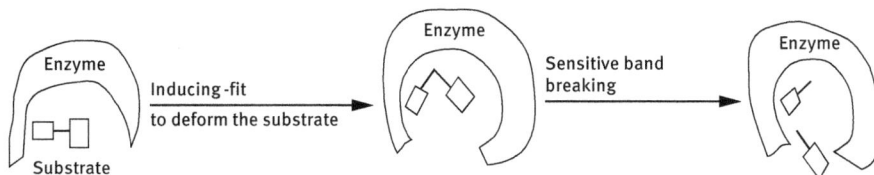

Figure 4.10: The schematic diagram of inducing fit and substrate deformation.

5) **Metal ion catalysis**

Metal ions are necessary cofactors in many enzymes. Its catalytic role is similar to that of acidic chemical catalysis, but multivalence metals play a strong role than that of proton, and also have a coordination function, easily making the substrates immobilized on the enzyme molecules.

6) **The effect of microenvironment**

Each protein enzyme has a particular spatial structure. This provides the environment for functional groups to act and is called microenvironment. The clap of two active sites is relatively nonpolar. This environment is characterized by a low dielectric constant and repellent to polar water molecule. In a nonpolar environment, the force between two charged objects increases significantly than that in a polar environment. Catalysis groups are in polarization in the low dielectric environment. When substrate is combined with an active site, the forces between catalyst groups and the sensitive bonds of substrate would be larger than that in polar environment. This hydrophobic microenvironment prompted the reaction rate to improve.

7) **Synergic catalysis**

The multiple catalysis of enzyme usually results from the synergy of a few elementary reactions, for example, chymotrypsin Ser195 as a nucleophilic for nucleophilic catalysis, while the side-chain groups of His57 play a role of alkaline catalysis.

4.2.3 Types of Biocatalysts

In the organic synthesis of biocatalysis, there are two main types of biological catalysts that can be used: whole cells and vitro enzyme. As recommended by the International Union of Biochemistry enzyme classification methods, biological catalysts are divided into six major categories (Table 4.5).

Table 4.5: Systems classification of biocatalyst enzyme.

Classification	Type of catalytic reaction	Typical examples
① Redox enzymes	Redox reaction with electron transfer: C–H, C–C, C=C bonds	Alcohol dehydrogenase
② Transfer enzymes	Transfer of functional groups: aldehyde, ketone, acyl, phosphonium, methyl, etc.	Hexokinase
③ Hydrolytic enzymes	Hydrolysis reactions: ester, amide, lactones, lactams, epoxides, nitrile	Trypsin
④ Lyase	Cracking reaction: C=C, C=N, C=O bonds	Pyruvate decarboxylase
⑤ Isomerase	Isomerization reactions: racemization, epimerization, rearrangement, etc.	Maleic acid isomerase
⑥ Synthetic enzymes	With cracking triphosphate C–O, C–S, C–N, C–C bonds	Pyruvate carboxylase

4.2.4 Typical Process of Biocatalysts Utilization

1. **Enzymatic production of L-amino acid**

   L-Amino acid is an important chiral compound with physiological activity and is widely used in medicine, food and feed industry. With enzyme catalysis, various substrates can be converted to L-amino acids, or split the amino acids into L-amino acids. A variety of enzymes have been used for production of L-amino acids, and some have undergone continuous production using immobilized enzyme. The production of L-amino acid from acyl amino acids by enzymatic optical dissolution is illustrated as an example.

   Amino acylation enzyme-catalyzed the asymmetric hydrolysis of racemic N-acyl amino acids, including L-acyl amino acids by hydrolysis to form L-amino acid and the remaining D-acyl amino acid racemization by chemical regeneration to give N-acyl amino acid, and it endures asymmetric hydrolysis again. And so forth, the N-acyl amino acid chemical synthesized can be transformed into L-amino acids. Amino acylation enzyme's optimal temperature is 60 °C, pH value is 7.5–8.5 and Co^{3+} plays a role for enzyme activation.

   In the industrial production, immobilized L-aminoacylase has been used on continuous production of L-phenylalanine, L-tryptophan and other amino acids.

2. **Biological production of acrylamide**

   Acrylamide (AM, $CH_2=CH\text{-}CONH_2$) is a versatile organic chemical raw material; its production contains chemical catalysis and microbial catalytic hydration. The traditional methods of chemical catalysis include sulfuric acid chemical hydrolysis and chemical hydration with reduced copper catalyst. Both of them have many problems such as complex process, difficulty in product refining, pollution and many by-products. Compared with chemical catalysis, microbial enzyme-catalyzed hydration has obvious advantages in feed conversion, reaction conditions, product quality, environmental protection, production safety and equipment investment. The microbial enzyme used in hydration is nitrile hydratase, a protein produced by some microorganisms in the process of oxime or urea metabolism. It can catalyze hydration of nitrile (−CN) of acrylonitrile (AN) into amide (-CONH$_2$). The catalytic process is shown in Figure 4.11.

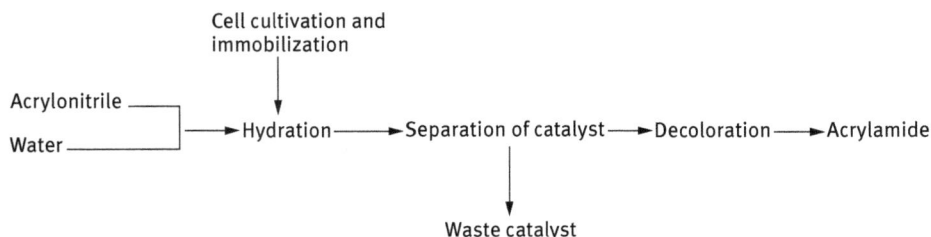

Figure 4.11: Biocatalytic production flow diagram of acrylamide.

4.3 Asymmetric Catalytic Synthesis

4.3.1 Overview

Chirality is an essential property of nature, especially of a variety of substances in life (such as amino acids, polysaccharides and nucleic acids) and plays a decisive role in the normal growth and health [12–21].

Chirality is the noncoincidence of the molecules or a compound and its mirror image, which means that certain types of groups in a compound molecule or molecule can be arranged in two forms that mirror each other, but they cannot overlap (such as a person's left- and right-hand relation).

Chiral compounds are in the same atomic composition, but with different arrangement of atoms in three dimensional, leading to differences in configuration. A pair of compounds having such an enantiomeric relationship is called enantiomers. If the pair of enantiomers is mixed together in equal amounts, they are called racemates.

Chirality has a rotation in the physical properties of matter, that is, they can rotate the plane-polarized light at a certain angle. The ability of the enantiomers to rotate the polarized light is equal, but in the opposite direction. Among them, the right-handed enantiomer (R) rotates the polarized light in the clockwise direction; the enantiomer which rotates the polarized light in the anticlockwise direction is called the left-handed enantiomer (S). While for racemates, since the rotational effects of the constituent enantiomers cancel each other without causing polarization rotation, there is no optical activity.

For the performance evaluation of chiral compounds, the ee values of the enantiomers are mainly used to characterize them. ee value is defined as follows: ee value refers to the percentage of the extra amount of an enantiomeric mixture of an isomer (such as R) than another isomer (such as S) of the total amount.

It reflects the optical purity of chiral compounds; the greater the ee value, the higher the optical purity. Since the absolute content of one enantiomer can be determined by analytical instruments at present, the optical purity of the compound cannot be expressed with ee value.

Although the composition of the chiral compounds and the number of atoms are equal, but because of the different arrangement of atoms, the properties of enantiomers are completely different; hence, chiral compounds in new drug development have an important status.

Chiral compounds are mainly derived from natural products and synthetic products. And synthetic product is an important way to obtain new chiral compounds. Synthetic methods include racemate dissolution, substrate-induced chiral synthesis and chiral catalytic synthesis, of which the chiral catalytic synthesis has been given the most attention in academic and industries.

Asymmetric reaction technologies can be divided into four ways: ① asymmetric reaction of chiral resources; ② asymmetric reaction of chiral assistants; ③ asymmetric reaction of chiral reagents; and ④ asymmetric catalytic reaction. Asymmetric catalytic synthesis is characterized by mild reaction conditions and good stereoselectivity. (R-)isomers or (S) isomers are also easy to produce, and prochiral substrate from a wide range of resources.

For mass production of chiral compounds, asymmetric catalytic synthesis is the most economical and most practical technology. Therefore, the development of high efficiency, high selectivity and high yield of chiral catalyst has become the core of the development of chiral technology.

4.3.2 Principle and Process Analysis of Asymmetric Catalytic Synthesis

1. **The principle of asymmetric catalytic synthesis**

   Conventional asymmetric synthesis involves the introduction of asymmetric elements in the symmetrical starting reactants or reactions with asymmetric reagents, thus requiring the consumption of stoichiometric amounts of chiral auxiliary reagents. While asymmetric catalytic synthesis is generally the use of reasonably designed chiral metal complex catalyst or biological enzyme as a chiral template to control the reactants of the textured surface, a large number of prechiral substrates are selectively converted into a specific configuration of the product, to achieve chiral amplification and proliferation.

   Most of the asymmetric catalytic reactions occur on the sites of functional group with the plane sp2 hybrid carbon atoms and tetrahedral sp3 hybrid carbon atoms. These functional groups include carbonyls, enamines, enols, imines and olefinic bonds, and the reaction includes asymmetric hydrogenation or asymmetric addition of other groups. For example:

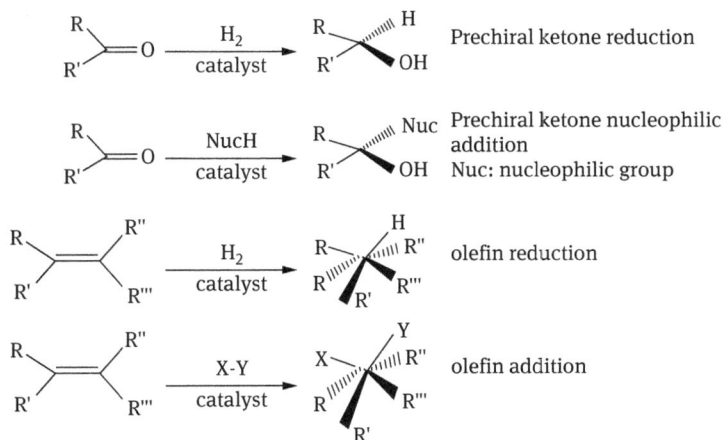

Enantiomerically enriched substrates can be converted to enantiomers by asymmetric catalysis, which disrupts the symmetry of the starting substrate. For example:

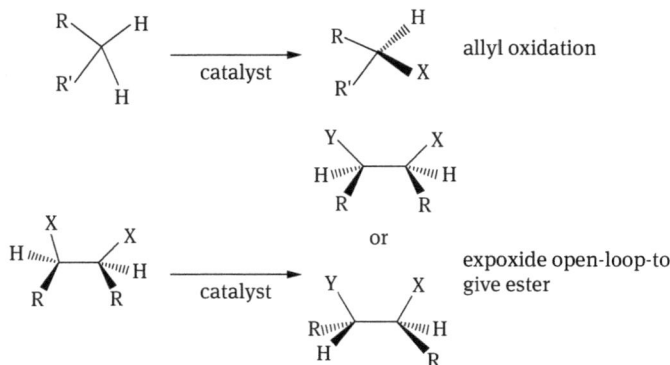

catalyst → allyl oxidation

catalyst → or → expoxide open-loop-to give ester

Asymmetric catalysis can also be achieved by kinetic dissolution of the racemic mixture of substrates. One of the enantiomers of the substrate is selectively converted to the product in the reaction, while the other does not respond and preserve. In certain reactions, both enantiomers are converted to the same enantiomer when the substrate is converted to the product. The following are examples of kinetic resolution of racemic substrates:

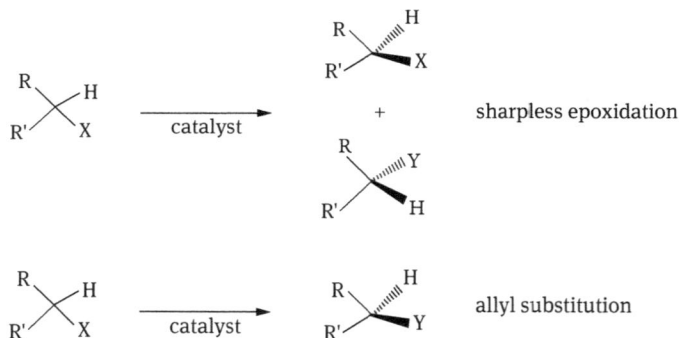

catalyst → + → sharpless epoxidation

catalyst → allyl substitution

In asymmetric catalytic reactions, the substrate and the catalyst with a chiral center combine to form a diastereomeric transition state. In an environment without a chiral center, the possibility of the formation of two enantiomers of the molecular structures mirrorring each other is equal. In the case of chiral centers, the difference in the activation energies of the two transition states with different spatial configurations will result in the preferential selection of an enantiomer. The difference in activation energy of the transition state is due to the interaction between the chiral catalyst and the substrate. The activation energies differ by 6.0 and 12 kJ/mol at 25 °C, respectively, resulting in 80% (90%:10%) and 98% (99%:1%) enantiomeric excesses, respectively.

Catalysts for asymmetric catalysis should have the function of controlling the activation of different substrates and the reaction products. On heterogeneous catalysis, the catalyst interacts with the adsorbed substrate and the adsorbed chiral auxiliary (modifier) to control the activation process. Experiments show that the substrates that can be converted into the optically active enantiomers with functional groups will interact with the active sites of the chiral centers of the catalyst.

2. **Process analysis of asymmetric catalytic synthesis**

Asymmetric catalytic synthesis reaction process is similar to the general organic catalytic reaction process, which mainly involves the ways of reaction operation (reactor form, the separation between the reactants and product) and the reaction process conditions (raw material than the reaction time, reaction temperature, catalyst dosage, feed rate, solvent and its addition, inert gas protection). When the reactants or products are temperature-sensitive substances or are susceptible to oxidation, protection is required using an inert gas (nitrogen or argon). When the reaction time is longer, intensification technology is needed to employ, such as the use of microwave heating. It is recommended that the reaction is carried out in an environmental-friendly solvent (ionic liquid and supercritical carbon dioxide) or in a solvent-free manner.

The following is a brief analysis of the advanced synthetic technologies.

1) **Asymmetric reaction catalyzed by ionic liquids**

Ionic liquids are excellent solvents for many organic reactions; for catalytic reactions, it can be recycled in most cases by extraction of the product, so that the use of ionic liquids can be considered as a ligand-free catalyst-immobilized method. If this catalyst immobilization method is used, the enantioselective reaction is catalyzed by a chiral catalyst, and the recycling of the catalyst can be achieved. In fact, asymmetric reactions in ionic liquids have been studied less frequently, and are currently carried out in asymmetric hydrogenation, hydroformylation, cyclopropanation, palladium-catalyzed allylalkylation and epoxide ring-opening reactions. Molecular Identification and Selective Synthesis Laboratory, Research Institute of Chemistry, Chinese Academy of Sciences developed a new type of highly efficient ionic liquid asymmetric small organic molecule catalyst – Baylis-Hillman. The catalyst not only has high catalytic activity and is equivalent to the currently reported optimal catalyst, but also can utilize the properties of the ionic liquid itself to achieve the purpose of recycling. The catalyst was highlighted by the American Chemical Society Heart Cut website as "a very efficient and reusable Baylis-Hillman catalyst."

2) **Asymmetric synthesis in supercritical media**

In 1995, Burk successfully catalyzed the hydrogenation of dehydrogenated amino acids in supercritical CO_2 with Ru-Duphos, and obtained products with optical purity of 96.8%.

3) **Solvent-free system in asymmetric catalytic reaction**

The use of a solvent is a conventional means of organic synthesis for dissolving reactants or carrying the small molecular by-products of the reaction while achieving heat exchange and equilibration of the process. However, in the production process, the use of solvents has also produced a corresponding negative effects, such as the inevitable loss of the recovery process and solvent pollution for the environment. Therefore, the solvent-free system is the most popular choice.

There are two ways to realize the solvent-free system: one is to use the reactants as the solvent; the other is the solid-phase reaction by solid reactants grounding mixture. The asymmetric reaction of aldol reaction as an example to illustrate the implementation of asymmetric reaction and the effect of solvent-free system.

1.1 mol aldehydes, 1.0 and 0.1 mol (S)-proline were put into a ball mill, and ball milled and mixed materials, and then heated to achieve solvent-free catalyzed asymmetric aldol reaction. In the reaction with a nucleophile as the reaction medium, a large excess of ketone is usually required in order to obtain a higher yield of the product and acceptable stereoselectivity. As shown in Table 4.6, aldol reaction of aldehydes and ketones occurs in almost the same amount in the absence of solvent. The ball milling technique (Method A) enables the catalytic reaction to proceed efficiently, resulting in a high yield of 99.0%, good stereoselectivity (ee value up to 99.0% or more).

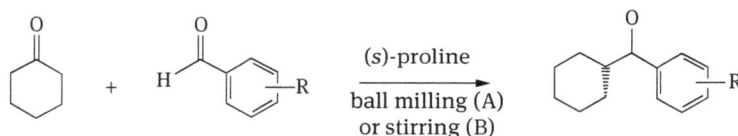

Compared with ball-milling method A and traditional magnetic stirring method B, the milling time can be shortened greatly. The same reaction conditions were applied to the whole solid reactant system. The diastereoselectivity of the product was excellent (anti/syn up to 99/1) and good enantioselectivity (ee value up to 98.0%).

4.3.3 Catalyst Systems in Asymmetric Catalytic Reactions

1. Asymmetric metal complex catalyst

Asymmetric metal complex catalysts are complexes formed by chiral ligands and transition metals. Asymmetric metal complex catalyst has the following advantages: ① high catalytic activity, so long as high stereoselectivity, can achieve the

Table 4.6: Asymmetric catalytic reactions by homogenous and heterogeneous metal catalysts.

Reaction types		Ni	Cu	Co	Rh	Pd	Pt	Ir	Ru	Mo	Tl	Fe	V
Oxidation	C=O			×	×	×	×		×				
	C=O	o	o	×o	×			×	×				
	C=N	o	o	o	o×				×				
Hydroformylation				×	×	×							
Hydrocyanation				×	×								
Hydrosilylation	C=C				×	×							
	C=O				×	×	×						
	C=N				×								
Hydroalkenation		×											
Intercoupling reaction		×				×							
Cyclopropanation			×	×									
Polymerization		×									×	×	
Epoxidation										×	×		
Isomerization				×						×			×
Amination						×							

Note: o presents heterogeneous catalyest, × presents homogeneous catalysts

desired catalytic effect; ② different metals and ligands combined with each other to get a wide range of catalysts for the different types of reaction; ③ few catalyst input, such as Rh/DEGPHOS as catalyst, for the production of L-phenylalanine, 1 ton reactants consume only 400 g catalyst; ④ the metal complex catalyst can be immobilized, recyclable and reusable.

Asymmetric metal complex catalysts can be represented by L (M), L represents a chiral ligand and M represents a central metal ion (see Table 4.6). From the table we can see that the complexes of each metal ion are suitable as catalysts for one or more reactions.

The excellent chiral ligands should have the following characteristics: ① When asymmetric centers of the substrate are formed, the chiral ligands should bind to the central ion without inducing a solvent effect; ② the catalytic activity should not be affected by the introduction of chiral ligands; ③ the ligand structure should be convenient for chemical modification, and can be used for the synthesis of different products. Different metal ions should select their appropriate ligands for different catalytic reactions.

Currently used ligands are divided into the following categories: chiral phosphine, chiral amines, chiral alcohols, chiral amides and hydroxyl amino acids, chiral dioxime, chiral sulfoxides and chiral crown ethers. The chiral phosphine ligand is most widely used, and it can coordinate with a variety of metal ions to form complex catalyst with excellent performance.

As chiral phosphine ligands are expensive, chiral ligands containing nitrogen are currently being developed, and the nitrogen-containing chiral ligands employ a lower cost, such as various substituted diphenylethylenediamines.

2. **Biocatalyst system**

Biocatalysts refer to microorganisms and enzymes used in catalysis with regional and stereoselectivity, mild reaction conditions and environmental-friendly features. However, due to the high price of most biocatalysts and the limitation of substrate adaptation, the combination of chemical synthesis and enzymatic synthesis is commonly used in industries. For some key steps, pure enzymes or microorganisms catalytic synthesis is employed, and the general synthetic steps employed chemical synthetic methods to achieve complementary advantages. At present, biocatalytic organic synthetic reactions include hydrolysis, esterification, reduction and oxidation.

1) **Enzyme-catalyzed asymmetric hydrolysis**

Most of the currently used enzymes for hydrolysis are hydrolases. These enzymes are versatile and require no coenzyme to directly catalyze the reaction. They are also highly resistant to organic solvents and have good stereochemistry selectivity in alcohols, carboxylic acids, esters, amides and amines. So in organic synthesis, racemic alcohols, acids and other substrates undergo kinetic dissolution, that is, the difference in reaction rate of hydrolysis or transesterification leads to two enantiomers in racemate compound by enzymes or micro-catalyst further gives two optically active products by dissolution. At present, chrysanthemic acid and cyanohydrin dissolution in the synthesis of pesticides is the most used. In 1988, Mintcuta applied for a patent for the industrial dissolution of chrysanthemic acid and its precursors. (±)-*cis-trans*-2,2-Dimethyl-3(2,2-dichlorovinyl)-cyclohexane in the presence of a buffer solution of NaOH or Na_2CO_3 at 50 °C and (+)-*trans*-2,2-dimethyl-3(2,2-dichloroethenyl) cyclopropanecarboxylate with optical purity of 100% were obtained after 48 h of separation.

2) **Biocatalytic asymmetric reduction**

L-Carnitine (L-carnitine), also known as vitamin BT, plays an important role in the metabolism of fatty acids in the body. It is an internationally recognized functional food additive, added to infant formula, athletes' drinks, nutritional supplements and weight loss foods. Usually the chemical synthesized D-carnitine is not only without physiological activity, but is also the L-carnitine antagonist. L-Carnitine is prepared with a-chloroacetyl acetoacetate reduction by baker's yeast as the key step, and then by two-step conventional reaction.

3. **Chiral solid catalyst system**

Chiral solid catalyst system includes chiral molecular sieve catalyst, chiral metal organic framework catalysts, chiral phosphonic acid catalyst and asymmetric phase transfer catalyst.

4.4 Organic Synthesis in Fluorine Biphase System

4.4.1 Working Principle of Fluorine Two-Phase System

In perfluorinated solvents, also known as fluorine solvents or perfluorocarbons, the hydrogen atoms in carbon atoms are all replaced by fluorine atoms in alkanes, ethers and amines. The density of perfluorinated solvents is greater than that of organic solvents, and perfluorinated solvents are colorless, nontoxic and have good thermal stability, named as a new green solvent. Perfluorinated solvents are excellent solvents for gases and can dissolve large amounts of hydrogen, oxygen, nitrogen and carbon dioxide, but are poorly soluble in organic solvents and organic compounds.

Fluorine biphase system (FBS) is a nonaqueous liquid–liquid two-phase reaction system, which consists of ordinary organic solvent and perfluorinated solvent. Due to the high electronegativity of fluorine atoms in the perfluorinated solvent molecules and the van der Waals radius is similar to that of the hydrogen atoms, the C–F bonds are highly stable and are nonpolar. At lower temperatures (e.g., at room temperature), the perfluorosolvent has very low miscibility with most common organic solvents and is divided into two phases (fluorine and organic). However, with the increase of temperature, the solubility of common organic solvents in the perfluorinated solvent increases sharply. At a certain temperature, some perfluorocarbon can dissolve with the organic solvent into a single phase, which provides good homogeneous conditions for a reaction. At the end of the reaction, once the temperature is lowered, the system reverts to the catalyst-containing fluorophase and the product-containing organic phase (OP). The reaction principle of the fluorine two-phase system is shown in Figure 4.12.

4.4.2 Applications of Fluorine Two-Phase System

1. **Hydroformylation of olefins**

   The hydroformylation of olefins is an important industrial process, and the reaction is usually carried out with HRh (CO) (PR$_3$)$_3$ as the catalyst, which is easily soluble in the OP [22–24].

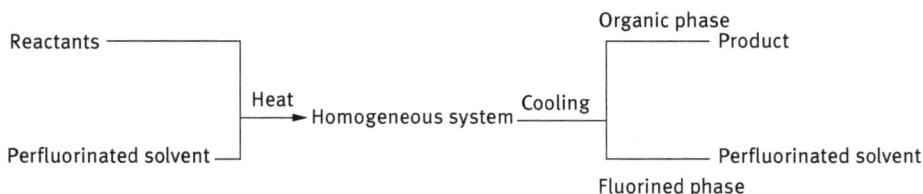

Figure 4.12: Working principle of fluorine two-phase system.

However, when the aldehyde product is separated from the catalyst by distillation in the posttreatment, the catalyst decomposes. In the hydroformylation of olefins with the fluorine-containing HRh (CO) {P [CH$_2$CH$_2$ (CF$_2$)$_5$CF$_3$]$_3$}$_3$ catalyst, the presence of a fluorine chain in its phosphine ligand makes it easily soluble in the fluorine phase, while the solubility of aldehyde is very small, so after the reaction, the simple liquid–liquid separation is used to separate the catalyst from fluorine phase, and is isolated the product from the OP.

2. **Oxidation**

The fluorine two-phase system is well adapted for the oxidation because the solubility of oxygen in the perfluorosolvent is high and the perfluorosolvent is not easily oxidized. In addition, the vast majority of oxidation products are polar, insoluble in perfluorinated solvents, and thus the separation of product is simple and convenient. The fluorinated solvent complex catalyst KRu (C$_7$F$_{15}$COCH$_2$COC$_7$F$_{15}$)$_2$ is an excellent catalyst for the epoxidation of olefins having two substituents in the presence of isobutanol. Even if monovalent olefins are present at the same time, only double-substituted olefins are selectively oxidized to the corresponding epoxy compounds. The fluorine-soluble catalyst may be used three times with negligible loss of activity.

4.5 Organic Synthesis by Phase Transfer Catalysis

4.5.1 Overview

Phase transfer catalysis (PTC) is a new method of chemical synthesis developed in the late 1960s and has been applied in industry since 1970 for many fine chemicals production [25]. In recent years, PTC has been successfully applied to various types of organic reactions, including halogenation, alkylation, acylation, carboxylation, esterification, sulfidation, condensation, addition, reduction, oxidation, polymerization and many base-catalyzed reactions.

PTC refers to the process by which a particular catalyst is used to accelerate the reaction of two substances in a mutually immiscible two-phase system (a liquid–liquid two-phase system or a solid–liquid two-phase system). In the reaction, the catalyst transfers an entity (such as an anion) that actually participates in the reaction from one phase to another, so that it reacts smoothly with the substrate.

PTC has the following advantages: ① faster reaction rate and shorter reaction time; ② less by-product and high product selectivity; ③ mild reaction conditions and lower energy consumption; and ④ ordinary phase transfer catalyst is inexpensive and easy to obtain. So PTC has gained wider applications both in the laboratory and in the industry, especially in fine chemical synthesis.

Commonly used phase transfer catalysts are quaternary ammonium salts, crown ethers, cryptand and open-chain polyethers.

4.5.2 Principle of PTC

In a neutral medium, the phase transfer catalyst provides a lipophilic cation which migrates the anions of the reactants from the aqueous phase (AP) or the solid phase to the OP. In an alkaline medium, one of the reactants loses the proton at the interface of the two phases and initiates the reaction. The role of the phase transfer catalyst is to rapidly transfer the anions produced at the interface to the OP. The reaction was allowed to proceed until it completes.

Quaternary ammonium salt is the earliest discovered and the most commonly used phase transfer catalyst, and quaternary ammonium salt action mechanism is as follows:

In the above immiscible two-phase system, the nucleophile $M^+ Nu^-$ is dissolved in the AP only. The organic reactant R-X is soluble only in the organic solvent and not in the AP. The two cannot be easily contacted to trigger a chemical reaction. If the quaternary ammonium salt $Q^+ X^-$ is added, its quaternary ammonium cation, Q^+, is lipophilic, so that it is soluble in the AP as well as in the OP. When the quaternary ammonium salt is contacted with the nucleophilic $M^+ Nu^-$ in the AP, the anion Nu^- in the nucleophile can be exchanged with the anion X^- in the quaternary ammonium salt to form a $Q^+ Nu^-$ ion pair, The AP migrates to the OP. Because Nu^- is naked with high activity, it reacts with R-X in the OP by rapid nucleophilic substitution to form the target product R-Nu. The simultaneous formation of $Q^+ X^-$ ions in the reaction can also migrate from the OP to the AP to complete the catalytic cycle of phase transfer. The quaternary ammonium cation, Q^+, is not consumed in the recycle, and only acts as a nucleophile Nu^- immigrater, so only a small amount of quaternary ammonium salt is needed to complete the reaction.

The main factors affecting the PTC are as follows: (1) the distribution of phase transfer catalyst (such as quaternary ammonium salt) between the AP and the OP, or the distribution of anions between the two phases; (2) structure and condensed state of the catalyst; (3) reactivity of ion pair in low polarity solvent; (4) reaction rate; (5) hydration degree of anion in OP.

4.5.3 Applications of PTC

1. Medical synthesis

PTC alkylation of Schiff base opens up a new path for the preparation of a physiological activity of amino acids. For example, the amination reaction of the benzophenone and the amino acid ester is carried out, followed by alkylation under phase transfer conditions, followed by hydrolysis to give a methylene-alkylated amino acid. The alkylation of the condensed amines is accomplished by PTC in very simple conditions without the use of a lithium catalyst in anhydrous conditions. The mixture was stirred overnight at room temperature with 1 mol of the amide and an alkylating agent RX (1.2–4.0 mol), 10% aqueous NaOH solution, (C_4H_9) N^+Cl^- catalyst overnight. The OP was separated, washed with saturated aqueous sodium chloride solution, wherein the alkylation product is separated. The residual benzophenone was removed and refluxed with concentrated hydrochloric acid for 6 h to give the amino acid. The reaction is as follows:

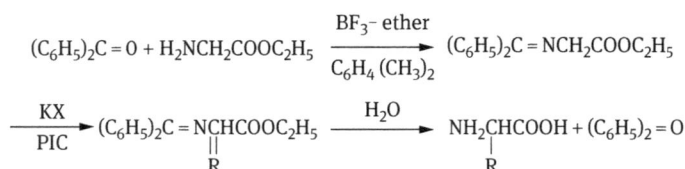

$$(C_6H_5)_2C=O + H_2NCH_2COOC_2H_5 \xrightarrow[C_6H_4(CH_3)_2]{BF_3-ether} (C_6H_5)_2C=NCH_2COOC_2H_5$$

$$\xrightarrow[PIC]{KX} (C_6H_5)_2C=\underset{R}{\overset{||}{N}}CHCOOC_2H_5 \xrightarrow{H_2O} NH_2\underset{R}{\overset{|}{C}}HCOOH + (C_6H_5)_2=O$$

4.6 Combinatorial Chemistry Synthesis

4.6.1 Overview

Combinatorial chemistry originates from the groundbreaking work of Merrifield R. B. in 1963 on the solid-phase synthesis of peptides, also known as combinatorial synthesis, combinatorial libraries and automated synthesis [26–29]. Combinatorial chemistry was originally developed to meet the needs of biologists for developing high-throughput screening techniques for a large number of new chemical libraries. Subsequently, the invention of a polypeptide synthesizer makes the process a conventional means. In the mid-1980s, Geysen established a multi pointed synthesis method. Houghton introduced the concept of peptide library in his established tea bag method. In 1991, the establishment of hybrid split synthesis indicates that combinatorial chemistry has entered a stage of rapid development. In 1996, in the United States, a "combinational chemistry" seminar was held. In the same year, two kinds of combination chemistry and closely related magazines "diversity" and "biological screening" were born. The American Chemical Society also founded the *Journal of Combinatorial Chemistry* in 2000.

Combinatorial chemistry is the combination of chemical synthesis, computeraided molecular design, and high-throughput screening. It can form a large number of related compounds (called chemical library) in a short time by the bonding systematically and repeatedly with different structures of the building block. By their rapid performance screening, researchers can find the best target performance or the structurally active compounds. Combinatorial chemistry is characterized by the formation of tens of thousands of molecules of the compound through a few steps of reaction, when compared with the traditional synthesis method. It can achieve the preparations of thousands and thousands of compounds at the same time to greatly accelerate the synthesis of new compounds. Combinatorial chemistry consists of three parts: the synthesis of compound libraries, library selection, library analysis and characterizations. According to the structure type of the compound library to be synthesized and the required chemical reaction conditions, the compound library can be synthesized by a solid-phase synthesis method, a liquid-phase synthesis method and a liquid-phase synthesis with a carrier. Microwave-assisted combinatorial synthesis technology is a new combinatorial chemistry technology for the preparation of chemical compound libraries. It can not only overcome the shortcomings of traditional solid-phase synthesis technology and liquid-phase synthesis technology, but also improve the product yield.

Combinatorial chemistry originated from the synthesis of drugs, and then moves to the organic synthesis of small molecules, molecular structure analysis, molecular recognition research, catalyst screening, receptor and antibody research and materials preparation and other fields. It is a new type of chemical technology. This is a cutting-edge discipline of molecular biology, pharmaceutical chemistry, organic chemistry, analytical chemistry, combinatorial mathematics and computer-aided design. It plays a more and more important role in the fields of pharmacy, organic synthetic chemistry, life sciences, materials science and catalysis.

4.6.2 Principles of Combination of Chemical Synthesis

Combinatorial chemistry breaks the conception of traditional synthetic chemistry, and it is no longer synthesized individually for individual compounds, but to choose a series of building blocks with similar structure and reactivity $(A_1, A_2, ..., A_n)$ to react with another building block $(B_1, B_2, ..., B_n)$ to obtain a mixture of all the combinations. A few dozens, or even tens of thousands of compounds can simultaneously be synthesized to form a compounds library.

In the conventional chemical synthesis mode, the reaction initiators are usually in pairs. Each synthesis reaction produces only one compound, the product after separation and purification, as the intermediate and then to proceed the second step reaction. This was repeated until the desired product was obtained.
Different from the traditional chemical synthesis, the reaction mode of combinatorial chemical synthesis is as follows:

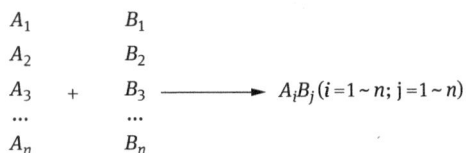

$$
\begin{array}{cc}
A_1 & B_1 \\
A_2 & B_2 \\
A_3 \quad + & B_3 \quad\longrightarrow\quad A_iB_j\,(i=1\sim n;\, j=1\sim n) \\
\cdots & \cdots \\
A_n & B_n
\end{array}
$$

Table 4.7: Relationship between the numbers of building blocks, numbers of reaction steps, and molecule numbers in a library.

Numbers of building blocks	Numbers of reaction steps	Molecule number in library
10	2	10^2
	3	10^3
	4	10^4
20	2	20^2
	3	20^3
	4	20^4
100	2	100^2
	3	100^3
	4	100^4

Such a step reaction can generate $n \times n$ compounds. If $AiBj$ is reacted with the building blocks (C_1, C_2, ..., C_n) and (D_1, D_2, ..., D_n), more compounds can be generated. Combinatorial chemical synthesis is characterized by the use of a few steps to obtain tens of thousands of compounds (see Table 4.7).

4.6.3 Applications of Combinatorial Chemical Synthesis

Combinatorial chemistry was originally developed to meet the needs of drug development. It has made remarkable achievements in research and development of new drugs, which has developed into a new scientific research technology in recent 10 years.

1. **New drug synthesis**

   Hydrogenated uracil is a precursor of herbicidal active material synthesized by solid-phase synthesis. Wang resin was selected as the carrier for solid-phase synthesis to construct hydrogenated uracil compound library. Acrylic acid or acryloyl chloride and Wang resin 1 to get acrylate 2; 2 reacts with primary amine by Michael addition to get secondary amine 3; 3 then adds with isocyanate to give urea 4; 4 dissociates from the resin in acid catalyst and simultaneously loop closed, thus obtaining the target hydrogenated uracil 5. 5 is confirmed by ^{1}H-NMR, high-performance liquid chromatography and mass spectrometry. On the basis of these results, a small chemical library containing nine kinds of hydrogenated uracil is prepared by changing substitutes and their structures were confirmed by gas chromatography-mass spectrometry. The reaction is as follows:

Wang resin

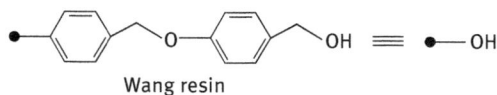

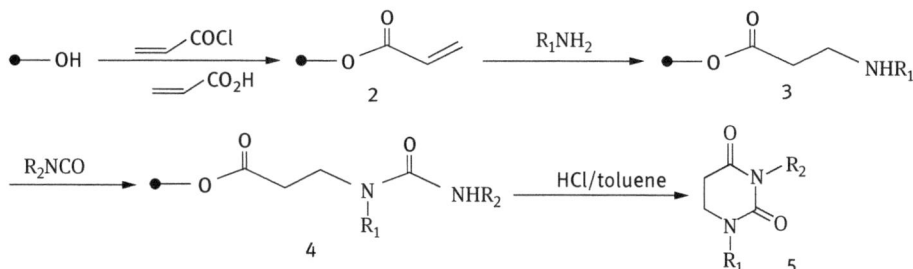

2. Heterogeneous catalyst preparation

Combinational catalytic technology is the combination of combinational chemical methods into the field of catalytic research. It is used in catalyst design, preparation, evaluation, screening and characterization. On the basis of computer-aided molecular design, advanced instrument analysis and high-speed screening technology, a combination of catalyst library and efficient evaluation system was designed. It can be used to reevaluate the present catalysts or used for research and development of new catalytic systems with promising applications. For example, the researchers have developed a unique catalyst preparation and screening system: an improved inkjet printer is used to print Pt, Ru, Rh, Os and Ir – five metals with various content on a conductive carbon fiber paper – and then reduction of metal ions to obtain 80 kinds of two-component, 280 three-component and 280 four-component composition, i.e., a total of 640 electrode material spots. To evaluate the activity of these dots, it is found that $Pt_{62}Rh_{25}Os_{13}$ and $Pt_{44}Ru_{41}Os_{10}Ir_5$ alloy catalysts show the best catalytic performance for the electro-oxidation of methanol.

Questions

1. Solid acid and base catalyst is suitable for what kinds of reactions?
2. What are the advantages and disadvantages of solid superacid as catalyst?
3. Mention the typical preparation method of solid superacids.
4. Mention the typical preparation method of zeolite and its catalytic properties.
5. Mention the typical preparation method of heteropoly acids(salt). Explain the procedure for the preparation of ammonium molybdophosphate by solid-phase method.
6. Explain briefly the properties of polymer acid and its representative products.
7. Sum up the typical preparation method of supported solid bases.

8. What are the properties of ionic liquid and typical preparation methods?
9. What are the characteristics of enzyme catalysis and application areas?
10. Illustrate the working principles of FBS.
11. Give examples to explain the applications of PTC in polymer synthesis.

References

[1] Yu Shitao, Wang Daquan. Solid Acids and Fine Chemical Industries. Beijing: Chemical Industry Press, 2006.
[2] Chen Zhongming, Tao Keyi. Research progress on solid bases catalysts. Chemical Engineering Progress, 1994, 3: 18–25.
[3] Takamiya, N., Koinuma, Y., Ando, K., et al. N-methylation of aniline with methanol over a magnesium oxide catalyst. Nippon Kagakukaishi, 1979, 125: 1452.
[4] Gong Changshen. Practical Technologies in Green Chemical Processes. Beijing: Chemical Industry Press, 2002.
[5] Xu Ruren, Pang Wenqin, Huo Qisheng. Zeolites and Porous Material Chemistry. Beijing: Science Press, 2015.
[6] Chen Weilin, Wang Enbo. Heteropolyacid Chemistry. Beijing: Science Press, 2013.
[7] Xiaoyejiafu, Fubuying. Solid Base Catalysis. Shanghai: Fudan University Press, 2013.
[8] Wang Limin, Tian He. New Methods for Fine Organic Chemical Synthesis. Beijing: Chemical Industry Press, 2004.
[9] Zhang Suojiang, Lv Xingmei. Ionic Liquids. Beijing: Science Press, 2017.
[10] Yuan Qinsheng, Zhaojian. Enzymes and Enzyme Engineering. Shanghai: Northeast University of Science and Technology Press, 2005.
[11] Ma Wu sheng, Ma Tongshen, Yang Shengyu. Research progress on the nitrile hydrolase and its applications in vinyl cyanide production. Chemistry Research, 2004, 15(1): 75–79.
[12] Zhang Yubin, Chiral Synthesis by Bio-catalysis. Beijing: Chemical Industry Press, 2002.
[13] Okuhara T, Mizuno N, Misono M. Advance in Catalysis. London: Academic Press, 1996.
[14] Dai Li xin, Jing Bihui. Catalytic asymmetrical synthesis. China Fundamental Science, 2005, 3: 15–17.
[15] Zhong Bangke. Catalysis in Fine Chemical Processes. China Petrochemical Press, Peking, 2002: 145–154.
[16] Qian Yanlong, Chen Xinzi. Metal organic chemistry and catalysis. Beijing: Chemical Industry Press, 1997: 224.
[17] Rodrigue, Z. B., Rantanen, T., Bolm, C. Solvent free Asymmetric Organocatalysis in a Ball mill. Angewandte Chemie Internmational Edition, 2006, 45(41): 6924–6926.
[18] Jiang Lijuan, Zhang Zhaoguo. Asymmetrical hydroxyl-aldehyde condensation by small organic compounds. Organic Chemistry, 2006, 26(5): 618–626.
[19] Sir John Meurig Thomas, Design and Applications of Single-site Heterogeneous catalyst, Translated by Zhang Long, Hu Jianglei, Beijing: Chemical Industry Press, 2014.
[20] Lin, G. Q., You, Q. D., Cheng, J. F. Chiral Drugs Chemistry and Biological Action. New Jersey: John Wiley & Sons, Inc., 2011: 14–21.
[21] Dalko, P. I., Moisan, L. In the golden age of organocatalysis. Angewandte Chemie Internmational Edition, 2004, 43(39): 5138–5175.
[22] Adrian, P. D., Meriel, R. K. Fluorous phase chemistry: A new industrial technology. Journal of Fluorine Chemistry, 2002, 118(12): 3–17.

[23] Pozzi, G., Colombani, J., Miglioli, M., et al. Epoxidation of alkenes under liquid-liquid biphasic conditions; synthesis and catalytic activity of Mn(III)-tetraarylporphyrins bearing perfluoroalkyl tails. Tetrahedron, 1997, 53(17): 6145–6162.

[24] Klement, I., Lutjens, H., Knochel, P. Transition metal catalyzed oxidations in perfluorinated solvents. Angewandte Chemie International Edition, 1997, 36(13): 1454–1456.

[25] Qu Rongjun, Sun Changmei, Wang Chunhua, et al. Applications of phase transfer catalysis in the production of polymers. Journal of Catalysis, 2003, 9(6): 716–724.

[26] Gallop, M. A., Barrett, R. W., Dower, W. J., et al. Application of combinatorial technologies to drug discovery: Background and peptide combinatorial libraries. Journal of Medicinal Chemistry, 1994, 37(9): 1233–1251.

[27] Thompson, L. A., Ellman, J. A. Synthesis and applications of small molecule libraries. Chemical Reviews, 1996, 96(1): 555–600.

[28] Reddington, E., Sapienza, A., Gurau, B., et al. Combinatorial electrochemistry: A highly parallel, optical screening method for discovery of better electrocatalysts. Science, 1998, 280(5370): 1735–1737.

[29] Fancis, M. B., Jacobsen, E. N. Discovery of novel catalysts for alkene epoxidation from metal-binding combinatorial libraries. Angewandte Chemie International Edition, 1999, 38(7): 937–941.

5 Green synthesis chemistry for polymer materials

5.1 Introduction

With the introduction of green chemistry as a subject by the end of the last century, people's understanding of green chemistry combining with various disciplines is deepening, and the green strategy for polymer material synthesis technologies and applications is gradually formed [1–3]. The research content of green chemistry for polymer involves environmental-friendly raw materials, nontoxic polymer synthesis process or catalyst, nontoxic polymer material itself and atom economic reaction characteristics without by-product formation.

The concept of "ecological polymer materials" in the synthesis of polymer materials relates to the ecological chemistry (mainly refers to raw materials and polymerization process), ecological production (mainly refers to production environment), ecological utility, ecological recovery, recycling and a profound effect for residues in the ecological environment. The ideal ecological study for polymer materials content should include the utility of nontoxic and harmless raw materials, harmless (waste gas, wastewater and waste residue) material production (zero emissions), no environmental pollution during high polymer molding, easy waste recycling and recovery for materials.

Therefore, the requirements for green synthesis of polymers are shown as follows:

(1) Raw materials friendly to the environment and scarce resources are forbidden to use. To select agricultural and sideline products as raw materials is the best choice. For example, in the synthesis of alkyd paint, mainly agricultural and sideline products, soybean oil, castor oil fatty acid anhydride and polyols are employed to prepare alkyd resin, and then the resin reacts with isophorone diisocyanate (IPDI), extended by chain extender, to prepare polyurethane-modified alkyd resin. In the synthesis of polyurethane elastomer or paint, people tend to give priority to the castor oil as preferred raw material to prepare the polymer which can meet the requirements.

(2) Nontoxic solvents are desirable in the polymerization process. For example, water as solvent, ionic liquid, supercritical fluid (SCF) as reaction medium and sometimes even some toxic solvents are to be used, but these should recycle and reduce the residual ratio in the product.

(3) Polymerization reaction process should be environmentally friendly. For example, the use of microwave-induced polymerization, photo-initiated polymerization, radiation cross-linking polymerization, plasma polymerization and other green processing technology will not cause a hazard to the environment.

(4) Highly efficient and nontoxic catalysts are trends of development. Highly efficient and nontoxic catalysts are usually used to improve the catalytic efficiency, shorten the polymerization time and reduce the energy required for the reaction. For example, enzyme-catalyzed polymerization is a typical polymerization reaction carried out with nontoxic catalysts.

https://doi.org/10.1515/9783110479317-005

(5) No by-products were produced; at least no toxic by-products are generated during the polymerization process.

This is exactly the embodiment of the atom economy proposed by Professor Trost in polymer polymerization. Based on the above several aspects, this chapter will give the elaboration of the contemporary development of the polymer green synthesis technology, understanding the typical polymer green synthesis technology from the principle and its application in the development of national economy.

5.2 Polymerization Technology Employing Water as Reaction Medium

Water is the origin of life; it is only a nontoxic liquid reaction medium in chemical solvents. Whether it is organic chemical reaction or polymer chemical reaction, the reaction in water or polymer products in the dispersion medium can greatly reduce the damage to environment [4]. To pursue the realization of water as the dispersion medium for industrial production is a dream of chemists or chemical science and technology workers.

Free radical polymerization of α-unsaturated monomers (monomers) in presence of water phase avoids the use of organic solvents besides solving the engineering problems of exothermic in aqueous phase, whether the process of reaction process or the application of product doesn't pose a hazard to the ecological environment.

5.2.1 The Advantages and Disadvantages

In fact, the free radical polymerization of the unsaturated monomer (vinyl monomer) in the aqueous phase is usually referred to the emulsion polymerization in polymer chemistry. The simplest recipe of emulsion polymerization mainly consists of four components, which are monomer, water, water-soluble initiator and water-soluble emulsifier. Emulsion polymerization has the following advantages when compared with bulk polymerization, solution polymerization and suspension polymerization.

1. **Solution of engineering problems for heat exchange**

   The main features of emulsion polymerization attribute to water as the reaction medium, polymerization reaction system with low viscosity, easy controlled internal heat exchange and the other main features. When compared with the bulk polymerization, its continuous phase is aqueous and reaction spots locate in latex particles in aqueous phase. Although the internal viscosity of latex is very high, the viscosity of reaction system is low, which is basically close to that of the continuous phase (water), and the viscosity of the system in the process of polymerization viscosity changes little.

2. **With high relative molecular weights and reaction rate**

   When compared with free radical bulk polymerization, solution polymerization, suspension polymerization and emulsion polymerization follow the similar reaction mechanism consisting of chain initiation, chain growth and chain termination reaction, and follow the similar reaction kinetics. However, emulsion polymerization also has its own unique mechanism which can improve the reaction rate and increase the average relative molecular weight of polymer in perfect harmony. Therefore, it follows another dynamic law.

   In the emulsion polymerization system, the initiator is dissolved in the aqueous phase and continuously decomposed into free radicals. Once the free radicals in aqueous phase are diffused into the micelles or latex particles, the polymerization occurs in isolated latex particles; free radicals in latex particles can maintain a longer life and the higher average relative molecular mass. The chain termination reaction occurs only when the extra active free radicals enter the latex particles from the aqueous phase, thereby the polymer with higher relative molecular weight can be obtained. In addition, the latex particle's surface with the same charge, owing to the electrostatic repulsion between the particles, there is no possibility for both to collide with each other for different latex and chain termination reaction between latex. So, the chain termination rate of emulsion polymerization is very low, the free radical active chain has sufficient time for chain growth reaction, and the polymer with superhigh average molecular weight is obtained.

   On the other hand, both the lower chain termination reaction rate and the longer life of free radical lead to the higher reaction rate and higher molecular weight, from which the special characteristics of the emulsion polymerization method are embodied.

3. **Representing the development trend of chemistry and chemical industry**

   Most emulsion polymerizations are based on water as the medium, both the use of expensive solvents and the trouble of solvent recovery are avoided; meanwhile the possibility of causing fire and polluting the environment is reduced. As water replaces the organic solvent medium, many favorable characteristics such as no burning, no explosion, nontoxic, tasteless, no environmental pollution, the safety production features, harmless to the human body and great improvement in the work environment are obtained. The water is cheap and easy to get; it can significantly reduce the production cost. With the introduction and strengthening of environmental protection law in the world, employing water instead of organic solvent as reaction medium is the inevitable trend of development. Therefore, employing water as the dispersion medium in the polymerization shows a strong vitality.

4. **Easy treatment of polymer products**

   After polymerization, through postprocessing the polymer emulsion is condensed into massive, granular or powdered polymer, and then processed into a variety of products. It is also directly used as adhesive, paint and other materials used in the construction industry directly, textile, papermaking, leather, etc. In the case of polymer latex application, it is particularly suitable for emulsion polymerization.

In addition to the above-mentioned advantages, some disadvantages of emulsion polymerization are also available: (1) when solid polymer from the emulsion is needed, available condensation, washing, dehydration, drying and a series of postprocessing process will lead to the higher production cost; (2) when the emulsifier used in the reaction system is difficult to be removed from the emulsion, all residues in the polymer will affect the electrical performance, transparency, water resistance of final product and surface gloss of workpiece; (3) when the effective space available for reactor is reduced, the rate of equipment utilization is reduced.

Despite the drawbacks of emulsion polymerization, however its many advantages determine the economic value, social value and environmental value of emulsion polymerization in polymer chemical industry.

5.2.2 Compositions of Aqueous Polymerization System and Their Function

Free radical polymerization of α-unsaturated monomers (monomers) in aqueous phase composition mainly comprises monomer(s), emulsifier, water-soluble initiator, dispersing medium, water, etc. The functions of each component of the four parts are described as follows:

1. **Monomers**

    The common monomers for emulsion polymerization is vinyl monomers (such as styrene, ethylene, vinyl acetate, vinyl chloride, vinyl chloride and other two partial), conjugated diene monomer and acrylic acid or methyl acrylate monomer (such as methyl acrylate, ethyl acrylate, acrylic acid hydrocarbon propyl ester, methyl methacrylate (MMA), methacrylic acid, acrylamide, acrylonitrile and butyl hydrocarbon acrolein). The dosage of monomer is mostly controlled from 40% to 60%

    Emulsion polymerization is presently the predominant process for the commercial polymerizations of vinyl acetate, chloroprene, various acrylate co-polymerizations and co-polymerizations of butadiene with styrene and acrylonitrile. It is also used for methacrylates, vinylchloride, acrylamide and some fluorinated ethylenes.

    In order to reach the desirable properties of emulsion polymers, co-polymerizations of a variety of monomers are often used. For the emulsion formulation design, based on the specific properties of polymer, selected monomers are necessary, such as MMA, acrylonitrile, acrylamide and polar groups can give emulsion polymer with good weatherability, transparency and antipollution; butadiene, chloroprene and vinyl chloride to give emulsion polymer water resistance, flame resistance and oil resistance; acrylic acid, itaconic acid and fumaric acid monomer-containing carboxyl groups can greatly improve the stability of polymer emulsion, and for use in the future, carboxyl groups can provide further curing cross-linking reaction potential. In case of high tensile strength and other mechanical properties for emulsion polymer, hard monomers leading to high T_g should be chosen in formulation, such as MMA, styrene, acrylonitrile and vinyl chloride In case of supertoughness or flexibility for emulsion polymer or using as coatings and adhesives for emulsion at room temperature, the soft monomers leading to the low T_g, such as butyl acrylate, butadiene chloroprene, should be considered.

In order to improve the hardness, tensile strength, wear resistance, solvent resistance and water resistance of emulsion polymer, a network structure polymer is often constructed from the linear polymer. The copolymer should contain reactive comonomers with cross-linking groups, such as (meth)acrylic acid and hydroxyethyl, (meth)acrylic acid, (methyl)acrylamide. Because of the cross-link reaction of reactive groups with itself or with the added cross-linking agent, the formed cross-link network can improve the properties of polymer significantly.

2. **Emulsifiers**

The emulsifier plays a very important role in emulsion polymerization system. The monomers can be dispersed into small monomer droplets by means of the emulsifier. The monomer droplets are stabilized by surfactant molecules absorbed on their surfaces. The monomer droplets are usually regarded as monomer "warehouse." The hydrophobic monomer-soluble micelle made of surfactant molecules in water is an important source of latex particles. The emulsion polymer is stabilized by surfactant molecules absorbed on the surface of latex particle; the polymer emulsion in polymerization, storage, transportation and application is kept stable; and demulsification can't be condensed, at the same time, the emulsifier also directly affects the reaction rate of emulsion polymerization.

According to the nature of the hydrophilic group of the emulsifier, the emulsifier in emulsion polymerization is divided into four types, which are anionic, cationic, ampholytic and nonionic. Anionic emulsifiers commonly used are carboxylic acid salts (such as potassium lauric acid, stearic acid, oleic acid and potassium salt), sulfate (12 sodium dodecyl sulfate and 16 sodium dodecyl sulfate), sulfonate (two succinic acid ester sulfonate and sodium sulfonate) and so on. Sulfate and sulfonate emulsifier can be used in acid and alkaline conditions, and organic carboxylic acid salt in weak alkaline medium can be used to maintain the ionic carboxyl group.

Cationic emulsifier can be used in the preparation of positively charged polymer emulsion. Cationic emulsifier is commonly used in emulsion polymerization with ammonium salt (e.g., 12 alkyl ammonium chloride) and quaternary ammonium salt type (e.g., 16 alkyl three methyl bromide).

The application of ampholytic emulsifier in emulsion polymerization is relatively less. However, the nonionic emulsifiers, such as polyol ester-type fatty alcohol polyoxyethylene ether, alkyl phenol polyoxyethylene ether, polyoxyethylene fatty acid ester, polyoxyethylene nonionic emulsifier, Span and Tween are widely applied in emulsion polymerization.

The type and concentration of emulsifier have significant effect on the size and number of emulsion particle, relative molecular weight of the polymer, the polymerization rate and the stability of polymer emulsion.

3. **Initiators**

According to the generation mechanism of free radicals, the initiator for emulsion polymerization is divided into two categories: one is thermal decomposition initiator and another is redox initiator.

1) **Thermal decomposition initiator**

The mostly used thermal decomposition initiators in emulsion polymerization are peroxide, such as potassium sulfate and ammonium peroxide, and the free radical generated by the cracking occurs in the case of heat.

2) **Redox initiator**

This kind of initiator is composed of oxidant and reducing agent. A free radical leads to the polymerization from a redox reaction between an oxidizing agent and a reducing agent. Redox initiators used in emulsion polymerization usually are referred to persulfate-mercaptan, persulfate-sodium bisulfite, chlorite bisulfite, hydrogen peroxide and ferrous salt, organic peroxide, hydrogen peroxide and ferrous salts of organic polyamine redox system.

With an increase of initiator in concentration, there will be an increase in the rate of free radical formation, the nucleation rate, the number of latex particles and the polymerization rate. Meanwhile, the increase of free radical generation rate also enhances the chain termination rate, and decreases the average relative molecular weight of polymer. In order to reflect the characteristics of high molecular weight and high reaction rate, the concentration of initiator should be controlled properly.

4. **Reaction medium**

The polymerization of unsaturated monomers (vinyl monomers) in aqueous solution has strict requirements on the water medium. Because the metal ions in the water (especially calcium, magnesium, iron, lead and plasma) can seriously affect the stability of polymer emulsion, and have inhibition effect on the polymerization process, the distilled water or deionized water are employed in the emulsion polymerization process, the conductivity value of water should be controlled below 10 mS/cm.

Reaction temperature for some monomers is required at –10 °C even at low temperature of –18 °C. As the freezing point of water is °C, the antifreeze agent is also added in water. One kind of antifreeze agent is the nonelectrolyte antifreeze agent, such as methanol, ethanol, ethylene glycol, acetone, glycerol, ethylene glycol alkyl ether, two oxygen six ring. Another kind of antifreeze agent is the electrolyte antifreeze agent, such as NaCl, KCl and K_2SO_4. Electrolyte antifreeze is cheap and easy to get; the appropriate antifreeze can reduce the critical micelle concentration of emulsifier and polymerization rate and can improve the stability of emulsion system, so the electrolyte antifreeze is only in a small range to reduce the freezing point of water. In particular, nonaqueous solvents as a medium for emulsion polymerization are rarely used.

5.2.3 Principle of Aqueous Polymerization

Emulsion polymerization process of α-unsaturated monomer (monomer) in aqueous solution is divided into four stages, that is, the stage of dispersion (emulsion phase), nucleation stage (phase I), latex particle growth stage (phase II) and polymerization stage (stage III).

1. **Dispersion (emulsion phase)**
   Before the addition of the initiator, the polymerization reaction does not occur in the system, the monomer is only dispersed in droplet form by the stabilization of emulsifier and mechanical stirring; therefore, the dispersion stage is also called emulsion phase.

   The micelle from emulsifier is formed in the water, and the critical micelle concentration (CMC) for a special emulsifier is a certain value at a certain temperature. The average number of molecules in a micelle is called the aggregation number, the general aggregation number of the emulsifier is between 50 and 200, and the particle size is 5–10 nm. In the normal emulsion polymerization system, there are 10^{18} micelles per 1 cm^3 water.

   Figure 5.1 is schematic diagram of emulsion polymerization system in dispersion stage. From Figure 5.1, emulsifier exists in the form of micelle, single molecule adsorbed on the surface of monomer droplets or a single molecule in water phase. Most monomer exists in monomer droplets, only a small amount

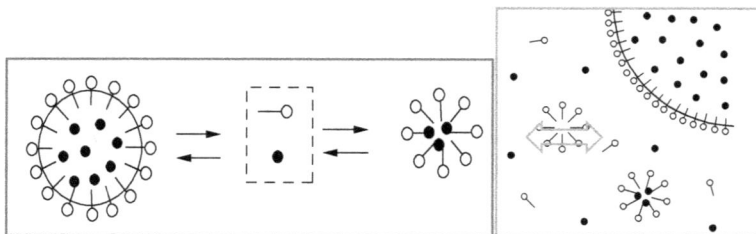

Figure 5.1: Schematic diagram of dynamic equilibrium among the emulsifier, monomer, monomer droplets and micelles. ● monomer, O emulsifying agent, water phase, ✳ micelle.

of monomer is distributed in hydrophobic micelle or dissolved in the water. To balance a dynamic relationship for emulsifier and monomer, they are built among water, monomer droplets and micelles.

2. **Nucleation stage (phase I)**

When the water-soluble initiator is added to the reaction system, the initiator is decomposed to generate free radicals at the reaction temperature. Before the beginning of the polymerization, the induction stage of the polymerization reaction is often passed. After the induction stage, reaction speedup stage occurred. As the formation of latex particles mainly occurred in this stage, stage I is also called latex particle formation stage. The generated radicals produced can diffuse into micelles, to monomer droplets, or dissolved in aqueous phase. Because the number of micelles is much more than that of monomer droplets, under normal circumstances, most of the free radicals will enter into soluble micelles. Once the free radicals entered into the hydrophobic micelles, they immediately initiate monomer to form living polymer in micelle, and the micelle polymer particles change into the monomer swelling (latex). This process is called micelle nucleation process. The micelle from the emulsifier is the main source of latex particles; the greater the concentration of the emulsifier, the more the number of micelles in the system and the more the number of latex particles, the higher is the rate of polymerization reaction.

The polymerization mainly occurred in the latex particles. In the polymerization reaction, the monomer in the latex particles is gradually depleted, and a free monomer molecule in water phase will continue to diffuse into the latex particles. The monomer exists in dynamic equilibrium among latex and aqueous and monomer droplets, with the continuous consumption of monomer in latex particles. The mobile equilibrium constantly moves along the monomer droplets, water phase and latex particle direction to keep the constant concentration in latex particles, as shown in Figure 5.2.

Although the free radicals in aqueous phase polymerize free monomer molecules dissolved in water or diffused into monomer droplets to form latex

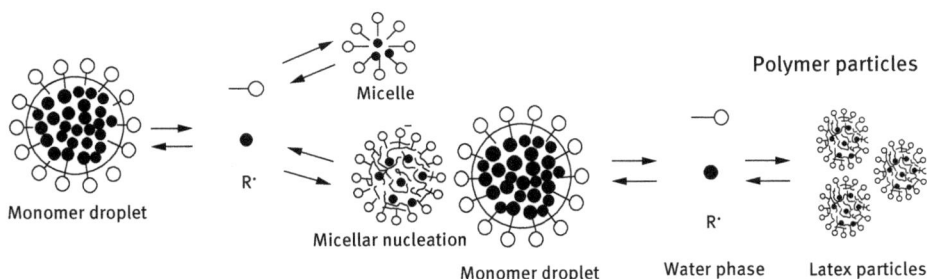

Figure 5.2: Schematic diagram of dynamic equilibrium of particles from micellar nucleation to latex particles formation.

particles, the research results show that the latex particles mainly occurred in solubilization micelles (i.e., by micellar nucleation mechanism), and the other two generation mechanisms of latex particle are very few, and even can be ignored.

3. **Latex particle growth stage (phase II)**
    The end of the acceleration phase is characterized by the disappearance of the micelles; all of the micelle particles in the system are transformed into the latex particles and the number of latex particles in the phase II is no longer increased, and the number of the particles is kept constant. The polymerization of the monomer in latex particles makes their size become larger. With the continuous consumption of the monomer in the latex particles, the balance of the monomer moves from the droplets to water phase and then to the latex particles, and the monomers in monomer droplets gradually decrease till the disappearance of the monomer droplets. This interval from deplete of micelles to disappearance of monomer droplets, and with the constant number and increasing size for the latex particles is called phase II or latex particle growth stage. In stage II, the emulsifier maintains a dynamic equilibrium among the monomer droplet, surfactants, free radical and polymer latex particles, as shown in Figure 5.3.

4. **Polymerization completion stage (stage III)**
    At the stage of polymerization, the characteristics are the disappearance of both the particle and the single body fluid droplet. Only the emulsion particles and water phase exist in the system. Emulsifier, monomer and free radicals are determined by their dynamic equilibrium in these two phases. The monomer is converted into a polymer under the action of free radical; hence this stage is also called polymerization completion stage (Figure 5.4).

    In this phase, the initiator in the aqueous phase continues to decompose the free radicals, and the free radicals also continue to diffuse into the latex particles and leads to polymerization in the latex particles. At this time, since the latex particles can only consume the stored monomers, the monomer can not be supplemented, so the concentration of monomer in the latex particles

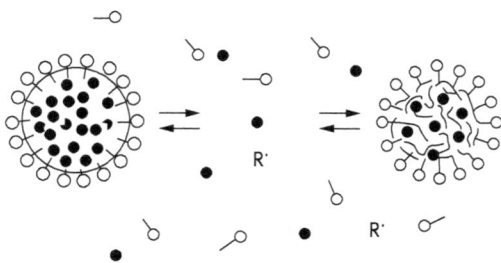

Figure 5.3: Schematic diagram of dynamic equilibrium among the monomer droplets, surfactants, free radical, and polymer latex particle.

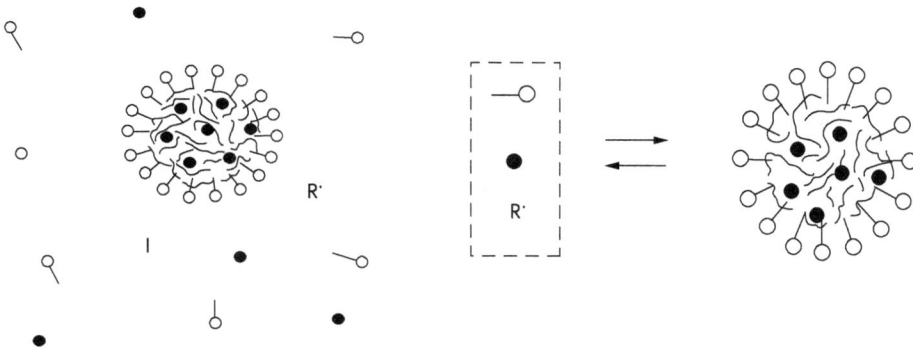

Figure 5.4: Schematic diagram of dynamic equilibrium among the monomer, surfactants, free radical, and polymer latex particle.

gradually decrease with an increase in reaction time. With the monomer concentration gradually reduced, polymer concentration increased gradually, the macromolecular chain tangled together, the rate of latex viscosity gradually increased and the rate of termination for free radicals in latex particles is decreasing rapidly. The life of the free radicals has grown longer. Therefore, in phase III, with the increase of conversion rate of monomers in latex particles, the reaction rate increased very much, This phenomenon is called Trommsdoff effect or gel effect.

The polymerization kinetics curve is shown in Figure 5.5.

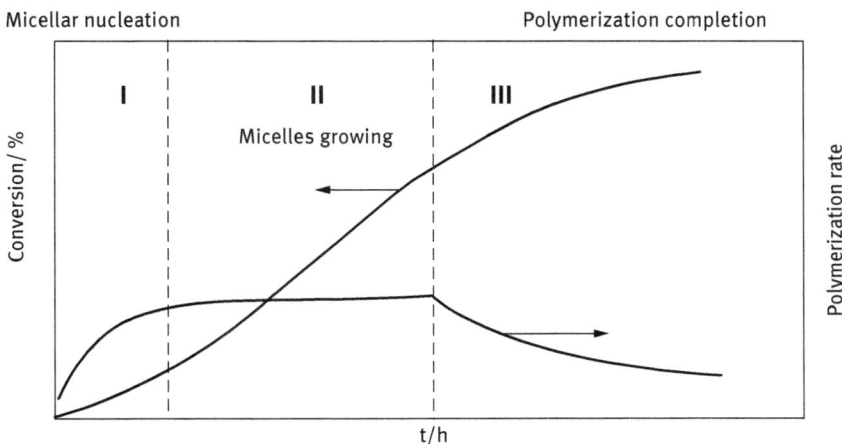

Figure 5.5: Kinetic curves of emulsion polymerization.

5.2.4 Application of Green Polymer Latex

1. **Application of polymer emulsion in textile**

   A lot of polymer emulsion show excellent adhesion properties, so besides a binder for printing paste in the textile, the polymer emulsions are also used as an adhesive in the following textile process, such as electrostatic flocking, as adhesive-coated cloth, short fiber can be planted in the cloth linters by electrostatic interaction, and after drying and baking short fiber vertical adhesive the cloth to form electrostatic flocking products.

2. **Application in construction industry**

   Emulsion paint (commonly known as latex paint) is one kind of a water-based paint, which employed synthetic polymer emulsion as the base material, pigments, fillers and additives were dispersed in the formation for water dispersion system. The latex paint has the advantage of quick drying, and the container can be used again. Most latex paints are odorless or slightly taste, and in the humid climate conditions it can work on the wet surface of the construction; the film has the characteristics of alkali resistance, and weathering, so the latex paints are quite popular to be used.

3. **Application of polymer latex in metal coating industry**

   Metal surface coating is a common process for metal materials processing. The coating can make the metal surface smooth and beautiful luster, and can play the role of internal corrosion resistant effect. It is widely used for ships, cars, bridges and other large metal structures; it can not only play the role of beauty but also can improve the service life and corrosion resistance of metal components. Electrophoretic paint is a common coating for metal surface which is briefly described as follows.

   The electrophoretic paint is a water-based coating, polymer particles of which can auto deposit on the anode plating in the electric field of. Electrophoresis (electro deposition) paint that is similar to ordinary electroplating, works in electrolytic tank. According to the types of charged particles, the particles are divided into anode electro-deposition and cathode electro-deposition. Usually the anodic electro-deposition often uses the latex particles with negative charge from carboxyl anion groups or particles around the adsorption of anionic emulsifier around latex particles. If the latex particles contain quaternary ammonium salt and other cationic groups, the latex particles with a positive charge can prepare cathode electrophoresis paint to be used. Its working principle is shown in Figure 5.6.

   After neutralization by the organic acid, cationic polymer resin under the action of electric field occur the following reactions:

   Polymer particles with positive charge make directional movement toward the cathode under the electric field, at the same time, the following reactions occur in the cathode region:

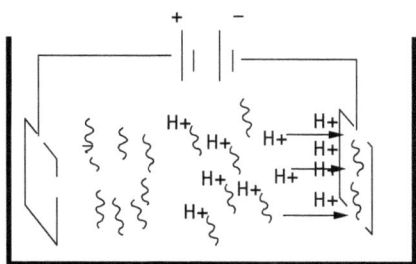

Figure 5.6: Schematic diagram for cathodic electrophoresis paint.

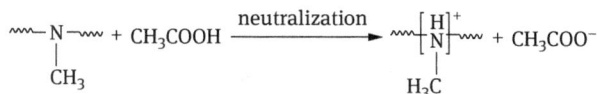

$$\text{---N---} + CH_3COOH \xrightarrow{\text{neutralization}} \left[\text{---N---}\right]^+ + CH_3COO^-$$

$$2H_2O + 2e \rightarrow 2OH^- + H_2\uparrow$$

The increase in hydroxyl ion concentration cathode interface enhanced the pH value, the electric neutralization reaction occurred with positively charged particles and deposition of polymer occurred on the surface of the cathode.

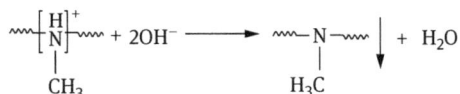

$$\left[\text{---N---}\right]^+ + 2OH^- \longrightarrow \text{---N---}\downarrow + H_2O$$

While, the following reaction also occurs in the anode area:

$$2H_2O \rightarrow O_2\uparrow + 4H^+ + 4e$$

A wet layer of resin film is formed on the surface of the cathode plating part, and the wet resin film is solidified via high temperature or other method to obtain the dense electrophoresis paint film.

5.3 Polymerization Technology in Ionic Liquids

As a substitute for the traditional organic solvents, the new field of green chemistry technology has been explored via investigation of the application of ionic liquids [5, 6]. Especially in polymerization, the application research for ionic liquids got significant development and many excellent achievements.

5.3.1 Radical Polymerization

In recent years, the study of the free radical polymerization in room temperature ionic liquids has got notable research effect that the traditional free radical polymerization cannot achieve. The typical characteristics are as follows:

1. **The faster polymerization reaction rate and the higher relative molecular weight**

   For example, MMA free radical polymerization can easily be carried out in ionic liquids of 1-butyl 3-methyl imidazole six fluoride phosphate ([bmim][PF$_6$]) (Figure 5.7). With an increase in concentration of ionic liquids, the chain propagation constant k_p enhances, and the k_t of chain termination rate constant decreases.

   By comparing the radical polymerization reaction for MMA in ionic liquid [bmim][PF$_6$] and benzene solvent, the polymerization reaction rate of MMA in ionic liquids is about 10 times in the benzene, and the prepared polymethyl methacrylate (PMMA) polymer has a larger relative molecular weight. Because of the high concentration of [bmim][PF$_6$] it makes ionic liquid polar enhancement, and the homogeneous phase for the reaction system has always been maintained in the polymerization process, so the faster reaction rate of polymerization was observed.

2. **Block copolymerization as target product**

   Generally, the traditional radical polymerization cannot realize the synthesis of block copolymer. Since the use of ionic liquids as solvents, synthesis of block copolymer by radical polymerization reaction became a reality. Co-polymerization of styrene (St) and MMA monomer in [bmim][PF$_6$] can get PSt-b-PMMA block copolymer with weight average molecular weight (Mw) from 2×10^5 to 8×10^5.

   Radical polymerization reactions in ionic liquids have two characteristics which are as follows:

   (1) Because of the higher viscosity for ionic liquids, with precipitation of polymer, the probability of termination for living growing chain radical that are controlled by diffusion became lower, and the life of free radical got extended; so both the relative molecular weight and polymerization rate got increased;

   (2) Because of the solubility of polystyrene (PS) in the ionic liquid is very small, and the solubility of PMMA in the ionic liquid is strong, St is added first and PS is prepared. To a certain degree of polymerization the residue monomer

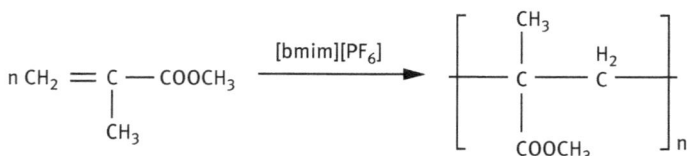

Figure 5.7: Free radical homopolymerization for MMA monomer in [bmim] [PF$_6$].

of St was removed from the reaction system by a vacuum, and then MMA monomer was added. Since MMA is completely dissolved in ionic liquids, the polymerization can proceed.

Ionic liquids can not only be used in the free radical polymerization, but also in the study of active free radical polymerization, and some achievements have been obtained.

Most of the studies on living radical polymerization are about atom transfer radical polymerization (ATRP) and reverse ATRP. The ATRP reaction is initiated by an organic halide (initiator), a transition metal low valent salt (catalyst) and a nitrogen-containing bidentate ligand. The reaction mechanism is shown in Figure 5.8.

It can be seen from the above reaction mechanism that the content of transition metal compounds plays an important role in controlling the polymerization reaction. The ability of ordinary organic solvents to dissolve transition metal compounds is weak, and the ATRP reaction is generally heterogeneous. In order to achieve the purpose of effective control of polymerization, it is often necessary to add a large number of metal catalysts, which not only increases the reaction cost but also inevitably lead to the pollution of the residual metal ions to the polymer.

Carmichael studied the ATRP reaction for MMA in [bmim][PF$_6$], as shown in Figure 5.9. It was found that the polymerization can obtain a very high reaction rate and low molecular weight distribution polymer (Mw/Mn = 1.30–1.40) at a low temperature.

Because the [bmim][PF$_6$] is insoluble in toluene, polymer can be extracted from the reaction mixture with toluene after reaction, and the vast majority of metal ions is still in ionic liquids and does not pollute the polymer; therefore, the problem of residual metal ions polluting polymer has been solved.

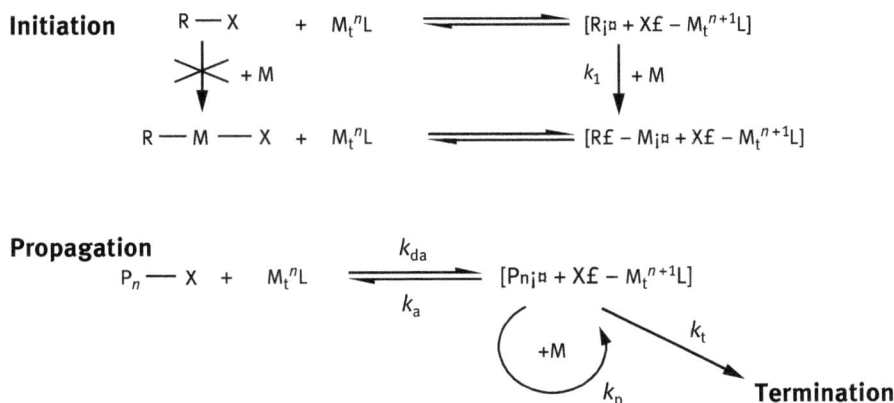

Figure 5.8: Atom transfer radical polymerization reaction mechanism.

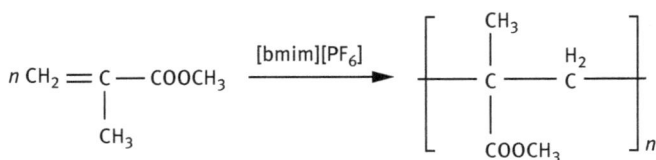

Figure 5.9: ATRP reaction of MMA in [bmim] [PF$_6$].

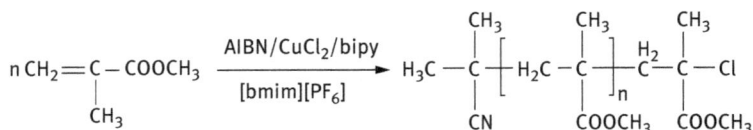

Figure 5.10: Reverse ATRP reaction for MMA in [bmim] [PF$_6$].

In addition to ATRP reaction for the MMA in ionic liquid, similar ATRP polymerization reactions for series of acrylate monomers in [bmim][PF$_6$] can also occur. Comparison of ATRP reactions in ionic liquids with that in organic solvent, the former showed the following advantages, such as high polymerization reaction rate and mild reaction temperature, easy separation from ionic liquid and recyclable utilization.

Like ATRP, the reverse ATRP also realize to control the polymerization by reducing the concentration of free radicals affected by the redox reaction and adjusted by the transition metal catalyst. In reverse ATRP, with conventional free radical polymerization initiators, such as Azobisisobutyronitrile azobidsisobutyronitrile (AIBN) and benzoyl peroxide (BPO) initiators, instead of alkyl halides, and with transition metal salt instead of low salt, the problems about toxicity and catalyst for ATRP have been effectively solved.

For example, the trigger system of AIBN/CuCl$_2$/bipy pyridine (al) was used to initiate the reverse ATRP reaction for MMA in [bmim][PF$_6$], as shown in Figure 5.10.

[bmim][PF6] has good solubility of metal compounds; homogeneous phase polymerization of ATRP system can be formed in [bmim][PF6], only a small amount of catalyst ([AIBN]/[CuCl$_2$] = 8/1) in the system can effectively control the polymerization reaction, thereby reducing the pollution of metal ions on polymer. After the polymer is separated from the reaction mixture, the ionic liquid can be recycled and reused by simple treatment. MMA and AIBN were added to the recovered ionic liquid, and the reaction was carried out in the same way again.

5.3.2 Ionic Polymerization

Successful radical polymerization reactions in the ionic liquid greatly inspired scientists' interests to use ionic liquids as solvents for other polymerization [7]. Theoretically, ionic liquids with high polarity are more beneficial for ion polymerization reaction; in fact, radical polymerization reactions in ionic liquids were investigated much, and research results of ionic polymerization reactions in ionic liquids are little reported.

Vijayaraghavan et al. [7] with a new type of Bronsted acid and double oxalic acid root boric acid (HBOB) as initiators studied the cationic polymerization of styrene in dichloromethane (DCM) and ionic liquid [PI⁴⁻] [Tf₂N] (*N*-methyl-*N*-butyl pyrrole,3-methyl sulfonate acid salt). The reaction principle is shown in Figure 5.11.

Comparing with the traditional organic solvent DCM, the polymer with high relative molecular weight and narrow molecular weight distribution has been prepared in [P₁₄] [Tf₂N]. A mixture of ionic liquids and the initiator can also be recycled.

5.3.3 Polycondensation and Addition Polymerization

So far, there have been relatively few reports about condensation and addition reactions in ionic liquids [8]. Vygodskii [8] found that in [R₁R₂im] (1 and 3 on the imidazolium cation substituent) ionic liquid without additional catalyst, polyamide has been prepared by addition reaction of hydrazine with four carboxylic acid biacyl or polycondensation of hydrazine with the biacylchloride. As shown in Figure 5.12, the polymer with superhigh relative molecular weight is obtained.

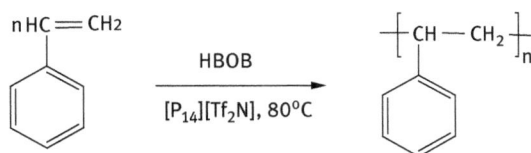

Figure 5.11: Cationic polymerization reaction of styrene in [P14] [Tf2N].

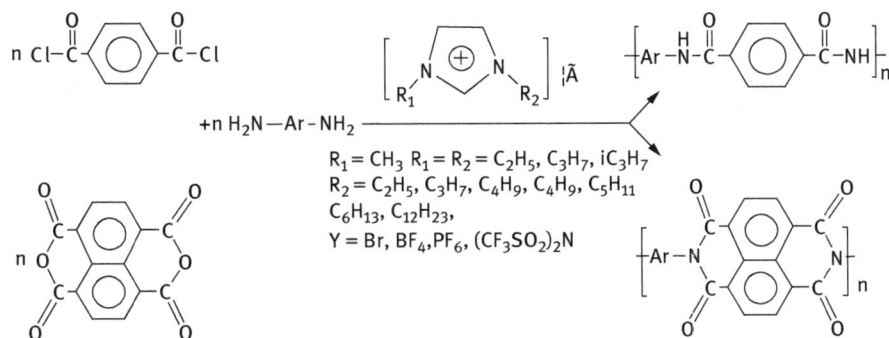

Figure 5.12: Pol-condensation reactions in ionic liquids.

$$n\ C_6H_5 \xrightarrow[\text{[bmim][BF}_4\text{] or [bupy][BF}_4\text{]}]{\text{[Rh(I)]/NEt}_3} \text{PPA}$$

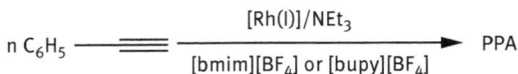

Figure 5.13: Coordination polymerization of phenyl acetylene in ionic liquids.

5.3.4 Coordination Polymerization

In the past, the coordination polymerization was carried out under the condition of high temperature and pressure in the presence of Ziegler-Natta catalyst. Pinheiro [9] fulfilled the coordination polymerization of ethylene in [bmim][AlCl$_4$] ionic liquid in the presence of two imine nick catalyst, under a mild condition to achieve.

Mastrorilli [10] studied the coordination polymerization of phenyl acetylene in [bmim][BF$_4$] and [bupy][BF$_4$] (N-butyl four fluoboric acid pyridine salt) in the presence of Bh(I) as catalyst, triethylamine as cocatalyst, respectively, as shown in Figure 5.13.

Results showed that in the above two kinds of ionic liquids, the yield of coordination polymerization was very high, the molecular weight of the prepared polymer has reached 55,000–200,000, the catalyst can be recycled, and the activity is not significantly reduced.

5.3.5 Electrochemical Polymerization

The application of ionic liquids in electrochemical polymerization was started earlier than that in other polymerization methods. As early as in 1978, the electrochemical reaction from benzene to poly(phenylene oxide) has been carried out in bupy/AlCl$_3$ ionic liquids by Osteryoung et al. [11].

Arnautov [12] tried to use bupy/AlCl$_3$(OC$_2$H$_5$) ionic liquid instead of the traditional chloroaluminate ionic liquid to realize the electrochemical synthesis of benzene. Recently, Zein [13] used [Hmim][CF$_3$SO$_3$]1-hexyl-3-methyl imidazole three trifluoromethylsulfonate and [P$_{14}$][Tf$_2$N] ionic liquid as electrolyte, which is stable to air and water, and to study the synthesis of PPP film. The results showed that the ionic liquid is nontoxic, odorless, and noncorrosive, and the PPP film has a good electrochemical activity and the polymerization rate is faster.

In addition, Sekiguchi [14] can also be used to undergo electrochemical polymerization in [emim][CF$_3$SO$_3$] ionic liquid. Taking [emim][CF$_3$SO$_3$] as medium, the morphology and structure of poly (vinyl chloride) film on the anode surface can be controlled well, and the rate of polymerization was also improved. Pyrrole [15] can also be used to synthesize polypyrrole film by electrochemical polymerization in [bmim][PF$_6$], [emim][Tf$_2$N], and [bmpy][Tf$_2$N] plasma liquid.

5.4 Polymerization Technology in SCFs

One of the most active research fields in the study of nontoxic and harmless solvents is the development of SCF, such as supercritical carbon dioxide as solvent [16]. Supercritical carbon dioxide is the carbon dioxide fluid which temperature and pressure are above critical point (311 °C, 7,477,179 Pa). It shows the density of liquid, and the solubility of the conventional liquid solvent. Under the same conditions, it has not only the gas viscosity, but also a higher mass transfer speed and a greater compressibility. The density, solvent solubility and viscosity for SCFs can be adjusted by pressure and temperature. Its biggest advantage is nontoxic, nonflammable, low price and so on.

SCF has unique physical and chemical properties and cannot cause environmental pollution or rarely. It can achieve both atomic economic reaction and highly selective reaction.

Supercritical chemical reaction is divided into two categories, that is reaction medium and reactants in the supercritical state.

The exploration of supercritical reaction medium has been studied. People are most interested in studying carbon dioxide and water for their environmentally friendly, abundant and inexpensive features on earth, while the recycling of carbon dioxide as medium (solvent) can reduce the impact of greenhouse gas on the atmospheric environment.

5.4.1 Polymerization in Supercritical Carbon Dioxide

The solubility of fluoropolymers is very small in the traditional organic solvent [16]. Chlorofluorocarbon organic solvents as medium are commonly used to prepare the fluoropolymers; however, chlorofluorocarbons organic solvents were verified for damaging the ozone layer. Fortunately, it was discovered that the solubility of fluoropolymer is high and can achieve homogeneous polymerization in supercritical carbon dioxide. Since De Simone [16] put forward the novel supercritical carbon dioxide polymerization in 1922, there have been more and more reports about the polymerization system, involving homogeneous polymerization, precipitation polymerization, dispersion polymerization, emulsion polymerization reaction and so on.

De Simone et al. [16] investigated the continuous precipitation polymerization of vinylidene fluoride (VF_2) in supercritical carbon dioxide. The conversion of VF_2 is 7–26% at 75 °C, 2.5 mol/L and its reaction rate is 27×10^{-5} mol/L·s. The relative molecular weight of the resulting polymer in solid power is about 150 kg/mol and its melting index is 3.0 at 230 °C.

In addition, polymerization of acrylic acid and mixed butane oligomerization have also been reported.

5.4.2 Depolymerization of Polymer in Supercritical Water

Treatment to waste plastics by SCFs is an innovative technology. Supercritical water has the performance of organic solvents in normal state; it can dissolve organic matter, but can't dissolve the inorganic matter and even has the oxidation. It can be completely miscible with air, nitrogen, oxygen, carbon dioxide and other gases, so it can be used as the medium of oxidation reaction and directly carry the oxidation reaction. It can also decompose and degrade macromolecules, recycle valuable products and recycle resources to meet the needs of environmental protection.

1. **Treatment of polyethylene (PE) or PS with supercritical water**
    The preliminary experimental results of supercritical water on the degradation of PS foam showed that the reaction efficiency was the highest within the initial period of 30 min. The additives can promote the degradation reaction and obtain the product with lower relative molar mass, and the ratio of efficiency to cost is the highest when the dosage of additives is about 5%. When the reaction time was shorten or there were no additives, an increase of reaction temperature can play a significant role in promoting degradation.

    Combination of PE with water at 400 °C, PE was degraded into oil composed of alkanes and olefins in 1–3 h. The distribution of products can be changed by optimization of temperature, water and reaction time. On the occasion of some conditions, aromatic hydrocarbons can also be generated. Emissions from this technology are oil and water, which are easy to be separated from each other, almost free of harmful substances, environmentally friendly and the wastewater can be recycled.

2. **Treatment of polyvinyl chloride (PVC) with supercritical water**
    Supercritical water can also be used to degrade PVC polymer to get different organic compounds. Experimental results indicate that it is possible to use the supercritical water technology in the recovery of PVC waste. Chlorine atoms in PVC are recycled in water in the form of HCl. The results show that these chlorinated compounds can be completely converted into environmental-friendly products during the process of supercritical hydrolysis without harmful by-products to be discharged. The approach provided a feasible method to solve the white pollution.

3. **The degradation of polyethylene(terephthalate) (PET) with supercritical water and supercritical methanol**
    The degradation of PET in supercritical water and supercritical methanol has been studied [17]. In supercritical water, the oligomer is first formed during degradation, followed by the formation of terephthalic acid. The recovery ratio of high purity terephthalic acid (about 97%) is above 90%. In supercritical methanol, the reaction degradation rate is fast, conditions are appropriate, ethylene glycol product can be recovered 100%, and no gas and by-products were formed. The depolymerization of PET by supercritical methanol showed that the supercritical

methanol could depolymerize the PET quickly, and the PET could be completely depolymerized in 30 min at 280 °C and 8 MPa. The depolymerized product has high purity and could be easily separated. Solid phase terephthalic acid has high purity and can be used as raw materials for production; Liquid phase is the mixture of methanol and ethylene glycol, and they can be easily separated by distillation in which ethylene glycol and methanol can be recycled. The reaction requires no catalyst and is easy to realize industrial continuous production. The reaction system has no corrosivity, and no pollution to the environment.

5.4.3 Supercritical Enzyme Catalytic Reaction

Enzyme is a kind of biocatalyst that has highly efficient catalysis effect under mild conditions and has high region-selectivity and stereo-specificity [18]. Enzyme-catalyzed biosynthesis from water-based media to nonaqueous organic solvent media is a major breakthrough in enzyme catalysis, it provides a new development opportunity for supercritical carbon dioxide-catalyzed reactions.

SCFs as the enzyme catalytic reaction medium can carry out the reverse reaction of enzymatic hydrolysis reaction, such as esterification, ester exchange and peptide synthesis, it also increases the thermal stability of the enzyme.

The polarity of the microenvironment near the enzyme activity center was studied in the esterification reaction of lauric acid and n-butanol catalyzed by lipase in supercritical carbon dioxide. The solvent properties of SCFs were characterized by its solubility parameter and dielectric constant. The solubility parameter and dielectric constant of carbon dioxide in different pressures were calculated. The maximum reaction rate under different pressures was measured, and the influence of the solubility parameter δ and the dielectric constant ε on the maximum reaction rate was studied and better correlated.

The advantages of the supercritical enzyme catalytic reaction is that it can improve the reaction rate of a chemical reaction, reduce reaction temperature, convert heterogeneous reactions into homogeneous reactions, reduce catalyst deactivation rate, enhance selectivity, improve the mass transfer rate of multiphase reaction, replace the harmful solvents with environmental-friendly solvents, combine chemical reaction and separation process together.

Supercritical reaction technology plays an important role in many fields, such as bioengineering, pharmaceutical engineering and material engineering. With the continuous development of supercritical reaction, new technologies will continue to emerge, which will create enormous economic and social benefits for mankind.

SCFs are different from conventional gases, liquids and solids, although a lot of researches on SCF has been made, there are still some limitations. Supercritical reaction technology, whether it is basic research or applied research, is still in the process of continuous exploration of specific stages. Supercritical reaction technology as an efficient reaction technology meets the requirements of sustainable development in today's society. With the deepening of research, supercritical reaction technology will be widely used.

5. 5 Synthesis of Waterborne Polyurethane with Low Residual Volatile Organic Compounds (VOCs)

In the process of polymerization, achieving nontoxic solvent medium such as the previously discussed with water, ionic liquids and SCFs instead of organic solvents are the research contents of green chemistry [19–22]. Using certain toxic solvents in the synthesis of polymer material, ensuring toxic solvents are recycled and low residual in the product are still the contents of polymer green synthesis. The preparation of water-based polyurethane resins is the typical example.

Based on the fact for polyurethane adhesives with superadvantages, including adjusted hardness, good low temperature resistance, better flexibility, strong adhesive strength and so on, they are widely used in many fields. At present, polyurethane adhesives is mainly based on solvent polyurethane resin. Organic solvents are flammable, explosive, volatile, odor, pollution of air and more or less toxic. Over the past 20 years, with the increasing public pressure on earth, for environment protection some developed countries have established the fire regulations and solvent laws; these factors urged researchers to spend considerable efforts to devote for the development of waterborne polyurethane resins.

Water-based polyurethane resin employing water as the medium showed the superadvantages, such as nonflammable, little odor, no-pollution to the environment, energy conservation and easy operation. At present, people pay more and more attention to the water-based polyurethane resins.

Comparing with solvent-based polyurethane adhesives, in addition to the above advantages, waterborne polyurethane adhesives also shows the following characteristics.

(1) Most of the waterborne polyurethane adhesive does not contain reactive NCO group, so the curing of resins mainly depends on the cohesion and adhesion of the intermolecular polar groups. The solvent or solvent free single component and two-component polyurethane adhesive can make the best use of the reaction of NCO groups to strengthen the adhesive performance in the curing process. The carboxyl, hydroxyl and other groups of waterborne polyurethane can also be involved in cross-linking reaction under proper conditions.

(2) Besides the extra polymer thickener, the important factors affecting the viscosity of waterborne polyurethane are ionic charge, core-shell structure, emulsion particle size and so on. The more ions and counterions on the polymer molecules chain (the free ions opposite to polarity of ion groups on main chain and the side chains of polyurethane in solution), the higher will be the viscosity of waterborne polyurethane. The weight percent and the molecular weight of polyurethane,

cross-linking agent and other factors have no obvious influence on the viscosity of water-based polyurethane. In contrast, the main influencing factors affecting the viscosity of the solvent based polyurethane adhesive are the molecular weight, the degree of branching, the concentration of the polymer weight percent and so on. With the same polymer weight percent, the viscosity of water-based adhesives is lower than that of solvent adhesives.

(3) Viscosity is an important parameter for adhesives. The viscosity of waterborne polyurethane dispersions (PUDs) is generally adjusted by water-soluble thickener and water. While the solvent-based adhesives can be adjusted by the solid content, the relative molecular weight of the polyurethane, or the choice of suitable solvent.

(4) Because of the lower volatility of water than that of organic solvent, the drying rate of aqueous polyurethane adhesive is slower, and due to its higher surface tension, the water-based PUD shows the poor wetting ability on the surface of the hydrophobic substrate. Owing to the majority of waterborne polyurethane resins containing hydrophilic groups is the main ingredient of the adhesive, and sometimes water-soluble polymer thickener was added into the recipe, so , the dried adhesive film without cross-linking structure showed the poor water resistance.

(5) Waterborne polyurethane resins can be mixed with a variety of waterborne resins to improve its performance or reduce costs. At this point, we should pay attention to the electrical properties of water-based resin and acid–base; otherwise, it may lead to the aggregation of water-based polyurethane resin.

(6) Water-based polyurethane resin showed the features of low odor, convenient operation and easy to clean. Solvent-type polyurethane resin used to consume a large amount of organic solvent and is not as convenient as aqueous polyurethane resin when needs to be cleaned.

5.5.1 Classifications of Waterborne Polyurethanes

Due to the diversity of raw materials and formulations, after 40 years' development of waterborne polyurethane, a variety of preparation methods and formulations people have been developed. A wide variety of water-based PUDs can be classified into several ways.

1. **Classification by appearance**
   Waterborne polyurethane can be divided into polyurethane emulsions, PUDs and polyurethane aqueous solutions. The widely applied waterborne polyurethanes are polyurethane emulsions and dispersions, which are collectively known as water-based polyurethane or polyurethane emulsion in this chapter. The classification by appearance is shown in Table 5.1.
2. **Classification by using form**
   Waterborne polyurethane adhesives can be divided into single-component and two-component types by using the form. Single-component type can be used

Table 5.1: Classification of waterborne polyurethane dispersions.

Index	Aqueous solution	Dispersion	Emulsion
State	Solution to colloid	Dispersing	Dispersing
Appearance	Transparent	Translucent milky white	Nebulous urine
Particle size/μm	<0.001	0.001–0.1	>0.1
Relative molecular weight	1,000–10,000	Thousands to 2 hundred thousands	>5,000

directly, or without added crosslinking agent, to get the desired performance for the PU film. If the required performance of polyurethane cannot be obtained when used alone, it is necessary to add cross-linking agent in the single-component PUD. In general, the adhesive performance for single-component waterborne polyurethane can be improved after cross-linking agent was added. In these cases, the PUDs consist of waterborne polyurethane resin, and cross-linking agent was called two-component waterborne polyurethane adhesives

3. **Classification by the properties of hydrophilic groups**
   According to the type of ionic groups contained in the side chain or main chain of the polyurethane, waterborne polyurethane can be classified as anionic, cationic and nonionic. Anionic or cationic water-based polyurethane is also known as ionomer-type water-based polyurethane.
   (1) Anionic waterborne polyurethane can be subdivided into sulfonic acid type and carboxylic acid type. The ionic groups that are in the side chain are the majority of anionic waterborne polyurethane. The carboxyl ions and sulfonic acid ions were introduced into the polyurethane chain by the carboxyl-containing chain extender or sulfonate-containing chain extender.
   (2) Cationic waterborne polyurethane generally refers to the main chain or side chain containing ammonium ions (usually quaternary ammonium ions) or sulfonium ions. The vast majority of cases is quaternary ammonium waterborne polyurethane extended by chain extender containing tertiary amine group. Tertiary amines and secondary amines can be formed to hydrophilic ammonium ions by the action of acids or alkylating agents.
   (3) Nonionic waterborne polyurethane is the aqueous polyurethane that does not contain ionic groups in molecule. The preparation of nonionic waterborne polyurethane is summarized as follows. Ordinary polyurethane prepolymer or polyurethane organic solution was emulsified by high shear force in the presence of emulsifier. The prepared polyurethane prepolymer contains hydrophilic segments or hydrophilic groups: the hydrophilic segments are generally PE oxide of low molecular weight and hydrophilic groups are generally hydroxyl methyl.
   (4) The molecule structure of hybrid waterborne polyurethane resin contains ionic and nonionic hydrophilic groups or segments.

5.5.2 Raw Materials of Waterborne Polyurethane

1. **Oligomer polyols**

   In the preparation of waterborne polyurethane adhesives, the most commonly used oligomer polyols are polyether glycol and polyester glycol. And some small varieties of oligomer polyols such as polyether triol, low branched degree polyester polyols and polycarbonate diol are sometimes used.

   Polyether polyurethane has good flexibility at low temperature and better water resistance. The price of polypropylene glycol (PPG) is lower than that of polyester diol. Therefore, most of the researches and development for waterborne polyurethane in China mainly use PPG as the main raw materials of oligomer polyols. Polyurethane made from polytetramethylene ether glycol has good mechanical properties and hydrolysis resistance, but its relatively high price limits its wide applications.

   Polyester polyurethane has high strength and good adhesion, but the hydrolysis resistance of polyester is poorer than that of polyether, so the most polyester-based waterborne polyurethane has a shorter storage stability period. The flexibility of the aliphatic irregular structure polyester is also good. The one-component polyurethane emulsion adhesive is prepared by the crystalline polyester diol with regular structure. The layer of this kind of adhesive is heat-activated and bonded, and the adhesive has the higher initial strength. The waterborne polyurethane made from aromatic polyester polyols showed the higher adhesion and cohesive strength on the metal, PET and other materials.

   Polycarbonate polyurethane has good hydrolysis resistance, weather and heat resistance but its high price limits its wide applications.

2. **Isocyanates**

   The common diisocyanates to prepare polyurethane emulsion are 2,4/2,6-tolylene diisocyanate (TDI), diphenylmethane diisocyanate (MDI) and other aromatic diisocyanate, as well as hexamethylene diisocyanate (HDI), isophorone diisocyanate (IPDI), hydrogenate diphenylmethane-4,4'-diisocyanate (H12MDI) and other aliphatic and alicyclic diisocyanate. Waterborne polyurethane made from aliphatic or alicyclic diisocyanate showed a better hydrolysis resistance than that from aromatic diisocyanate, so the products of waterborne polyurethane have good storage stability. The foreign high-quality polyester-based waterborne polyurethanes are generally made from aliphatic or alicyclic isocyanate. By the restrictions of raw materials of varieties and prices, domestic waterborne polyurethanes have to employ TDI as the raw material of diisocyanate.

3. **Extenders**

   The chain extenders are often used in the preparation of waterborne polyurethane, and hydrophilic chain extenders can introduce various ionic groups. In addition to this special kind of chain extender, 1,4-butylene glycol, ethylene glycol, diethylene glycol, ethylene glycol, ethylene diamine, two ethylene three amine chain extender are often used in reaction system.

Due to the higher reactivity of the amine with isocyanate than that of water, the diamine chain extenders can be mixed in water or made into ketimine, and their extension reactions are carried out in emulsifying and dispersing process.

(4) Water

Water is the main medium of water-based polyurethane adhesive, in order to prevent the influence of calcium, magnesium and other impurities in water on the stability of anionic water-based polyurethane. The distilled water or deionized water was used to prepare water-based polyurethane resin. In addition to be used as solvent or dispersion medium for waterborne polyurethanes, water is also an important reaction material. The synthesis of water-based polyurethane is mainly based on prepolymer method; herein, water is also involved in chain extension while the polyurethane prepolymer is dispersed with water. Due to the chain extended by water and diamine, in fact, most water-based polyurethanes are polyurethane–urea emulsions (dispersions). Polyurethane–urea has the better cohesion and bonding strength than that of the pure polyurethane, and the water resistance for urea is better than that of ammonia ester bond:

$$2R\text{-NCO} + H_2O \rightarrow RNHCONHCONHR + CO_2$$

5.5.3 Preparation of Waterborne Polyurethane Resin

Waterborne polyurethane resin preparation consists of two steps: (1) preparation of the polyurethane with high molecular weight or high molecular weight prepolymer by the reaction diisocyanate with oligomer glycol; (2) dispersing prepolymer in water under the action of strong shearing force.

Polyurethane is generally hydrophobic. In order to prepare waterborne polyurethane, one approach is that polyurethane prepolymer or polyurethane solution is dispersed in water under the action of shearing force by external emulsification method. The stability of the waterborne polyurethane resin is poor. Another method is to introduce hydrophilic components or groups in the process of preparing polyurethane, it is dispersed in water under the action of shearing force without addition of emulsifiers. This is called a self-emulsifying method, which greatly improved the stability of waterborne polyurethane resin.

The most commonly used method for preparation of waterborne polyurethane is the "pre polymer dispersion method," that is to prepare NCO-based polyurethane prepolymer containing hydrophilic groups first (NCO wt% is generally below 10%), and then the polyurethane prepolymer is emulsified under the action of shear force. As the molecular weight of prepolymer is relatively low, only a little amount of solvent was added or even without solvent, herein, water can participate in the extending reaction of the prepolymer as extender to make the prepolymer chain growth and to prepare waterborne polyurethane with higher molecular weight.

Diamine (or hydrazine) chain extenders were also to carry out the chain growth reaction rapidly.

$$2 \text{ -\!\!\!-\!\!\!- } NCO + H_2O \longrightarrow \text{ -\!\!\!-\!\!\!- } NHCONH \text{ -\!\!\!-\!\!\!- }$$

$$2 \text{ -\!\!\!-\!\!\!- } NCO + H_2N\text{---}R\text{---}NH_2 \longrightarrow \text{ -\!\!\!-\!\!\!- } NHCONH\text{---}R\text{---}NHCONH \text{ -\!\!\!-\!\!\!- }$$

As diamines are with high activity, it is generally necessary to protect the active -NH_2 by the reaction of ketones with amine to produce ketimine. The diamine is regenerated by the reaction of ketimine with water in the prepolymer emulsification, and the chain can be extended smoothly.

In order to improve the properties of the emulsion, blocking the residue-terminated NCO groups is a good choice, and the blocked polyurethane emulsion was prepared. After filming, the film was treated by high-temperature process, NCO is stripped and reacted with active hydrogen of the polyurethane itself and the substrate to form crosslinking. The commonly used sealants are ketoxime, caprolactam, sodium bisulfite and so on.

This chapter mainly introduces the synthesis technology of waterborne polyurethane by self-emulsification process.

5.5.4 Preparation of Anionic Waterborne Polyurethane Resin

The anionic waterborne polyurethane resin is the most common waterborne polyurethane, and the following is a brief introduction of its synthetic technology [20]. Herein, the commonly used carboxyl-containing chain extender is dihydroxymethylpropionic acid (DMPA).

The prepolymer was synthesized in the following two ways:

(1) The oligomer diol reacted with excessive diisocyanate to form prepolymer first, then extended with DMPA to form the prepolymer containing carboxyl groups:

(2) Diisocyanate, oligomer polyol and chain extender DMPA are mixed together and react each other to prepare carboxyl group-containing prepolymer:

$$2HO \text{~~~~} OH + 4OCN - R - NCO + HOCH_2 - \overset{\overset{\displaystyle CH_3}{|}}{\underset{\underset{\displaystyle COOH}{|}}{C}} - CH_2OH$$

$$NCO - R - NH - \overset{\overset{\displaystyle O}{\|}}{C} - O \text{~~~~} H_2C - \overset{\overset{\displaystyle CH_3}{|}}{\underset{\underset{\displaystyle COOH}{|}}{C}} - CH_2 \text{~~~~} O - \overset{\overset{\displaystyle O}{\|}}{C} - HN - R - OCN$$

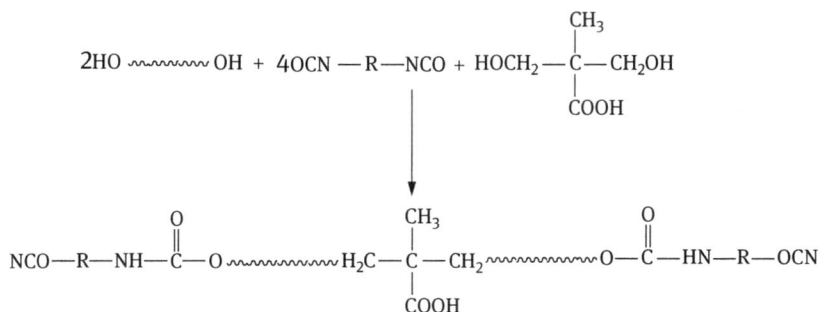

Two kinds of emulsification methods are employed.

(1) The triethylamine (NEt$_3$) which is called salt agent is added to prepolymer, and the carboxyl group is neutralized to carboxylic acid ammonium salt group. The neutralized prepolymer is the highly viscous liquid due to the interionic forces. A small amount of solvent is necessary for dilution to facilitate the shear emulsification:

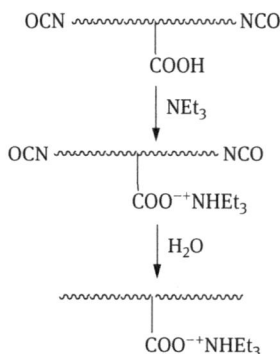

$$OCN \text{~~~~} \underset{\underset{\displaystyle COOH}{|}}{} NCO$$

$$\downarrow NEt_3$$

$$OCN \text{~~~~} \underset{\underset{\displaystyle COO^{-+}NHEt_3}{|}}{} NCO$$

$$\downarrow H_2O$$

$$\text{~~~~} \underset{\underset{\displaystyle COO^{-+}NHEt_3}{|}}{} \text{~~~~}$$

(2) Making salt-forming agent such as sodium hydroxide, ammonia water and triethylamine into dilute aqueous alkali solution, and the prepolymer is poured into the above aqueous solution to emulsify. As the viscosity of unionized (nonneutralized) prepolymer is lower than that of the ionized pre polymer, it can be emulsified by water with little solvent. Or, the water-containing salt-forming agent may be poured into the prepolymer with vigorous stirring to make the prepolymer emulsion and extend the chain:

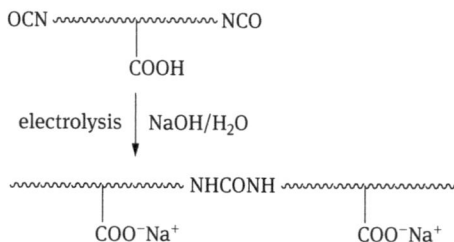

$$OCN \text{~~~~} \underset{\underset{\displaystyle COOH}{|}}{} NCO$$

$$\text{electrolysis} \downarrow NaOH/H_2O$$

$$\text{~~~~} \underset{\underset{\displaystyle COO^-Na^+}{|}}{} NHCONH \text{~~~~} \underset{\underset{\displaystyle COO^-Na^+}{|}}{} \text{~~~~}$$

During the emulsion process, besides water being used as chain extender of prepolymer in prepolymer particles, dibasic amine can also be used as chain extender. The aqueous solution of the diamine can be added to the polyurethane emulsion which has just been dispersed, or the ketoimine from dual primary amine and methyl ethyl ketone mixed with prepolymer, dispersing in water and extending chain at the same time. Finally, by film evaporation decompression process, the solvent was removed from the dispersion to obtain a water-based polyurethane resin with very low content of VOCs.

5.5.5 Preparation of Cationic Waterborne Polyurethane Resin

The cationic groups in cationic waterborne polyurethane chain were organic ammonium (quaternary ammonium ion) group and sulfhydryl group, while the latter is not of application value in preparation of waterborne polyurethane [21, 22]. Actually cationic waterborne polyurethane is just the waterborne polyurethane whose backbone (or side chains) contains quaternary ammonium ions.

The commonly used extender in the preparation of cationic waterborne polyurethane should be the diols with tertiary amino groups (*N*-methyldiethanolamine). Such extender agents are applied to synthesize NCO-terminated prepolymer-containing tertiary amino groups, then the prepolymer was quaternized, or neutralized by acid, then emulsified in water. Finally, polyurethane resin with low VOCs was obtained after the solvent is removed from the dispersion by film evaporation decompression process. The preparation principle is shown in Figure 5.14.

5.5.6 Performance of Waterborne Polyurethane

Compared with solvent-based polyurethane adhesives, waterborne polyurethane (mainly emulsion) adhesives have some special properties as shown in Table 5.2.

5.5.7 Applications of Waterborne Polyurethane

Waterborne polyurethane resins are widely used in the fields of coatings, adhesives and additives.

1. **Adhesives**

   Like solvent-based polyurethane adhesive, waterborne polyurethane adhesives also showed superproperties. It can be used for bonding a variety of substrates.

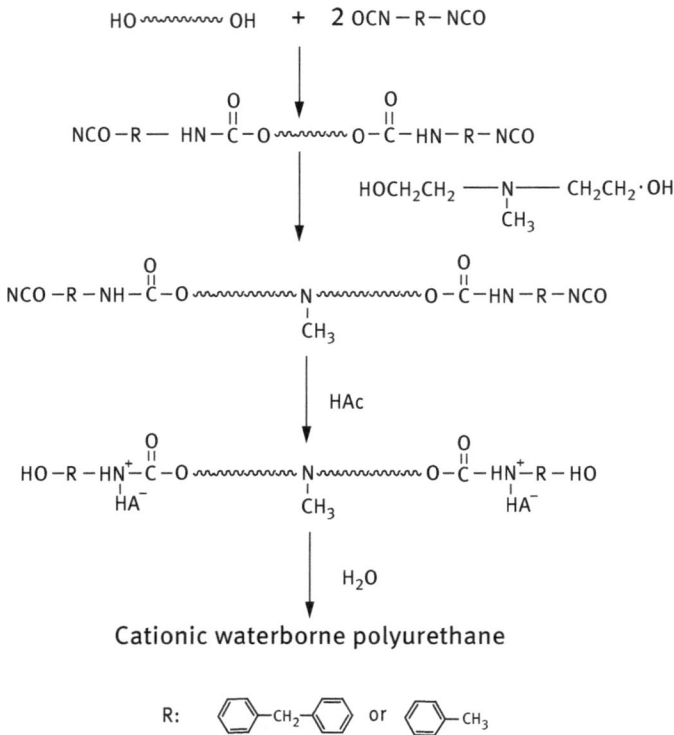

Figure 5.14: The principle for preparation of cationic waterborne polyurethane.

(1) Manufacture of a variety of laminates, including plywood, food packaging composite plastic film, fabric laminates and laminates of various thin materials, such as flexible PVC plastic films or plastic sheets and laminated products of other materials (such as wood, fabric, paper, leather and metal).

(2) Flocking adhesives, artificial leather adhesives, fiberglass and other fiber bundle adhesives, and ink adhesives.

(3) Bonding of common materials, such as bonding of automotive interior decoration materials.

2. **Coating agent**
   Due to the superflexible and wear-resistant properties, polyurethane material is widely used as coating of natural leather, artificial leather, canvas, clothing fabric, conveyor, plastic, floor, paper, automotive interior decoration and so on.

3. **Wood processing**
   Wood processing is the largest application for water-based adhesives.

Table 5.2: Comparison of performance for emulsion-based and solvent-based polyurethane adhesive.

Performance index		Emulsion type	Solvent type
Liquid properties	Appearance	Translucent to milky white dispersion	Homogeneous transparent liquid
	Solid content	20–60% (has nothing to do with M_w)	20–100% (has nothing to do with M_w)
	Solvent type	Water (sometimes with a small amount of solvent)	Organic solvent
	Viscosity	Low, independent of molecular weight, can be thickened	High molecular weight viscosity, relating to the solvent and concentration
	Viscous flow characteristics	Non-Newtonian type (general constant degeneration)	Newtonian type
	Wettability	Surface tension is higher, the low energy surface moisture run bad, can add leveling agent change	Good wetting of low-energy surface by different type of visual solvent
	Drying property	Slow (the evaporation energy of water is high)	Fast
Construction performance	Film-forming properties	Must be above 0 °C, dependent on temperature and humidity	Small temperature dependence
	Blending properties	The same ionic properties of different polymers can be mixed	Related to the polymer and solvent system
	Mechanical properties	Bad to good	Good
Film properties	Water resistance	A little bad to good (adding cross-linking agent to enhance)	Good
	Solvent resistance	A little bad to good (adding cross-linking agent to enhance)	Single component is bad, two components are good
	Heat resistance	Slightly different in thermal plasticity, cross-linking type of good	

5.6 Radiation Cross-linking Polymerization Technology

As a new and high technology processing method, the radiation technology has been paid more and more attention in the world. Radiation processing industry has developed rapidly in recent 30 years, which is widely used in many fields, such as industry, agriculture, national defense, medical and health, food and environmental protection. Radiation method has its originality in reducing energy consumption, environmental pollution control and product quality, which has become dominant in the processing of polymer materials, and is a trend of green processing technology of polymer materials.

Radiation cross-linking polymer technology is the use of high-energy or ionizing radiation to cause ionization and excitation of polymers and some secondary reactions, which cause chemical reactions and achieve chemical cross-linking between the macromolecules. This led to the formation of intermolecular cross-linking network. It is one of the effective means for polymer modification to prepare new materials.

After radiation cross-linking, the structure and properties of the polymer have been changed, and the application range of the polymer has been widened. Dozens of radiation processed products have been put into industrial production at present. These products are widely used in some industrial sectors, and some products have been necessary for the people's daily life. The industrial scale or batch production varieties are increasing, which breaks the past traditional point of view that "radiation on polymer materials is only destructive." Radiation cross-linking has opened up a new research direction to realize the performance modification of polymer materials by radiation. At present, the technology of radiation cross-linking has become a high-tech industry which can't be ignored in the developed countries of the world.

5.6.1 The Basic Principles of Radiation Cross-linking and Pyrolysis

Under the action of ionizing radiation, molecular chain of the polymer can be cross-linked, and the molecular weight increases. Radiation cross-linking results in the increase of the molecular weight of the polymer with an increase in radiation dose, and finally the polymer forms a three-dimensional network structure, which leads to the normal melting point not to occur again. Generally speaking, the radiation cross-linking of the polymer does not need to add any additives, and the purpose of crosslinking can be achieved at room temperature.

The radiation cross-linking of polymers is a complex process, which can lead to the cross-linking of the main chain and the degradation of the main chain.

The basic principle of polymer radiation cross-linking is that polymer macromolecules in effect of high-energy or radioactive isotopes (such as Co-60 ray) results in ionization and excitation which generates macromolecular free radicals and free radical reaction. Radiation also results in some secondary reactions as well as a variety of chemical reactions. The following reaction mechanisms exist during polymer radiation cross-linking.

(1) The dehydrogenation of the neighboring molecules generated by irradiation, and the two free radicals formed cause the coupling reaction to produce cross-link:

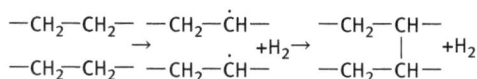

$$
\begin{matrix}
-CH_2-CH_2- \\
\rightarrow \\
-CH_2-CH_2-
\end{matrix}
\quad
\begin{matrix}
-CH_2-\overset{\cdot}{C}H- \\
\qquad +H_2 \rightarrow \\
-CH_2-\overset{\cdot}{C}H-
\end{matrix}
\quad
\begin{matrix}
-CH_2-CH- \\
\ \ \ \ \ \ \ | \qquad +H_2 \\
-CH_2-CH-
\end{matrix}
$$

(2) Independently generated two mobile free radicals combine to produce cross-links:

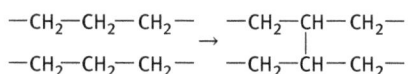

$$
\begin{matrix}
-CH_2-CH_2-CH_2- \\
\rightarrow \\
-CH_2-CH_2-CH_2-
\end{matrix}
\quad
\begin{matrix}
-CH_2-CH-CH_2- \\
\ \ \ \ \ \ \ \ \ \ \ | \\
-CH_2-CH-CH_2-
\end{matrix}
$$

(3) Ion and molecular directly react to produce cross-link:

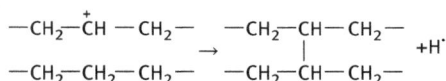

$$
\begin{matrix}
-CH_2-\overset{+}{C}H-CH_2- \\
\rightarrow \\
-CH_2-CH_2-CH_2-
\end{matrix}
\quad
\begin{matrix}
-CH_2-CH-CH_2- \\
\ \ \ \ \ \ \ \ \ \ \ | \qquad +H^{\cdot} \\
-CH_2-CH-CH_2-
\end{matrix}
$$

(4) Free radical and double bond react to produce cross-link:

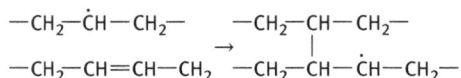

$$
\begin{matrix}
-CH_2-\overset{\cdot}{C}H-CH_2- \\
\rightarrow \\
-CH_2-CH=CH-CH_2
\end{matrix}
\quad
\begin{matrix}
-CH_2-CH-CH_2- \\
\ \ \ \ \ \ \ \ \ \ \ | \qquad \overset{\cdot}{\ } \\
-CH_2-CH-CH-CH_2-
\end{matrix}
$$

(5) Free radical produced by the main chain cracks recombines to produce cross-link:

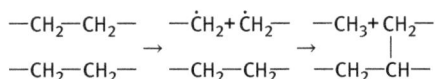

$$
\begin{matrix}
-CH_2-CH_2- \\
\rightarrow \\
-CH_2-CH_2-
\end{matrix}
\quad
\begin{matrix}
-\overset{\cdot}{C}H_2+\overset{\cdot}{C}H_2- \\
\rightarrow \\
-CH_2-CH_2-
\end{matrix}
\quad
\begin{matrix}
-CH_3+CH_2- \\
\ \ \ \ \ \ \ \ \ | \\
-CH_2-CH-
\end{matrix}
$$

(6) Cyclic reaction leads to cross-linking:

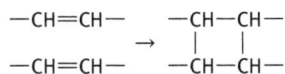

$$
\begin{matrix}
-CH=CH- \\
\rightarrow \\
-CH=CH-
\end{matrix}
\quad
\begin{matrix}
-CH-CH- \\
\ \ | \ \ \ \ | \\
-CH-CH-
\end{matrix}
$$

Radiation cracking means that under the action of ionizing radiation, the main chain of polymer is broken, and molecular weight decreased. Radiation cracking results in the molecular weight of the polymer decreases with an increase in radiation dose, and cracks into a monomer finally.

The main feature of the radiation pyrolysis is the formation of two shorter polymer molecules when the chain cracks, and the average molecular weight decreased. Under the action of high energy beam, both the cross-linking and the cracking of polymer were carried out simultaneously. So some polymers are mainly cross-linked, and some are mainly cracked. Some polymers are cross-linked, while others are dominated by cracking. Even in the same polymer, the cross-linking and pyrolysis behaviors are different under different conditions. Under usual conditions, the cross-link or cracking is mainly closely related to the molecular structure of the polymer itself.

According to a large number of experimental results, the following experience is followed, which judged the tendency to be cross-linked or cracked from structure of the polymer.
(1) The polymer-containing CH_2-$C(R_1)R_2$ – in the structural unit, or the main chain containing the quaternary carbon atoms, radiation cracking mainly happens, where R_1 stands for alkyl, Cl and F.
(2) The low polymerization heat of polymers is generally dominated by radiation degradation. Thermal cracking tends to generate single polymer, mainly by radiation cracking.
(3) The polymer, whose main chain is -C-O- as the repeating unit structure (such as formaldehyde) or whose branched chain connects to the backbone with -C-O- (such as poly(vinyl formal)), is easy to radiation crack.

5.6.2 The Main Features of Radiation Polymerization

Compared with the general thermal chemical polymerization, radiation polymerization showed the following characteristics:
(1) The radiation polymerization does not need any addition of initiator or catalyst, and only relies on the power radiation on monomers to initiate polymerization. So the polymer is pure. It is more favorable for the preparation of biomedical materials or optical materials.
(2) With a strong penetration of gamma rays, the reactions can be carried out uniformly and continuously to prevent local overheating, and the reactions are easy to control.
(3) The radiation is independent of the physical state of the monomer. Therefore, the radiation can be carried out in liquid phase polymerization, but also solid phase polymerization or gas phase polymerization.
(4) The initiation rate of polymerization is only related to the radiation intensity (dose rate) because the formation rate of free radicals or ions in initiating the reaction is only related to the dose rate. Therefore, the polymerization reaction can be controlled easily by adjusting the dose rate.
(5) The general chemical method is difficulty or unable to cause polymerization of monomers (such as hexafluoropropylene, α-methyl styrene, butadiene, perfluoropropylene ethyl monomer). Radiation could induce their polymerization reaction, and even allows ketones and CO_2 to initiate polymerization, so new methods for the preparation of new polymer has been established.

(6) Radiation polymerization provides a novel and special processing method for some industrial production. Especially for the complex form and difficult parts, the method of in situ radiation polymerization of monomer can be considered.

(7) The effect of radiation on the monomer is independent of temperature. Radiation polymerization can be carried out in conditions of low temperature or cold. It makes certain enzyme or biological cells to be immobilized easy, not easy to lose activity at low temperature.

5.6.3 Effect of Radiation Cross-linking on the Properties of Polymer

1. **Effect on the structure of the molecule**
   When the radiation is acting on the polymer, the polymer will form a chemical cross-linking bond, so that the average molecular weight of the polymer increases, and the solubility of the polymer decreases. When the radiation reaches a certain dose, the cross-linked network structure is formed between molecules. When irradiation is at a certain dose, a cross-linked network structure that is no longer dissolved or melted was formed.

2. **Effect on the mechanical properties of polymer**
   After radiation cross-linking, generally speaking, the tensile strength, hardness, wear resistance and elastic modulus of the material will increase, and the elongation at break will decrease. The polymer materials with excellent comprehensive properties can be obtained by choosing proper blending system and radiation conditions.

3. **Effect on thermal and thermomechanical properties of polymers**
   Irradiation enhanced the cross-linking density of polymer materials, such as PE, PS, PVC and PVDF (polyvinylidene fluoride), which can improve the thermal stability and heat resistance. For example, the long-term working temperature region of PE increased from the temperature range of 60–70 °C to the range of 125–135 °C, and the short-term temperature increased from 140 to 300 °C. After irradiation, the temperature for PVDF increased from 150 to 175 °C.

4. **Effect on the flame retardant properties of polymer materials**
   Compared with the chemical cross-linking, radiation cross-linking is carried out under the normal temperature and pressure, and the cross-linking time is short. The appropriate dose control can reduce the internal decomposition concentration, and improve cross-link density (chemical method is 40%, and the radiation cross-linking degree is more than 70%). This cross-linking can effectively improve the gas diffusion during polymer materials combustion, the heat resistance, reduce smoke and the melt dripping.

5. **Improvement of electrical performance**
   In PE cable curing with steam, the high pressure steam will inevitably penetrate into the PE layer to cause many micropores and high contaminant concentration, and cable in use easily dissociate and branch aging; however, the introduction of

cross-linking agent causes damage of high-frequency characteristic for the material. By means of radiation cross-linking can avoid or eliminate the pores, filth or bulging, and eliminate water treeing and electrical tree phenomenon, ensure uniformity and high purity of the insulation layer, show the higher frequency characteristic and better long-term performance.

5.6.4 Industrial Application of Radiation Cross-linking Technology

Radiation cross-linking can improve the thermal and mechanical properties of polymer materials. In the actual production, it is usually the first forming, the post cross-linking; the processing technology is simple, and the cross-linking degree is easy to be controlled by the dose.

Since the discovery of the cross-linking phenomenon of PE radiation in the 1950s, a powerful radiation processing industry in the field of polymer radiation cross-linking and product development has been formed. Polymer materials are mainly PE and PVC cross-linked products. Domestic research in this area started in the 1960s and 1970s; radiation cross-linked PE and PVC wire and cable insulation and heat shrinkable joint jacket and other products have been developed in the 1980s, and the radiation processing industry in China has been formed.

At present, the main products of radiation cross-linking are wire, cable, heat shrink materials, rubber and latex radiation cross-linked products (such as tires and medical latex tubes), radiation cross-linked foam, etc.

The main production process of radiation cross-linking products is shown in Figure 5.15.

5.7 Plasma Polymerization Technology

Plasma is the fourth state of matter besides solid, liquid and gas. It is the ionized gas with the basic equivalent positive and negative charged particles, which can be divided into thermal equilibrium plasma and low temperature plasma [23–25]. When thermal equilibrium plasma electron temperature and gas (ion) temperature reach a balance, not only the electron temperature is high, the heavy particle temperature is also high. It is also called high temperature plasma (ionization above 5,000 °C). General organic compounds and polymers are split at this temperature, and difficult to generate the polymer, which is commonly used to generate the high-temperature-resistant inorganic substances. Low-temperature plasma (ionization at 100–300 °C) is characterized by the electron and gas temperature not reaching thermal equilibrium, also known as nonequilibrium plasma, its electron temperature can reach up to more than 10^4 K, while heavy particles such as ions and atomic temperature can be as

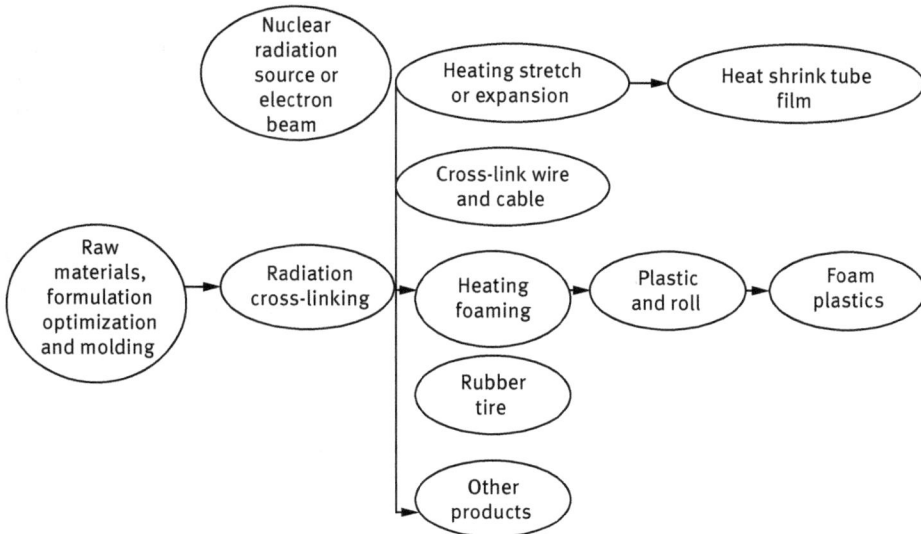

Figure 5.15: Schematic diagram of processing technology for radiation cross-linked products.

low as 300–500 K, usually in the 1.33×10^4 Pa by direct current (DC) glow discharge, radiofrequency or microwave discharge method.

Due to the big difference between electron temperature and gas temperature, low-temperature plasma can generate the stable polymer, which is commonly used in plasma polymerization.

Plasma polymerization technique is a novel synthetic technology, which means the plasma generated by gas ionization to activate the monomer, the activation of plasma electrons, ions, free radicals, photons and excited state molecules (the outermost electron in anti-bond molecular orbital) can be referred to as living sites and growth center for the polymerization, which initiates monomer polymerization.

Almost all organic compounds and organic compounds are polymerized by plasma. This unique synthetic technology is of great significance in the preparation of membrane materials and surface modification of polymer materials.

5.7.1 Types and Characteristics of Plasma

The plasma is divided into two types, one is reactive plasma, such as excited states of O and N atoms. They can not only initiate the organic monomer to generate polymerization active sites (such as alkyl radical R·), and also participate in the polymerization reaction.

Another is nonreactive plasma, such as H, He, Ne atomic in excited state. They are in high energy to impact on the surface of the material and motivate surface molecules of material to produce larger free radicals. These radicals result in larger surface cross-linking polymerization, forming dense surface layer, while H, He, Ne plasma themselves are not involved in this reaction, and only plays a role in conveying an energy.

Plasma polymerization has the following features:

(1) Plasma polymerization does not require monomer with a single unsaturated unit, almost all organic compounds and organic metal compounds are polymerized by plasma polymerization, regardless of double bonds, or the characteristics of the two or more functional groups from the chemical structure; in normal circumstances, monomers do not or difficult to carry out polymerization in the system can quickly become easy to polymerization and the rate can be very fast, such as CO_2 and styrene polymerization, CO, H_2 and N_2 polymerization.

(2) The generated polymer membranes with high density network structure, and the network size and branched degree can be controlled to some extent. The formed plasma polymer film has excellent mechanical properties, chemical stability and thermal stability, such as the plasma polymerized film of ethane.

(3) Plasma polymerization process is very simple; both inner electrode type and outer electrode type are generally react to certain vacuum first, and then filling the monomer vapor, or filling the gas mixture of carrier gas and monomer, and keep the set pressure value (usually in 1. 3×10^{-1} –1.3×10^{-2} Pa). The traffic is usually about 10–100 mL/m, under the appropriate choice of the discharge power producing plasma, and then can generate polymer film on the substrate surface.

5.7.2 Mechanism of Plasma Polymerization

The plasma polymerization mechanism is a free radical polymerization mechanism consisting of the formation of free radicals, chain initiation, chain propagation and chain termination, including the deposition mechanism of the film formation, which also relates to the mechanism of cross-linking process.

The mechanism of the deposition process and the mechanism of the formation of the film have been studied by scientists. A relatively consistent opinion that has been formed at present is generally believed that deposition mechanism is referred to the reaction that takes place simultaneously on the surface of the gas and the substrate, and the formed polymer in gas phase deposits on the substrate surface. The mechanism of the cross-linking process is described as follows. The free electrons are generated by the discharge of the monomers with higher energy; and a large number of hydrogen atoms, free radicals and derived monomers are generated by collision. These groups of chemical activity are very high, participating in various reactions, in the growth process of the polymer chain. The main chain will continue to be charged by the impact of electrons, randomly generated at a position in the main chain of free radicals, leading to the formation of branching or cross-linking.

5.8 Enzyme-catalyzed Polymerization

The enzyme-catalyzed polymerization (enzymatic polymerization) is a new trend of polymer science [26–34]. Its importance is gradually increasing, and the enzyme-catalyzed reaction provides a new synthetic strategy for the synthesis of polymer.

In recent years, the study of enzymatic catalysis technology is focused on small molecule reactions, such as hydrolysis, including carboxylic ester, phosphate ester, amide and nitrile amide hydrolysis reaction, esterification. Research reports of enzyme catalysis and the small molecules of the enzyme-catalyzed reaction accounted for more than 95% of the total reported. However, few studies of enzyme-catalyzed polymerizations have been reported, and it is still in the initial stage. It needs to be further studied for chemical scientists and enrich the research content of enzyme catalysis technology.

In the use of nonfossil renewable resources as the starting material of functional polymeric materials, enzyme-catalyzed polymerization has an important advantage over the use of scarce resources, and contributes to global sustainable development. In enzymatic polymerization, the polymerization product can be obtained under mild conditions without the use of toxic reagents. Therefore, the enzyme-catalyzed polymerization has great application potential in the environmental-friendly synthesis of polymer materials, which provides a good example for the realization of "green polymer chemistry."

5.8.1 Enzyme-catalyzed Ring-opening Polymerization

The study of enzyme-catalyzed polymerization technology mainly concentrated in ring-opening polymerization and polycondensation reaction [26, 27]. The preparation of polycarbonate has been developed by ring-opening polymerization. Synthesis of polysaccharide, aliphatic polyester, polyphenylene ether and its derivatives are typical poly-condensation reactions.

The typical polymerization technology of enzyme catalysis is described as follows.
1. **Synthesis of polycarbonate**
   The enzyme-catalyzed ring-opening polymerization of six-membered and seven-membered cyclic carbonates was initiated by lipase. In 1997, Matsumura et al. first reported the lipase-catalyzed polymerization of cyclic carbonate.

the sixmembered   poly(trimethylene carbonate)
1,3-dioxan-2-one (TMC)   [P(TMC)]

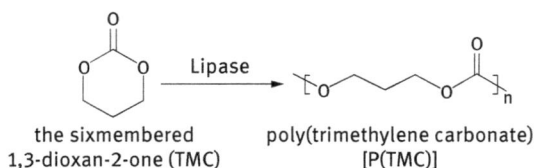

Figure 5.16: Enzyme-catalyzed ring-opening polymerization of the six membered 1,3-dioxan-2-one (TMC).

The reaction mechanism was shown in Figure 5.16. The reactions catalyzed by various enzymes under different conditions were studied. The results showed that cyclic carbonate was easy to realize polymerization wither the temperature range of 60–100 °C, and the relative molecular weight of the polymer reached up to 169,000. In the absence of enzyme blank samples in 24 h, no change for trimethylene carbonate (TMC) was found. It was proved that polymerization reaction was caused by enzymes.

Poly(trimethylene carbonate) (PTMC) with a number average molecular weight of 15,000 has been obtained by using Novozym-435 enzyme as catalyst (see Figure 5.16).

Reaction temperature rises from 55 to 85 °C, the conversion of the reaction is almost constant while the molecular weight of the polymer decreased. When the water content decreased, the polymerization reaction rate decreased, and the molecular weight increased.

Through the analysis of small molecule products, the mechanism of the ring-opening polymerization of TMC catalyzed by lipase is described as follows.

Initiation reaction

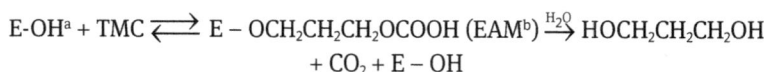

$$E\text{-}OH^a + TMC \rightleftarrows E - OCH_2CH_2CH_2OCOOH\ (EAM^b) \xrightarrow{H_2O} HOCH_2CH_2CH_2OH$$
$$+ CO_2 + E - OH$$

Formation of TMC dimer

$$EAM + HOCH_2CH_2CH_2OH \longrightarrow HO(CH_2)_3OCOO(CH_2)_3OH + E\text{-}OH$$

Formation of polymer

$$nHO(CH_2)_3OCOO(CH_2)_3OH \xrightarrow{EAM} HO(CH_2)_3O\text{-}[\text{-}COO(CH_2)_3O\text{-}]_n\text{-}COO(CH_2)_3OH$$

where E-OH^a is the enzyme and EAM^b is the complex of enzyme and TMC.

The polymerization of enzyme catalytic carbonate in the carbon chain containing carboxyl substituent has been reported. The synthesis of 5-methyl-5-carbobenzoxy-1, 3-dioxane-2-ketone is shown in Figure 5.16. When using lipase as catalyst, polymer with high average molecular weight of 5,900 has been obtained.

2. **Enzyme-catalyzed ring-opening polymerization of aliphatic ester**

For fatty lactone ring-opening polymerization, more researches on ε-caprolactone have been investigated. Uyama studied the ring-opening polymerization of

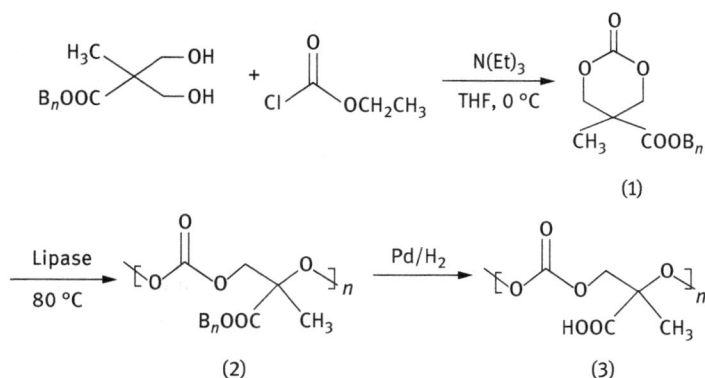

Figure 5.17: Enzyme-catalyzed ring-opening polymerization of 5-methyl-5-benzyl oxygen carbonyl-1, 3-dioxane-2-ketone.

Initiation reaction

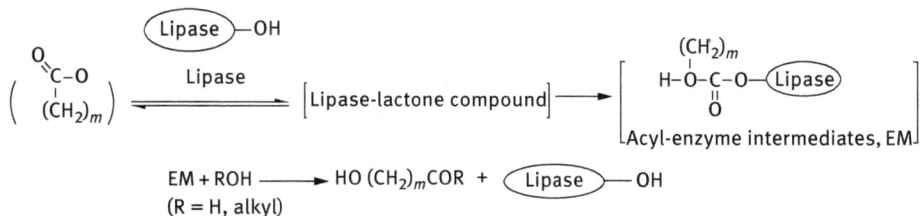

Propagation reaction

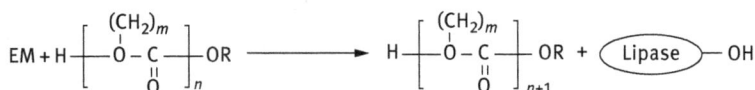

Figure 5.18: Enzyme-catalyzed ring-opening polymerization mechanism of aliphatic ester.

ε-caprolactone and δ-valerolactone catalytic polymerization by different lipase as catalyst, and the possible reaction mechanism was put forward.

The control step in the ring-opening polymerization process of fatty lactone catalyzed by lipase is reaction of lactone with lipase, which produces acyl enzyme reaction intermediates (enzyme-activated monomer, EM). The initiation step is the acyl carbon atom intermediates nucleophilic attack by the water in enzyme to produce omega-hydroxy carboxylic acid ($n = 1$). In the chain growth stage, intermediate growth by nucleophilic attack of terminal hydroxyl polymer on EM, a polymer chain unit extended. Polymerization kinetics showed that the control step of polymerization process is enzyme-activated monomer formation.

Therefore, the polymerization may be based on the mechanism of monomer activation (Figure 5.18).

The result for ^{13}C-NMR and ^{1}H-NMR proved that the terminated groups for the products are carboxyl and hydroxyl groups. With an increase in temperature, polymerization rate of ε-caprolactone increased; with an increase in conversion, the molecular weight also increased. Polymerization of δ-valerolactone showed the similar results. Although the conversion of delta valerolactone is higher than that of ε-caprolactone, the polymer molecular weight is less than that of ε-caprolactone.

If the reactants are the mixture for ε-caprolactone and δ-valerolactone in the system, the copolymers ε-caprolactone and δ-valerolactone have been obtained. With an increase in the content of ε-caprolactone, the relative molecular weight of copolymer increased. The analysis result for ^{13}C-NMR indicates that the polymerization products are random copolymers.

In addition, the large ring lactone, such as ω-pentadecane lactone also can take the ring-opening polymerization reaction.

5.8.2 Enzyme-catalyzed Stepwise Polymerization

1. Synthesis of polysaccharides

Employing two β-cellulose sugar fluoride as glycosyl donor, the cellulose was synthesized by the action of cellulase [28–30]. In acetonitrile acetic acid buffer solution, the polycondensation reaction of β-cellulose difluoride with cellulose under the action of cellulase is shown in Figure 5.19.

In the methanol phosphate buffer solution, α-amylase as a catalyst, α-D-maltose takes a condensation reaction to produce a maltose oligomer. In the process of polymerization, 1,4-α-glycosidic bond was formed by region selection and stereoselection. However, other substrates, such as D-maltose, β-D-maltose fluoride and α-D-gluconate, did not get a condensation product.

Employing two β- wood sugar fluoride as substrate monomer, and xylanase as catalyst, the artificial xylan was first synthesized by the action of transglycosidation. Polymerization of substrate monomer smoothly generates the corresponding condensation product, with ^{13}C-NMR, the structure of natural xylan and the effect of xylanase on the natural xylan were studied. By regional selection and stereoselective effects of fluoride monomer, polycondensation between wood two sugar units was carried out to generate 1,4-β-bond stereoregular artificial xylan.

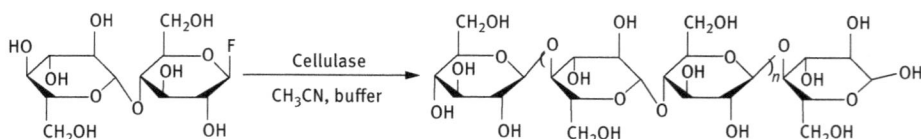

Figure 5.19: The enzyme-catalyzed condensation reaction of β-cellulose difluoride.

In addition, the enzyme-catalyzed reaction is also used to synthesize a number of nonnatural polysaccharides.

2. **Synthesis of polyaniline and its derivatives**

   Polyaniline (PANI) and its derivatives are important conductive materials [31, 32]. Because of their great potential electronic properties and excellent thermal stability, the development of synthesis and application of PANI and its derivatives have been widely concerned [31]. Formaldehyde and other toxic substances were used in the ordinary chemical polymerization method. They are harmful to the environment, and the enzymatic polymerization technology can overcome the disadvantages. In catalyzed polymerization, PANI and its derivatives are usually catalyzed by horseradish peroxidase (HRP). In the presence of H_2O_2, HRP can catalyze and oxidize a series of aromatic amines and phenols [32].

   Generally, the catalytic mechanism can be described as

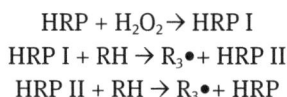

$$HRP + H_2O_2 \rightarrow HRP\ I$$
$$HRP\ I + RH \rightarrow R_3\bullet + HRP\ II$$
$$HRP\ II + RH \rightarrow R_3\bullet + HRP$$

HRP was oxidized by H_2O_2 to get two valent intermediate HRP I, then HRP I oxidized the substrate RH, and the partial oxidation intermediate HRP II was obtained, and then HRP II oxidized the substrate RH. HRP was returned to its initial state after two steps of single electron reaction. The free radical R_3. reacted with each other to form dimer, and the dimer continued the oxidation chain growth reaction to get the polymer finally.

3. **Synthesis of Polyphenylene ether and its derivatives**

   Polyphenylene ether (PPO) is a kind of high-performance engineering plastics with excellent thermal stability and chemical stability [33, 34]. In 1996, for the first time, it was found that at room temperature, the oxidoreductase initiates two methoxy-hydroxybenzoic acid to form PPO in water-soluble organic solvent, and a high yield was obtained, as shown in Figure 5.20.

   In the mixed solution of water-miscible organic solvent and buffer solution, compound of 2,6-diphenoxy phenol can take the enzyme polymerization and the polymer with molecular weight of thousands can be obtained.

   Enzyme-catalyzed polymerization is a multidisciplinary research field, which provides a bridge for the communication among polymer chemistry, organic chemistry and biochemistry. At present, the study of enzyme-catalyzed

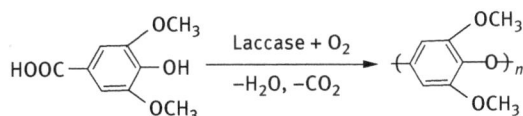

Figure 5.20: Enzyme catalytic polymerization of 3, 5-dimethoxyhydroxy benzoic acid.

polymerization is still in the exploratory stage, and the reaction mechanisms are not very clear. There are still a lot of works to carry out in the control of reaction conditions and the optimization of enzyme. However, as a new polymerization method, enzymatic polymerization has established a novel and environmental-friendly approach to synthesize polymer. It is an effective method to prepare new functional polymer materials, and also has a wide application prospect in medicine, environmental protection and national defense. With the deepening of the research, the enzyme-catalyzed polymerization will achieve a breakthrough in the polymerization technology, and becomes one of the main methods for the preparation of green polymers.

Questions

1. Please explain the reasons why the micelles are the reaction sites of α-unsaturated monomer in the presence of water medium.
2. What are the characteristics of emulsion polymerization with water as the dispersion medium?
3. What kinds of polymerization can be carried out in ionic liquids? Please illustrate.
4. What is polymer radiation cross-linking technology?
5. Please sketch the characteristics of radiation polymerization.
6. Please sketch the effect of radiation cross-linking on properties of polymers.
7. Please explain the principle of self-emulsifying water-based polyurethane.
8. What are the characteristics of water-based polyurethane?
9. Please explain the definition and concept of plasma polymerization? What are the characteristics of plasma polymerization?
10. Please sketch the mechanism of ring-opening polymerization of caprolactone catalyzed by enzyme.

References

[1] Feng, X. D. Prospect of polymer chemistry in twenty-first Century. Polymer Bulletin, 1999 (3): 1–9.
[2] Ge, M. L. Research progress of green polymers. Aging and Application of Synthetic Materials, 2002 (4): 22–26.
[3] Zhan, M. S. Research status and development of green polymer materials. Plastic Additives, 2003 (1): 12–17.
[4] Odian, G. Principles of Polymerization, 4th edition. Hoboken: A John Wiley & Sons, Inc., 2004: 350–356.
[5] Wang, Y. Y., Sun, H., Dai, L. Y., et al. Application of ionic liquids in polymers. Polymer Bulletin, 2006 (5): 20–25.
[6] Hong, K. L., Zhang, H. W., Mays, J. M., et al. Conventional free radical polymerization in room temperature ionic liquids: a green approach to commodity polymers with practical advantages. Chemical Communications, 2002, 13: 1368–1369.

[7]   Vijayaraghavan, R., Mac Farlane, D. R., Living cationic polymerisation of styrene in an ionic liquid Electronic supplementary information (ESI) available: GPC results for the two-step living polymerisation of styrene by HBOB in the IL. Chemical Communications, 2002, 13: 1368–1369.

[8]   Vygodskii, Y. S., Lozinskaya, EI., Shaplov, A. S. Ionic liquid as novel media for the synthesis of condensation polymer. Macromol Rapid Communication, 2002, 23: 676.

[9]   Carlin, R. T., Wikes, J. S. Complexation of Cp2 MCl 2 in a chloroaluminate molten salt: Relevance to homogeneous Ziegler-Natta catalysis. Journal of Molecular Catalysis, 1990, 63: 125–129.

[10] Mastrorilli, P., Nobile, C. F., Galb, V., Suranna, G. P., Farinola, G. Rhodium(I) catalyzed polymerization of phenylacetylene in ionic liquids. Journal of Molecular Catalysis A: Chemistry, 2002, 184: 73–78.

[11] Gale, R. J., Osteryoung, R. A. Raman spectra of molten aluminum chloride: 1-butylpyridinium chloride systems at ambient temperatures. Inorganic Chemistry, 1978, 19: 2728–2729.

[12] Arnautov, S. A. Electrochemical synthesis of polyphenylene in a new ionic liquid. Synthetic Metals, 1997, 84: 295–296 .

[13] El Abedin S. Z., Borissenko, N., Endres, F. Electropolymerization of benzene in a room temperature ionic liquid. Electrochemistry Communications, 2004, 6(4): 422–426.

[14] Sekiguchi, K., Atobe, M., Fuchigami, T. Electropolymerization of pyrrole in 1-ethyl-3-methylimidazolium trifluoromethanesulfonate room temperature ionic liquid. 2004, 4(11): 881–885.

[15] Jennifer, M., Pringle, J. E., MacFarlane, R., et al. Polyelectrolyte in ionic liquid electrolytes. Polymer, 2004, 45: 1447.

[16] De Simone, J. M., George, W. Continuous precipitation polymerization of vinylidenefluoride in supercritical carbon dioxide: Modeling the rate of polymerization. Industrial and Engineering Chemistry Research, 2000, 39(12): 4588–4596.

[17] Cao, W. L., Zhang, J. C. Application of supercritical fluid technology in the depolymerization of PET. Journal of Beijing University of Chemical Technology, 1999, 26(4): 73–74.

[18] Liu, S. L., Zong, M. H. Research progress of enzymatic catalysis in supercritical fluids. Microbiology, 2001, 28(1): 81–85.

[19] Dieterich D. Aqueous Emulsions, Dispersions and Solutions of Polyurethanes: Synthesis and Properties. Elsevier Sequoia S.A., Lausanne, Progress in Organic Coating, 9 (1981): 281–340.

[20] Blank, W. J., Tramontano, V. J. Properties of crosslinked polyurethane dispersions. Progress in Organic Coatings, 1996, 27(1): 1–15.

[21] AL-Salah, H. A., Xiao, H. X., McLean, Jr. J. A., Frisch, K. C. Polyurethane cationomers. I. Structure-properties relationships. Polymer Science Part A: Polymer Chemistry, 1988, 26: 1609–1620.

[22] Chars, Wu Chungchen, Show An, Polyurethane ionomers: Effects of emulsification on properties of hexamethylenediisocyanate-based polyether-polyurethane Cationomers. Polymer, 1988, 29(11): 1995.

[23] Liu, Z. J. Introduction to plasma polymerization. Modern Physics Knowledge, 1998, 10(4): 9–10.

[24] Tan, C. Plasma polymerization technology and its application. New Chemical Materials, 1994(7): 37–38.

[25] Wen, G. A., Zhang, W. G., Lin, C. Y. Mechanism of plasma initiated polymerization. Polymer Bulletin, 1999 (6): 67–70.

[26] Uyama, H., Namekawa, S., Kobayashi, S. Mechanistic studies on the lipase-catalyzed ring-opening polymerization of lactones. Polymer Journal, 1997b, 29,299–301.

[27] Matsumura, S., Tsukada, K., Toshima, K. Enzyme-catalyzed ring-opening polymerization of 1,3-dioxan-2-one to poly(trimethylene carbon-ate). Macromolecules, 1997, 30(10): 3122–3124.

[28] Kobayashi, S., Kashiwa, K., Kawasaki, T., et al. Novel method for polysaccharide synthesis using an enzyme: The first in vitro synthesis of cellulose via a nonbiosynthetic path utilizing cellulase as catalyst. Journal of American Chemical Society, 1991, 113: 3079–3084.

[29] Kobayashi, S., Shimada, J., Kashiwa, K., et al. Enzymatic polymerization of α-D-maltosyl fluoride utilizingα-amylase as the catalyst: A new approach for the synthesis of maltooligosaccharides. Macromolecules, 1992, 25: 3237–3241.

[30] Kobayashi, S., Wen, X., Shoda, S. Specific preparation of artificial xylan: A new approach to polysaccharides synthesis by using cellulase as catalyst. Macromolecules, 1996, 29: 2698–2700.

[31] Alva, K. S., Kumar, J., Marx, K. A., et al. Enzymatic synthesis and characterization of a novel water- soluble polyaniline. Macromolecules, 1997, 30: 4024–4029.

[32] Liu, W., Kumar, J., Tripathy, S., et al. Enzymatically synthesized conducting polyaniline. Journal of American Chemical Society, 1999, 121: 71–78.

[33] Ikeda, R., Uyama, H., Kobayashi, S. Novel synthetic pathway to a poly(phenylene oxide): Laccase-catalyzed oxidative polymerization of syringic acid. Macromolecules, 1996, 29: 3053–3054.

[34] Ikeda, R., Sugihara, J., Uyama, H., et al. Enzymatic oxidative polymerization of 2 ,6-dimethyphenol. Macromolecules, 1996, 29: 8702– 8705.

6 Green technology in fine chemical industry

Fine chemical industry is a special chemical industry which produces fine chemicals. Fine chemical industry is important in modern chemical industry, and it is a symbol for the level of not only science and technology but also comprehensive strength of a country. The implication of green fine chemical includes using the principle and technology of green chemistry, employing nontoxic and harmless raw materials, developing green synthetic technique and environmental-friendly chemical procedures, and producing harmless to human health and environmental-friendly fine chemical products. The connotation of green fine chemical industry covers the greenization of chemical raw materials, the greenization of chemical industry technique, and the greenization of fine chemical products.

The greenization of fine chemical raw materials requires the selection of nontoxic and harmless raw materials as more as possible to decrease the environmental pollution in the procedures of fine chemicals production.

The greenization of chemical industry techniques requires the employment of new chemical technology such as catalytic technology and biotechnology to develop an efficient, high-selective, atom-economical reaction and green synthetic process, reduce or eliminate the harmful wastes from fountainhead, or improve the chemical reaction technology, reduce or avoid the use of harmful materials, reduce the emissions of by-product, and achieve zero discharge of wastes ultimately.

The greenization of fine chemical products requires research and development of pollution-free alternatives for the traditional chemical products, and design and synthesis of safer chemicals; and the use of environmental-friendly ecological materials for realizing a harmonious combination between human and nature, according to the new method and concept of green chemistry.

6.1 Greenization of Pharmaceutical Industry

6.1.1 Introduction

The pharmaceutical industry belongs to the typical fine chemical industry which is beneficial to the people's livelihood and health, and mankind's quality of life is closely related to the pharmaceutical industry. The characteristics of pharmaceutical industry are the wide varieties, fast replacement, multi-synthetic steps, and often-employing complex raw materials with lower yield and massive wastes which cause environmental pollution problems.

The objective of green pharmaceutical industry is to make the chemical principles and technologies used in the pharmaceutical industry to meet the requirements

https://doi.org/10.1515/9783110479317-006

of green technology during the process of medicine production. The traditional processes are improved and the yield is increased when the environment-friendly technologies are adopted in the chemical producing procedures, or the raw materials are fully transformed, toxic and harmful emissions are cut down even and are reduced to zero emissions in the reaction process. Green pharmaceutical industry has not only important economic benefits but also significant social and environmental benefits.

Green pharmaceutical industry is an important research direction in green chemistry; its main features are the pollution treatment being designed as the first condition for selecting production process, and the cleaner production technology being implemented. The standards of greenization include low energy consumption, no pollution, renewable resources, waste recycling and separation, or degradation. Green pharmaceutical industry includes green chemical pharmacy, green biological pharmacy, and green natural pharmacy according to the sources of materials and production method.

6.1.2 Green Chemical Pharmacy

The contents of green chemical pharmacy are using principle and technology of green chemistry for improving the atomic utilization rate, reducing or eliminating the harmful byproducts, recycling the solvent and reagent, using environment-friendly technology, and achieving the harmless production process.

The utilization of catalytic technology is an important method to promote the development of "green pharmacy." The catalytic processes include various forms of chemical catalysis and biological catalysis, which are important ways to achieve high atomic economic reaction. The utilization of catalytic methods can help accomplish some reactions which cannot be achieved by conventional methods, shorten the synthesis steps and improve the reaction yield.

The screening and optimization of the catalyst is very important in catalytic synthesis procedure. At present, there are many important green catalysts researched, such as nanozeolite, nano composite oxide lattice oxygen, heteropoly acid, conjugated solid superacid, supported transition metal nitrides, carbides, and water-soluble homogeneous organic metal complexes.

An important research content of green chemical technology is to obtain the optical active substances by asymmetric catalytic synthesis method, which is the raw material that has to be fully transformed to reduce the toxic and harmful emissions. The use of asymmetric catalytic synthesis to obtain optical active substances let the raw materials to be fully converted and reduce the emissions of toxic and harmful substances in the reaction process, which is one of the important research contents of green chemical technology.

1. Synthesis of Ibuprofen

Ibuprofen is a medicine in the nonsteroidal anti-inflammatory drug (NSAID) class that is used for treating pain, fever, and inflammation. This includes painful menstrual periods, migraines, and rheumatoid arthritis [1, 2].

The original synthetic pathway of ibuprofen was named after Brown method developed by Boots Company, which employed isobutyl benzene as a raw material, through Friedel–Crafts reaction producing the intermediate *p*-isobutylacetophenone, then through the Darzens condensation and hydrolysis reaction to obtain 1-(4-isobutyl phenyl) propionic aldehyde, and finally through oxidation ibuprofen is obtained. Ibuprofen can also be produced by oximation reaction of 1-(4-isobutyl phenyl) propionaldehyde, followed by hydrolysis, the method to get the products through six-step reaction is shown in Figure 6.1.

The atom utilization of raw materials is only about 40% in this route. Ibuprofen was synthesized only by three steps in BASF and BHC in Germany (Figure 6.2).

In the carbonylation reaction, $PdCl_2(PPh_3)_2$ was used as catalyst. When IBPE/ $PdCl_2(PPh_3)_2$ equaled to 1,500, reaction temperature 130 °C, CO pressure 16.5 MPa,

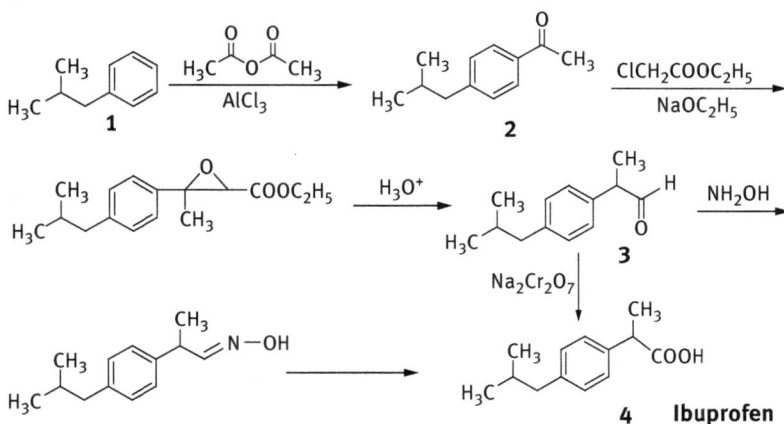

Figure 6.1: The route for synthesis of ibuprofen by Brown method.

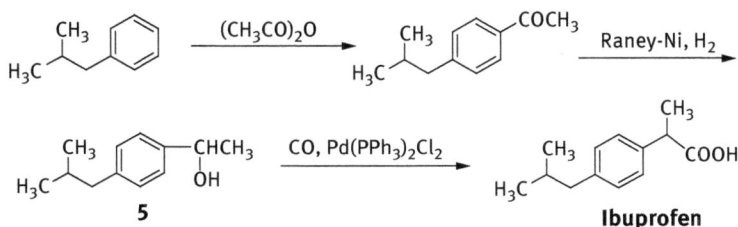

Figure 6.2: The route for synthesis of ibuprofen by BASF and BHC.

IBPE or methyl ethyl ketone (MEK) as the solvent in 10–26% hydrochloric acid, then the conversion and selectivity will reach 99% and 96%, respectively, after 4 h.

Compared with the traditional Boots method, BHC process is a typical atomic economic reaction with simple synthesis, high atom utilization of raw materials, uses less number of solvent and avoids producing a large number of wastes, which leads to less pollution. In Boots process, the oximation includes a six-step reaction, where only part of the substrate was converted into products(40.03% atom utilization), while the BHC process takes only three steps (77.44% atom utilization). If the recovery of by-product acetic acid is considered, the effective atom utilization of BHC process is up to 99%. BHC has won the Presidential Green Chemistry Challenge Award for the year 1997.

2. **Synthesis of naproxen**

Naproxen is an NSAID of the propionic acid class (the same class as ibuprofen) that relieves pain, fever, swelling, and stiffness [3, 4]. It is a nonselective COX inhibitor, usually sold as the sodium salt. Naproxen is an excellent nonsteroidal anti-inflammatory and analgesic drug, mainly used for the treatment of rheumatoid arthritis, ankylosing spondylitis, various types of rheumatism and rheumatoid disease, periarthritis of shoulder tendinitis, and so on. The chemical name of naproxen is (S)-(+)-1-(6-methoxy-2-naphthyl)propionic acid. The traditional synthetic method used beta naphthol as materials, through etherification, Friedel–Crafts propionylated reaction, bromination, ketone condensation, hydrolysis, hydrogenation, and chiral separation. The process is shown in Figure 6.3.

The characteristics of this method are the long route, the high cost, and the serious pollution, and large amount of concentrated sulfuric acid and sodium hydroxide being used in the reaction, which causes substantial wastewater and does not meet the requirements of environmental protection.

A new method to synthesize racemic naproxen was developed by Monsanto Company. Through electrolytic carboxylation and catalytic hydrogenation, 6-methoxy-2-acetyl naphthalene was converted to racemic naproxen. The total yield reaches to 83% catalyzed by phase transfer catalyst, based on aluminum anode in DMF solution with the pressure of CO_2 at 0.253 MPa. The electrolytic product 2-hydroxy-2-

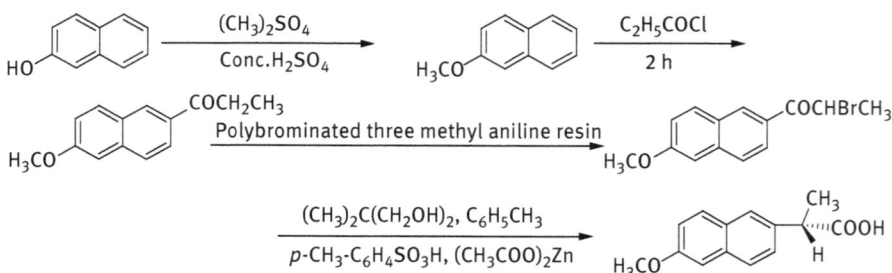

Figure 6.3: The traditional synthetic method of Naproxen.

(6′-methoxy-2′-naphthyl) propionic acid was converted into 2-(6′-methoxy-2′-naphthyl methacrylate) acrylic acid after dehydration.(see Figure 6.4).

The anti-inflammatory and analgesic effect of (S)-naproxen is 28 times higher than that of (R)-naproxen, but half of its by-products will be generated through enantioseparation of racemic naproxen, which does not meet the requirements of atom economy. High yield and enantioselectivity can be obtained when preparing (S)-naproxen through asymmetric synthesis while employing chiral phosphine ruthenium complexes as catalyst.

A new chiral diphosphine ligand was synthetized by Qiu, the yield and enantioselectivity catalyzed by [RuCI (p-cymene)]Cl in the catalytic hydrogenation of 2-(6-methoxy-2-naphthyl) acrylic acid in methanol solvent was 100% and 97%, respectively.(see Figure 6.5)

In another research at DuPont, Markovnikov addition reaction of 6-methoxy-2-naphthalene ethylene with HCN in the presence of 1,2-glycol phosphinate ligand Ni(0) complexes, the yield of cyanide was 100% and the enantioselectivity was 85%. In the next step reaction of cyanide hydrolyzation to (S)-naproxen in acid, the total yield exceeds 90% and e.e. value is 99% after crystallization, which shows a significant greenization.

3. **Synthesis of L-DOPA**

L-DOPA (L-3, 4-dihydroxyphenylalanine) is an amino acid that is made and used as part of the normal biology of humans, some animals, and plants. Some animals and humans make it via biosynthesis from the amino acid L-tyrosine. L-DOPA is

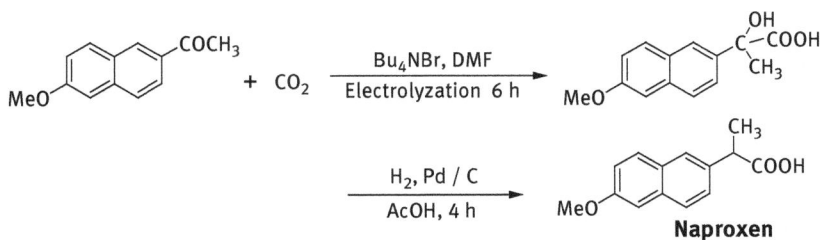

Figure 6.4: The synthetic method of naproxen by Monsanto.

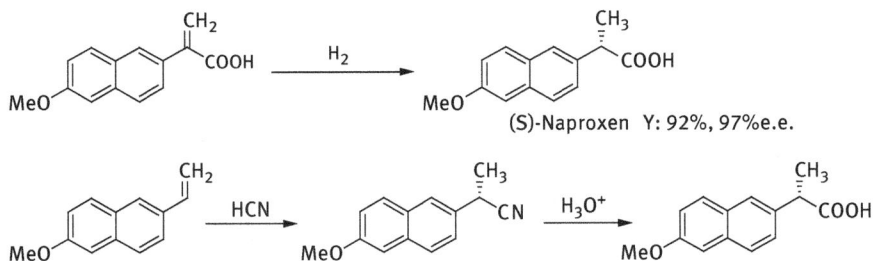

Figure 6.5: The modified synthetic method of naproxen.

the precursor to the neurotransmitters dopamine, norepinephrine (noradrena-line), and epinephrine (adrenaline) collectively known as catecholamines. Fur-thermore, L-DOPA itself mediates neurotrophic factor release by the brain and CNS [5, 6]. L-DOPA is used for the treatment of neurological Parkinson syndrome, which was once produced by enzyme-catalyzed procedure with complex opera-tion. A large-scale production of L-DOPA employing Rh-DIPAMP as the catalyst in Monsanto is a significant method with 95% optical yield.

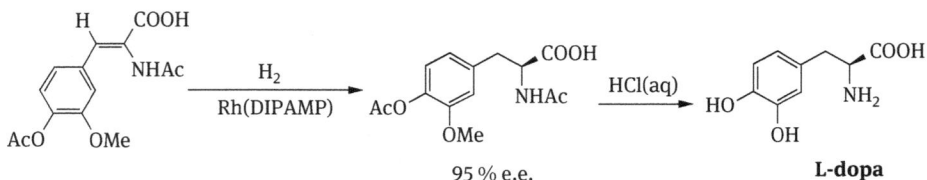

A chiral binaphthyl ligand (BINAP) metal catalyst Rh-BINAP produced by Kyoto University in Japan was used in the asymmetric catalytic hydrogenation of C=C to produce the intermediate of L-DOPA with optical yield of 65%.

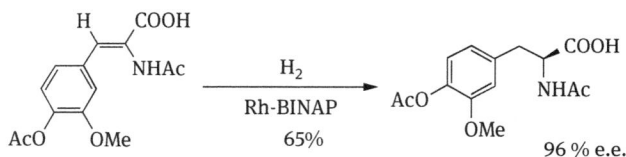

4. Synthesis of sitagliptin

Sitagliptin is dipeptidyl peptidase-4 (DPP-4) inhibitors for the treatment of type 2-diabetes, possessing good curative effect and little side effect [7]. The greener synthetic pathway award of US Presidential Green Chemistry Challenge Award in 2006 was awarded to Merck Company for the preparation of Januvia™ containing sitagliptin active ingredients from beta amino acids. In the route, waste can be reduced dramatically, and the total yield increased nearly 50%.

Januvia™ is a new drug for the treatment of type 2-diabetes. More than 200 pounds of clinical trial medicines were prepared through the first-generation synthetic sitagliptin route in Merck which is enlarged to large-scale production after a simple adjustment. However, the synthesis procedure still requires eight steps, including many complex operations. In addition, several macromolecular reagents are needed, which do not exist in the terminal product molecule and eventually become a waste.(see Figure 6.6).

Researchers in Merck Company with cooperative colleagues in Solvias Company, specialized in the study of asymmetric catalyst reaction, found that the beta amino acid derivatives can be obtained with high optical purity and high yield catalyzed by metal rhodium salt with two ferrocenyl as ligands.

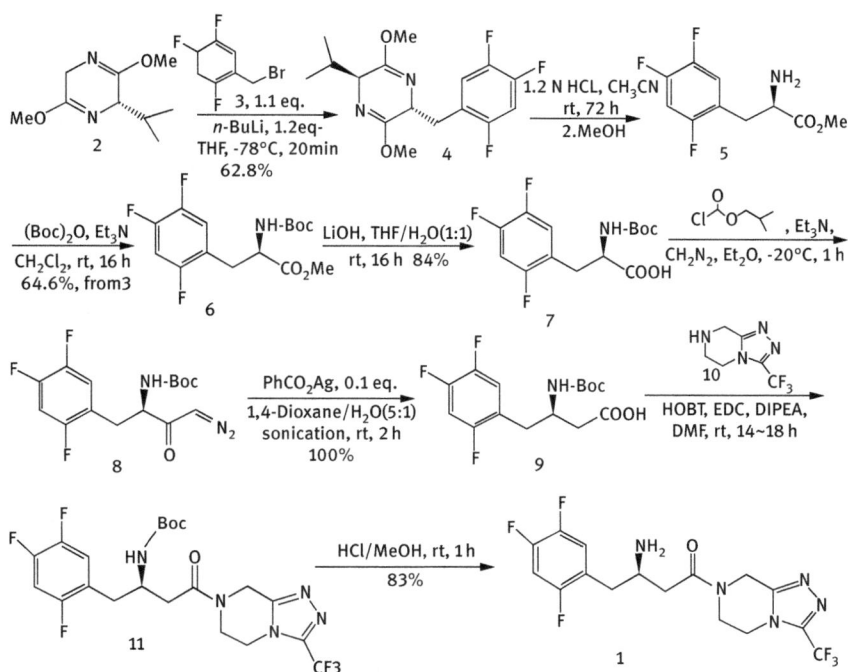

Figure 6.6: The traditional synthetic method of Sitagliptin by Merck.

The findings provided a common method for the synthesis of biologically active beta amino acids. Because of the expensive chiral catalysts, scientists and engineers in Merck used the catalyst only in the final synthesis step, the reaction yields have been increased significantly.

The whole reaction process of sitagliptin from the precursor dehydrogenation to the asymmetric hydrogenation is achieved in a container. After the completion of the hydrogenation reaction, 95% noble metal rhodium can be recovered and reutilized. Since the reactive amino group in the sitagliptin molecule is exposed until the last step, no group requires protection. The new synthetic route has three steps, which decreases raw material consumption, reduces energy consumption, cuts down waste emissions, and increases the total yield. (see Figure 6.7)

6.1.3 Green Biopharmaceutical

Biopharmaceutical is a kind of pharmaceutical technology combining biological principle and pharmaceutical engineering, and is widely used of modern biotechnology in the field of medical and pharmaceutical. Biotechnology can provide new means for research and innovation in medicine, can be used for the preparation of a large number of bioactive substances which are difficult to obtain, endogenous proteins, and peptides and makes

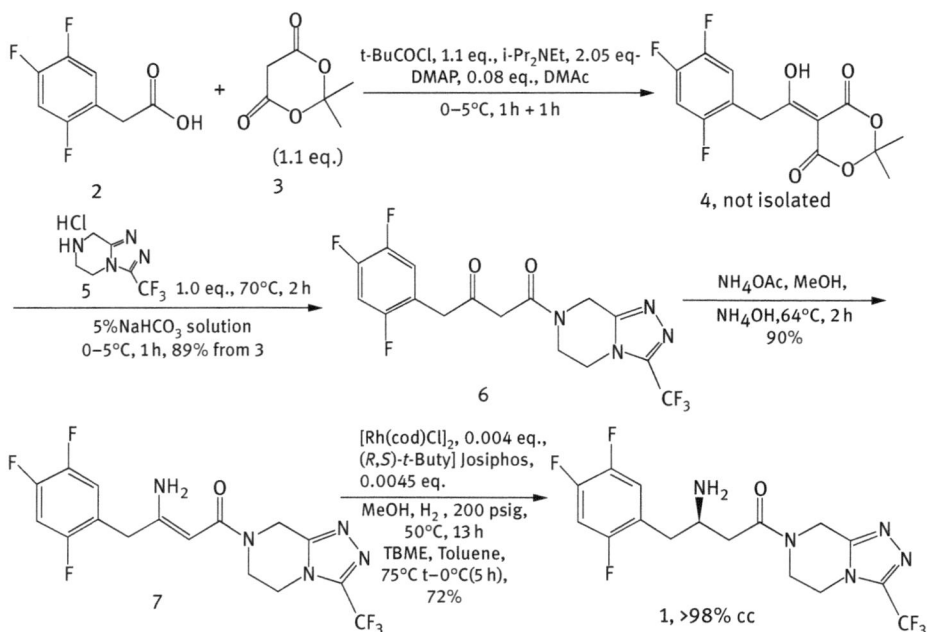

Figure 6.7: The synthetic method of Sitagliptin by Merck and Solvias.

them become new drugs. The new drugs can be used to improve vaccines to prevent the disease, and to establish a new diagnostic method which plays more and more important role in the prevention of diseases which seriously endanger people's health.

The modern biotechnology mainly includes:

(1) Gene engineering, including recombinant DNA technology and its transgenic technology
(2) Cell engineering, including technology of cell, protoplast fusion, animal, and plant cell large-scale culture
(3) Enzyme engineering, including immobilization technology of enzyme or cell, as well as enzyme chemical modification, physical modification, enzyme gene cloning, and enzyme gene mutation technology
(4) Fermentation technology, including high-density fermentation, continuous fermentation, and other new fermentation technologies
(5) Modern biological reaction engineering and modern separation technology of biological products

Using modern biotechnology not only optimizes the process conditions, the reaction conditions become mild, but also reduces energy consumption and avoids pollution. Therefore, the new biotechnologies have shown stronger vitality in the green synthesis and screening of new pharmaceutical direction.

1. **Preparation of human erythropoietin**

   Erythropoietin (EPO), also known as hematopoietin or hemopoietin, is a glyco-protein that controls erythropoiesis, or red blood cell production. It is a cytokine (protein signaling molecule) for erythrocyte (red blood cell) precursors in the bone marrow. Human EPO has a molecular weight of 34 kDa.

   EPO is produced by interstitial fibroblasts in the kidney in close association with peritubular capillary and proximal convoluted tubule. It is also produced in perisinusoidal cells in the liver. While liver production predominates in the fetal and perinatal period, renal production is predominant during adulthood.

   Exogenous erythropoietin can be provided to people whose kidneys cannot make enough EPO. Recombinant human erythropoietin (rhEPO) is produced by recombinant DNA technology in cell culture. Several different pharmaceutical agents are available with a variety of glycosylation patterns and are collectively called erythropoiesis-stimulating agents (ESA). Major examples are epoetin alfa and epoetin beta. The specific details for labeled use vary between the package inserts, but ESAs have been used in the treatment of anemia in chronic kidney disease, myelodysplasia, and cancer chemotherapy. Boxed warnings include a risk of death, myocardial infarction, stroke, venous thromboembolism, and tumor recurrence. rhEPO has been used illicitly as a performance-enhancing drug; it can often be detected in blood due to slight differences from the endoge-nous protein, for example, in features of posttranslational modification [8].

   1) **Large-scale cultivation of rhEPO engineering cells**

      After enlarging cultivation, the engineering cell lines were inoculated into the packed bed bioreactor under suitable pH and dissolved oxygen condi-tions. The inoculated engineering cell lines were firstly allowed to grow in the fetal bovine serum contained culture medium, which was then replaced by serum-free perfusion culture. Culture supernatant was collected, in which the expression of EPO quantity therein was about 5000 IU/ml. After ion exchange, chromatography, reverse phase chromatography and molecular sieve chromatography treatment, the EPO with high purity and high specific activity was obtained.

   2) **Purification technology**

      The purification process was: culture supernatant→ion exchange chromatog-raphy→desalination→C_4 reverse phase chromatography→ultrafiltration con-centration→molecular sieve chromatography. The fermentation supernatant was added to the ion exchange chromatography column balanced by buffer solution, then eluted by NaCl solution, EPO component is collected; and then desalinated by the Sephadex column. The collected liquid is eluted discontin-uously by anhydrous alcohol from C_4 column; the collected EPO component is diluted and concentrated by ultrafiltration. After processing with ion exchange chromatography, purification by the reversed phase chromatography, and molecular sieve chromatography, EPO activity component is obtained.

The characteristic of the above procedure was that the operation time was short. The purification period was only 48 h, and avoiding the bacterial contamination caused by long time treatment may result in higher pyrogens and excessive degradation of the EPO molecules in the products. After repeated chromatography, the purity of EPO samples was higher than 90%.

The total yield of purified biological activity EPO was 46%, which was proved to have immune properties same as the natural EPO. Therefore, the technology is suitable for large-scale production of rhEPO with high purity and activity.

2. **Preparation of new calcitonin**

Calcitonin (also known as thyrocalcitonin) is a 32-amino acid linear polypeptide hormone that is produced in humans primarily by the parafollicular cells (also known as C-cells) of the thyroid, and in many other animals in the ultimopharyngeal body. It acts to reduce blood calcium (Ca^{2+}), opposing the effects of parathyroid hormone [9, 10].

Calcitonin has been found in fish, reptiles, birds, and mammals. Its importance in humans has not been as well established as its importance in other animals, as its function is usually not significant in the regulation of normal calcium homeostasis. It belongs to the calcitonin-like protein family.

Calcitonin (nCT) and its analogs have been widely used in clinical treatment of osteoporosis disease and hypercalcemia. The bioactivity of salmon calcitonin is the highest among the calcitonin, whose activity is 50 times higher than of human calcitonin, but the long-term use of nonhuman sources of salmon calcitonin will produce antibodies. The new calcitonin (nCT/pGEX-2T/BL E.coli 21) with high activity, long half-life, and low antigen was obtained by employing human and salmon calcitonin as the lead compound. The specific process is as follows [11].

1) **Fermentation of engineering bacteria**

The gene engineering bacteria nCT/pGEX-2T/E.coli BL21 single mycelium were inoculated in a Luria-Bertani culture medium. After being shocked and cultured at 37 °C for 15 h, the bacteria were then inoculated into a shaking flask with culture medium. By ventilation and addition of isopropyl thio-β-d-galactosidase (IPTG), the concentration of engineering bacteria gradually turned to 0.1 mmol/L. Then further cultivation for 4 h, the bacteria were collected via centrifugalization,The collected bacteria were used for mass preparation of target bacteria by fermentation in a tank.

2) **Purification of fusion protein**

The collected centrifugal fermentation bacteria was placed in centrifuge tubes, and put the centrifugal tube in the refrigerant (dry ice and ethanol) after adding a certain amount of succinic acid-1,4-butylene glycol polyester soliquoid. After intermittent centrifugation, the supernatant was collected.

The concentration of fusion protein was adjusted to 10 mg/mL by chromatography column, elution, dialysis, and polyethylene glycol concentration.

3) **The purification of calcitonin precursor**

Through sulfonation, dissociate cracking by bromine hydride of the fusion protein; put the liquid into the fast gel column, wash to absorbance $A_{280\ nm} < 0.05$ under 10 mmol/L HCl; then wash by eluent (10 mmol/L HCl, 100 mmol L NaCl), detected by UV absorption; collect the outflow, then frozen dried and purified by RP-HPLC.

4) **Preparation of new calcitonin**

The new calcitonin precursor was added into ammonia solution (pH = 9.5, regulated by HCl), added DMSO to make polypeptide completely dissolved; then added 50 mol/L carboxypeptidase solution, and reacted at 37 °C for 1 h, then added trifluoroacetic acid to the reaction solution with swing to terminate the reaction (final concentration 1%). pH was adjusted to 8 with 2 mol/L NaOH and add cysteine to make the final concentration of 5 mmol/L, react for 1 h at 37 °C to restore disulfide bonds, then purified using RP-HPLC and collect the effluent liquid. After vacuum distillation, freeze-drying, analyzed by amino acid composition analysis and mass spectrometry, the components of the new calcitonin were determined.

6.1.4 Green Natural Medicine

Green natural medicine makes full use of modern science and technology, according to medical standards, researches and develops safe, efficient, stable and controllable modern Chinese medicine products. The keys to realize the modernization of the natural pharmacy by means of modern science and technology are the studies of modernization of natural medicine production technology, industrialization of manufacturing technology, and large-scale production by adopting modern science and technology. In the procedure of the manufacture of natural medicine, the pharmaceutical industry makes use of the supercritical fluid extraction, ultrasonic extraction, and resin adsorption and separation .

1. **Extraction of *ginkgo flavones***

There are many active ingredients in ginkgo biloba, such as flavonoids, terpenoids, and phenolic acids, which have strong physiological activity of the central nervous system, circulatory system, respiratory system, and digestive system; they can increase cerebral blood flow, improve the circulation function of cerebral vessels, protect brain cells, expand coronary artery, prevent angina and myocardial infarction, prevent thrombosis, and improve the immune capacity. In addition, it has antibacterial, anti-inflammatory, anti-allergic effects, which is beneficial to coronary heart disease, angina pectoris, cerebral arteriosclerosis, senile dementia, and hypertension patients.

At present, a solvent extract has been adopted to obtain ginkgo biloba flavones. For example, ginkgo biloba flavone was obtained when employing 60% acetone as the extract solvent, after extraction, separation, and purification. But there were disadvantages of the process: long time extraction, multiple wash, filtering, and extraction process; consuming much organic solvent, high production cost, low yield, and poor quality of products; producing a large number of wastewater and waste residue, pollution to natural environment, and residual containing heavy metals and organic solvents in products that will bring toxic side effects.

Application of supercritical fluid extraction of active components of ginkgo biloba is an effective method to overcome above disadvantages. The process of supercritical extraction is shown in Figure 6.8.

After drying and crushing, the green ginkgo leaf has been graded and loaded to the extractor and sealed, open the extractor, separator, and other heating device to preheat the whole system. At the same time set the required extraction temperature; open the air inlet switch of CO_2 and start the compressor, increase the pressure to achieve the required range, then open the valve, and input carbon dioxide. When the required temperature and pressure of extraction is achieved, keep a certain time and then start the separation operation; when the pressure is stable at 10 MPa, remove ginkgolic phenolic acid, chlorophyll, and other

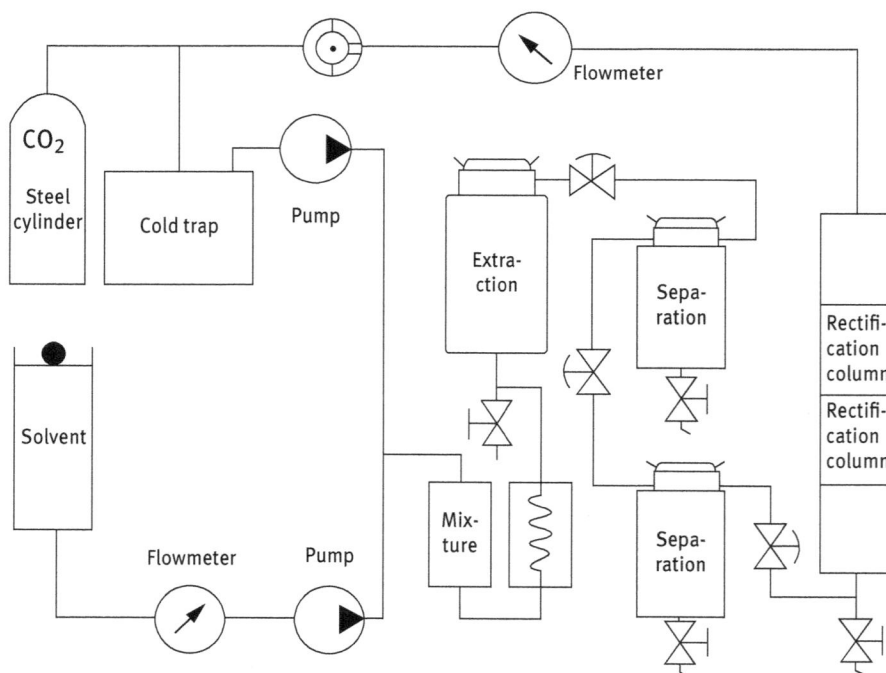

Figure 6.8: Extraction process of flavonoids from *Ginkgo biloba* leaves by supercritical CO_2.

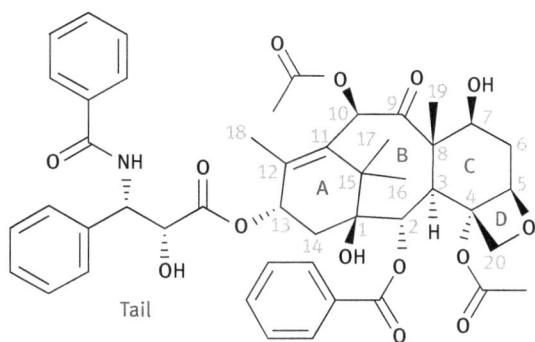

Figure 6.9: The structure of paclitaxel.

impurities; when the pressure is above 10 MPa and stable, the effective components of ginkgo biloba leaves are extracted, separated, collected, and the structure of extraction products are determined [12].

Carbon dioxide is used as the extraction medium in the supercritical fluid
extraction process with safe, nontoxic, and mild extraction condition. The natural
quality of the effective components of ginkgo biloba leaves is maintained and no
residual of heavy metals and toxic solvents is present. Compared to the solvent
extraction, supercritical fluid extraction is a greener extraction technology.

2. **Extraction of paclitaxel**

Paclitaxel (PTX, see Figure 6.9), sold under the brand name Taxol among others,
is a chemotherapy medication used to treat a number of types of cancer. This
includes ovarian cancer, breast cancer, lung cancer, Kaposi's sarcoma, cervical
cancer, and pancreatic cancer. It is given by injection into a vein. There is also an
albumin-bound formulation [13].

The nomenclature for paclitaxel is structured on a tetracyclic 17-carbon
(heptadecane) skeleton. There are a total of 11 stereocenters. The active stereoi-
somer is (−)-paclitaxel and chemical name is β-(benzoylamino)-α-hydroxy-(2aR,
4S, 4aS, 6R, 9S, 11S, 12S, 12aR, 12bS)-6, 12b-bis(acetyloxy)-12- (benzoyloxy)-2a,3,
4,4a,5,6,9,10,11,12,12a,12b-dodecahydro-4,11-dihydroxy-4a,8,13,13-tetramethyl-
5-oxo-7,11-methano-1H-cyclodeca[3,4]benz[1,2-b]oxet-9-yl ester, (αR,βS)-benzene-
propanoic acid.

The discovery was made by Monroe E. Wall and Mansukh C. Wani at the
Research Triangle Institute, Research Triangle Park, North Carolina, in 1971. These
scientists isolated the natural product from the bark of the Pacific yew tree, *Taxus
brevifolia*, determined its structure and named it as "taxol," and arranged for its
first biological testing [14]. The compound was then developed commercially by
Bristol-Myers Squibb (BMS), who had the generic name assigned as "paclitaxel."
Paclitaxel was approved for medical use in 1993. It is in the World Health Organization's list of essential medicines, the most effective and safe medicines needed
in a health system.

Paclitaxel is approved in the United Kingdom for ovarian, breast and lung, bladder, prostate, melanoma, esophageal, and other types of solid tumor cancers as well as Kaposi's sarcoma. It is recommended in National Institute for Health and Care Excellence (NICE) guidance of June 2001 that it should be used for nonsmall cell lung cancer in patients unsuitable for curative treatment, and in first-line and second-line treatment of ovarian cancer. In September 2001, NICE recommended paclitaxel should be available for the treatment of advanced breast cancer after the failure of anthracyclic chemotherapy, but that its first-line use should be limited to clinical trials. In September 2006, NICE recommended paclitaxel should not be used in the adjuvant treatment of early node-positive breast cancer. In 2005, its use in the United States for the treatment of breast, pancreatic, and nonsmall cell lung cancers was approved by the Food and Drug Administration (FDA).

Paclitaxel is one of several cytoskeletal drugs that target tubulin. Paclitaxel-treated cells have defects in mitotic spindle assembly, chromosome segregation, and cell division. Unlike other tubulin-targeting drugs such as colchicine that inhibit microtubule assembly, paclitaxel stabilizes the microtubule polymer and protects it from disassembly. Chromosomes are thus unable to achieve a metaphase spindle configuration. This blocks the progression of mitosis and prolonged activation of the mitotic checkpoint triggers apoptosis or reversion to the G-phase of the cell cycle without cell division. The ability of paclitaxel to inhibit spindle function is generally attributed to its suppression of microtubule dynamics, but recent studies have demonstrated that suppression of dynamics occurs at concentrations lower than those needed to block mitosis. At higher therapeutic concentrations, paclitaxel appears to suppress microtubule detachment from centrosomes, a process normally activated during mitosis. Paclitaxel binds to beta-tubulin subunits of microtubules.

It is more complex and difficult to extract taxol from the natural yew trees. The main reason is that the taxol content in plants is too low, the highest taxol content is less than 0.02%, and there are 200 kinds of taxol analogues; the chemical structures and properties are similar with paclitaxel, therefore the separation of taxol is very difficult.

The principle of supercritical CO_2 fluid extraction of taxol is shown in Figure 6.10. Put the sample in the extraction tank, carbon dioxide and methanol were pumped to the fluid mixer through CO_2 pump and modifier pump respectively; after mixing in manifold in the extractor at pressure 27.6 MPa and temperature 31°C, the mixture was transferred to enter the extraction pool. When the dynamic extraction is started, the supercritical CO_2 flows into the collection bottle through a restrictor and then undergoes decompression. The extracts were brought out by the fluid to be collected in the collecting liquid. Methanol was used as an absorption solution, and the collected liquid was concentrated by vacuum rotary evaporation and dried for 2 h at 50°C. The content of taxol was detected when collected at 30, 60, 90, and 120 min, respectively (see Figure 6.10) [15].

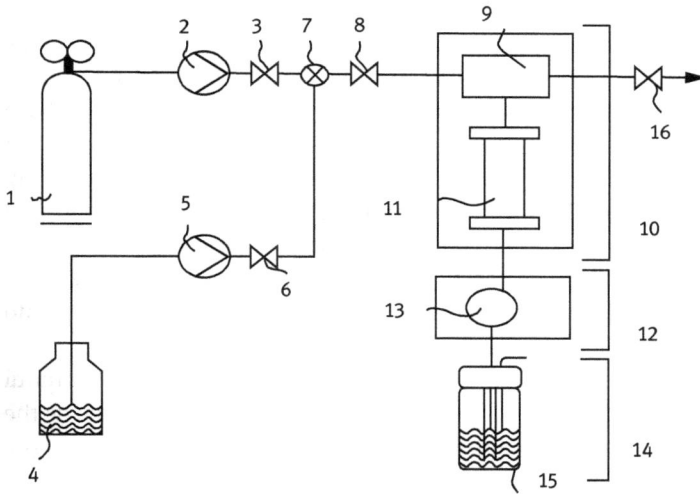

Figure 6.10: The principle of supercritical fluid extraction of taxol. 1, CO_2 cylinder; 2, CO_2 pump; 3, 6, 8, valves; 4, modifier containers; 5, modifier pump; 7, fluid mixer; 9, manifold; 10, extraction unit; 11, extraction tank; 12, current limiting unit; 13, current limiter; 14, collection unit; 15, collection bottle; 16, exhaust valve.

By adding ethanol in CO_2 as modifier and maintaining the extraction of taxol in the appropriate temperature and pressure, most of the bark of paclitaxel can be an effective extraction, and the selectivity to paclitaxel is better than the traditional ethanol extraction. The extraction of taxol by supercritical fluid is a green process with low solvent consumption, high yield, and less waste.

6.2 Greenization of Pesticide Industry

6.2.1 Introduction

Green pesticide is a kind of high efficient, stable, convenient, and environmental-friendly pesticide product to use, which is manufactured by pollution-free materials and procedures without harmful by-products. The use of green pesticides cannot only protect the normal growth of crops, to ensure a stable crop yield, but also reduce environmental pollution. In accordance with green agricultural products, pesticide products should have the characteristics of high efficiency, low toxicity, low residue, and good selectivity in the future, which include green biological pesticides, green chemical pesticides, and green pesticide formulations.

6.2.2 Green Biological Pesticides

Biological pesticides are a kind of formulations which can prevent and control harmful organisms by employing living organisms or its metabolites, which have the advantages of good selectivity, no pollution, no drug resistance, wide range of raw materials, and so on. Biological pesticides can be divided into microbial living, microbial metabolites, mixture of living and metabolites, botanical pesticides, and biochemical pesticides. Moreover, transgenic plant with antiviral and disinfection function can also be included in the category of biological pesticide. According to the prevention of different objects, green biological pesticides can also be divided into pesticides, fungicides, herbicides, plant growth regulators, etc.

Green biological pesticides include active pesticides of natural enemies of harmful organisms, some of the secondary metabolites of biological metabolism, such as the gene introduction of resistance to insect crops or disease crops. Among which, the most widely used category is microbial pesticides.

1. **Microbial pesticides**

   Microbial pesticides include microbial pesticides, microbial herbicides, and agricultural antibiotics. The microbial pesticide is a kind of formulations using insect pathogens as insecticide; these pathogens can be used without modification by the application of near natural state, standard genetic operations, or recombinant DNA technology to improve the insecticidal activities. Insect pathogens included viruses, bacteria, fungi, protozoa, and so on; all of them have pathopoiesis and lethal effects on their host insects.

   Microbial herbicides employ natural enemy living organisms, such as herbivorous and pathogenic microorganisms by way of ecology use, which control weed population in the economic, ecological, and aesthetic acceptable level in the natural state. The research and development of microbial herbicides began in mid-twentieth century, and used microorganism itself for weed control directly.

   1) **Avermectins**

      The avermectins (see Figure 6.11) are a series of drugs used to treat parasitic worms. They are a 16-membered macrocyclic lactone derivatives with potent anthelmintic and insecticidal properties [16, 17]. These naturally occurring compounds are generated as fermentation products by *Streptomyces avermitilis*, a soil actinomycete. Eight different avermectins were isolated in four pairs of homologue compounds, with a major (a-component) and minor (b-component) component usually in ratios of 80:20 to 90:10. Other anthelmintics derived from the avermectins include ivermectin. It is fermentation metabolites of soil microorganism of gray *Streptomyces*.

      Avermectin is an agricultural antibiotic insecticide and acaricide, which can prevent and cure the phytophagous mites and Lepidoptera, Hymenoptera, Coleoptera, Hemiptera, and other pests effectively; it is a new, high efficient, low toxic, and nonpolluting biological pesticide. Merck Sharp

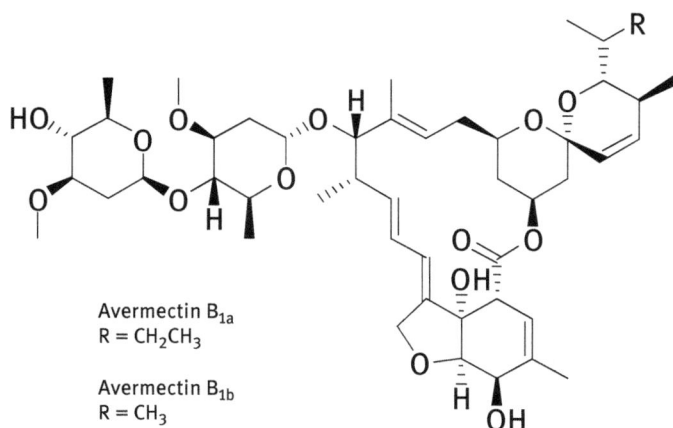

Figure 6.11: The structure of avermectin.

and Dohme Agvet had employed the mixture of two kinds of avermectin Bla and Blb as insecticides and acaricides, which had been industrialized for production.

Avermectin belongs to insect nerve poison. The main function is to interfere neuro physiological activities of the insects with contact and stomach toxicity; it has no internal absorption but has a strong penetration, and can be a transverse conduction in the plant. The insecticidal and acaricidal activities are 5–50 times higher than those of the commonly used pesticides. The dosage of pesticide is only 1–2% of common pesticides. Avermectin is nontoxic to egg in which the embryo is undeveloped; however, it has a strong ovicidal toxicity in the late stage of embryo development.

The medicament has good controlling effect against the resistant insect, and no cross-resistance to organophosphorus, pyrethroid, and carbamate pesticides.

Currently, the production process of avermectin is mainly the following steps.

(1) Strain improvement. The qualities of avermectin strain have important influence on the avermectin production. The fermentation unit of original strain of avermectin is very low; the fermentation unit of the first discovered strain MA-4680 was only 9 µg/mL. Though the changes of fermentation conditions have greatly improved the yield, it is still only 120 µg/mL, which is not suitable for large-scale fermentation. The strain was mutated and obtained by UV, and the fermentation unit reached to 500 µg/mL. The yield has been greatly improved compared with the original strain. A strain of streptomycin-resistant mutant were obtained through ultraviolet (UV) mutagenesis of an original strain by Feng Jun,

the fermentation unit increased 116 times than that of the original strain, and the fermentation unit of another mutant increased 215 times; the fermentation efficiency was increased 116 times, and the ratio of B1a and B1b in the product was increased from 8 to 20 by using the nitrosoguanidine to induce mutation.

(2) Fermentation medium: fermentation medium is very important in the industrialization of fermentation products. At present, many culture mediums have been used for the fermentation, due to a variety of different components, especially the natural components have great influence on the yield of fermentation; it is necessary to choose a good combination of culture medium and culture condition. The commonly used carbon sources are sugar, oil, organic acid, and low carbon alcohol. Using glucose as carbon source can obtain the highest yield of fermentation, and the mycelial growth amount is the largest; however, the yield of dry silk weight was less than when using wheat flour as carbon source, therefore wheat flour is an ideal carbon source. In the medium optimization, more than 170 kinds of fermentation media with different proportions and starch were selected as the most suitable carbon source.

(3) Crystallization: Crystallization is the key step for avermectin purifying process. Bagner Oarl studies a direct toluene extraction method of abamectin in fermentation. The fermentation liquid without filtering was directly adjusted to a pH level of 2.5 by adding sulfuric acid, and then it was extracted under heat by a mixture of toluene and fermentation broth in the ratio 2:1. The toluene extract solution was collected and concentrated to obtain the crude product. Using the direct crystallization method for twice, the total extraction yield and purity reached to 97.3% and 95%, respectively.

2) **Bialaphos**

Bialaphos (see Figure 6.12) is a natural herbicide produced by the bacteria *Streptomyces hygroscopicus* [18] and *Streptomyces viridochromogenes*. Bialaphos is a protoxin but nontoxic as well. When it is metabolized by the plant, the glutamic acid analog glufosinate is released which inhibits glutamine synthetase. This results in the accumulation of ammonium and disruption of primary metabolism [19].

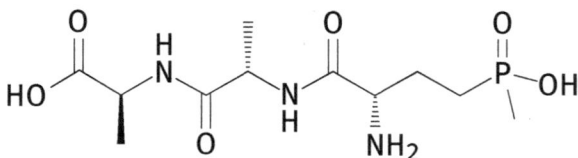

Figure 6.12: The structure of bialaphos.

The sodium salt of bialaphos can be used to control the annual weeds on the cultivated land, and also be used for the prevention of the annual and perennial weeds on the uncultivated land. As a biologically active acid, bialaphos is an efficient and irreversible inhibitor of glutamine synthetase, which can cause the accumulation of ammonia and inhibit the phosphorylation of photosynthesis.

In addition, bialaphos also inhibits the photosynthetic phosphorylation in photosynthesis. It is absorbed by the stem and leaves, and has the function of internal absorption and contact. It is used for the removal of a variety of annual and perennial single leaf and dicotyledonous weeds in grapes, apples, citrus orchards, as well as for no-tillage, nonarable land weeding. The product has the properties of easy loss of activity, easy metabolism, and biological degradation in soil, so it is safety for use.

3) **Kasugamycin**

Kasugamycin (Ksg, see Figure 6.13) is an aminoglycoside antibiotic that was originally isolated from *Streptomyces kasugaensis*, a *Streptomyces* strain found near the Kasuga Shrine. Kasugamycin was discovered by Hamao Umezawa [20], who also discovered kanamycin and bleomycin, as a drug which could prevent growth of a fungus causing rice blast disease. It was later found to inhibit bacterial growth also. It exists as a white, crystalline substance and it is also known as kasumin.

Like many of the known natural antibiotics, kasugamycin inhibits proliferation of bacteria by tampering with its ability to make new proteins, the ribosome being the major target. Kasugamycin inhibits protein synthesis at the step of translation initiation. Kasugamycin inhibition is thought to occur by direct competition with initiator transfer RNA. Recent experiments suggest that kasugamycin indirectly induces dissociation of P-site-bound fMet-tRNAf-Met from 30S subunits through perturbation of the mRNA, thereby interfering with translation initiation [21].

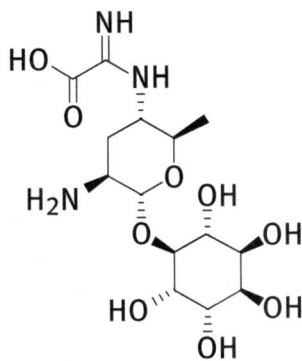

Figure 6.13: The structure of kasugamycin.

Kasugamycin specifically inhibits translation initiation of canonical but not of leaderless mRNA. For initiation on leaderless mRNA, the overlap between mRNA and kasugamycin is reduced and the binding of tRNA is further stabilized by the presence of the 50S subunit, minimizing Ksg efficacy. Kasugamycin also induces the formation of unusual 61S ribosomes in vivo, which are proficient in selectively translating leaderless mRNA. 61S particles are stable and are devoid of more than six proteins of the small subunit, including the functionally important proteins S1 and S12.

Kasugamycin is an ideal agent for the prevention and treatment of a variety of bacterial and fungal diseases, which has the function of prevention, treatment, growth, and regulation. It is not only a special antibiotic against rice blast but also has a good controlling effect on rice thin disease, citrus flow plastic disease, sand skin disease, kiwi ulcer disease, pepper bacterial scab, celery early blight, and bean curd. Komycin may cause slight harm to peas, beans, soybeans, grapes, citrus, and apples but no harm to rice, potatoes, beets, tomatoes and, other vegetables. Komycin is less toxic to mammals, and has good environmental compatibility, and no adverse effect on nontarget organisms and environment.

2. **Plant source pesticide**
The active ingredients of plant-derived pesticides are naturally occurring, which are generally low in toxic and has no residue. They are safe on humans and animals, and easy for degradation. The possibility of accumulation of toxicity in the environment is small, the environmental compatibility is good, and the pests are difficult to produce resistance: they do not kill insect's natural enemies, and have a broad development prospects in agriculture.

1) **Azadirachtin**
Azadirachta indica is widely planted in tropical and subtropical regions. The extract of azadirachtin (see Figure 6.14) contained in seeds, leaves, and bark has strong insecticidal properties. Azadirachtin has a variety of effects on herbivorous insects, and has a strong food repellent effect. Many insects are not eating azadirachtin-treated crops. Azadirachtin can interfere the molting of insects by antagonizing ecdysone, resulting in morphological defects in insects when contacted with the crop sprayed by azadirachtin; they can also reduce their fecundity by disturbing the mating behavior of phytophagous insects. Azadirachtin is a broad-spectrum, high-efficient, low-toxic, easy-to-degrade, nonresidual pesticide and nonresistant insecticide to almost all plant pests, and has no pollution to humans and animals and to the surrounding environment.

Azadirachtin is a chemical compound belonging to the limonoid group and a secondary metabolite present in neem seeds. It is a highly oxidized tetranortriterpenoid which boasts a plethora of oxygen bearing functional groups, including an enol ether, acetal, hemiacetal, tetrasubstituted epoxide, and a variety of carboxylic esters. Azadirachtin has a complex molecular structure;

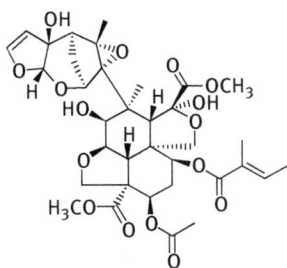

Figure 6.14: The structure of azadirachtin.

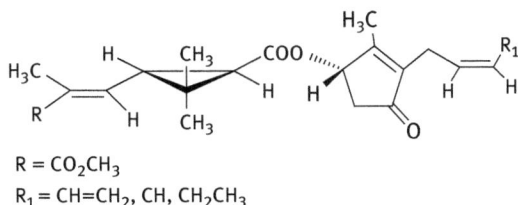

R = CO$_2$CH$_3$
R$_1$ = CH=CH$_2$, CH, CH$_2$CH$_3$

Figure 6.15: The structure of pyrethrum.

it presents both secondary and tertiary hydroxyl groups and a tetrahydrofuran ether in its molecular structure, alongside 16 stereogenic centers, seven of which are tetrasubstituted. These characteristics explain the great difficulty encountered when trying to prepare this compound from simple precursors, using methods of synthetic organic chemistry. Hence, the first total synthesis was published over 22 years after the compound's discovery: this first synthesis was completed by the research group of Steven Ley at the University of Cambridge in 2007 [22, 23]. The described synthesis was a relay approach with the required heavily functionalized decalin intermediate being made by total synthesis on a small scale, but it has been derived from the natural product itself for the gram-scale operations required to complete the synthesis.

The best solvent in ultrasonic assisted extraction of azadirachtin is methanol with the material to liquid ratio of 1:2, ultrasonic power of 200 W, extraction time of 15 min, and the best conditions of extraction yield is 0.38%; in microwave extraction, employing methanol as solvent has microwave power of 210 W, material to liquid ratio of 1:3, radiation time of 3 min, total extraction for 3 times, and extraction yield up to 0.52%. Compared with the traditional magnetic stirring method, the method is simple and the extraction time is short [24].

2) **Pyrethrum**
   Pyrethrum (see Figure 6.15) was a genus of several Old World plants now classified as *Chrysanthemum* or *Tanacetum* (e.g., *Chrysanthemum coccineum*) which are cultivated as ornamentals for their showy flower heads [25]. *Pyrethrum* is used as a common name for plants formerly included in the genus *Pyrethrum*. *Pyrethrum* is also the name of a natural insecticide made

from the dried flower heads of *Chrysanthemum cinerariifolium* and *Chrysanthemum coccineum*.

Pyrethrum has high economic value and ornamental value. The poisonous substances which existed in roots, stems leaves, and flowers can be extracted as the pyrethrum materials for the preparation of various pesticides, for killing aphids, mosquitoes and flies, cabbage caterpillar, cotton bollworm, etc. The leaves of *Chrysanthemum cinerariifolium* are often used to produce mosquito-repellent incense for mosquito killing and driving flies, and have special effects on bugs, lice, and fleas. *Pyrethrum* can also be directly used to kill pests such as aphids, mosquitoes and flies, cabbage caterpillar, and cotton bollworm. It has advantages of not polluting environment, not destroying ecological balance, no resistance and no harmful to human, livestock, poultry, and mammals.

Pyrethrin is a nonintramuscular contact insecticide, which can be combined to sodium ion channel of insect cell membrane, extending its opening time, causing insect shock and death. Pyrethrin is generally used together with synergist, a kind of synergistic ether. Pyrethroids are unstable to light, and synergistic ether can inhibit the decomposition of pyrethroids. *Pyrethrum* has relatively low levels of toxicity and many pyrethrins are used in combination with pesticides.

The extraction of pyrethrin by microwave has the advantages of short time, high yield, low solvent consumption, better product quality, and shallow color [26]. Microwave extraction of pyrethroids can be carried out by continuous microwave irradiation. When the total time of microwave irradiation is equal to the traditional extraction time and the operating temperature is below to the boiling point of the solvent, higher yield and higher content of pyrethrin in the extract can be obtained along with the microwave irradiation time.

3. **Genetical engineering pesticides**

Genetic engineering, also called genetic modification, is the direct manipulation of an organism's genome using biotechnology. It is a set of technologies used to change the genetic makeup of cells, including the transfer of genes within and across species boundaries to produce improved or novel organisms. New DNA may be inserted in the host genome by first isolating and copying the genetic material of interest using molecular cloning methods to generate a DNA sequence, or by synthesizing the DNA, and then inserting this construct into the host organism. Genes may be removed, or "knocked out," using a nuclease. Gene targeting is a different technique that uses homologous recombination to change an endogenous gene, and can be used to delete a gene, remove exons, add a gene, or introduce point mutations.

Genetic engineering techniques have been applied in numerous fields including research, agriculture, industrial biotechnology, and medicine. Enzymes used in laundry detergent and medicines such as insulin and human growth hormone are now manufactured in GM cells, experimental GM cell lines and GM animals such as mice or zebra fish are being used for research purposes, and genetically modified crops (GMCs) have been commercialized. Humans have altered the genomes of species for thousands of years through selective breeding, or artificial selection [27, 28] as contrasted with natural selection, and more recently through mutagenesis. Genetic engineering as the direct manipulation of DNA by humans outside breeding and mutations has only existed since the 1970s.

GMCs, GM crops, or biotech crops are plants used in agriculture, the DNA of which has been modified using genetic engineering techniques. In most cases, the aim is to introduce a new trait to the plant which does not occur naturally in the species. Examples in food crops include resistance to certain pests, diseases, or environmental conditions, reduction of spoilage, or resistance to chemical treatments (e.g., resistance to aherbicide), or improving the nutrient profile of the crop. Examples in nonfood crops include production of pharmaceutical agents, biofuels, and other industrially useful goods, as well as bioremediation [29].

Genetically modified foods or GM foods, also known as genetically engineered foods, are foods produced from organisms that have had changes introduced into their DNA using the methods of genetic engineering. Genetic engineering techniques allow for the introduction of new traits and have greater control over traits than previous methods such as selective breeding and mutation breeding [30].

Commercial sale of genetically modified foods began in 1994, when Calgene first marketed its unsuccessful Flavr Savr delayed-ripening tomato [31]. Most food modifications have primarily focused on cash crops, such as soybean, corn, canola, and cotton in high demand by farmers. GMCs have been engineered for resistance to pathogens, herbicides and for better nutrient profiles. GM livestock have been developed, although as of November 2013, none were on the market [32].

It seems that there is no difference between GM crops and ordinary plant; the only difference is that GM crops have an extra gene that makes them produce many additional properties. Biologists have known how exterior genes can be transplanted into a plant's DNA so that it has some new features, such as antiremoval characteristics of herbicides, resistance to plant virus, and pest resistance. This gene can come from any life: bacteria, viruses, or insects. So, a crop can be planted into a characteristic that is not e obtained by hybridization through biological engineering technology, which will greatly promote the quality and yield of crops.

The first GMC was tobacco which contains antibodies to antibiotics planted in 1983, and the first marketed food came out in the United States ten years later. In 1996, the tomato cake which was made from genetically modified tomato was permitted to be sold in supermarkets. Genetically modified cattle and sheep, fish and shrimp, foodstuff, vegetables, and fruits have been cultivated internationally and have been put into market. The total number of GMCs in the world has increased by 40 times in the last seven years. Genetically modified organisms are commonly plant, animal, and microbes, where plants are the most.

Papaya was genetically modified to resist the ringspot virus. "SunUp" is a transgenic red-fleshed sunset papaya cultivar that is homozygous for the coat protein gene PRSV; "Rainbow" is a yellow-fleshed F1 hybrid developed by crossing "SunUp" and nontransgenic yellow-fleshed "Kapoho.". The *New York Times* stated, "In the early 1990s, Hawaii's papaya industry was facing disaster because of the deadly papaya ringspot virus. Its single-handed savior was a breed engineered to be resistant to the virus. Without it, the state's papaya industry would have collapsed. Today, 80% of Hawaiian papaya is genetically engineered, and there is still no conventional or organic method to control ringspot virus" [33]. The GM cultivar was approved in 1998 [34]. In China, a transgenic PRSV-resistant papaya was developed by South China Agricultural University and was first approved for commercial planting in 2006; as of 2012, 95% of the papaya grown in Guangdong province and 40% of the papaya grown in Hainan province were genetically modified [35].

1) **Genetically modified insect-resistant plants**

Transferring insecticidal genes of bacteria to crops can make crops have insecticidal feature. At present, in genetically modified insect-resistant plants, the most widely used is Bacillus thuringiensis toxin protein gene, abbreviated as Bt gene. Genetically modified soybean, cotton, and potato with insect resistance have been developed and applied in America. The cultivated genetically modified Bt cotton were not only having strong resistance to Lepidoptera pests, but also had no negative effects on the yield and fiber qualities. Some properties are even better than those of their parents.

The extensive researches of modification of Bt ICP (insecticidal crystal protein), construction of expression vector, transformation of tissue, cultivation of insect resistant plants had been carried out and 50 different transgenic plants had been obtained, such as anti-potato beetle potatoes, anti-Lepidoptera tomatoes, anti-lepidopteran cotton, anti-lepidopteran, and *Coleoptera* corn. The advantages of transgenic insect-resistant plants are obvious. They can not only resist the harm of pests but also reduce the use of pesticides, which further reduce the pesticide poisoning and the death of natural enemies. So the transgenic crops protect the environment.

Cotton is one of important crops in China. The researches of transgenic cotton have made a series of important achievements in insect resistance, herbicide resistance, disease resistance and, cotton fiber quality improvement. Bt insecticidal gene was synthesized for the first time in Chinese Academy of Agricultural Sciences Institute of Biotechnology in 1992. The gene was introduced into the cotton plant to obtain GK series transgenic Bt cotton. Recombining cowpea trypsin inhibitor gene (CpTI) and Bt gene, then putting them into cotton, bivalent genetically modified (CpTI + Bt) SGK insect- resistant cotton is obtained. The toxin in Bt transgenic cotton can be expressed persistently, put the cotton bollworm in high pressure of Bt toxic protein for the whole cotton growing period, thus reducing the pesticide usage of about 60%, and the yield of transgenic insect cotton increased to an average of 10% than that of the common cotton.

2) **Genetically modified herbicide resistant plants**
The effect of genetically modified herbicide is that it mainly lets crops obtain or enhance the resistance genetic traits to herbicides through the transgenic technology, and many excellent herbicides have been widely used to promote the research and development of new herbicides.

Nearly 300 species of plants had been cultivated as anti-herbicide varieties since 1998. The herbicides involved glyphosate, glufosinate, sulfonylurea, imidazolinones, bromoxynil, 2,4-d, and so on. The commercialized crops were corn, soybeans, wheat, rape, sugar beet, flax, tobacco, rice, cotton, and so on. For example, anti-glyphosate corn developed by Monsanto and anti-imidazoline wheat developed by Cyanamid and Agripro.

The use of genetically modified method is an effective way to improve the trait of corn. A large number of herbicide-resistant transgenic corn species appeared since 1988, and the introduced herbicide resistance genes are glyphosate, glufosinate, glufosinate, 2,4-D, and sethoxydim, and a significant progress has been acquired in herbicide resistant transgenic corn, such as anti-glyphosate corn in Monsanto, anti-mimicolinone corn in Cyanamid, anti-glyphosate corn in Agripro, and anti-sethoxydim corn in BASF. As to anti-glyphosate corn, the mechanism is to inhibit the activities of 5-enolpyruvylshikimiate acid-3-phosphate in the synthesis of essential amino acids, which can kill weeds. Shikimic acid is an important intermediate produced in the catalyzed process of enolpyruvylshikimate phosphate synthase (EPSPS), which is sensitive to glyphosate resulting in EPSPS protein over expression in anti-glyphosate herbicides crops.

6.2.3 Green Chemical Pesticides

1. **Research area of green chemical pesticides**
Chemical pesticide is a kind of chemical agent which has the functions of insecticidal, bactericidal, virus killing, weeding, etc. According to the different targets, chemical pesticides can be divided into insecticides, fungicides, and herbicides. Chemical pesticide is still the main method to control pests and diseases due to the characteristics of rapid effects, low consumption, and easy production. The characteristics and requirements of green chemical pesticide are to target high biological activities, which are nearly nontoxic to humans and animals, harmless to natural enemies and pests, and easily degradable or low residue in nature.

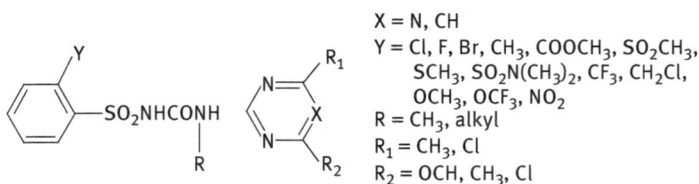

$$X = N, CH$$
$$Y = Cl, F, Br, CH_3, COOCH_3, SO_2CH_3,$$
$$SCH_3, SO_2N(CH_3)_2, CF_3, CH_2Cl,$$
$$OCH_3, OCF_3, NO_2$$
$$R = CH_3, alkyl$$
$$R_1 = CH_3, Cl$$
$$R_2 = OCH, CH_3, Cl$$

Figure 6.16: The structure of sulfonylureas.

2. Examples of the synthesis green chemical pesticides
1) Sulfonylurea herbicides

Sulfonylureas (see Figure 6.16) have been commercialized for weed control: amidosulfuron, azimsulfuron, bensulfuron-methyl, chlorimuron-ethyl, ethoxysulfuron, flazasulfuron, flupyrsulfuron-methyl-sodium, halosulfuron-methyl, imazosulfuron, nicosulfuron, oxasulfuron, primisulfuron-methyl, pyrazosulfuron-ethyl, rimsulfuron, sulfometuron-methyl sulfosulfuron, terbacil, bispyribac-sodium, cyclosulfamuron, and pyrithiobac-sodium [36]. Nicosulfuron, triflusulfuron methyl [37], and chlorsulfuron are broad-spectrum herbicides that kill plants, weeds, or pests by inhibiting the enzyme acetol acetate synthase.

The first sulfonylurea herbicide, chlorsulfuron, was developed in DuPont in 1982. After that a series of structures have been modified and developed. Sulfonylurea herbicide has a high efficiency of weeding and the dosage is about 10–100 g/km², which is 100–1,000 times higher than the traditional herbicide weeding. Sulfonylurea herbicide has low toxicity to animals; it is not accumulated in the nontarget organisms, and can be degraded by chemical and biological processes in soil. The structure of the sulfonylurea herbicide is as follows:

The sulfonylurea herbicides act on the acetyl lactate synthase (ALS) in plants, which is absorbed through the roots and leaves and conduction in plants is bidirectional. Duo to the inhibition of ALS, the synthesis of branched-chain amino acids was inhibited, and the cell division was inhibited, which led to the destruction of the normal growth of weeds. The herbicidal activity was highly correlated with the inhibition of ALS inhibition. Owing to the lack of a biosynthetic pathway of branched-chain amino acids (valine, leucine, and isoleucine) in animals which is obtained from plant products, the safety of sulfonylurea herbicides is high in animals because it inhibits the biosynthesis of branched-chain amino acid.

3) Amino acid pesticide

Glyphosate (N-(phosphonomethyl)glycine, see Figure 6.17) is a broad-spectrum systemic herbicide and crop desiccant. It is an organophosphorus compound, specifically a phosphonate. It is used to kill weeds, especially

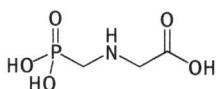

Figure 6.17: The structure of glyphosate.

annual broadleaf weeds and grasses that compete with crops. It was discovered to be an herbicide by Monsanto chemist John E. Franz in 1970. Monsanto brought it to market in 1974 under the trade name Roundup, and Monsanto's last commercially relevant United States patent expired in 2000.

Farmers quickly adopted glyphosate, especially after Monsanto introduced glyphosate-resistant Roundup Ready crops, enabling farmers to kill weeds without killing their crops. In 2007, glyphosate was the most used herbicide in the US agricultural sector and the second-most used in home and garden (2,4-D being the most used), government and industry, and commerce. By 2016 there was a 100-fold increase from the late 1970s in the frequency of applications and volumes of glyphosate-based herbicides (GBHs) applied, partly in response to the unprecedented global emergence and spread of glyphosate-resistant weeds [38].

Glyphosate is absorbed through foliage, and minimally through roots [39], and transported to growing points. It inhibits a plant enzyme involved in the synthesis of three aromatic amino acids: tyrosine, tryptophan, and phenylalanine. Therefore, it is effective only on actively growing plants and is not effective as a pre-emergence herbicide. An increasing number of crops have been genetically engineered to be tolerant of glyphosate (e.g., Roundup Ready soybean, the first Roundup Ready crop, also created by Monsanto) which allows farmers to use glyphosate as a post emergent herbicide against weeds. The development of glyphosate resistance in weed species is emerging as a costly problem. While glyphosate and formulations such as Roundup have been approved by regulatory bodies worldwide, concerns about their effects on humans and the environment persist [40].

3) **Fluoride pesticide**
The number of N-containing heterocyclic compound is about 70% in highly efficient pesticides, and the number of fluorine-containing compounds in nitrogen-containing heterocyclic pesticides is nearly 70%.

Based on the theory of bioisosterism, the fluorine-containing pesticides had been obtained instead of H, Cl, Br, CH_3, and OCH_3 in the original structures with fluorine or fluorine groups (such as CF_3, OCF_3, $OCHF_2$), such as fungicides fluoroquinazolinone, instead of H atom with fluoride atom in quinazolinone, or CH_3 with CF_3 in diphenyl ether herbicide, or H or Cl with F or CF_3 in pyrethroid insecticide such as cypermethrin and fenvalerate.

The introduction of fluorine atom in the structure of pesticide will increase the lipophilicity of compounds, and fluoride and hydrogen are not

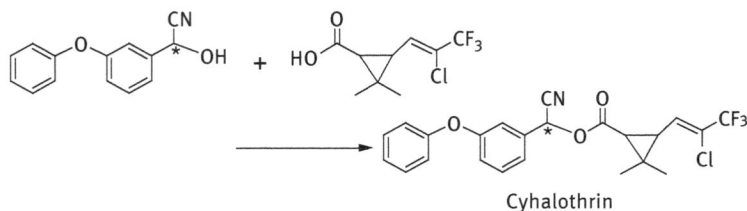

Cyhalothrin

Figure 6.18: The preparation of glyphosate.

easily identified by the receptor resulting in irreversible inactivation of the receptors, or even prevent the metabolism of organisms. So, the bioactivities were higher than the corresponding fluorine free compound.

Cyhalothrin (see Figure 6.18) is an organic compound that is used as a pesticide [41]. It is a pyrethroid, a class of man-made insecticides that mimic the structure and insecticidal properties of the naturally occurring insecticide pyrethrum which comes from the flowers of *Chrysanthemum*. Synthetic pyrethroids, like lambda-cyhalothrin, are often preferred as an active ingredient in insecticides because they remain effective for longer periods of time. It is a colorless solid, although samples can appear beige, with a mild odor. It has a low water solubility and is nonvolatile. It is used to control insects in cotton crops [42].

6.2.4 Green Pesticide Preparations

Generally, pesticides cannot be used directly; it must be prepared from various types of formulations and be used. Usually the pesticide preparations are processed by the original pesticide and additives. Pesticide additives are mainly carriers and surfactants. The carrier of liquid pesticide preparation is mostly with aromatic hydrocarbon as organic solvent, and solid pesticide formulations of the carrier is mainly porous material with large specific surface. Pesticide preparation needs to add surfactants to play the role of emulsification, dispersion, adhesion, wetting, penetration, etc.

The new research area of pesticide formulation is water-based, ultramicro, dust-free, and controlled-release. Due to the toxic aromatic solvents, the pesticide formulations are no longer registered using toluene and xylene as solvent. In the design of green pesticide formulation, it is necessary to take into account the wetting, dispersing, enhancing spreading, and improving the efficacy of the surfactant, as well as the sustainable development. Therefore, many natural or botanical surfactants (alkyl glucoside, sorbitol ester, plant source surfactant) are gradually replacing the mineral surfactants. Environmental-friendly pesticide formulations include aqueous formulations, particulate formulations, high concentration butter, a high content of powder preparation, vegetable oil suspension agent, etc.

1. **Microemulsion of pesticides**

    Microemulsion pesticide is a kind of liquid pesticide produced by liquid or solvent, which is dispersed in water by 10–100 nm particles under the action of surfactant such as emulsifier and dispersant. Microemulsion does not contain or contain a small amount of organic solvents. A mixture of nonionic surfactant or nonionic surfactant and anionic surfactant is added in the microemulsion without or containing a small amount of organic solvent. Microemulsion in the appearance of a transparent or translucent homogeneous system looks like the real solution; in fact, it is essentially oil dispersed in water emulsion. Compared with other dosage forms, microemulsion has the advantages of good stability, good adhesion, strong penetration, high safety, and high efficiency.

    Microemulsion can prevent harm to the human body and the environment, as well as improve the efficacy of pesticides. So it is regarded as green pesticide formulations. The research on the microemulsions of organophosphorus pesticides such as malathion, parathion, diazinon, and phorate was carried out in the United States and Japan, and the thermal stability of the active component was disposed.

2. **Microcapsule of pesticides**

    Microcapsule is a new pesticide formulation which packs the pesticide active component in polymer bag, the particle size of the capsule is about several microns to several hundred microns. When the microcapsules are spread on fields or exposed to the insects surface, the capsule breaks, dissolves, hydrolyzes, or spreads through the wall, then the encapsulated pesticide is released slowly, which prolongs the drug residue life, reduces the frequency of spraying, and reduces environmental pollution.

    A modified spinosad has been synthesized in Clarke Company, which is effective for killing mosquito larvae. Spinosad is a kind of environmental safe insecticide but unstable in water, which limited its application and is ineffective for killing the larvae of mosquito in water. Clarke developed a method of embedding spinosad in gypsum matrix, so that the spinosad can release slowly into the water to control mosquito larvae effectively. The efficacy of the larvicide is 2–10 times higher than that of the traditional insecticide and the toxicity is 15 times less than organic phosphorus compounds. In addition, it leaves no residue in the environment and is nontoxic to wild animals.

    The substrate of the microcapsules is gypsum which is formed by water-insoluble calcium sulfate and water. By adding different amount of hydrophilic polyvinyl alcohol, the release time of pesticides can be adjusted. Polyvinyl alcohol dissolves slowly, exposing the pesticide and calcium sulfate to water, and calcium sulfate absorbs water to form gypsum and releases the pesticide. Due to the achievement, Clarke received the Green Chemicals Award of 2010.

6.3 Green Functional Materials

6.3.1 Polyaniline Materials

Polyaniline (PANI, see Figure 6.19) is a conducting polymer of the semi-flexible rod polymer family. Although the compound itself was discovered over 150 years ago, only since the early 1980s polyaniline has captured the intense attention of the scientific community. This interest is due to the rediscovery of high electrical conductivity. Among the family of conducting polymers and organic semiconductors, polyaniline has many attractive processing properties. Because of its rich chemistry, polyaniline is one of the most studied conducting polymers in the past 50 years [43].

Polyaniline was prepared by oxidation polymerization of aniline by chemical oxidation or electrochemical anodization. The most extensive oxidation system is persulfate system. In the medium of strong acidic solution, aniline reacts in a "head-to-tail" connection to obtain a polymer production with conductivity and electrochemical activity, the kind of anion in the solution, the reaction temperature, the concentration of aniline, and the reaction environment will have impacts on aniline polymerization. In the preparation of polyaniline using electrochemical methods, the product structure and performance also depend on the electrode material and surface morphology [44].

Electrochemical anodic oxidation reaction requires selecting an appropriate electrochemical condition in the electrolytic solution containing aniline. In the electrolytic cell which contains sand core, the platinum electrode of 1 cm × 1 cm is used as the electrode and the graphite rod is used as the auxiliary electrode and controls the platinum potential at 0.8 V and the aniline will polymerize in an acidic aqueous solution.

The oxidative polymerization of aniline occurred on the anode and is deposited on the electrode surface to form powder. The reaction process follows the electrochemical→chemical→electrochemical mechanism. The commonly used oxidation system is a persulfate, such as aluminum persulfate. The oxidative polymerization of aniline is processed in accordance with the principle of free radicals.

Polyaniline conductive material can show many unique advantages such as mechanics, electrical and optical properties, and also great potential properties in antistatic, electromagnetic shielding, electrode materials, and antifouling anticorrosive coatings. Polyaniline has not only good conductivity and electrochemical properties but also good environmental stability, unique optical and catalytic properties, and electrochromic properties. The application of polyaniline is very extensive, such as secondary

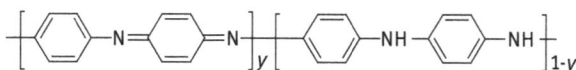

Figure 6.19: The structure of polyaniline.

lithium battery, electrochromic devices, sensor for a volatile organic compounds, anti-static packaging materials, light emitting diodes, and gas separation film [45].

6.3.2 Graphene

Graphene is a new kind of two-dimensional materials with carbon atoms tightly packed into honeycomb lattice structure. It is a planar thin film composed of six corners of a honeycomb lattice by carbon atoms with sp2 hybrid orbitals. Graphene has been once considered as a hypothetical structure, cannot exist independently. Until 2004, physicist Andre Geim and Konstantin Novoselov Sholov have successfully isolated graphene from graphite in the University of Manchester; the excellent electrical, optical and mechanical properties have attracted many attentions, and the two men have won the Nobel Prize in Physics in 2010 for their "pioneering work on 2D graphene materials" [46].

Graphene is currently the thinnest but also the most hard nano materials in the world, the thermal conductivity is higher than that of carbon nanotubes and diamonds. Studies show that as graphene has high conductivity, it is the smallest resistivity material in the world under room temperature. Its electron transfer is extremely fast and be expected to develop a kind of thinner, faster-conductive, new-generation electronic components, or transistors. Graphene is essentially a transparent, good conductor, also suitable for the production of transparent touch screen, light plate, and even solar cells [47].

1. **Structures of graphene**

   Graphene is a crystalline allotrope of carbon with 2-dimensional properties. Its carbon atoms are densely packed in a regular atomic-scale chicken wire (hexagonal) pattern [48]. It can be bent into zero dimensional (0D) fullerene, curled one-dimensional (1D) carbon nanotubes, or stacked into three-dimensional graphite. Therefore, graphene is the basic unit of other graphite materials.

   The basic structure of graphene is the most stable six-membered ring of benzene in organic structure which is the most optimal two-dimensional nano-material. The ideal graphene structure is a planar hexagonal lattice, which can be regarded as a layer of exfoliated graphite molecules, where each carbon atom is sp2 hybrid, and contributes a residual p orbital electron for the formation of large π bond, electrons in the large π bond can move freely endowing graphene a good electrical conductivity. The two-dimensional graphene structure can be viewed as the basic unit of the formation of all sp2 hybrid carbonaceous materials (Figure 6.20).

2. **Physical properties of graphene**

   Graphene is one of the strong materials in the world, harder than the diamond and the strength is 100 times higher than the best steel. If physicists produce a

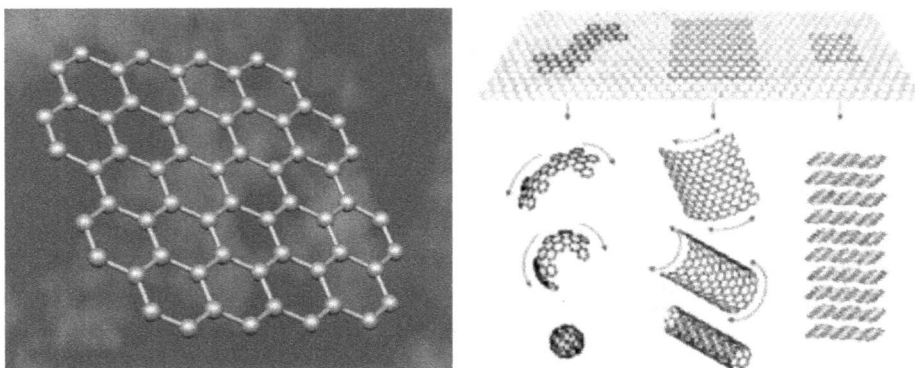

Figure 6.20: The structure of graphene.

king of graphene with the thickness equivalent to ordinary plastic food packaging bag (thickness of about 1 million nm), it needs almost 2×10^4 N pressure for breaking it. In other words, if the bag is made of graphene, it will be able to carry about 2 tons of goods [49].

Graphene is the best conductive material in the world, the electrons can move efficiently in graphene, while for conventional semiconductors and conductors, such as silicon and copper, the conductivity are not better than graphene. Due to the collision of electrons and atoms, the traditional semiconductor and conductor release some energy in the form of heat, but graphene is different; the electrons don't loss energy, which makes them have an excellent electrical characteristics.

3. **Applications of Graphene**

   1) **Application in nanoelectronic devices**

      Geim and Kim found that graphene had 10 times higher carrier mobility (about 10 amps/Vs) than commercial silicon devices in 2005, which is hardly affected by temperature, doping, showing ballistic transport properties of sub-micron scale at room temperature (up to 0.3 K under 300 m), demonstrating prominent advantage as a nanoelectronic device of graphene, and making the room temperature ballistic effect transistor which is more attractive in the field of electronic engineering become possible [50].

      Larger fermi velocity and low contact resistance will help to further reduce the switching time of the device, and the super high frequency operating response characteristic is another significant advantage of the graphene based electronic devices. In addition, the scale of graphene reduced to nanoscale or even a single benzene ring also maintains good stability and electrical properties, which make it possible to explore single-electron devices.

   2) **Solar cells**

      By covering graphene on the surface of traditional single crystal silicon, it is found that the crystal silicon has an excellent photoelectric performance. As a simple solar cell model, the photoelectric conversion efficiency can reach more than 10% after being optimized by graphene. The graphene-silicon model can be further extended to the structure of graphene and other semiconducting materials. This model, which can combine graphene with traditional materials, plays an important role in the practical application of graphene [51].

3) **Photon sensor**

Graphene can also be the appearance of photon sensors; this sensor is used to detect the information carried in the fiber. Up to now, the role has been played by the silicon, but the silicon era seems to end. In 2012, a research team at IBM revealed for the first time their graphene photodetectors, graphene-based solar cells, and liquid crystal displays. Because graphene is transparent, an electric panel manufactured therewith has a superior light transmittance than other materials.

A graphene/n-type silicon heterojunction has been demonstrated to exhibit strong rectifying behavior and high photoresponsivity. By introducing a thin interfacial oxide layer, the dark current of graphene/n-Si heterojunction has been reduced by two orders of magnitude at zero bias. At room temperature, the graphene/n-Si photodetector with interfacial oxide exhibits a specific detectivity up to 5.77×10^{13} cm Hz 1/2 W^2 at the peak wavelength of 890 nm in vacuum. In addition, the improved graphene/n-Si heterojunction photodetectors possess high responsivity of 0.73 A/W and high photo-to-dark current ratio of $\approx 10^7$. These results demonstrate that graphene/Si heterojunction with interfacial oxide is promising for the development of high detectivity photodetectors [52].

4. **Preparation of graphene**

There are many methods for the preparation of graphene such as mechanical stripping, chemical vapor deposition, oxidation-reduction, organic synthesis, and carbon nanotube stripping. In 2004, Geim obtained single layer graphene using micro mechanical stripping method from highly oriented pyrolytic graphite for the first time. Through this method high quality graphene can be prepared, but it has the defects of low yield and high cost, and it cannot meet the requirements of industrialization.

Chemical vapor deposition (CVD) is a chemical process used to produce high quality, high performance, solid materials deposited on a heated solid substrate surface. The process is often used in the semiconductor industry to produce thin films. High quality and large area graphene can be prepared by CVD method. CVD method can meet the requirement of large-scale preparation of high quality graphene, but the cost is higher and the process is complex [53].

Oxidation-reduction process is one of the best methods for preparing graphene with low cost and wild conditions. The process oxides natural graphite with strong acid and oxidizing agents to graphite oxide (GO), then reduced by reducing agent to remove oxygen groups of graphene on the surface [54].

The high-quality graphene is obtained by the reduction of graphite oxide using environmental-friendly reductant VC or sodium citrate to avoid environmental pollution and reduce the use of chemical reagent [55].

6.4 Green Electronic Chemicals

Electronic chemicals usually refer to special fine chemicals in supporting electronics industry. The main products include packaging materials, photoresist, super pure chemical reagents, specialty gases, silicon wafer polishing material matched for integrated circuits, substrate resin, resist dry film and cleaning agent matched for printed circuit boards, liquid crystal, polarizing plate matched for liquid crystal display, and so on. Electronic chemicals have strict quality requirements on the cleanliness of environmental requirements, packaging, transportation, and storage. The product replacement is fast, the investment is high, the research is difficult, and the products are high value-added.

6.4.1 Photoresist

The radiation anticorrosion additive (photoresist) refers to an anticorrosion thin film material that changes its solubility by irradiation or irradiation with UV light, electron beams, excimer laser beams (KrF248 nm and ArF193 nm), X-rays, and ion beams. A positive resist is increased by exposure and development, and a negative-type photoresist is reduced by solubility [56].

The main varieties of photoresist are polyvinyl alcohol cinnamic acid ester gum, polyethylene terephthalate malonic acid ethylene glycol polyester rubber, cyclized rubber-type purchase glue (equivalent to OMR-83 plastic), and diazonaphthoquinone sulfonyl chloride as sensitizer, the main body of the UV-positive photoresist. UV-negative plastic has been made in China; UV glue can meet the 2 μm process requirements, deep UV-positive and -negative plastic (polymethyl isopropenyl ketone, chloromethyl polystyrene, resolution 0.5–0.3 μm). Electron beam is poly(methyl methacrylate-glycidyl methacrylate-co-ethyl acrylate) (resolution 0.25–0.1 μm) and x-ray positive gel (polybutylene sulfone-poly (1,2-dichloroacrylic acid). Resolution is 0.2 μm.

In order to meet the requirements of submicron graphics processing technology in microelectronics industry, in addition to the near UV (g-line, 436 nm), the far UV (i-line, 365 nm), excimer laser beam (KrF 248 nm and ArF 193 nm), electron beam, X-ray, and ion beam and etc. are also applied in the development of new radiation resistant materials. The developed materials are called UV engraved rubber (UV positive photoresist, ultraviolet negative photoresist), carved deep UV glue, X-ray beam glue, ion beam glue et al., respectively. Current emphasis are focused on the development of i-line photoresist and electron beam chemically amplified resist (CAR) glue.

CAR is based on poly (4-hydroxystyrenes) as the optimal chemical platform, adding the appropriate photoacid generator, cross-linker, and other components. In the radiation source exposure, the photochemical gain is up to 10^2–10^8, by which CAR can achieve a high sensitivity (1–50 mJ/cm^2), and get high-precision graphics.

In recent years, photoresist in the microelectronics industry has another new type of use that is the use of photosensitive media materials for multi-chip components (MCM). MCM technology can significantly reduce the size of the electronic system, reduce its weight, and improve its reliability. MCM technology has been widely used in advanced military electronic and aerospace electronic equipment in abroad. In the middle of 1990s, BCB (styrene-propylene) is a unique thermosetting resin. BCB is mainly used for MCM processing and IC Ga/As interconnection, its dielectric constant has reached 2.65; It is better than polyimide dielectric material, and has a lower water absorption, curing temperature, better flatness, heat resistance, and other characteristics. Therefore, BCB is considered to be the second-generation silicon-based MCM dielectric material after SiO$_2$ and polyimide [57].

6.4.2 Polyimide Materials

Polyimides (PI, see Figure 6.21) are a class of polymers with imide functional groups in the molecular main chain structure, including aliphatic PI and aromatic PI. PI is a polymer of imide monomers. Polyimides have been in mass production since 1955. With their high heat resistance, polyimides enjoy diverse applications in roles demanding rugged organic materials, e.g., high-temperature fuel cells, displays, and various military roles. A classic polyimide is Kapton, which is produced by condensation of pyromellitic dianhydride and 4,4'-oxydianiline [58].

Polyimides are often prepared by the condensation of two aromatic anhydride and two aromatic amines. The polyamic acid precursor was obtained first and then polyimide by imidization or cyclodehydration was obtained. Polyimide material is heat stable polymer material with excellent properties. The synthesis process is as shown in Figure 6.22.

Polyimides have excellent electrical properties and mechanical properties, high thermal stability, thermal oxidation and chemical stability, thermal expansion coefficient, good solvent resistance, dimensional stability and good processability, high precision and easy molding of complex shape, and many other excellent properties,

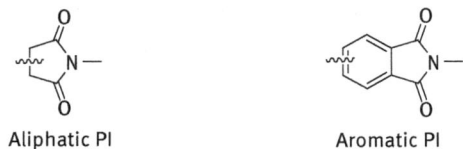

Aliphatic PI   Aromatic PI   **Figure 6.21:** The structure of polyimides.

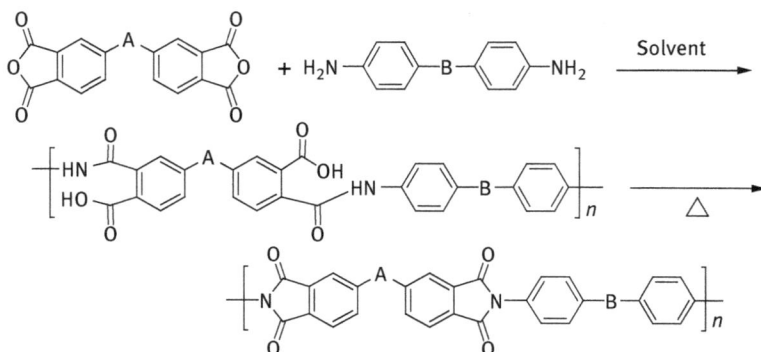

Figure 6.22: The preparation of polyimides.

the product can be used for the preparation of high performance plastic products in aerospace, electronics, nuclear power, communications, automobile, and other cutting-edge technology field, and it has become the main material with good prospect of development.

In the microelectronics industry, polyimide materials are mainly used in packaging materials, in contact coating film, to brush the substrate and adhesive materials, and in other aspects of the substrate. With the deepening of the traditional polyimide modification work, many emerging technologies and industries continue to emerge. Development of polyimide materials research has become a hot area.

6.4.3 Epoxy Molding Compound

Epoxy molding compound (EMC) is composed of epoxy resin and its curing agent (phenolic resin) and other components of the molding powder. When kept under the influence of heat it will cross-link curing into a thermosetting plastic; in the injection molding process, it will package the semiconductor chip, give it a certain structure shape, and make it become a semiconductor device. The method of plastic packaging which is used to product transistors, integrated circuits (IC), large-scale integrated circuits (LIC), very large-scale integrated circuits (VLIC), and other LICs have been widely used and become mainstream in the domestic and foreign. The diagram of curing reaction of epoxy resin and curing agent is as shown in Figure 6.23.

The epoxy-molding compound is composed of o-cresol formaldehyde epoxy resin, linear phenolic resin, silica powder, accelerator, coupling agent, modifier, mold release agent, flame retardant, and colorant. o-Cresol formaldehyde epoxy resin is used as an adhesive, the curing agent is linear phenolic resin, with other components in a certain proportion. Under the action of heat and curing accelerator, the epoxy group of epoxy resin has a high reactivity. The solidified epoxy-molding compound

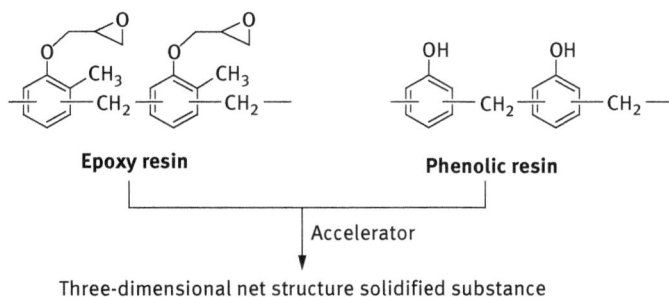

Epoxy resin　　　　　　**Phenolic resin**

Accelerator

Three-dimensional net structure solidified substance

Figure 6.23: The curing reaction of epoxy resin and curing agent.

has excellent adhesion, excellent electrical insulation properties, high mechanical strength, heat resistance, good chemical corrosion resistance, low water absorption, and low molding shrinkage.

According to the difference in packaging materials, electronic packaging materials can be divided into plastic packaging, ceramic packaging, and metal packaging. The latter two are airtight packaging, mainly used in aerospace, aviation, and military fields, and plastic packaging is widely used in civilian field. Nowadays, 90% of the semiconductor devices are packaged in plastic, and the plastic packaging materials are more than 90% in epoxy resin, which indicate that epoxy-molding compound has become an important support of the semiconductor industry.

6.4.4 Green Battery Materials

1. Lithium iron phosphate

Lithium iron phosphate, also known as LFP, is an inorganic compound with the formula $LiFePO_4$. It is a gray, red-gray, brown, or black solid that is insoluble in water. The material has attracted attention as a candidate component of lithium iron phosphate batteries, which are related to Li-ion batteries. It is targeted for use in power tools, electric vehicles, and solar energy installations [59].

Most lithium batteries (Li-ion) used in 3C (computer, communication, consumer electronics) products use cathodes of other lithium materials, such as lithium cobalt oxide ($LiCoO_2$), lithium manganese oxide ($LiMn_2O_4$), and lithium nickel oxide ($LiNiO_2$). The anodes are generally made of carbon. Lithium iron phosphate exists naturally in the form of mineral triphylite, but such material has insufficient purity for use in batteries.

Compared to the traditional cathode material of Li-ion secondary battery, $LiFePO_4$ has a wider range of raw material sources, lower cost, and no environmental pollution. Lithium iron phosphate is the most secure lithium-ion battery cathode material, does not contain any harmful heavy metals. Iron lithium iron

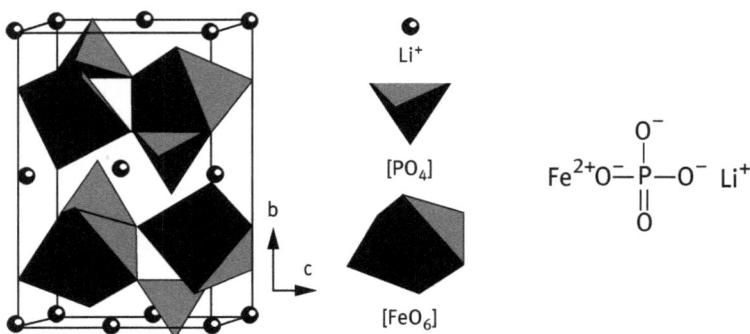

Figure 6.24: The crystal structure of lithium iron phosphate ($LiFePO_4$).

phosphate crystal lattice is stable, under the condition of 100% DOD, it can charge and discharge more than 2,000 times, which has good reversibility. Lithium iron phosphate materials have high capacity, high discharge power, long cycle life, good thermal stability, and high temperature performance advantages, has become the powerful lithium-ion battery of choice for high-security cathode material.

1) **The structure of $LiFePO_4$**

$LiFePO_4$ is a polyanionic lithium salt with olivine structure as shown in Figure 6.24. The spatial position of Li ions, PO_4 tetrahedra, and FeO_6 octa-hedra is shown in $LiFePO_4$ crystals. The crystal has a hexagonal compact packing of oxygen atoms, the lithium atom and the iron atom are located at the 4c and 4a positions of the oxygen octahedron respectively. LiO_6 octa-hedra in the (010) surface and the surrounding four FeO_6 octahedra are in a co-edge way to form a zigzag-shaped plane. The phosphorus atom is located in the 4C of the tetrahedral site, each PO_4 with a FeO_6 octahedron has a common point, with another FeO_6 having a common edge and a common point. Li^+ is in the 4a position to form a straight-line linear chain, parallel to the c-axis, which makes Li^+ to be free from the embedded and embedded in charge-discharge process. The covalent bond of P-O in $LiFePO_4$ crystal has strong thermodynamic and kinetic stability during the charge-discharge procedures.

2) **Electrochemical properties of $LiFePO_4$**

LFP batteries have an operating voltage of 3.3 V, charge density of 170 mAh/g, high power density, long cycle life, and stability at high temperatures. In $LiFePO_4$, lithium has +1 charge and iron has +2 charge balancing the 3-charge for phosphate. Upon the removal of Li, the material converts to the ferric form $FePO_4$ [60].

The iron atom and 6 oxygen atoms form an octahedral coordination sphere, described as FeO_6, with the Fe ion at the center. The phosphate groups, PO_4, are tetrahedral. The three-dimensional framework is formed by the FeO_6 octahedra sharing O corners. Lithium ions reside within the octahedral channels in a zigzag manner. In crystallography, this structure is thought to belong to the P_{mnb} space group of the orthorhombic crystal system. The lattice constants are as follows: $a = 6.008$ Å, $b = 10.334$ Å, and $c = 4.693$ Å. The volume of the unit cell is 291.4 Å.

In contrast to two traditional cathode materials $LiMnO_4$ and $LiCoO_2$, lithium ions of $LiMPO_4$ migrate in the lattice's one-dimensional free volume. During charge/discharge, the lithium ions are extracted concomitant with oxidation of Fe:

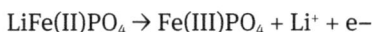

$$LiFe(II)PO_4 \rightarrow Fe(III)PO_4 + Li^+ + e-$$

The process is reversible:

$$Fe(III)PO_4 + Li^+ + e- \rightarrow LiFe(II)PO_4$$

Extraction of lithium from $LiFePO_4$ produces $FePO_4$ with a similar structure. $FePO_4$ adopts a P_{mnb} space group with a unit cell volume of 272.4, only slightly smaller than its lithiated precursor. Extraction of lithium ions reduces the lattice volume, as is the case with lithium oxides. $LiMPO_4$'s corner-shared FeO_6 octahedra are separated by the oxygen atoms of the PO_4^{3-} tetrahedra and cannot form a continuous FeO_6 network, and thus reducing conductivity.

A nearly close-packed hexagonal array of oxides centers provides relatively little free volume for Li^+ ions to migrate within. For this reason, the ionic conductivity of Li^+ is relatively low at ambient temperate. The details of the lithiation of FePO4 and the delithiation of LiFePO4 have been examined. Two phases of the lithiated material are implicated [60, 61].

Although LFP has 25% less capacity than other lithium batteries due to its material structure, it has 70% more than nickel-hydrogen batteries. The major differences between LFP batteries and ordinary lithium batteries are that LFP batteries do not have safety concerns such as overheating and explosion, they have 4–5 times longer cycle lifetimes than the lithium batteries, and 8 to 10 times higher discharge power than lithium batteries.

LFP batteries have drawbacks including, higher costs given less time on the learning curve. The energy density is significantly lower than $LiCoO_2$. Lithium cobalt oxide has multiple disadvantages. It is one of the most expensive components of traditional li-ion batteries. The cobalt in $LiCoO_2$ is listed as a possible human carcinogen by IARC. $LiCoO_2$ can experience problems with runaway overheating and outgassing, particularly in lithium polymer battery packs. LFP batteries do not need the sophisticated charge monitoring in traditional Li-ion, but still benefit from balance charging on a regular basis, especially in repeated high-current discharge rate uses like electric-powered flying model aircraft.

One important advantage over other lithium-ion chemistries is thermal and chemical stability, which improves battery safety. LiFePO4 is an intrinsically safer cathode material than LiCoO2 and manganese spinel. The Fe–P–O bond is stronger than the Co–O bond, so that when abused, (short-circuited, overheated, etc.) the oxygen atoms are much harder to remove. This stabilization of the redox energies also helps fast ion migration.

As lithium migrates out of the cathode in a $LiCoO_2$ cell, the CoO_2 undergoes nonlinear expansion that affects the structural integrity of the cell. The fully lithiated and unlithiated states of $LiFePO_4$ are structurally similar which means that $LiFePO_4$ cells are more structurally stable than $LiCoO_2$ cells. No lithium remains in the cathode of a fully charged $LiFePO_4$ cell – in a $LiCoO_2$ cell, approximately 50% remains in the cathode. $LiFePO_4$ is highly resilient during oxygen loss, which typically results in an exothermic reaction in other lithium cells.

As a result, lithium iron phosphate cells are much harder to ignite in the event of mishandling (especially during charge) although any fully charged battery can only dissipate overcharge energy as heat. Therefore, failure of the battery through misuse is still possible. It is commonly accepted that LiFePO4 battery does not decompose at high temperatures. The difference between LFP and the LiPo battery cells commonly used in the aeromodelling hobby is particularly notable.

3) **Synthesis of LiFePO$_4$**

There are many methods for the preparation of $LiFePO_4$ for lithium ion batteries, including: solid-phase synthesis, emulsion drying, sol–gel process, solution coprecipitation, vapor phase deposition, electrochemical synthesis, electron beam irradiation, microwave process, hydrothermal synthesis, ultrasonic pyrolysis, and spray pyrolysis [62].

High-temperature solid-phase method is characterized by mixing the raw materials according to the chemical ratio. In order to improve the homogeneity of mixing of the various materials, the raw material mixture should be wet-milled in the presence of an organic solvent, or the raw material mixture should be ball-milled in dry state under an inert gas atmosphere. The raw material mixture after ball milling is decomposed into $LiFePO_4$ precursor by heating in an inert atmosphere at 300–400 °C and then pressed into a ball or into pieces, then the solid-phase reaction takes place at 600–800 °C in an inert gas atmosphere to form $LiFePO_4$.

High-temperature solid-state method has disadvantages such as high cost of raw materials, high technology, high energy consumption, equipment corrosion, and so on. At present, the improvement of high-temperature solid-state method mainly focuses on the influence of reaction materials on the electrochemical performance of $LiFePO_4$, optimization of process parameters, in situ surface coating process, or in situ doping structure modification process.

Sol–gel method is the first preparation of the corresponding material of the aqueous solution, in the presence of complexing agent and stabilizer conditions, controlling of the reaction between the material solutions forms a stable sol system. A solid gel is made by heating, adding electrolyte, adding complexing agent, and destroying the stability of the sol. The olivine structure of $LiFePO_4$ was prepared by heating the solid gel through one-step heating or two steps. Sol–gel method has the shortcomings of high production cost, long cycle time, and large-scale production difficulty.

The chemical coprecipitation method first coprecipitates the in-situ elements of each material into a solid-state mixture precursor in the form of a plurality of insoluble compounds. Compared with the direct ball-milling method, the coprecipitation method is used to coprecipitate the in-situ elements in each material in the form of coprecipitated insoluble compounds, and the mixing degree is high. However, compared with the sol-gel method, the coprecipitation method precursors in the various elements of the mixing level are still less than molecular-level mixing.

The formation process of the coprecipitation mixture is a rapid reaction process, and the production cycle of the precursors is short, which is suitable for large-scale industrial production. The insoluble compounds of the respective elements in the coprecipitate mixture may be either inorganic or organic compounds and are not only suitable for the ferrous iron compound material but also for the trivalent iron compound raw material. It is no longer required that all of the extraneous elements in the feed material must eventually escape from the reaction system in the form of gaseous products, extending the range of choice for the iron, phosphorus, and phosphonium containing compounds. These extraneous ions do not affect the purity or structure of the final product as long as the extraneous ions in these compounds can exist in the solution in soluble state and can be separated from solid phase. Since the process of preparation of the coprecipitated precursor does not require a high-temperature heating process, this aspect reduces the energy consumption and simplifies the production equipment. The preparation process of the precursor is carried out in solution, and the oxidation of the ferrous iron does not affect the purity of the final product, and the inert gas protective property can be omitted; thus simplifying the process complexity.

4) **Application of LiFePO$_4$ battery**

$LiFePO_4$ battery will be widely used in many area such as large electric vehicles: buses, electric cars, and hybrid vehicles; light electric vehicles: electric bicycle, golf cart, small flat battery car, forklift, cleaning cars, and electric wheelchairs; power tools: electric drills, saws, lawn mowers; remote control cars, boats, aircraft toys, etc.; solar and wind power energy storage equipment: UPS electric source and emergency lights, warning lights and Miner's lamp small medical equipment, portable instrument, etc.

Lithium iron phosphate as a lithium battery material is developed in recent years, the safety and cycle life is better than other materials: Charge-discharge cycle life of up to 2,000 times, a single 30 V battery charging over-voltage does not burn, puncture no explosion. All those are also the most important technical parameters of power battery. Large capacity lithium-ion battery made by lithium iron phosphate cathode material is easier to use in series connection. And it meets the needs of electric vehicles frequent charge and discharge. It is a nontoxic and nonpolluting material. It has a good and safe performance, wide range of raw material sources, cheap price, and long service life as advantages. It is an ideal cathode material for the new generation of lithium ion battery.

2. **Lithium hexafluorophosphate**

Lithium hexafluorophosphate is an inorganic compound with the formula LiPF6 widely used in lithium battery electrolyte, has good conductivity and electrochemical stability in power batteries, energy storage batteries, and other daily batteries. $LiPF_6$ is an irreplaceable electrolyte for lithium-ion batteries because of its conductivity, cost, safety, and environmental impact [63].

$LiPF_6$ is in white crystal or powder, the relative density is of 1.50, with strong deliquescence and solubility in water, it is also soluble in low concentrations of methanol, ethanol, acetone, carbonates, and other organic solvents. When in air or heating, $LiPF_6$ will be decomposed rapidly and will release PF_5 and white smoke. At present, the preparation of $LiPF_6$ are mainly gas-solid methods, hydrogen fluoride solvent methods, complex methods, and solution methods.

1) **Gas–solid method**

In a nickel vessel, PF_5 gas and LiF solid were directly reacted to obtain solid $LiPF_6$. The reaction formula was [64]

$$PF_5(g) + LiF(s) \rightarrow LiPF_6(s)$$

The whole reaction is carried out under high temperature and high pressure without solvent, but the reaction is insufficient and the yield is low, so it is difficult to achieve large-scale production. In order to solve the problem, it was proposed that HF gas is reacted with LiF solid to produce $LiHF_2$ solid, then HF in Li-HF$_2$ is removed under reduced pressure at 600–700 °C to obtain highly active porous LiF. And then LiF and PF_5 gas reacted at 200 °C to produce $LiPF_6$, the reaction equation is as follows:

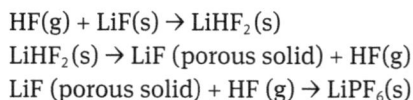

$$HF(g) + LiF(s) \rightarrow LiHF_2(s)$$
$$LiHF_2(s) \rightarrow LiF \text{ (porous solid)} + HF(g)$$
$$LiF \text{ (porous solid)} + HF(g) \rightarrow LiPF_6(s)$$

The purity of the product can be reached to 99.9%. Due to more production steps and high cost, it is difficult to prepare homogeneous porous LiF during the reaction. Therefore, the application of gas-solid method is limited, and it is difficult to produce large-scale continuous production.

2) **Hydrogen fluoride solvent method**

As PF_5, LiFs are easy to dissolve in HF, so use HF solvent as a medium for homogeneous reaction. The steps are as follows:

(1) LiF is dissolved in anhydrous HF to form LiF solution.

(2) PF_5 gas is blown into the solution so that LiF and PF_5 react to form $LiPF_6$ solution; the solution was recrystallized by cooling to low-temperature, filtered, and dried to obtain high purity LiPF6 crystals. The equation is as follows:

$$PF_5(g) + LiF(s) \rightarrow LiPF_6(s)$$

Since PF_5 is more expensive, cheaper PCl_5 is usually used as a raw material, and the gaseous PCl_5 is reacted with HF to produce high purity PF_5, and then reacted with LiF dissolved in HF to produce $LiPF_6$. The reaction equation is as follows:

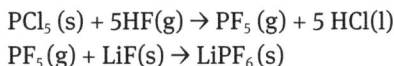

$$PCl_5(s) + 5HF(g) \rightarrow PF_5(g) + 5\,HCl(l)$$
$$PF_5(g) + LiF(s) \rightarrow LiPF_6(s)$$

Another advantage of preparing PF_5 by reaction of PCl_5 with HF is that the product PF_5 is gas, so the metal impurities are easily separated from PCl_5 to obtain highly pure PF_5. Use HF as solvent, the start materials HF and PF_5 are highly corrosive; therefore, it is necessary for high material quality requirements of the equipment and safety measures. In addition, the process needs to be carried out at low temperatures with larger energy consumption. However, the reaction can be controlled easily and high purity $LiPF_6$ products can be obtained after crystallization. At present, the producing of $LiPF_6$ mainly employs anhydrous HF as solvent in the industry.

3) **Complexing method**

LiF suspended in organic solvent reacted with PF_5, and the resulting $LiPF_6$ dissolved immediately in the organic solvent to form stable complexes, promoting the reaction to continue. At the end of the reaction, the unreacted LiF and other impurities are insoluble in the organic solvent, so the content of impurities are reduced. The organic solvent that can be used to form complexes with $LiPF_6$ is mainly low carbon alkyl ethers, low carbon alkyl esters, and acetonitrile.

The advantage of using organic solvent as the medium is to avoid the corrosion of the equipment; the price of organic solvent is cheap and easy to obtain. The disadvantage is that the raw material PF_5 is easy to react with the organic solvent to produce other impurities, and the complex formed by $LiPF_6$ and organic solvent is difficult to decompose, which increases difficulties in the subsequent purification and affects the final product purity [65].

4) Solution method

The nonaqueous organic solvents commonly used in lithium ion battery electrolytes include dimethyl carbonate, diethyl carbonate, ethyl methyl carbonate, and 1,2-dimethoxy- ethane.

By solution method, LiF is suspended in the organic solvent first, which is used for the electrolyte of lithium ion battery, then blow the PF_5 gas into the solution. In the gas–solid reaction, the product $LiPF_6$ can be dissolved in organic solvent in time, so that the interface is continuously updated which improved the reaction efficiency. The obtained solution is directly used as the electrolyte of lithium ion battery. The reaction temperature is generally controlled between −40 and 100 °C. If the temperature is low, the solvent will froze and cannot carry out the reaction effectively; if the temperature is high, the solvent is easy to react with PF_5, resulting in the change of color and the increase of viscosity of the solution. In general, the yield of the reaction is high and easy to be controlled.

Solution method avoids the use of highly corrosive anhydrous HF as solvent, but the use of PF_5 still has a strong corrosive property, and it easily reacts with the solvent to produce other impurities, affecting the product purity. In addition, the use of pure $LiPF_6$ solids can be used to prepare different electrolyte according to the needs of customers, and the electrolyte species is relatively simple when employing solution method, which also limits the range of use of the product. Because of the requirements of high temperature, anhydrous and oxygen free conditons in the production, the manufactures of $LiPF_6$ faced with high difficulties in strong corrosion and high purity demands. At present, only Japan and South Korea has achieved large-scale production and built patent barriers [66].

Questions

1. Describe the connotation of green fine chemical industry briefly.
2. Briefly describe the application of catalytic technology in green chemical pharmaceutics.
3. Describe the progress of green synthesis of anti-inflammatory and analgesic ibuprofen.
4. Brief introduction of green biopharmaceutical technology and its principle.
5. Briefly describe the preparation process of L- tryptophan by enzymatic conversion.
6. Briefly describe the characteristics of supercritical fluid extraction technology in the extraction of natural drugs.
7. Consult the literature and write out the molecular structure of the green anticancer drug taxol.
8. Briefly describe the characteristics of pesticide preparations and the direction of green development.

9. Briefly describe the development direction and trend of green pesticides.
10. Briefly describe the structural features of graphene and its green synthesis methods.

References

[1] Perera, D. R., Seiffert, A. K., Greeley, H. M. Ibuprofen and meningoencephalitis. Annals of Internal Medicine, 1984, 100: 619.
[2] Bouland, D. L., Specht, N. L., Hegstad, D. R. Ibuprofen and aseptic meningitis. Annals of Internal Medicine, 1986, 104: 732.
[3] Rossi, S. Australian Medicines Handbook. Adelaide: The Australian Medicines Handbook Unit Trust. 2013, ISBN 978-0-9805790-9-3.
[4] Joint Formulary Committee British National Formulary (BNF). London: Pharmaceutical Press, pp. 665, 673. 2013, ISBN 978-0-85711-084-8.
[5] Lopez, V. M., Decatur, C. L., Stamer, W. D., Lynch, R. M., McKay, B. S. L-DOPA is an endogenous ligand for OA1. PLoS Biol. 2008, 6(9): 1861–1869
[6] Hiroshima, Y., Miyamoto, H., Nakamura, F., et al. The protein Ocular albinism 1 is the orphan GPCR GPR143 and mediates depressor and bradycardic responses to DOPA in the nucleus tractus solitarii. British Journal of Pharmacology, 2014, 171(2): 403–414.
[7] Herman, G. A., Stevens, C., van Dyck, K. Pharmacokinetics and pharmacodynamics of sitagliptin, an inhibitor of dipeptidyl peptidase IV, in healthy subjects: Results from two randomized, double-blind, placebo-controlled studies with single oral doses. Clinical Pharmacology and Therapeutics, 2005, 78(6): 675–688.
[8] Momaya, A., Fawal, M., Estes, R. Performance-enhancing substances in sports: A review of the literature. Sports Medicine, 2015, 45(4): 517–531.
[9] Boron, W. F., Boulpaep, E. L. Endocrine system chapter. Medical Physiology: A Cellular and Molecular Approach. Elsevier/Saunders. 2004, ISBN 1-4160-2328-3.
[10] Costoff, A. Sect. 5, Ch. 6: Biological Actions of CT. Medical College of Georgia. July 5, 2008. Retrieved 2008-08-07.
[11] Erdogan, M. F., Gursoy, A., Kulaksizoglu, M. Long-term effects of elevated gastrin levels on calcitonin secretion. Journal of Endocrinological Investigation, 2006, 29(9): 771–775.
[12] Zhang, Y. X., Qiu, Y. F. Study on supercritical extraction of Ginkgo biloba leaves by CO_2. Chinese Journal of Traditional Medical Science and Technology, 2006, 13(4): 255–256.
[13] Peltier, Sandra; Oger, Jean-Michel; Lagarce. Enhanced oral paclitaxel bioavailability after administration of Paclitaxel-loaded lipid nanocapsules. Pharmaceutical Research, 2006, 23(6): 1243–1250.
[14] Wall, M. E., Wani, MC. Camptothecin and taxol: discovery to clinic – thirteenth Bruce F. Cain Memorial Award Lecture. Cancer Research 1995, 55(4): 753–60.
[15] Li Hua, Li Ming, Zheng Zhijian, et. al. The application of supercritical fluid extraction for extracting and separating taxol. Journal of Fudan University (Natural Science), 2003, 42(3): 453–456.
[16] Ōmura, Satoshi; Shiomi, Kazuro . Discovery, chemistry, and chemical biology of microbial products. Pure and Applied Chemistry, 2007, 79(4): 581–591.
[17] Pitterna, T., Cassayre, J., et al. New ventures in the chemistry of avermectins. Bioorganic & Medicinal Chemistry, 2009, 17(12): 4085–4095.
[18] Murakami, T., Anzai, H., et al. The bialaphos biosynthetic genes of Streptomyces hygroscopicus: Molecular cloning and characterization of the gene cluster. Molecular & General Genetics, 1986, 205(1): 42–53.

[19] Duke, S. O., Dayan, F. E. Modes of action of microbially-produced phytotoxins. Toxins (Basel), 2011, 3(8): 1038–1064.

[20] Okuyama, A., Machiyama, N., Kinoshita, T., Tanaka, N. Inhibition by kasugamycin of initiation complex formation on 30S ribosomes. Biochemical and Biophysical Research Communications, 1971, 43: 196–199.

[21] Schluenzen, F., Takemoto, C., Wilson, D.N., et. al. The antibiotic kasugamycin mimics mRNA nucleotides to destabilize tRNA binding and inhibit canonical translation initiation. Nature Structural & Molecular Biology, 2006, 13: 871–878.

[22] Veitch, G. E., Beckmann, E., Burke, B. J. Synthesis of azadirachtin: A long but successful journey. Angewandte Chemie International Edition in English, 2007, 46(40): 7629–7632.

[23] Sanderson, K. Chemists synthesize a natural-born killer. Nature, 2007, 448(7154): 630–631.

[24] Zhao Shu-ying, song Zhan-qian, gao Hong, wang Qiu-fen. An effective method of microwave assisted process for extraction of azadirachtin from kernel of Azadirachta indica A. Chemistry and Industry of Forest Products, 2003, 23(4): 47–50.

[25] Cheng Xuan sheng l Zhao ping Z, yu yong. Natural insecticidal pyrethrum. Chinese Journal of Pesticides, 2005, 44(9): 391–394.

[26] Cheng Shouhong. Research on the use of microwave-aid technique for the extraction of pyrethrins. Pesticide Science and Administration, 2003, 24(9): 31–33.

[27] Root, C. Domestication. Westport, CT: Greenwood Publishing Groups, 2007.

[28] Zohary, D., Hopf, M., Weiss, E. Domestication of Plants in the Old World: The origin and spread of plants in the old world. Oxford: Oxford University Press, 2012.

[29] ISAAA 2013 Annual Report Executive Summary, Global Status of Commercialized Biotech/GM Crops: 2013 ISAAA Brief 46-2013.

[30] GM Science Review First Report Archived October 16, 2013, at the Wayback Machine., Prepared by the UK GM Science Review panel (July 2003). Chairman Professor Sir David King, Chief Scientific Advisor to the UK Government, p. 9.

[31] James, C. Global review of the field testing and commercialization of transgenic plants: 1986 to 1995. The International Service for the Acquisition of Agri-biotech Applications, 1996.

[32] Weasel, L. H. Food Fray. Nashville, TN: Amacom Publishing, 2009.

[33] Ronald, P., McWilliams, J. Genetically Engineered Distortions. The New York Times, May 14, 2010.

[34] The Rainbow Papaya Story. Hawaii Papaya Industry Association. Retrieved April 2015.

[35] Li, Y., et al. Biosafety management and commercial use of genetically modified crops in China. Plant Cell Reports, 2014, 33(4): 565–573.

[36] Appleby, A. P., Müller, F., Carpy, S. Weed Control, in Ullmann's Encyclopedia of Industrial Chemistry. Weinheim: Wiley-VCH, 2002.

[37] EFSA September 30, 2008 EFSA Scientific Report (2008) 195, 1–115: Conclusion on the peer review of triflusulfuron.

[38] Myers, J. P., Antoniou, M. N., Blumberg, B. Concerns over use of glyphosate-based herbicides and risks associated with exposures: a consensus statement. Environmental Health, 2016, 15(19): 13.

[39] The agronomic benefits of glyphosate in Europe .Monsanto Europe SA. February 2010. 39.

[40] Cressey, D. Widely used herbicide linked to cancer. Nature, (March 25, 2015).

[41] Cyhalothrin. Cameo Chemicals. United States National Oceanic and Atmospheric Administration.

[42] Robert, L. M.. Insect Control in Ullmann's Encyclopedia of Industrial Chemistry. Weinheim: Wiley-VCH, 2002.

[43] Okamoto, Y., Brenner, W. Ch. 7: Polymers, in Organic Semiconductors, New York: Reinhold Publishing Corporation. 1964, pp. 125–158.

[44] Chiang, J. C., MacDiarmid, A. G. Polyaniline': Protonic acid doping of the emeraldine form to the metallic regime. Synthetic Metals, 1986, 1(13): 193–205.

[45] Fehse, K., Schwartz, G., Walzer, K., Leo, K. Combination of a polyaniline anode and doped charge transport layers for high-efficiency organic light emitting diodes. Journal of Applied Physics, 2007, 101(12): 143507.

[46] The Nobel Prize in Physics 2010 was awarded jointly to Andre Geim and Konstantin Novoselov for groundbreaking experiments regarding the two-dimensional material graphene, The Nobel Foundation.

[47] Global Demand for Graphene after Commercial Production to be Enormous, says Report. Azonano.com. 28 February 2014.

[48] Cooper, D. R., D'Anjou, B., et al. Experimental review of graphene. ISRN Condensed Matter Physics, 2001, 2012: 1–56.

[49] Bonaccorso, F., Colombo, L., Yu, G., et. al. Graphene, related two-dimensional crystals, and hybrid systems for energy conversion and storage. Science, 2015, 347(6217): 1246501.

[50] Park, D.-W., et. al. Graphene-based carbon-layered electrode array technology for neural imaging and optogenetic applications. Nature Communications, 2014, 5: 5258.

[51] Research Hints at Graphene's Photovoltaic Potential, Newly observed properties mean graphene could be a highly efficient converter of light to electric power, by Mike Orcutt, MIT. March 1, 2013.

[52] Xinming Li, Miao Zhu, Mingde Du. High Detectivity graphene-silicon heterojunction photodetector. Small, 2016, 12(5): 595–601.

[53] Gall, N. R., Rut'Kov, E. V., Tontegode, A. Ya. Influence of surface carbon on the formation of silicon-refractory metal interfaces. Thin Solid Films, 1995, 266(2): 229–233.

[54] Chakrabarti, A., Lu, J., Skrabutenas, J. C., et.al. Conversion of carbon dioxide to few-layer Graphene. Journal of Materials Chemistry, 2011, 21(26): 9491–9493.

[55] Yuan Wen-hui, Gu Ye-jian1, Li Li. Green synthesis of Graphene/Pt nanocomposites with vitamin C. Journal of Chemical Engineering of Chinese Universities, 2014, 28(2): 258–263.

[56] Zheng Jin-hong, Huang Zhi-qi, Hou Hong-sen. Evolution and Progress of Deep UV 248 nm Photoresists, 2003, 21(5): 346–356.

[57] Wu jian. The status and development of electronic chemicals in China (part 1). Advanced Materials Industry. 2002, 106(9): 12–15.

[58] Wright, W. W., Hallden-Abberton, M. "Polyimides", in Ullmann's Encyclopedia of Industrial Chemistry. Weinheim: Wiley-VCH, 2002.

[59] Park, O. K., Cho, Y., Lee, S., Yoo, H.-C. who will drive electric vehicles, Olivine or Spinel? Energy and Environment Science, 2011, 4(5): 1621–1633; Ozawa, Ryan. New Energy Storage Startup to Take Hawaii Homes Off-Grid. Hawaii Blog.

[60] Love, C. T., Korovina, A., Patridge, C. J., et. al. Review of LiFePO4 phase transition mechanisms and new observations from X-ray absorption spectroscopy. Journal of the Electrochemical Society. 2013, 160: A3153–A3161.

[61] Malik, R., Abdellahi, A., Ceder, G. A critical review of the Li insertion mechanisms in $LiFePO_4$ electrodes. Journal of the Electrochemical Society 2013, 160(5): A3179–A3197.

[62] Song Yang, Zhong Benhe, Liu Heng, Guo Xiaodong. A review of recent developments in the synthesis of lithium iron phosphate. Materials Review, 2010, 24(16): 292–296

[63] Goodenough, J. B., Kim, Y. Challenges for rechargeable Li batteries. Chemistry of Materials, 2010, 22(3): 587–603.

[64] Simmons, J. H. Practice of fluorine science in lithium hexafluorophosphate. Fluorine Chemistry, 1950, 22(1): 164–166.

[65] William Novis Smith, Jr Exton, pa. Preparation of lithium Hexafluorophosphate. US:3607020, 1971-09-21.

[66] Sato Keiji, Oe Meguru. Method for producing electrolyte solution for lithium ion battery and lithium ion battery using same. Europe: 1976048, 2008-01-10.

7 Green technologies for intermediate product synthesis

7.1 Introduction

Green chemical synthesis process emphasizes consciously applying the principles of green chemistry in the process of chemical production and fully considering the effective use of resources and environmental protection. In other words, green chemical synthesis process considers recycling energy and resources at the source of the chemical production and controlling the waste before the generation of pollution in the various aspects of the process, such as the sources of energy, raw materials, process technologies and equipment.

The characteristics of the green chemical synthesis process mainly reflect the following aspects [1]:

(1) It is better to use clean raw materials and energy. For example, none or less poisonous and harmful materials, biomass raw materials, clean solvents and green catalysts are always adopted in the process.

(2) It is better to employ advanced and effective technologies in synthesis process in order to reduce the rigor degree of the production process and improve the utilization of energy and materials.

(3) It is better to adopt the new reaction technology, the reaction and separation coupling technology, the multiproducts integration technology in order to strengthen the equipment production capacity, improve energy efficiency and reduce waste emissions.

(4) The raw materials and the products should be greenization. For example, the products can biodegrade as harmless materials after its application in order to achieve sustainable production.

7.2 Green Technologies in Intermediate Product Synthesis

In this section, we choose 1,3-propanediol (PDO) as an important intermediate product for green technologies in intermediate product synthesis to illustrate.

7.2.1 Application of PDO

PDO is an important chemical raw material to synthesize plasticizes, detergent, antiseptic and emulsifying agent. In addition, it can also be used in the food, cosmetics and pharmaceutical industries. Among these applications, the important application is that monomer PDO can be used in the production of new polyester material

https://doi.org/10.1515/9783110479317-007

polypropylene glycol terephthalate (PTT). PTT is a new type of polyester material, which has the characteristics of excellent resilience, stain resistance, dirt resistance, and so on. As a result, PPT is widely used in the carpet, engineering plastics, clothing materials and other application fields. However, in practice, the application of PTT is limited because of the high price of PDO. Then, the industrialized production of PDO is the key for the production of PTT [2, 3].

The industrial production methods of PDO are divided into chemical synthesis method and biosynthesis method, which are monopolized by Degussa Company of Germany, Shell Company of America and DuPont Company of America. Among them, epoxyethane carbonylation method (EO synthesis method) is used by Shell Company, acrolein hydration hydrogenation method (AC synthesis method) is adopted by Degussa Company and innovative microorganism fermentation (MF method) is employed by DuPont Company.

1. **AC synthesis method**

   AC synthesis method is mainly divided into two steps. The first step is the hydration reaction of acrolein to produce 3-hydroxypropyl aldehyde (HPA) with the acidic catalyst, such as acidic ion exchange resin and acid molecular sieve [4–6]. The second step is the catalytic hydrogenation reaction of HPA to produce PDO with Raney Ni catalyst. The reaction mechanism is shown as follows:

$$CH_2=CHCHO + H_2O \rightarrow HOCH_2CH_2CHO \tag{7.1}$$

$$HPA + H_2 \rightarrow HOCH_2CH_2CH_2OH \tag{7.2}$$

   The yield of PDO depends on the acrolein hydration reaction, and the quality of the final product is determined by the hydrogenation effect of HPA. The key factor of the two step reactions is the catalysts in the reaction.

   1) **Catalyst system**

      In the early research, the inorganic acid is used as catalyst in the acrolein hydration reaction, which can result in a lower yield and lower selectivity of PDO along with the occurrence of side reactions. Based on the principle of not affecting the activity of the subsequent hydrogenation reaction catalyst, the development of the new catalyst systems of acrolein hydration reaction is shown as follows: using chelating ion exchange resin as catalyst, using inorganic carrier-containing active center as catalyst and using acid/alkali buffer system as catalyst.

   2) **Technological process**

      Based on the patent of Degussa Company, the production of PDO technological process is shown in Figure 7.1.

      The mixtures of acrolein and water are passed to the hydration reactor that has two catalyst beds filled with sodium ion exchanger resin. There is an exchange heater with cooling water between the two catalyst beds to remove the reaction heat. After the reaction, the conversion of acrolein of 89.1% and

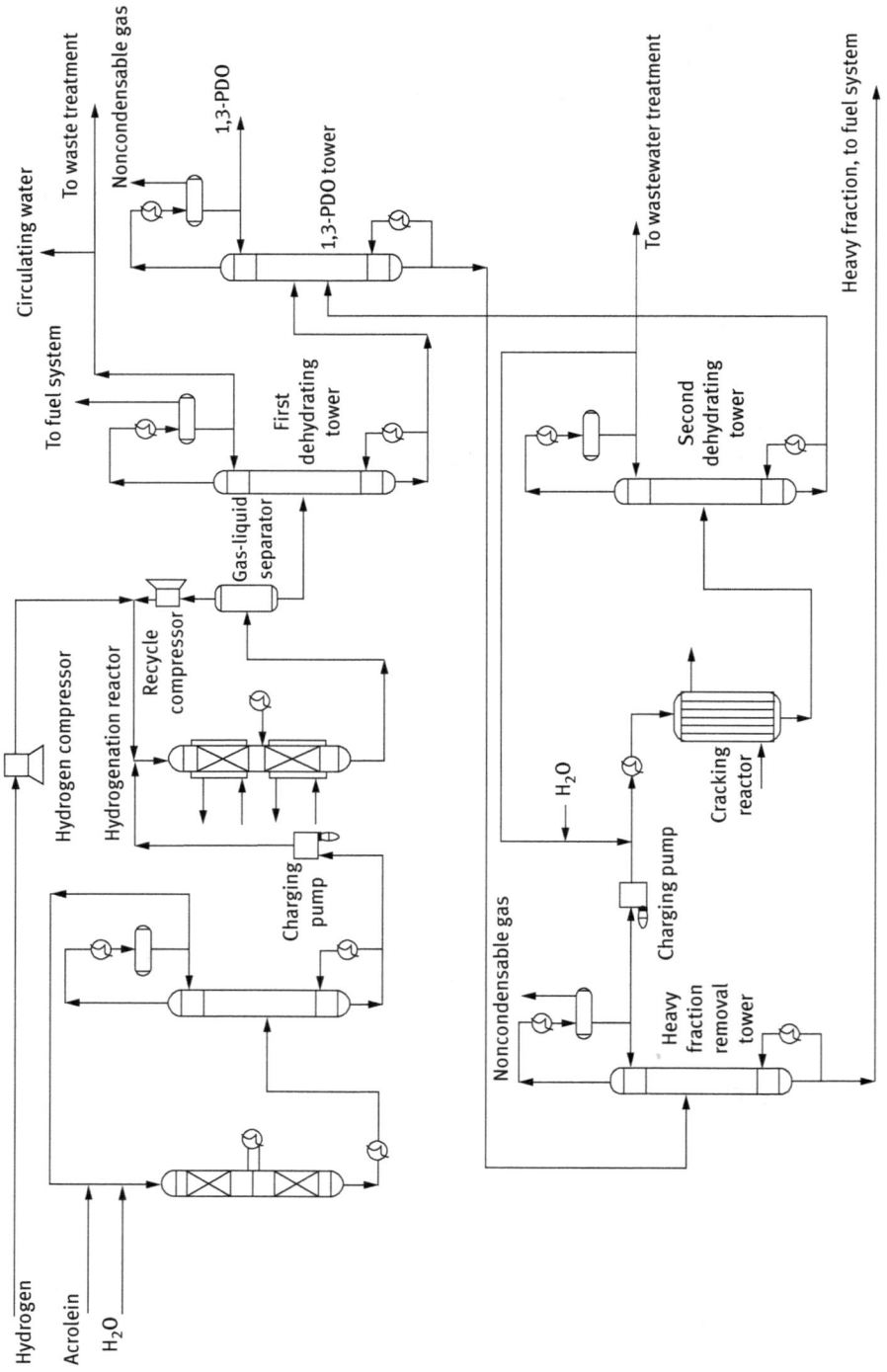

Figrue 7.1: PDO production process.

the selectivity of HPA of 85.1% can be obtained. The product of the hydration reactor is preheated and then is pumped into acrolein circulating tower. On the top of acrolein circulating tower, azeotrope of acrolein and water is returned to the hydration reactor. At the bottom of acrolein circulating tower, the discharge is forced to the pressure of 10 MPa and then mixes with the fresh hydrogen and the circulated hydrogen with the same pressure. Subsequently, the mixture gas is introduced into the hydrogenation reactor with two series of catalyst beds from the top to the bottom. In the two catalyst beds, the height of first catalyst bed is about 85% of the total height, and the temperature is maintained at 50 °C by using cooling water as the cooling agent. Then the product from the first catalyst bed is preheated and then into the second catalyst bed of the hydrogenation reactor under the reaction temperature of 125 °C. After the above reaction process, HPA conversion and PDO selectivity can be reached nearly 100%.

The product from the hydrogenation reactor is then introduced to the gas–liquid separator to separate out excess hydrogen which is recycled in the process, and the liquid product introduced the first dehydration tower. Stream from the top of the first dehydration tower is cooled to water which is almost reused in the system except a small part is discharged to the wastewater treatment system, and noncondensable gas from the top of the first dehydration tower is discharged as fuel. The liquid from the bottom of the first dehydration tower is sent to 1,3-PDO tower. PDO product can be obtained from the top of 1,3-PDO tower; at the same time, 3,3'-oxybis-propanol and acrolein polymer from the bottom of 1,3-PDO tower are put into the heavy fraction removal tower. 3,3'-Oxybis-propanol can be obtained from the top of the heavy fraction removal tower, and the heavy fraction can be obtained from the bottom to be used as fuel. Then, 3,3'-oxybis-propanol is compressed to 5.0 MPa and is mixed with the fresh and the recycled water to form 20.0% mixture solution. Subsequently, the mixture solution is entered in the cracking reactor filling dealuminated Y-type zeolite catalyst to conduct liquid hydrolysis reaction under the reaction temperature of 250 °C. In the cracking reactor, the conversion of 3,3'-oxybis-propanol of 73.0% and the selectivity of PDO of 72.0% can be obtained. The liquid product from the cracking reactor is then being sent to the second dehydrating tower. Water from the top of the second dehydrating tower is almost recycled in the system except a small part is discharged to the wastewater treatment system. The liquid mixtures of PDO from the bottom of the second dehydrating tower, unreacted 3,3'-oxybis-propanol and nonselective cracking products, are introduced to PDO tower to recycle the product.

2. **Epoxy ethane carbonylation method (EO synthesis method)**
   The raw materials of EO synthesis method are ethylene oxide and syngas, which is easy to obtain and transport. EO synthesis method can result in lower hydroxyl

content, a lower product cost price, a higher equipment investment and a relatively difficult technology than that of AC synthesis method [7, 8]. Especially, the preparation and the selection of the catalyst play the important roles in the industrialization production of PDO. EO synthesis method is divided into the one-step method and the two-step method.

1) **Two-step method**

The two-step method developed by Shell Company involves the following steps. First of all, HPA can be obtained from the hydroformylation reaction between epoxy ethane with CO and hydrogen. Then, PDO can be obtained from the hydrogenation reaction of HPA over a fixed bed catalyst under the reaction temperature of 100–200 °C and the pressure of 7.5–15.0 MPa. Between the two steps, the hydroformylation reaction and the catalyst used in the reaction are the key of EO synthesis method. Co-based catalyst in industrialized production is proposed by Shell Company.

The principle of the two-step method is illustrated as follows:

$$CH_2OCH_2 + CO + 2H_2 \rightarrow HPA \tag{7.3}$$

$$HPA + H_2 \rightarrow HOCH_2CH_2CH_2OH \tag{7.4}$$

Catalyst $Co_2(CO)_8$ used in the hydroformylation reaction (7.3) is prepared by the reaction between metallic cobalt salt and syngas using quaternary ammonium salt as promoter and ethers as solvent. In eq. (7.3), the separation between the reaction product HPA and the catalyst is very easy due to the use of ether solvent. As a result, the concentration of HPA of 35% can be obtained with $Co_2(CO)_8$ catalyst. In addition, by controlling the content of water and the concentration of HPA in the hydroformylation reaction, the conversion of EO can reach 100% and the selectivity of HPA can be up to 90%. Furthermore, the circulation utilization rate of cobalt-based catalyst can achieve 99.6% using water extraction technology of HPA, which can lead to lower catalyst consumption.

According to the influence of self-condensation of HPA to reduce the yield of PDO, a new patent put forward a new production process of PDO. In the new production process, the first step is the production of 3-hydroxymethyl propionate from EO and methanol, and the second step is the production of PDO from the hydrogenation reaction of 3-hydroxymethyl propionate. In the process, the selectivity and the activity of the hydrogenation reaction should be improved. In fact, the two steps in the reaction can be combined in one step by improvement of the catalyst, which is the origin of the one-step method.

2) **One-step method**

Compared with the process of two-step method, one-step method is the combination of the carbonylation reaction of EO and the hydrogenation of HPA. From the research, the catalyst and the catalyst promoter play important roles in one-step method. Under a certain reaction temperature and pressure, the yields of PDO and HPA of 65–78% can be obtained by using Ru-P complex as catalyst and water and a variety of acid as promoter; the selectivity of PDO of above 73% can be obtained by using Rh-based catalyst and triethanolamine as promoter; the conversion of EO of only 21–34% and the selectivity of PDO of 85–90% can be obtained by using tert-P/carbonyl cobalt composite catalyst and acetaldehyde is the main by-product; the yield of PDO of 87.2% and non-HPA can be obtained by using Co-tertiary phosphorus ligands-Ru complex as catalyst and its acid and metal salt catalytic system as accelerator. Shell Company has been applied to a series of patent in the production of PDO by one-step method.

The one-step method is carried out with a lot of improvement in new bimetallic catalysts and the appropriate ligands in order to obtain a higher yield of PDO by Shell Company. The one-step method of Shell Company is shown in Figure 7.2.

In Figure 7.2, the mixtures of the fresh and the recycled EO is compressed to 10.3 MPa and then is transported to the first section hydrogen formylation reactor. In the first section hydrogen formylation reactor, under the action of cobalt-ruthenium-phosphine and methylbenzene solvent, the mixed EO conducted the contact reaction with the mixture of synthesis gas from the second section hydrogen formylation reactor with H_2/CO mole ratio of 1 by adding hydrogen. The reaction heat is maintained at 90 °C by the way of cooling water in the reactor jacket and the dish tube in the reactor.

The liquid product from the first section hydrogen formylation reactor is cooled and introduced to the second section reactor. In the second section hydrogen formylation reactor, the gas from the top is mainly compressed to 10.3 MPa and mixed with the fresh synthesis gas except a few gas is discharged as fuel, and then the mixed gas and the liquid product from the first section is reacted in bubble contact. In the total hydrogen formylation reactor, the total residence time is about 3 h, and the reaction results are the conversion of EO of 58.1% and the selectivity of PDO of 85.7%. The liquid product from the upper part of the second section of reactor is decompressed and entered into the cycling EO tower and then is cycled to the first section hydrogen formylation reactor. The liquid product from the bottom of the second section of reactor is sent to the light component separation tower. The lighter component from the top of the light component separation tower is used as fuel, and the liquid from the bottom is cooled by the cooling and freezing water and then is introduced to the extraction tower. In the extraction tower with the cooling process water as extracting agent, the roughing 1,3-PDO can be obtained from the bottom of the extraction tower and then is introduced to the purification and refining process in order to obtain the purification of 1,3-PDO product. The toluene–benzene solvent

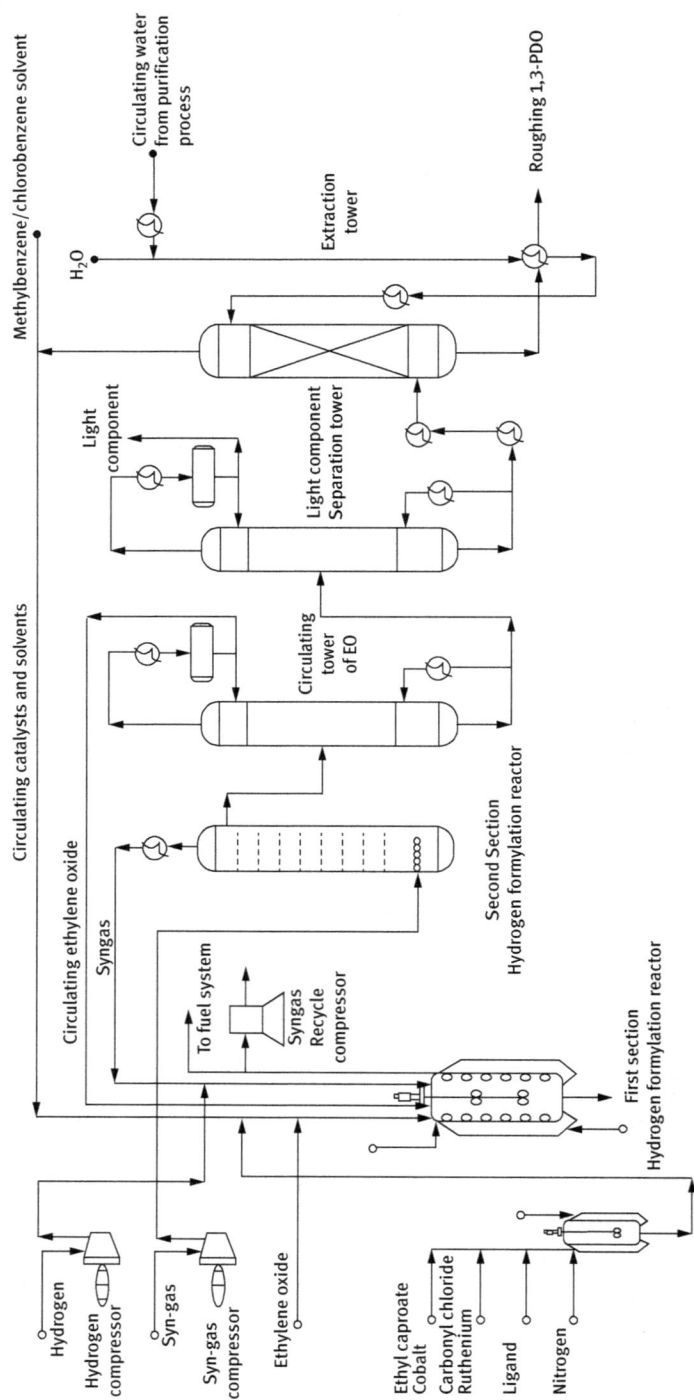

Figure 7.2: One-step method for the production of PDO fro EO: (a) reaction and distillaton, and (b) recuperation and purification.

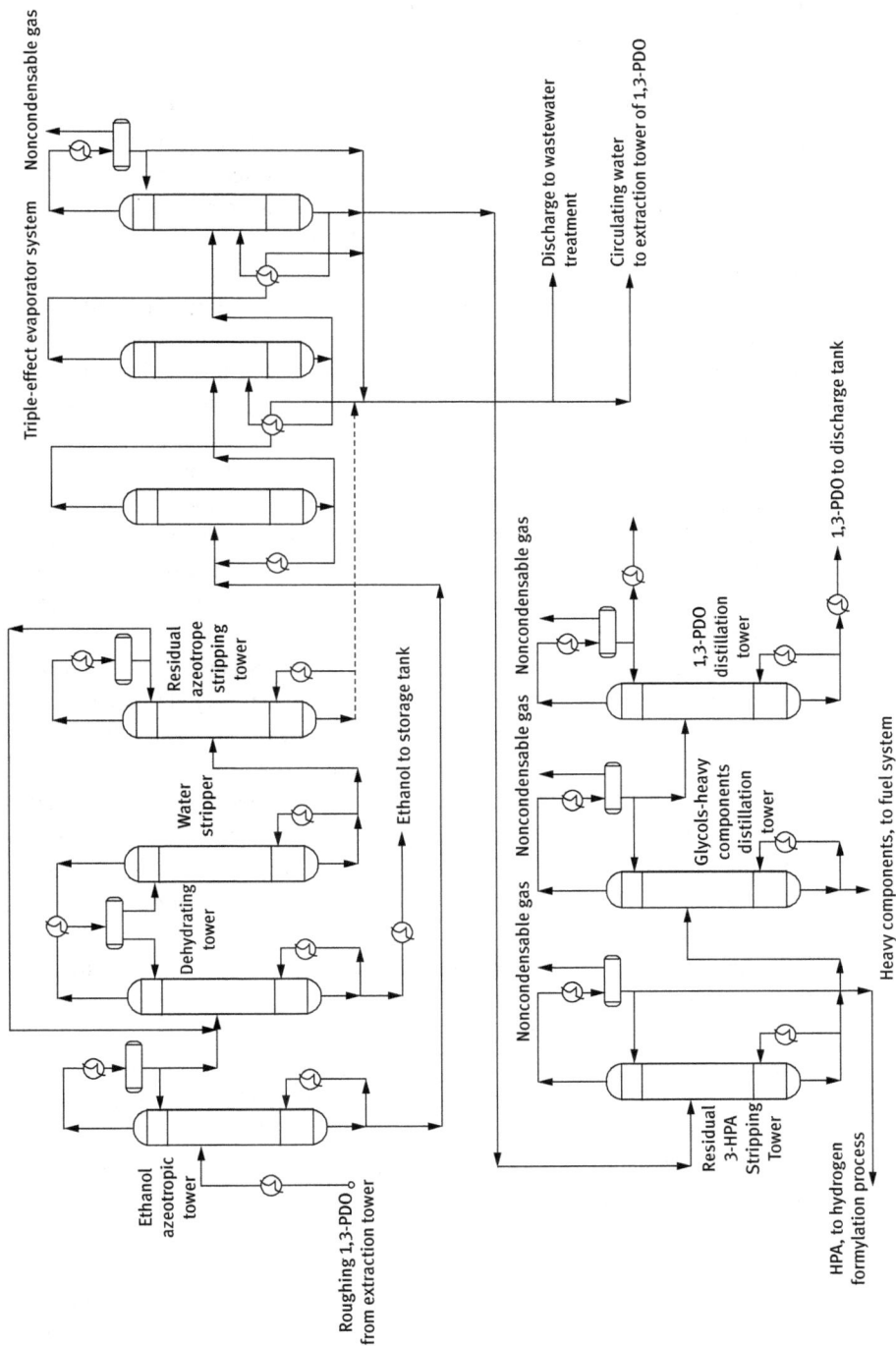

Figure 7.2: continued

and the catalyst can be recycled from the top of the extraction tower, and then are recycled to the first section hydrogen formylation reactor with the supplements of the solvent and catalysts.

The roughing 1,3-PDO is entered into the ethanol azeotropic tower. In the ethanol azeotropic tower, the azeotrope at the top is sent to the dehydrating tower. Water and water carrying agent from the top of the dehydrating tower is introduced to the water stripper, and the by-product ethanol from the bottom of the dehydrating tower can be obtained. Water-carrying agent from the top of the water stripper is recycled to the hydrating tower, and water containing trace ethanol from the bottom of the water stripper is entered into the residual azeotrope stripping tower. The azeotrope of residual ethanol and water from the top of the residual azeotrope stripping tower is sent back to the dehydrating tower, and water from the bottom of the residual azeotrope stripping tower is sent back to PDO extraction tower as the process water except part of water is discharged to the waste water treatment system. The mixtures of 1,3-PDO, water, 1,2-PDO and impurities from the bottom of the ethanol azeotropic tower are sent to the triple-effect evaporator to get rid of water. The operation condition of the triple-effect evaporator is in a reduced pressure to reduce the operation temperature and use of the steam heat energy effectively. The dehydrated 1,3-PDO from the bottom of the third evaporator is then introduced to the distillation process.

In the distillation process, the roughing 1,3-PDO is sent to the residual HPA stripping tower to strip the residual HPA, which is recycled to the hydrogen formylation reactor. The liquid of the bottom of the residual HPA stripping tower is sent to the glycols' heavy component distillation tower, where the glycol material can be obtained at the top and the heavy component can be obtained at the bottom. Then, the heavy component is sent to burn and recycle metal catalysts, and the glycol material is sent back to 1,3-PDO distillation tower. In PDO distillation tower, 1,2-PDO can be obtained at the top and 1,3-PDO can be obtained at the bottom.

Compared with two-step method for production of 1,3-PDO, due to the direct hydrogenation of HPA which has less stability generated from EO carbonylation, one-step method has the characteristics of a higher yield of product, a simplified production process and a lower production price. The only downside is that it is difficult for the distillation of 1,3-PDO because of high content of HPA in the liquid. Hence, in the production of high purity 1,3-PDO, aldehyde material from the decomposition of HPA plays a negative effect on the quality of 1,3-PDO which leads to a lower viscosity and a poor degree of color and luster for PPT polyester.

3. **Microorganism fermentation (MF method)**

MF is provided by DuPont Company of the United States, which is based on its innovation biological engineering method [9, 10]. MF methods are divided into three ways based on the cooperation with Genencor Company: (1) 1,3-PDO can be produced from glycerin disproportion reaction using intestinal bacteria as catalyst; (2) 1,3-PDO can be obtained from genetic engineering bacteria using glucose as substrate; and (3) 1,3-PDO can be obtained from corn syrup as raw material.

1) **Microbial fermentation process with glycerol as raw material**

The process of microbial fermentation is the process using glycerol as the raw material to produce PDO with the nature of *Klebsiella pneumonia* bacillus and butyric acid *Clostridium* under anaerobic conditions. During the fermentation process of bacteria, glycerol is consumed in two ways. One is that glycerol is dehydrated to produce HPA with glycerol dehydration enzyme, and then HPA is converted into PDO through reduction reaction. The other is that the by-product can be obtained from glycerol with the function of dehydrogenase. Due to the consumption of glycerol for the growth and oxidation metabolism of the bacteria, the conversion of glycerol to PDO is only 0.5%. In addition, in order to get a purity of 99.9% PDO, a relativity complicated production process should be adopted. As a result, the production price of MF synthesis method is difficult to compete with the chemical synthesis method. The microbial fermentation process using glycerol as the raw material is shown in Figure 7.3. In the joint development technology of Institute of Biological Science in Dalian University of Technology and PetroChina Jilin Petrochemical Company, the production of PDO is adopted through *Klebsiella* fermentation with glycerol as raw material. During the production of PDO pilot experiment, the purity of 99.0% and the yield of 85.0% PDO can be obtained by using fermented alcohol sink pretreatment and the distillation, and the quality of product PDO can meet the requirement of the polymerization reaction of PPT.

2) **Microbial fermentation process with glucose as raw material**

In order to reduce the production price of PDO, microbial fermentation process with glucose as raw material is proposed (see Figures 7.3 and 7.4). The genetic engineering technology is used in one-step production technology of PDO from glucose. In the technology, the glycerol is converted from glucose with *Escherichia coli* added *Saccharomyces cerevisiae*, and then PDO

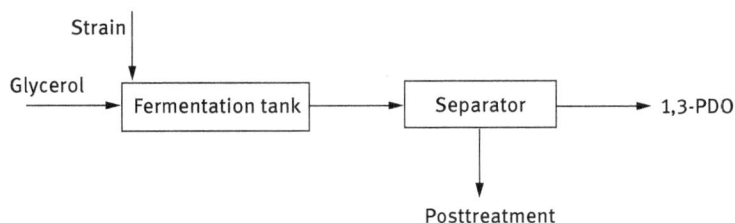

Figure 7.3: Microbial fermentation process using glycerol as raw material.

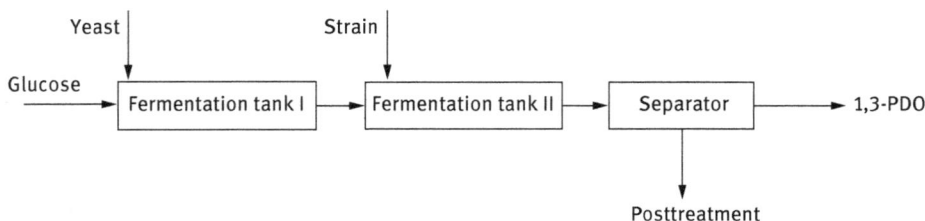

Figure 7.4: Microbial fermentation process with glucose as raw material.

can be obtained from glycerol with the existing citric acid coli and klebsiella gene, which can improve the yield of PDO effectively. DuPont and Tate & Lyle Cooperation Company in England verified this technology and the results showed the technology succeeded in the production of PDO in 45.4 t/a pilot plant founded in 2000. The results also demonstrate that the production price of MF synthesis method with glucose has the obvious advantage than that of chemical synthesis method.

4. **Comparison of 1,3-PDO production methods**

Compared with the MF synthesis method, the chemical synthesis of EO and AC synthesis method has industrialized the production of PDO due to the mature and feasible of the technology. Between EO and AC synthesis method, EO synthesis method uses ethylene oxide as raw material or the cheap syngas and ethylene, which can result in a lower production price. However, EO synthesis method has the disadvantages of a higher equipment investment and a relativity difficult technology, especially the preparation of the catalyst. AC synthesis method uses propylene as raw material to produce PDO, and as a result, the production price is relativity higher than that of EO synthesis method, and has the advantages of a relatively mild reaction condition, a relatively easy technology development and a very mature technology. In the above two methods, the product is mainly PDO and the by-product 1,2-propylene glycol and its dimer and trimer; as a result, there is a difficult separation and purification in the products. MF synthesis method is the method of genetic engineering bacteria production based on the gene recombination technology which used biomass as raw material, and has the advantages of an abundant resource, a regenerate raw material and nonpollution. Compared with the chemical synthesis methods, MF synthesis method has the advantages of a mild reaction condition, an easy operation, less by-products and green and environmental protection.

7.3 Green Technologies in Typical Product Synthesis

In this section, we choose the production of H_2O_2 as a typical green production process to illustrate.

Hydrogen peroxide is a multipurpose green oxidant and the traditional production is mainly by anthraquinone method. In practice, anthraquinone method has the advantage of the mature technology, but it also has the disadvantages of complex process, higher equipment investment, higher production price and pollution to the environment. Then, the new technologies are proposed for the production of hydrogen peroxide [11].

1. **Oxyhydrogen direct synthesis method**

It is a typical atom economy reaction for direct synthesis of hydrogen peroxide from hydrogen and oxygen. The production process is simple, cleaning, low production cost [12, 13]. And the important thing of the production process is the choice of the suitable catalyst.

In the process of direct synthesis of hydrogen peroxide from hydrogen and oxygen, the possible reactions are as follows:

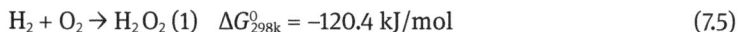

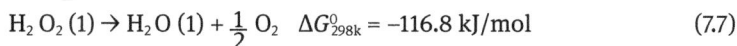

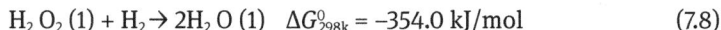

$$H_2 + O_2 \rightarrow H_2O_2 \ (1) \quad \Delta G^0_{298k} = -120.4 \ kJ/mol \tag{7.5}$$

$$H_2 + \frac{1}{2} O_2 \rightarrow H_2O_2 \ (1) \quad \Delta G^0_{298k} = -237.2 \ kJ/mol \tag{7.6}$$

$$H_2O_2 \ (1) \rightarrow H_2O \ (1) + \frac{1}{2} O_2 \quad \Delta G^0_{298k} = -116.8 \ kJ/mol \tag{7.7}$$

$$H_2O_2 \ (1) + H_2 \rightarrow 2H_2O \ (1) \quad \Delta G^0_{298k} = -354.0 \ kJ/mol \tag{7.8}$$

The catalysts adopted by the reactions are listed as eqs (7.5–7.9) are different. Some noble metal-loaded catalysts show the better catalytic performance in the above reactions. The catalytic active in the reaction eqs (7.6–7.9) can be limited by adding promoter in the catalyst to decrease the formation of by-products. The common carriers of the above catalysts are γ-Al_2O_3, SiO_2, C and so on.

In addition, in order to obtain a higher concentration of hydrogen peroxide, the reaction is processed in the strong acidity and medium containing halogen ions. Avoiding the corrosion of the reactor and the dissolution of the catalyst in the acidity medium, the solid superacid carrier is used in the neutral or weak acid medium reaction. The solid superacid carrier is composed of two kinds of metal oxides or the metal oxide-loaded sulfuric acid. Figure 7.5 shows the production of hydrogen peroxide in direct reaction from hydrogen and oxygen.

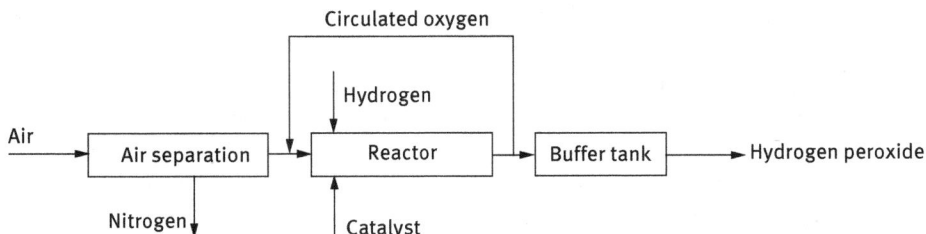

Figure 7.5: Process of hydrogen peroxide in direct reaction.

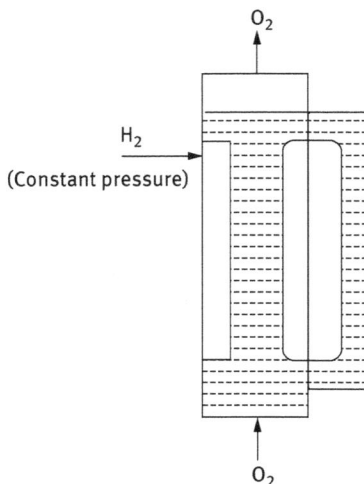

Figure 7.6: Diagram of membrane reactor in production of hydrogen peroxide.

In direct synthesis reaction, the controlling step of the reaction is the mass transfer process from the gas phase to the liquid phase of hydrogen; hence, it is advantage for adding the concentration of hydrogen in the organic solvent. Except for the traditional organic solvents, supercritical carbon dioxide is also a common solvent in direct synthesis reaction due to the high solubility of hydrogen and oxygen in supercritical carbon dioxide and a better mass transfer process.

The explosion limitation of hydrogen and oxygen is very wide and becomes wider along with the increasing of pressure. Hence, the safety is one of the important problems in designing the reactor. The fixed bed tubular reactor and the high pressure reactor are the main reactors for the production of hydrogen peroxide in the production. Membrane reactor is a safer reactor than the others because the alloy membrane in the reactor can separate hydrogen from oxygen to avoid the generation of explosive mixtures and to improve the safety of the production. In addition, the conversion of hydrogen of 100% can be obtained safely in membrane reactor, and the selectivity of hydrogen peroxide can be greatly improved. The diagram of the membrane reactor is shown in Figure 7.6.

In recent years, in the research of direct synthesis method for the production of hydrogen peroxide using hydrogen and oxygen as raw materials, the improvement is mainly focused on the following three aspects [14–20]:

1) **Improvement of catalysts**

   The researches on the catalysts include the composition and structure of catalyst, the carrier and the promoter of catalysts, and the surface modification methods, in order to improve the activity and selectivity of the catalyst.

2) **Development of solvents and additives**

   The purposes of the researches of solvents and additives are on the improvement of the generation rate of hydrogen peroxide and the stability of the catalyst. For example, it is better to adopt the mixture solvent of the lower alcohol and water by adding a small amount of inorganic acid and halogen as promoter.

3) **Modification of the reaction system**

In the production process, the security of operation is important. As a result, the researches on the modification of the reaction system are most focused on the development of the safety operation condition, such as controlling the molar ratio of hydrogen and oxygen, using the selective permeability organic or inorganic membrane device in the reactor, and so on.

2. **Integration of direct synthesis of hydrogen peroxide and the production of propylene epoxide**

According to the stoichiometric ratio, in the catalytic production of propylene epoxide from propene and hydrogen peroxide over TS-1 catalyst, the production of 1.7 tons of propylene epoxide need to consume 1 ton hydrogen peroxide. In addition, if the concentration of 30% commercialized hydrogen peroxide is used in the reaction system, the problems, such as the purification of raw material, storage, transportation, separation of water and high energy consumption of distillation, can be generated in the production. Hence, the combination of the production of hydrogen peroxide and the production of propylene epoxide is developed in order to achieve the goals of decreasing production cost, saving energy and reducing consumption.

1) **Integration of the production of propylene epoxide by propylene epoxidation and the production of hydrogen peroxide by anthraquinone method**

Integration of using methanol/water as extraction agent: methanol/water solvent separated from the epoxidation process is used as the extraction agent for extracting hydrogen peroxide, and then hydrogen peroxide can be used for the integration of the epoxidation process. The integration process can be illustrated in Figure 7.7.

In the production, alkyl anthraquinone dissolved in the appropriate solvent can produce alkyl hydrogen anthraquinone in the hydrogenation reactor, and then alkyl hydrogen anthraquinone is introduced in the oxidation reactor, where it is mixed with the air and processed the oxidation to produce hydrogen peroxide. Hydrogen peroxide can be extracted from the oxidation liquid with methanol/water solvent in the extraction tower. Soon afterward, the methanol/water solvent containing hydrogen peroxide is sent to the epoxidation reactor and is reacted with propylene to produce propylene epoxide with the action of TS-1 catalyst. The mixture from the epoxidation reactor is then separated in the flash tank to obtain propylene, and then the remaining part is distilled in the distillation tower. The product propylene epoxide can be obtained from the top of the distillation tower. At the bottom of the distillation tower, the mixture of methanol and water is

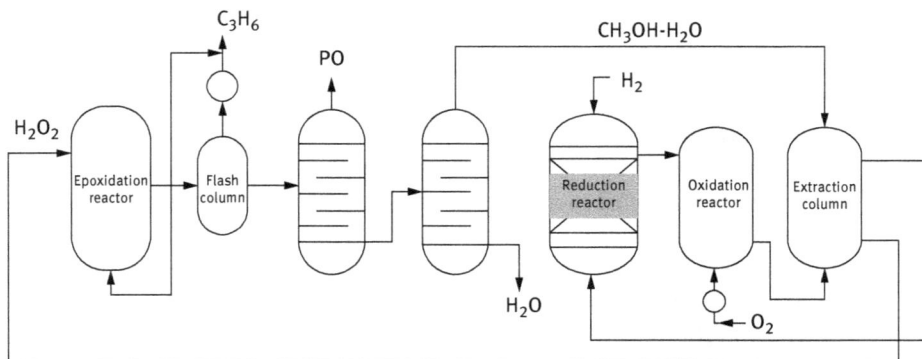

Figure 7.7: Integration of the production of hydrogen peroxide using methanol/water as extraction agent.

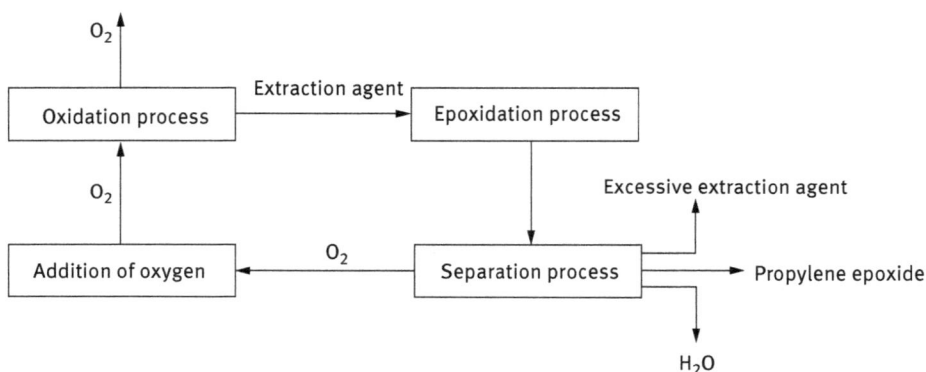

Figure 7.8: Integration process of using water-soluble anthraquinone as working medium.

divided into three parts: the first part is introduced to the extraction tower to extract hydrogen peroxide, the second part is recycled to the epoxidation reactor and the third part is sent to the distillation tower to remove water produced in the epoxidation.

2) **Integration of using water-soluble anthraquinone as working medium**
This integration is the improvement of the integration above. The integration process is shown in Figure 7.8. In the integration process, first, anthraquinone sulfonic acid alkyl ammonium salt is dissolved in the methanol/water solvent, and then is produced hydrogen anthraquinone sulfonic acid alkyl ammonium salt through hydrogenation reaction. Compared with the integration process ①, the integration process ② adopts methanol/water solvent which is similar with that of the epoxidation of propylene; as a result, there is no extraction process. In addition, due to a higher solubility in methanol/water solvent of anthraquinone sulfonic acid alkyl ammonium and the

addition of the water resistance hydrogenation catalyst, the structure dimensions of hydrogenation reactor and oxidation reactor can be shrinked.

3. **Catalysis process with manganese catalyst**

   The direct production of hydrogen peroxide with bivalent manganese ion as catalyst was proposed by the Herefordshire University. In the production, raw materials are air and hydroxylamine, and the final products are hydrogen peroxide, nitrogen and water. One of the advantages is that the reaction occurs in aqueous solution which can produce hydrogen peroxide in situ production. In the reaction, hydroxylamine and oxygen is converted to hydrogen peroxide, nitrogen and water under the reaction temperature of 20 °C and pH value of 8 with bivalent manganese ion exchange of montmorillonite as catalyst. As a result, the conversion of hydrogen peroxide of 75% can be obtained in less than 1 h. In practice, due to the higher price of hydroxylamine, an alternative material should be found in order to make the technology industrialized.

4. **Vacuum enrichment method**

   The problem of the lower purification efficiency of hydrogen peroxide in direct production can be solved by the patent of the production of hydrogen peroxide proposed by Kvaerner Company. In the patent, the reaction of the mixture is reacted in the organic solvent instead of water. When the reaction is conducted the time that the content of hydrogen peroxide is just below the saturation of that in the organic solvent, the mixture materials of the reaction are put into a vacuum place to evaporate hydrogen peroxide. Then, the evaporated hydrogen peroxide can be condensed into pure hydrogen peroxide with the quality of high concentration and low production cost. Now, the project is under the development stage.

5. **Discussion**

   The direct synthesis of hydrogen and oxygen to produce hydrogen peroxide is a typical atom economy reaction. During the production process, there are no damage products and by-products. But in the process, the safety is the important problem. The integration of the production of hydrogen peroxide and the production of propylene epoxide is the green chemical process, which pays attention on the full use of the materials and energies. However, the choice and the recycled of catalysts should pay more attention in order to guarantee a high yield of hydrogen peroxide and a low production cost.

7.4 Green Chemical Process

Green engineering refers to the engineering of the application of green chemistry and green ecological industrial technology to achieve the green engineering. Green engineering is the solution for the multiobjective optimization and combination problems including ecological balance with the point of view of system engineering. The objective of green engineering is that raw materials should be completely converted into the objective products, and also the whole life cycle of production and the service cycle of products should be safe, clean, efficient, pollution free and environment friendly.

The aim of the chemical process is developed a whole life cycle of the chemical process based on the principle of the green engineering. In other words, it should be environmentally friendly in the production of product from raw materials, in the process of usage and waste treatment, in the operation and the scrap cycle of the equipments and devices, and in the production process.

During the development of green chemical engineering, on the one hand, it is better to focus on the technology of green chemical engineering, such as the coupling of chemical technology, physical field synergy, miniaturization of devices, integration systems, recycling technology of waste; on the other hand, it is better to consider the development of renewable biomass resources utilization and integration of new process and so on. In addition, it also established the ecological industrial park, advocated green consumption, pursued the multiobjective of the environmental benefit, economic benefit and social benefit in order to meet the requirements of sustainable development.

Questions

1. What is the core of the green chemical process?
2. What is the characteristic of the green chemical process?
3. How to improve the utilization ratio of atoms in the green chemical process?
4. What issues should be payed attention in the green chemical process?
5. What are the characteristics of green chemical process in the production process of hydrogen peroxide?
6. What is the green engineering? What is the relationship between the green engineering and the circular economy?

References

[1] Li, D. H. Introduction of Green Chemistry and Chemical. Beijing: Science Press, 2005: 11.
[2] Yang, J. Q., Wang, X. Y. Process progress of 1,3-propanedial. Journal of Chemical Industry & Engineering in China, 2002 2(23): 11–14.
[3] Rita, L. D. Aquino. Three routes vie for the 1,3-propaneediol market. Chemical Engineering, 1999, 106(5): 56–62.
[4] Arntz, Dietrich. Process for producing Poly(1,3-Propylene Terephthalate). EP 412 337 A2, 1991.
[5] Haas, Tromas. Process for Producing Poly(1,3-Propylene Terephthalate). EP 0 572 812 Al, 1993.
[6] Arntz, Dietrich. Process for Producing Poly(1,3-Propylene Terephthalate)[P]. EP 0 487 903 A2, 1992.
[7] Shell Oil Co.(US). Process for making 1,3-propanediol and 3- hydroxypropanal. US 5 304 691, 1994.

[8]   Shell Oil Co.(US). Process for preparing 1,3-propane-diol. US 5 463 144, US 5 463 145, US 5 463 146, 1995.

[9]   Menzel, K., Zeng, A. P., Deckwer, W. D. High concentration and productivity of 1,3-propanediol from continuous fermentation of glycerol by Klebsiella pneumoniae. Enzyme and Microbiol Technology, 1997, 20(2): 82–86.

[10] Zhang, W. D., Meng, D. D., Wang, R. M., Ma, C. L. Research progress in 1,3-propanediol microbial production. Journal of Shandong Polytechnic University in China. 2011, 25(4): 11–14.

[11] Zheng, S. J., Xu, J., Ma, Z. Research progress of hydrogen peroxide production techniques. Chemical Propellants and Polymeric Materials in China, 2016, 14(2): 28–38.

[12] Lunsford, J. H. The direct formation of H_2O_2 from H_2 and O_2 over palladium catalysts. Journal of Catalysis, 2003, 216(1): 455–460.

[13] Bertsch-Frank. Process for producing hydrogen peroxide by direct synthesis. US6387346, 2002.

[14] Ma, Z. L., Jia, R. L., Liu C.J. Production of hydrogen peroxide from carbon monoxide, water and oxygen over alumina supported Ni catalysts. Journal of Molecular Catalysis A, 2004, 210(1): 157–163.

[15] Berglin. Method in the production of hydrogen peroxide. US4552748, 1985.

[16] Bengtsson. Process in the production of hydrogen peroxide. US5063043, 1991.

[17] Kato. Process for producing hydrogen peroxide. US5399333, 1995.

[18] Drackett. Electrolytic production of hydrogen peroxide using bipolar membranes. US5358609, 1994.

[19] Chuang. Production of hydrogen peroxide. US5338531, 1994.

[20] Ishii. Process for producing hydrogen peroxide. US6375922, 2002.

8 Green processes for carbon dioxide resource utilization

8.1 Overview of Global Carbon Dioxide Emissions

8.1.1 The Source of Carbon Dioxide

Carbon dioxide is an abundantly available substance in nature. It mainly exists in the gaseous form. Air contains about 0.03% of carbon dioxide. When carbon dioxide dissolves in water, it forms carbonic acid and its salt; it also reacts with various substances to form some solids such as the mountain rock and pebbles in the bottom of the sea.

It is estimated that the carbon present in the form of carbon dioxide on the earth, around 10^{14} tons, is ten times that of the carbon present in oil, coal and natural gas; in addition, the potential resource of carbonate in nature is estimated to be up to 10^{16} tons.

The main sources of carbon dioxide are respiration of animals and plants; degradation and transformation of plant and animal bodies; and volcanic eruption and fuel combustion, which have been one of the largest sources of carbon dioxide in the recent 100 years, about 2.6×10^{10} tons of carbon dioxide emissions to the atmosphere each year, while thermal power contributes around 40% of global carbon dioxide emissions. Other sources include cement production, vehicle exhaust and other human activities.

In addition to being present in earth's atmosphere, carbon dioxide is present in the form of dry ice on the surface of Mars, the Moon, Jupiter and other stars, in greater amounts than that of the earth's atmosphere.

8.1.2 The Present Situation and the Trend of Global Carbon Dioxide Emissions

1. **The status of emission of carbon dioxide around the world**

   In the past century, the global average temperature has increased by 0.6 °C since the pre-industrial revolution, and the average temperature in most industrialized European countries has increased by 0.9 °C [1, 2].

   Based on the 2009 greenhouse gas emissions released by the British risk assessment company, the top ten largest emitters are as follows. 1. China: the emissions of carbon dioxide into the atmosphere every year are more than 6.0×10^9 tons, but per capita emissions are not much. 2. The United States: its emission is 5.9×10^9 tons, with per capita emissions of 19.58 tons per year, second to Australia. 3. Russia: its carbon dioxide emissions every year surged to 1.7×10^9 tons. 4. India was 1.29×10^9 tons, of nearly 1.2 tons per capita emissions. 5. Japan: annual carbon dioxide emissions have fallen to a bit of 1.247×10^9 tons.

https://doi.org/10.1515/9783110479317-008

6. Germany: carbon dioxide emissions are 8.6×10^8 tons. 7. Canada, greenhouse emissions each year is of 6.1×10^8 tons. 8. The United Kingdom: greenhouse gas emission is 5.86×10^8 tons in 2008, so the UK government issued and implemented a climate-change bill which is the world's first country for greenhouse gas emissions legislation. 9. South Korea: annual greenhouse gas emission is 5.14×10^8 tons. 10. Iran: greenhouse gas emission is 4.71×10^8 tons.

8.2 The Separation and Fixing of CO_2

8.2.1 The Properties of Carbon Dioxide

1. **Physical properties**

   Under the normal temperature, carbon dioxide is a colorless, odorless gas; its relative molecular mass is 44.01, and 1.5 times higher than that of air in the same conditions. Some of the main physical properties are as follows [3].

   Carbon dioxide has a melting point of 216.6 K, boiling point of –194.27 K, gas density of 1.974 kg/m^3, liquid density of $1,022$ kg/m^3, solid density of $1,565$ kg/m^3, triple-point pressure and temperature of 0.518 MPa and 216.6 K, respectively, and liquid surface tension of about 3.0 dyn/cm.

   Liquid carbon dioxide and supercritical carbon dioxide can be used as a solvent; CO_2 gas can be liquefied to liquid by pressured cooling. Liquid carbon dioxide transformed to becomes a solid carbon dioxide (dry ice) by deep cooling. When carbon dioxide is present in the environment in high concentrations, it can cause serious physiological discomfort and even death.

2. **Chemical properties**

   Carbon dioxide is chemically stable and dissolves in water to produce carbonic acid, which is a weak acid. It also has weak oxidative capacity, and when it reacts with metal, a flammable gas is produced.

8.2.2 Separation Technologies of CO_2

Carbon dioxide separation is also called carbon dioxide capture and storage; it refers to the process of separating carbon dioxide from the mixture of gases in the fuel burning process, followed by compression, dehydration and transformation, and finally sealing on the geological layer of the earth forever. Chemical absorption method is a reaction of carbon dioxide with sorbent in an absorbing tower to generate a weakly combined intermediate. Then the carbon dioxide-rich liquid is desorbed in the tower to release carbon dioxide, and at the same time to regenerate the absorbent. The method of carbon dioxide capture, transportation and storage is shown in Figure 8.1.

Table 8.1: Key technologies for carbon dioxide separation.

Technologies		Working pressure	Industrial application	Key problems in industrialization
Absorption	Chemical (MEA)	$Pa \geq 3.5–17.0$ kPa	Extraction of CO_2 from Nature gas (NG) and smoke effluent	Energy consumption in the regeneration pretreatment of the other acidic gases during optimization of the regeneration
	Physical (cool, methanol glycols)	$Pa \geq 525$ kPa		
Adsorption	Nonisopressure adsorption	High pressure	Separation of CO_2 from H_2 mixture, NG, and smoke effluent;	Low adsorption capacity Low selectivity
	Nonisothermal adsorption	High pressure	separation of CO_2 from H_2 mixture and NG	Low CO_2 purity
Inorganic membrane (ceramic, Pd membrane)		High pressure	Separation of CO_2 from H_2 mixture and NG	Smaller specific area of ceramic membrane
Polymer membrane		High pressure		Low selectivity to CO_2 and membrane degradation

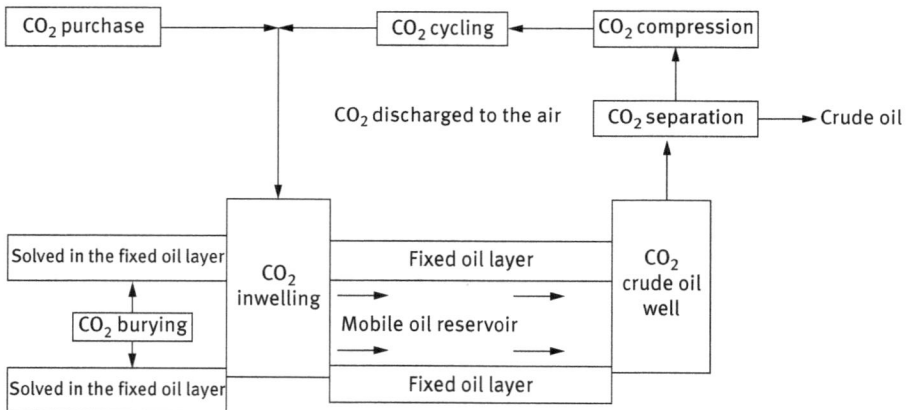

Figure 8.1: Concept representation for carbon dioxide capture, transportation and storage.

CO_2 separation technologies include absorption, adsorption, membrane separation technology and membrane absorption. Characteristics of these methods are listed in Table 8.1.

1. **Absorption method**

Absorption methods include both physical absorption and chemical absorption.

Physical absorption methods use the solubility difference of carbon dioxide and other gases in the mixture in the same absorber to remove carbon dioxide from the mixture. Commonly used physical absorption methods in industries include the flour method, rectisol method and selexol method. The choice of absorbent is crucial, and the absorbers should possess the properties of higher solubility of carbon dioxide, high selectivity to carbon dioxide, high boiling point, free corrosion and nontoxic and stable properties; usually, water, methanol and propene carbonate are used as absorbents. Physical absorption method has the advantage of less absorbent usage, low temperature and high absorption and absorbent regeneration efficiency at normal temperature, usually in the process of low-pressure adsorption or gas stripping with low energy consumption. This method is applied in the case of separate gas mixture with higher partial pressure of carbon dioxide with low removing yield of carbon dioxide. Typical flowchart of physical absorption of carbon dioxide is shown in Figure 8.2.

Chemical absorption methods involve reacting the feed gas with a chemical solvent in the absorption tower; carbon dioxide is absorbed in the solvent and a carbon dioxide-rich liquid is formed, which enters into an adsorption tower and carbon dioxide is released. The key step in this method involves accurately controlling the temperature and pressure of the absorption and adsorption tower. Commonly used chemical absorbents include alcohol amine, steric hindrance amine and carbonate aqueous solution with a concentration of less than 50% (when the concentration is too high, serious corrosion can occur); using a variety of alcohol amines can increase the absorption rate, alleviate corrosion, produce volatile products and reduce cost. A typical chemical absorption process is shown in Figure 8.3. It is worth mentioning that ionic liquids are an important class of CO$_2$ absorbers.

2. **Adsorption method**

Adsorption methods use solid adsorbent (e.g., natural zeolite, activated carbon, molecular sieve, activated alumina and silica gel) to selectively and reversibly adsorb CO$_2$ in the feed to recover carbon dioxide. The adsorbents adsorb carbon dioxide at low temperature (or high pressure), which is released at higher temperature (or

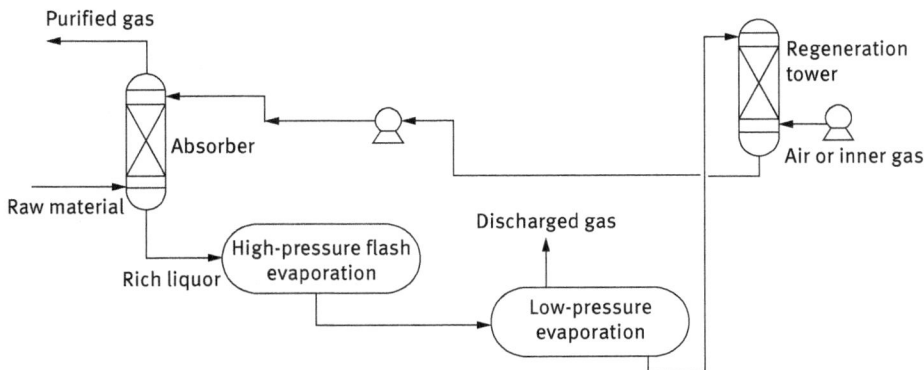

Figure 8.2: Typical flowchart of carbon dioxide physical absorption process.

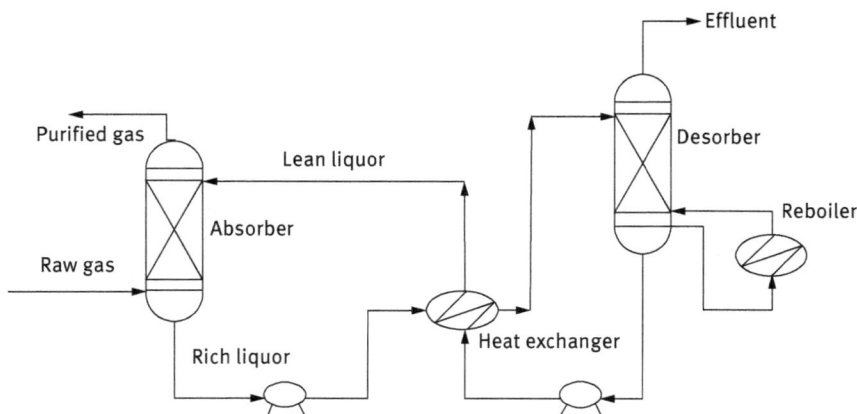

Figure 8.3: Typical flowchart of chemical absorption process.

low pressure); through periodic change of temperature (or pressure), the separation of carbon dioxide from other gases has been achieved. Chemical adsorption generally needs more adsorption towers to be installed in parallel to ensure that the whole process is continuous, and the key is the absorption capacity of the adsorbent. Adsorption capacity of solid adsorbents depends on the temperature and pressure of carbon dioxide; generally, the higher the CO_2 partial pressure and the lower the temperature, the more of carbon dioxide is absorbed. Because there is water vapor in exhaust gases, and water vapor will completely absorb with solid absorbent carbon dioxide to reduce CO_2 adsorption capacity, microparticles will enter the inner of adsorbent to deactivate the adsorbent. Based on these factors, chemical adsorption is more competitive than physical adsorption.

Although there have been reports that molecular sieve, activated charcoal and zeolite can help to achieve 100% CO_2 recovery, it is accomplished only under the condition of no water vapor and microparticles. Further research should focus on the new solid adsorbents with high selectivity and practical absorption capacity of carbon dioxide.

3. **Membrane separation**

Membrane separation is the use of polymer films, such as acetic cellulose fiber, polyimide, polysulfone membranes, with different gas permeability to separate carbon dioxide from a gaseous mixture. The driving force in the membrane separation process is a pressure difference: when pressure difference exists on both sides of a membrane, gas of high permeability will quickly pass through the membrane to form permitted gas flow, while low-permeability gas stays largely in the film inlet side to form residual gas; the two gas flows elute, respectively, so as to achieve the separation.

When membrane separation is used to process high CO_2 concentration mixture, no matter what kinds of film, except for high selectivity to CO_2, largest

CO_2 permeability is also expected. In addition, there still need to consider film life, film cost of maintenance and replacement, etc.

4. **The combination of membrane separation–absorption**

   The combination device of membrane separation–absorption is simple, lower investment cost and low separation efficiency. The two together can complement each other, the former conducts preseparation, the latter finished fine separation; this can achieve the goal of effective separation at low investment cost. For example, Norway's Statoil Company recovered CO_2 from natural gas, the original system employed amine solution as absorbent, installing huge absorption tower. After improvement, fluorine polymer membrane is used to proceed the pretreatment, making the investment of the absorption tower, scrubbing tower reduce 70–75%, covering area of plants being reduced of 65%.

8.3 Chemical Conversion Principles of Carbon Dioxide

8.3.1 The Structure of Carbon Dioxide

Carbon atom in CO_2 molecule is in *sp* hybrid orbital bonding. The *sp* hybrid orbitals in two carbon atoms, respectively, combined with two O atoms to form sigma bond. The other two *p* orbitals unhybridized to C atoms are at right angles to the sp hybrid orbitals, and overlapped side by side from the side with the oxygen *p*, respectively, to generate two three-center, four-electron delocalization PI bond. The C=O bond length in CO_2 is 0.116 nm, between that of double and triple bonds, and closer to that of a triple bond. The dipole distance of CO_2 is zero, which indicates that it has a linear structure, O=C=O. The CO_2 bond structure is shown in Figure 8.4.

8.3.2 CO_2 Activation Methods

1. **The chemical activation**

   Free energy of carbon dioxide is very small, and it is a very stable compound; therefore, converting it into other compounds containing carbon is very difficult [4–8].

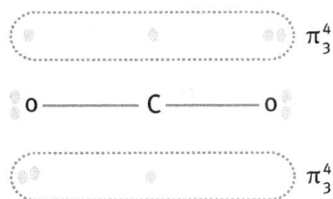

Figure 8.4: Diagram of CO_2 bonding structure.

Carbon dioxide can be converted into chemical raw materials, but, the key problem is the activation of carbon dioxide by catalytic process.

The molecular structure of CO_2 is linear; the C/O bond length is shorter than that of C=O in ketone (0.122 nm), which is composed of the three regular structures:

$$O=C=O \leftrightarrow O^+ \equiv C^- - O \leftrightarrow O - C^- \equiv O^+$$

From the CO_2 molecular structure, it is seen that the CO_2 molecule has two active sites, namely a Lewis acid site (C) and Lewis base site (O), which gives rise to the properties of electrophilicity and nucleophilicity in the activation reaction; This gives hint that CO_2 activation catalyst can be searched from compounds which can provide electrons or an empty orbital. By calculation, the first ionization energy of CO_2 is of 13.79 eV, being very difficult to give electron and hard to form CO_2^+ ion, but there is lower energy empty anti-bonding orbital in CO_2, it is easily to accept an electron to form a bending anion. Therefore, the most practicable way to achieve CO_2 activation is to enter an electron in the antibonding orbital of CO_2 to form a bending anion, making the C–O bonding order in CO_2 molecule from 2 to 1.5, reducing the activation energy of carbon dioxide molecules.

The CO_2 activation catalyst systems include alkaline catalyst, heteropoly acid catalyst, transition metal and rare earth metal catalysts. An inorganic alkaline compound combined with Lewis acid site in CO_2 molecules to make a C–O bond rupture of CO_2 and formed activated CO_2. Its representation reaction is the synthesis of dimethyl carbonate (DMC), with potassium hydroxide, magnesium oxide, calcium oxide, halide potassium, potassium carbonate, sodium carbonate, potassium hydroxide and sodium hydroxide, etc. as catalysts. The catalysis mechanism of DMC synthesis is

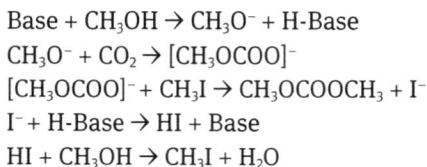

$$Base + CH_3OH \rightarrow CH_3O^- + H\text{-}Base$$
$$CH_3O^- + CO_2 \rightarrow [CH_3OCOO]^-$$
$$[CH_3OCOO]^- + CH_3I \rightarrow CH_3OCOOCH_3 + I^-$$
$$I^- + H\text{-}Base \rightarrow HI + Base$$
$$HI + CH_3OH \rightarrow CH_3I + H_2O$$

In the process, CH_3I is used as assistant catalyst to provide CH_3I^- that reacts with hydrogen generated to form hydrogen iodide, and then it reacts with methanol to form CH_3I again, completing a catalytic cycle. The activity of sodium compounds is far lower than that of the potassium compounds. Alkali strength enhancement can decrease the reaction temperature, and increase the reaction rate, but the selectivity is decreased. Moderate strength of alkali can improve catalytic activity, and organic alkali works better than inorganic base does, but the activity sequence is not consistent with alkaline strength.

In addition, research on the active adsorption of CO_2 on the surface of transition metal crystal has revealed dissociative adsorption of CO_2 on clean crystal

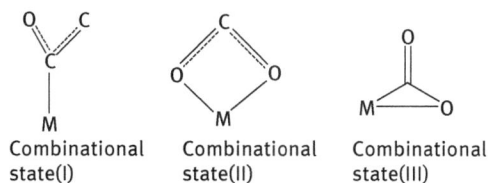

Combinational state(I)   Combinational state(II)   Combinational state(III)

Figure 8.5: The adsorption combination states of CO_2 with metal.

Cu(110), Fe(110) and Ni(110) surface. The adsorption combination state of CO_2 with metal may have three configurations (see Figure 8.5), with combinations of II and III being easier.

A precursor of the CO2 may further dissociate occurred on the transition metal surface, while oxidation occurs commonly in precious metal surface except palladium.

From the above analysis, the transition metals can activate CO2 by donating electron. if there exist at the same time a Lewis base, effective CO2 activation can beachieved. Synergical effect can be displayed when transition metal composite oxide is used; on the other hand, the crystal type of the catalyst has effect on its activity.

2. **Electrochemical activation**

The study of electric catalytic activation of CO_2 has been carried out at room temperature and atmospheric pressure with Pt as electrode and dialkyl imidazole, alkaline compound and methanol as electrolytes. For example, the electrochemical synthesis of DMC is accomplished with CH_3OK as alkaline. During electrochemical activation, CO_2 is adsorbed on the cathode by the adsorption of K^+ in the cathode; it obtains an electron from the electrode surface to give CO_2^-, and CO_2^- reacts with methanol adsorbed on the electrode to generate CH_3OCO^+ and adsorbed KOH. This cation then combines with CH_3O^- to form DMC. The adsorbed KOH further reacts with plenty of CH_3OH to generate CH_3OK, completing a catalytic cycle.

3. **Light activation**

In recent years, it has been proven that UV and visible light can break the thermodynamic limit or improve thermodynamics. The photocatalytic synthesis of DMC by methanol and CO_2 can enhance the activation of CO_2 and improve the yield of DMC. With easy process control and excellent atom economy, and with continuous-flow fixed bed reactor, the water formed can be removed simultaneously to enhance the formation of DMC. The yield of DMC was thus increased by 57%.

8.4 Utilization Examples of Carbon Dioxide Resources

Carbon dioxide can be used as raw material to produce many inorganic and organic chemical products. These products are used in almost all industries [9–15]. Chemical transformation of carbon dioxide can achieve the recycling of carbon dioxide, especially large-scale chemical production which consumed huge amount of carbon dioxide, plays an important role in reducing emissions.

Energy-saving and emission reduction are building a new era of harmonious ecology of mankind with nature. The green technology of carbon dioxide utilization plays a key role in the goal. The introduction of carbon dioxide in chemical synthesis is a very efficient approach. At present, with carbon dioxide as raw material, alcohol, alkanes, esters, amines, polyesters, polyamines and other chemicals have been prepared, while synthetic urea and sodium salicylate have been put into large-scale industrial production. The products tree from carbon dioxide is shown in Figure 8.6.

8.4.1 Application of Carbon Dioxide in Inorganic Synthesis

The preparation of inorganic compounds, such as lithium carbonate, sodium carbonate and strontium carbonate, from carbon dioxide has led to many industrial applications [16]. The green chemistry of sodium carbonate production is briefly discussed here.

Sodium carbonate (soda ash, soda) is widely used in glass, paper, soap, detergent, textile, leather and other important industries, and also as hard water softener and in the manufacture of sodium compounds. It is the first alkaline inorganic compound to realize industrial preparation. In 1892, Belgian industrial chemist E. Ernest Solvay invented an all-new alkali method with salt, limestone and ammonia as main raw materials; it is called the "ammonia alkali" or "Solvay soda" method. This method is of high yield, excellent quality, low cost and continuous production. However, the technology has been patented and protected, so until 1942, China's chemical engineering expert Hou Debang invented the "combined soda method" (also known as "Hou's soda method"), to lead soda ash industry into a new revolutionary period, and continues to now.

1. **Ammonia alkali method**

   With sodium chloride, limestone (calcinated to produce lime and carbon dioxide) and ammonia as raw materials, ammonia first passes into the saturated salt water to form ammonia salt solution, and then flows into the carbon dioxide to generate less-soluble sodium bicarbonate precipitation and ammonium chloride solution. The main process can be expressed as follows.

   (1) Preparation of CO_2: Calcination of limestone produces CO_2 gas.

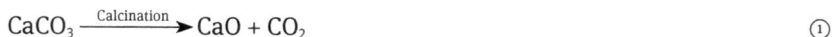

   $$CaCO_3 \xrightarrow{\text{Calcination}} CaO + CO_2 \qquad ①$$

   (2) Preparation of $NaHCO_3$: Refined NaCl and ammonia react with carbon dioxide to generate sodium bicarbonate:

   $$NaCl + NH_3 + CO_2 + H_2O \Leftrightarrow NaHCO_3 + NH_4Cl \qquad ②$$

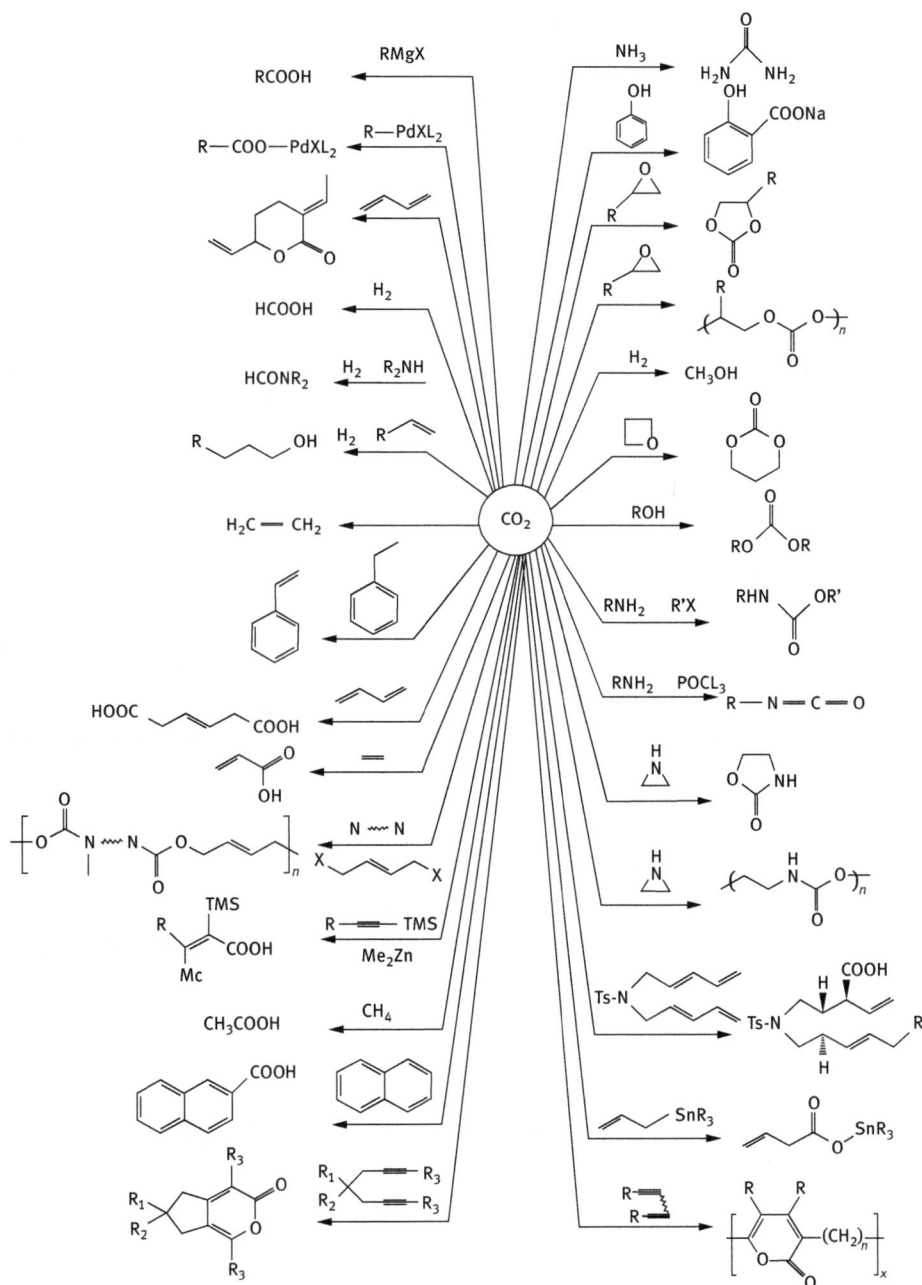

Figure 8.6: Chemical transformation pathways of carbon dioxide.

(3) Preparation of soda ash by the calcination of $NaHCO_3$: The product of reaction ② is filtered to give $NaHCO_3$ solid and NH_4Cl solution. The $NaHCO_3$ solid is calcined and decomposed to give the final product, soda ash:

$$2NaHCO_3 \xrightarrow{\text{Calcination}} Na_2CO_3 + H_2O + CO_2 \qquad ③$$

The product CO_2 is recovered and used for reaction ②.

(4) Preparation of lime emulsion: CaO from reaction ① reacts with water to yield $Ca(OH)_2$ emulsion for the following reaction ⑤:

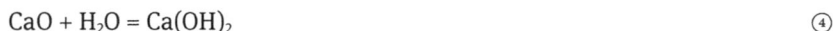

$$CaO + H_2O = Ca(OH)_2 \qquad ④$$

(5) Decomposition of NH_4Cl to recover ammonia: The NH_4Cl solution obtained by filtering reaction ② product was added with lime milk to decompose NH_4Cl to form NH_3:

$$2NH_4Cl + Ca(OH)_2 \Leftrightarrow 2NH_3 + CaCl_2 + 2H_2O \qquad ⑤$$

The process flowchart is shown in Figure 8.7. The total reaction for the ammonia-alkali method can be expressed as

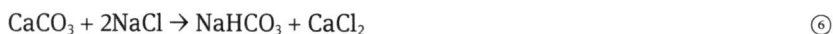

$$CaCO_3 + 2NaCl \rightarrow NaHCO_3 + CaCl_2 \qquad ⑥$$

It is seen that sodium ions from raw materials NaCl, CO_3^{2-} from limestone, while Cl^- in the raw material NaCl and Ca^{2+} in the $CaCO_3$ are useless, and finally discharged as by-product $CaCl_2$. It is possible to calculate the atomic efficiency and the E-factor of the ammonia alkali method:

Atomic utilization r = (the mass of the atoms entering the target product/ the mass of the atoms in all the reactant) × 100%
= [Relative molecular mass of sodium carbonate /(Relative molecular mass of calcium carbonate + 2 × Relative molecular mass of sodium chloride)] × 100%
= (106/217) × 100%
= 48.8%

E-factor = By-product/the quality of the product
= Relative molecular mass of calcium chloride/Relative molecular mass of sodium carbonate
= 111/106
= 1.05

It shows that the atomic efficiency of ammonia-alkali method is only 48.8%, E-factor is greater than 1, and in the production of 1 ton products, there are 1.05 ton of waste emissions. This method does not meet the criteria of green chemistry.

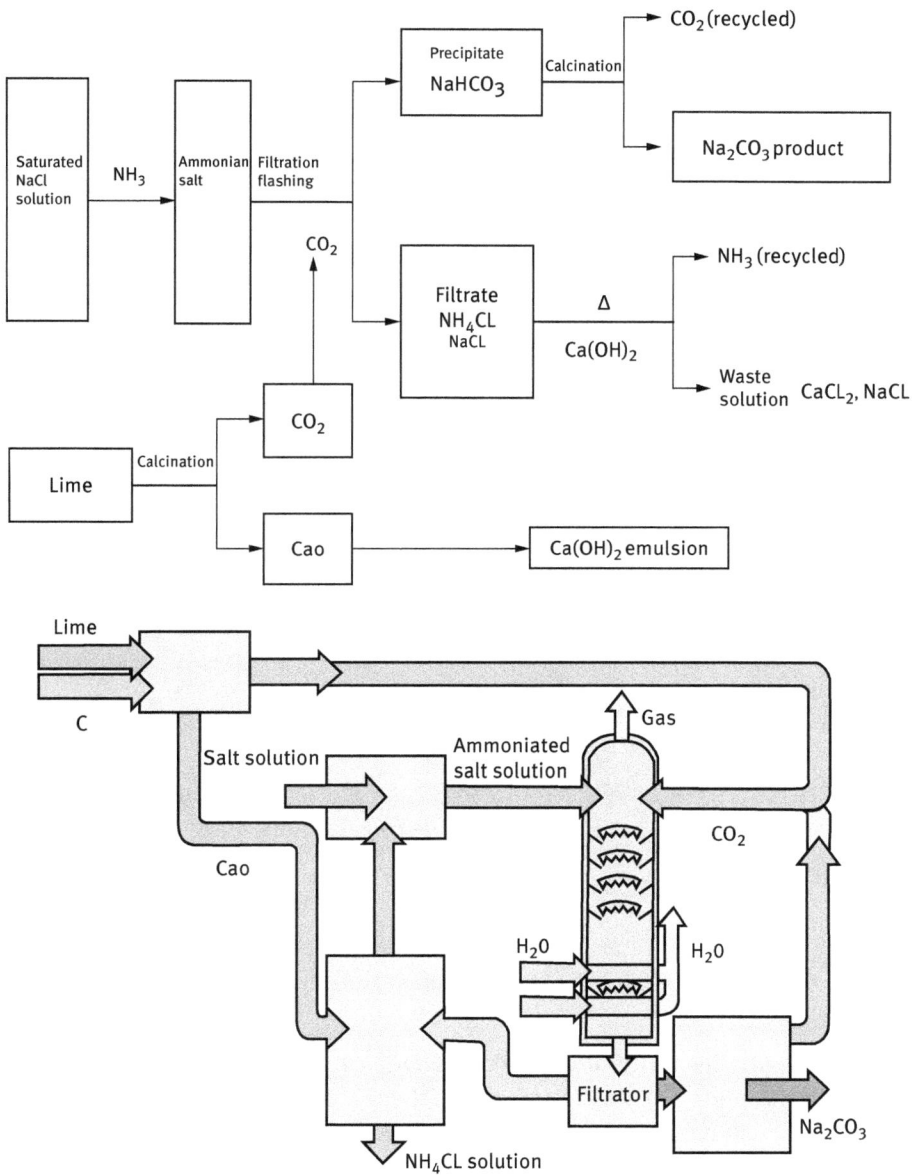

Figure 8.7: The process flowchart for the ammonia-alkali method.

2. Combined soda preparation method (Hou's soda production method)

In view of the shortcomings of ammonia–alkali method, Hou Debang invented this method. It is the combination of ammonia-alkali and synthetic ammonia process to produce soda ash and ammonium chloride simultaneously with salt, ammonia and carbon dioxide (waste gas from hydrogen production of coal-water shift reaction in ammonia plant) as raw materials; hence it is also called jointed alkali method. NH_4Cl was produced from NH_3 in the ammonia plant and Cl^- in the original salt, and soda ash was produced by using the by-product CO_2 of the ammonia plant and Na^+ in the original salt. From the chemical reaction point of view, the main difference between Hou 's and the ammonia alkali method is shown below:

(1) The raw material gas of the ammonia-alkali process is derived from the calcination of limestone, and CO_2 in Hou's process is derived from the by-product of the ammonia plant.

(2) Raw material ammonia in the ammonia-alkali method is recycled ($NH_3 \rightarrow NH_4Cl \rightarrow NH_3$), while ammonia in Hou's alkali method is not recycled, but transformed into the product NH_4Cl. The main steps of Hou's process are as follows:

(1) NaCl, ammonia and carbon dioxide react to produce $NaHCO_3$ and NH_4Cl:

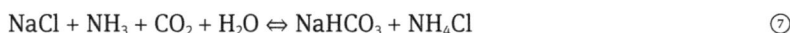

$$NaCl + NH_3 + CO_2 + H_2O \Leftrightarrow NaHCO_3 + NH_4Cl \qquad \text{⑦}$$

(2) The product is filtered to give $NaHCO_3$ solid and NH_4Cl liquid. The NH_4Cl solution is digested with ammonia to precipitate NH_4Cl solids. The $NaHCO_3$ solid is calcined to give soda ash:

$$2NaHCO_3 \xrightarrow{\text{Calcination}} Na_2CO_3 + CO_2 + H_2O \qquad \text{⑧}$$

The total reaction formula of Hou's method is

$$2NaCl + CO_2 + H_2O + 2NH_3 \rightarrow Na_2CO_3 + 2NH_4Cl \qquad \text{⑨}$$

The reaction products Na_2CO_3 and NH_4Cl are the target products. The atomic efficiency and E-factor of Hou's method are calculated as

$$\text{Atomic utilization} = \frac{106 + 107}{117 + 34 + 44 + 18} \times 100\% = 100\%$$

$$\begin{aligned} E\text{-factor} &= \text{By-product quality/Quality of products} \\ &= 0/213 \\ &= 0 \end{aligned}$$

It can be seen that all the atoms in the reactant in combined alkali reaction finally enter the product, the atomic economy of the reaction is 100% and the E-factor is 0, which conforms to the idea of green chemistry. The process flowchart is shown in Figure 8.8.

8.4.2 Applications of Carbon Dioxide in Organic Synthesis

Organic synthesis with carbon dioxide is of great concern to chemists. Therefore, there are many studies and reports in this area, and there are many successful examples of industrial applications [17–23].

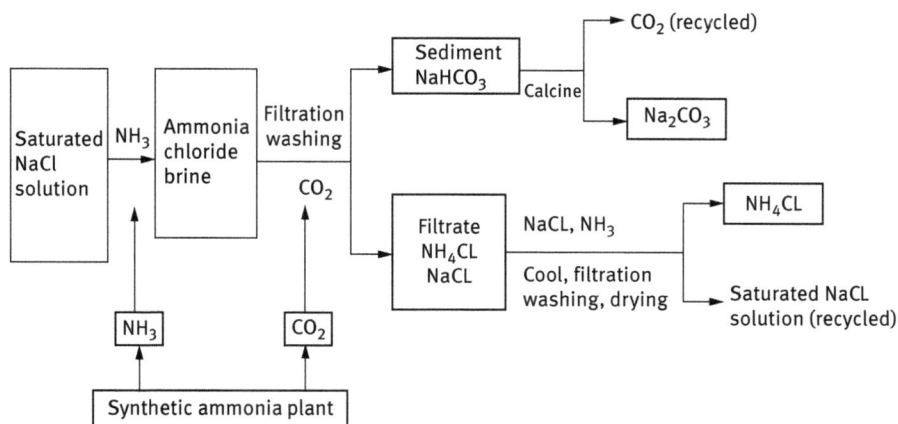

Figure 8.8: The process flowchart of combined soda preparation method.

1. **Synthesis of cyclic carbonates from carbon dioxide and epoxy compounds**

   Cyclic carbonate is mainly used in electrolytes, polymer monomers and pharmaceutical industry; it is also an important pharmaceutical intermediate, with high industrial value. The reaction for the synthesis of cyclic carbonate from CO_2 and epoxy compounds is described as

   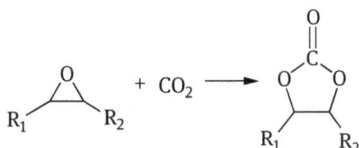

   This is considered to be one of the most successful CO_2 utilization processes, its atomic economy is 100% and has achieved commercial production. The catalytic system has been developed, including metal complex catalyst, ionic liquid catalyst, acid-base bifunctional heterogeneous catalyst, lanthanide oxychloride, supported heteropolyacid catalyst, supported Schiff base catalyst and supported ionic liquid catalyst. Among them, the homogeneous catalyst systems demonstrate better performance, and the yield of cyclic carbonate is very high, but the disadvantages of the homogeneous catalyst constrained its large-scale industrial application. In order to overcome the shortcomings of a homogeneous catalyst system, the development of highly efficient heterogeneous catalytic system has become a hot spot. In recent years, research has focused on the ionic liquid catalytic system with low toxicity and high activity, but it is difficult to separate from the product and cannot be reused in a continuous-flow reaction device such as a fixed bed. One way to overcome this problem is to immobilize the ionic liquid, that is, to support the ionic liquid on the surface of a certain carrier, so as to make the homogenous reaction into a heterogeneous one. Tables 8.2 and 8.3 show the

Table 8.2: Reaction data of various SiO_2-grafted ionic liquids.

Catalyst	Conversion (%)	S (%)
Silica gel	–	–
[Bmim]Cl	98.8	98.2
[Bmim]Cl/SiO_2	97.2	98.6
[Bmim]Cl-$AlCl_3$/SiO_2	90.1	99.7
[Bmim]Cl-$ZnCl_2$/SiO_2	98.0	98.6
[Bmim]BF_4	94.9	97.5
[Bmim]BF_4/SiO_2	93.2	99.0
[Bmim]OH	100	99.9
[Bmim]OH/SiO_2	99.1	100

Table 8.3: Reusability of SiO_2-grafted ionic liquid catalyst.

Used times	1	2	3	4
Conversion (%)	99.1	97.3	94.6	89.0

reaction data and repeatability of SiO_2-grafted ionic liquids in the cycloaddition of propylene oxide and CO_2. These catalysts have good reactivity and can be reused, showing further development potential.

2. **Other organic chemicals from carbon dioxide**
   Carbon dioxide has been used as raw material for the preparation of many valuable chemicals. Polycarbonate (PC) plastics are amorphous, thermoplastic, high impact, transparent and degradable plastics, can be used continuously at 135–145 °C and has been successfully applied in disposable packaging materials, tableware, preservation materials, disposable medical materials, plastic film and so on; its output has become the first of five general engineering plastics. China's PC is expected in the next few years to remain in the annual growth rate of 15–20%.

   The production of PC presently concludes the phosgene method. Phosgene is a kind of extremely toxic gas that seriously contaminates the environment, and extremely harmful to human health. Every ton of the product needs to consume chlorine of 368 kg, caustic soda of 415 kg and carbon monoxide of 145 kg, and produces 7.84 ton of wastewater.

   DMC is synthesized using carbon dioxide (CO_2) which reacts with diphenyl carbonate to prepare PC. The synthetic procedure is a clean route with no "three wastes" production, and has become the first choice for PC production.

   The most environment-friendly method to synthesize isocyanate is by thermal decomposition of carbamate. Carbamate can be synthesized from DMC instead of phosgene.

As a new type of green fuel, dimethyl ether is synthesized by hydrogenation of CO_2. The chemical reaction equation is as follows:

$$2CO_2 + 6H_2 \rightarrow CH_3OCH_3 + 3H_2O$$

This groundbreaking research has made great progress in energy revolution and the most promising application of CO_2 in chemical industry is to produce methanol. However, obtaining hydrogen at low cost has been an obstacle for massive production of methanol from CO_2 so far.

Another important application of CO_2 involves carbamate production with high yield by reacting organic amine and CO_2 using crown ether as catalyst.

Carboxylic acid can also be synthesized from CO_2 and hydrocarbons under photo-irradiation. For example, propane reacts with CO_2 to produce methacrylic acid, and methane reacts with CO_2 to yield formic acid.

Amine products can be synthesized through reaction of CO_2 with ammonia or amine. For example, methylamine can be synthesized from CO_2, H_2 and NH_3. N-Phenylformamide can be obtained from CO_2, H_2 and aniline, while diphenylurea can be produced from CO_2 and aniline.

The yield and selectivity have room for improvement in the cases mentioned above. Thus, researchers need to find a more suitable catalytic system for industrial production.

Questions

1. Illustrate the principles of carbon dioxide activation.
2. What are the essential problems involved in carbon dioxide utilization?
3. Summarize the green processes involved in the transformation of carbon dioxide.
4. Elucidate the key methods for the separation of carbon dioxide from industrial effluent.
5. Calculate the atomic efficiency of the reaction: $2CO_2 + 6H_2 = CH_3OH_3C + 3H_2O$.
6. Illustrate the advantages of the green preparation of DMC from carbon dioxide and methanol.
7. Elucidate briefly the key pathways of carbon dioxide emission reduction.
8. Design a new process for the transformation of carbon dioxide.
9. Explain how the chemical intensification technology is used in the chemical transformation of carbon dioxide.

References

[1] Xinbang Net, the first tenth carbon emission countries in the world, 2009-12-16, 23:13:28
http://koubei.xooob.com/gj/200912/395865.html

[2] Zhao Qinming. Recovery of carbon dioxide to benefit mankinds. Energy and Environment, 2010 (4): 25–28.

[3] Zhou Zhongqing, Qian Yanlong. Progress of carbon dioxide chemistry. Chemistry Communications, 1984 (5): 4–11.

[4] Li Hansheng, Zhong Sunhe. Research progress on the preparation of DMC from carbon dioxide and methano. Chemistry Progress, 2002, 14(5): 368–373.

[5] Bai Rongxian, Tan., Yisheng. Synthesis of C_2^+ alkanes by the hydrogenation of carbon dioxide over composite catalyst.chemistry Process, 2003, 15(1): 47–50.

[6] Zhu Yuezhao, Liao Chuanhua, Wang Chongqin, et al. Carbon Dioxide Emission Alleviations and Resource Utilization. Beijing: Chemical Industry Press, 2011.

[7] Jiang Qi, Li Tao, Liu Feng. Production of DMC from carbon dioxide and methanol over methoxyl compound of main group metals.chemistry Communications (internet edition), 1999, 9(1): C99094

[8] Zhong Sunhe, Li Hansheng, Wang Jianwei. Direct preparation of DMC from carbon dioxide and methanol over Cu-Ni/V_2O_5-SiO_2 catalyst. Physical Chemistry Bulletin, 2000, 16(3): 26–231.

[9] Tan Tianwei, Wang Fang, Deng Li etal. Research status and future developments of biomass energy[J].Modern Chemical Industry ,2003,23(9): 8–12.

[10] Jin Zhiliang, Qian Ling, Lv Gongxuan, Carbon dioxide-present and future. Chemistry Progress, 2010, 22(6): 1102–1115.

[11] Tian Hengsui, Li Feng, Lu Wenglong, et al. To develop the high quality CO_2 based fine chemicals to promote the optimization of industries distribution and energy-conservation and emission alleviation. Chemical Industry Progress, 2010, 29(6): 977–983.

[12] Wu Ying. Green organic synthesis from CO_2. Tianjing: Master thesis of Nankai University, 2009

[13] Qi Chaorong, Jiang Huanfeng. Organic reactions in supercritical carbon dioxide. Chemistry Progress, 2010, 22(7): 1274–1285.

[14] Wu Suxiang, Fan Honglei, Cheng Yan, etal. Organic catalytic reactions in CO_2/H_2O mixed medium. Chemistry Progress, 2010, 22(7): 1286–1294.

[15] Zhou Zongqing. Technological progress on the synthesis of polymer materials from CO_2. Natural Gas Chemical Engineering, 1994, 19(3): 41–45.

[16 Bai Yushan, Yu Yan. Green chemistry conception in the jointed soda production method. Chemistry Education, 2004(9): 48–50.

[17] Tomishige, K., Kuninori, K. Catalytic and direct synthesis of dimethyl carbonate starting from carbon dioxide using CeO_2-ZrO_2 solid solution heterogeneous catalyst: Effect of H_2O removal from the reaction system. Applied Catalysis A, 2002, 237(1–2): 103–109.

[18] Carnes, C. L., Klabunde, K. J. The catalytic methanol synthesis over nanoparticle metal oxide catalysis. Journal of Molecular Catalysis A: Chemical, 2003, 194: 227–236.

[19] Shiflett, M. B., Yokozeki, A. Solubilities and diffusivities of carbon dioxide in ionic liquids: [bmim][PF_6] and [bmim][BF_4]. Industrial & Engineering Chemistry Research, 2005, 44(12): 4453–4464.

[20] Toshiyasu Sakakura, Jun-Chul Choi, Hiroyuki Yasuda. Transformation of Carbon Dioxide. Chemical Reviews 2007, 107(6): 2365–2387.

[21] Roosen, C., Ansorge-Schumacher, M., Mang, T., et al. Gaining pH-control in water/carbon dioxide biphasic systems. Green Chemistry, 2007, 9: 455–458.

[22] Mengxiang Fang, Shuiping Yan, Zhongyang Luo, et al. CO_2 chemical absorption by using membrane vacuum regeneration technology. Energy Procedia, 2009, 1: 815–822.

[23] Cheng, H. Y., Meng, X. C., Liu, R. X., et al. Cyclization of citronellal to p-menthane-3,8-diols in water and carbon dioxide. Green Chemistry, 2009, 11: 1227–1231.

9 Green chemistry and chemical processes for biomass utilization

Biomass resources are considered to be the best alternative to fossil fuels currently. The comprehensive utilization of biomass is an important sustainable development strategy [1–4].

9.1 Introduction

9.1.1 Natural Conditions of Biomass

A large variety of plants on the earth make up giant chemical plants. They use solar energy to continuously synthesize inorganic substances such as water and carbon dioxide into various organic substances, further providing abundant and renewable biomass resource for human.

China is a large agricultural country with abundant biomass resources. In addition to crop land, it also possesses wasteland, grassland, saline land and marshland, which if managed effectively, can enormously increase the potential for agricultural development.

9.1.2 Biomass Concept

Biomass refers to various organisms that are produced through photosynthesis using the atmosphere, water, and land, etc.. That is, all living organisms with capable of growing are collectively referred to as biomass. It includes plants, animals and microorganisms [5–7].

9.1.3 Classification of Biomass

The main components of biomass fuels are C, H and O, while the main constituents of fossil fuels are C and H. Typical biomass resources include cellulose, hemicellulose, lignin, oil, starch and chitin. This book will focus on the chemical process of the former four biomass resources [10–11].

1. **Cellulose**
   Cellulose is the most abundant and widely distributed organic matter in nature, with 500 billion tons synthesized each year on earth, only 7 million tons of which undergoes chemical modification. It is a large molecule formed by connecting glucose unit with β-1,4-glycosidic bond.

2. **Hemicellulose**
   In the cell wall of plants, a part of plant polysaccharide is the hemicellulose, which is symbiotic with cellulose and soluble in alkaline solution, and far more

https://doi.org/10.1515/9783110479317-009

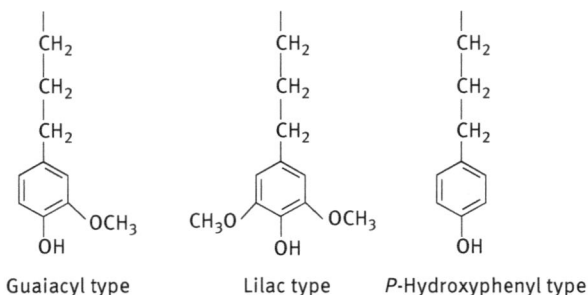

Figure 9.1: Basic structural unit of lignin.

soluble than cellulose in case of acid. Hemicellulose is a heterogeneous polymer composed of several different types of monosaccharides, namely pentose and hexose, xylose, arabinose and galactose.

3. **Lignin**

    The total amount of lignin synthesized is only less than cellulose, i.e., 150 billion tons of lignin per year. China's annual crop straw merely contains 700 million tons of lignin. From a paper industry by-product utilization perspective, lignin has not been fully exploited, causing environmental pollution.

    The basic structural units of lignin, as shown in Figure 9.1, contain hydroxyl groups and double bonds with active hydrogen, which can form hydrophilic group involved in the synthesis of various chemical products.

4. **Lipid**

    Lipid is a general term used to refer to oil and fat obtained from animal and plant, the main ingredient of which is glycerol with three fatty acids, simply known as triglycerides. Generally speaking, "oil" refers to the liquid state at room temperature, and "fat" refers to the semisolid or solid state at room temperature, but it is customary not to distinguish the two terms. From the structure chart, triglycerides can be considered as a glycerol molecule condensed with three fatty acid molecules (Figure 9.2).

    If the three fatty acids are the same, the resultant is a homotype triglyceride, otherwise, generating heterotype triglyceride. Most natural oils are mixed triglycerides containing acids, in addition to small amounts of phospholipids, wax, sterols, vitamins, hydrocarbons, fatty alcohols, free fatty acids, pigment and odorous volatile fatty acids, aldehydes and ketones.

9.1.4 Use of Biomass

Through photosynthesis, plants convert about 200 billion tons of CO_2 into carbohydrates each year and store about 3.1×10^{13} J of solar energy. The stored energy is about

$$\begin{array}{ccc}
\begin{array}{l}
CH_2\!-\!OH \\
| \\
CH\!-\!OH \\
| \\
CH_2\!-\!OH
\end{array}
+
\begin{array}{l}
R^1COOH \\
\\
R^2COOH \\
\\
R^3COOH
\end{array}
\longrightarrow
\begin{array}{l}
\qquad\quad O \\
\qquad\quad | \\
CH_2\!-\!O\!-\!C\!-\!R^1 \\
| \qquad\quad O \\
| \qquad\quad | \\
CH\!-\!O\!-\!C\!-\!R^2 \\
| \qquad\quad O \\
| \qquad\quad | \\
CH_2\!-\!O\!-\!C\!-\!R^3
\end{array}
+ 3H_2O
\end{array}$$

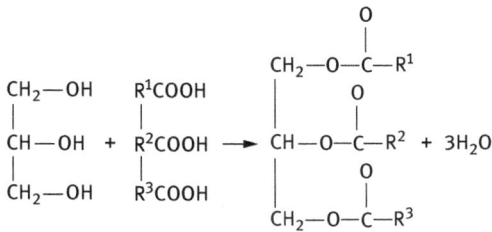

Figure 9.2: Preparation process of oil.

10–20 times of the total energy consumption in the world, but the utilization rate is less than 3%.

Among the main components of plant resources, what humans use most is cellulose, the least is hemicellulose, and the most difficult to use is lignin. In addition to cellulose, which has been used in the manufacture of paper and textiles, recent applications include the following:

1. **Preparation of Biodegradable Polymer Materials**
   Cellulose, lignin and starch of plant polysaccharide can be used to make valuable biodegradable plastics, such as in the development of American Wamer-Lambert Company's Novon brand of biodegradable plastics, which is a typical example of this kind of material. These products have not been widely used because of their high price, but have shown potential to solve the problem of "white pollution".
2. **The Conversion of Cellulose to Glucose and Alcohol**
   It is not difficult to further split cellulose into glucose when the crystal structure is destroyed and lignin and hemicellulose are separated. The conversion rate is above 95%. At present, there are some effective biological methods, which can directly convert cellulose into alcohol. The resulting sugar ethanol can replace the fossil oil and coal as the basic raw materials to develop a variety of organic chemical products, including gasoline and other basic energy sources.
3. **Utilization of Lignin**
   Lignin is a renewable plant fiber resource and stores the highest solar energy in plant fibers. It is the best substitute for petroleum. In the current paper industry, it is discharged as waste and causes serious pollution to the environment. In recent years, there have been many advances in the application of lignin, including its liquefaction directly and then used as a fuel, adhesives, and some low molecular weight chemicals such as phenols and organic acids.

9.1.5 Biomass Distribution

Of the 240,000 known vascular plants, about 25% of them are edible. The world's food comes from about 100 species, of which about 3/4 are crops such as wheat, rice, corn, potato, barley, sweet potato and cassava.

China has a vast territory, and has a wide range of biomass resources and a huge amount of resources. It provides abundant raw materials for the development and utilization of biomass.

In agricultural biomass resources of China, the distribution of crop straw is consistent with the distribution of crop planting. China's crop straw is mainly distributed in the eastern region, including the North China Plain and the northeast plain provinces and cities. Hebei, Inner Mongolia, Liaoning, Jilin, Heilongjiang, Jiangsu and other major grain-producing areas are also the major straw producers. The provinces with high straw resources of unit national land area are Shandong, Henan, Jiangsu, Anhui, Hebei, Shanghai and Jilin provinces.

The agricultural by-products include rice husk, corncob, bagasse and other sources, which are produced from grain processing factory, food processing factory, sugar factory and brewery. These resources are present in huge amounts, are of relatively concentrated origin and easy to collect. Among them, rice husk is mainly produced in Northeast China, Hunan, Sichuan, Jiangsu and Hubei provinces; corn cob in Northeast China and Hebei, Henan, Shandong and Sichuan provinces and bagasse in Guangdong, Guangxi, Fujian, Yunnan and Sichuan provinces.

The main types of forestry biomass resources in China include the harvested residues of mature or over-mature forests in forests, the clean-up of dead wood, the tending and cultivation of near-mature forests, and the thinning of middle-aged forests. According to the results of the sixth national forest resources inventory, forest areas of the northeast and Inner Mongolia, North China and Central Plains, the southern forest region and the tropical region of Southern China are the regions where the forestry biomass resources are concentrated.

9.1.6 Comprehensive Utilization of Biomass

Most of the organics in plant on the earth is of the fiber type (including rice straw, wheat straw, weeds, and all trees, etc.), among which the useful plant cellulose is difficult to be degraded by microorganisms or digested by humans due to crystallization and symbiosis with lignin.

Cellulose constitutes the skeletal framework of plants, while the structures of hemicellulose and lignin are more complex and are dispersed in and around the fibers as inclusions. The bonding force of lignin condenses the fibers together (Figure 9.3). The separation of cellulose, hemicellulose and lignin is a key step in a broad range of applications.

The invention of papermaking in ancient China is one of the early results of the artificial separation of cellulose, and the most important industrial sector dependent on plant resources is still pulp and paper industry. At present, alkaline pulping is widely used in the paper industry to remove lignin. The black liquor produced by the waste water can cause great pollution to the environment, which is particularly

Lignocellulose

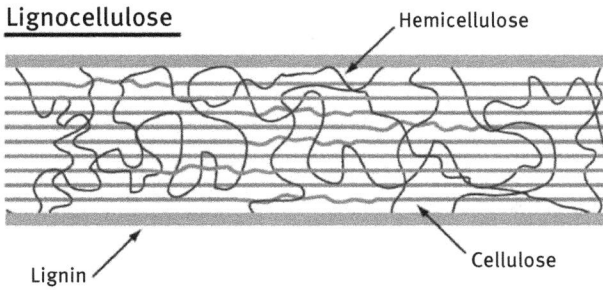

Figure 9.3: Lignocellulose structure.

prominent in the Huaihe River basin. At present, new green pulping technology and biochemical pulping technology are being developed. It is always the goal of human beings to transform plant resources into energy and basic raw materials. With regard to this many new methods of green chemistry have made some breakthroughs in this field.

1. **Biomass Conversion**

As shown in Figure 9.4, biomass conversion consists of two main processes: pretreatment and enzymatic hydrolysis. Through the pretreatment of biomass raw materials, cellulose and hemicellulose are exposed, and thus the surface area of substrate and enzyme is increased. Using water, acid, alkali and organic solvent pulping processes, compared with the traditional pulping methods, in the pretreatment of wood has yielded better results. Different types of lignocellulosic raw materials should be treated differently. Pretreatment is followed by enzymatic hydrolysis with cellulose, usually produced by fungi such as *Trichoderma*, *Penicillium* and *Aspergillus*, which can effectively decompose the cellulose microfibrils into different carbohydrate components. The enzymatic hydrolysis process can be carried out separately or with other biomass conversion processes. Separate hydrolysis and fermentation (SHF) offers greater flexibility compared to other process. Simultaneous saccharification and fermentation (SSF) has been considered as an efficient method for the production of bioethanol.

New technology is being developed to realize effective separation of cellulose, hemicellulose and lignin, which can be used for industrial processing. The most effective separation can be achieved by the combination of pretreatment and enzyme hydrolysis. In recent years, the basic research on the kinetics of biomass conversion has focused on reducing the cost of hydrolytic enzymes. The research project carried out jointly, by Novozymes, Genencor and American National Renewable Energy Laboratory, after 4 years of hard work, has finally succeeded in showing that costs involved in the enzyme hydrolysis of the model substrate can be reduced by 30 times.

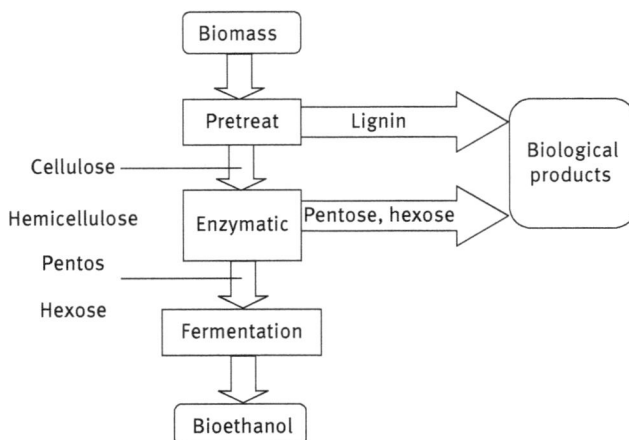

Figure 9.4: Process flow diagram of biomass conversion.

2. **Thermochemical treatment**

   This method uses the thermochemical process to produce synthetic gas through gasification. The chemical treatment method includes pretreatment, gasification, purification and regulation, and can produce a mixture of hydrogen, carbon monoxide, carbon dioxide and other gases. The products produced by the above two methods are intermediate products, which can be used as industrial chemicals after further synthesis and processing.

   If pyrolysis is used, the process of drying, grinding and screening will be needed, so that the raw material easily enters into the reaction vessel. This technology is already viable in industry. Biomass gasification is generally carried out in two stages. In the first stage, temperature is 450–600 °C. The volatile components in the biomass are pyrolyzed (burned under anoxic conditions). Rapid pyrolysis occurs at a lower temperature range of 450–550 °C, resulting in liquid pyrolysis of oil and a small amount of gas. The oil produced in the process of fast pyrolysis is 60–75% of the original fuel, and can be used as raw material for producing value-added chemicals or directly as biomass fuel. At higher temperature, pyrolysis vapors can be mainly converted into carbon monoxide, hydrogen, methane, volatile tar, carbon dioxide and water vapor. High-temperature pyrolysis will leave the solid residue of carbon, which accounts for 10–25% of the original fuel. A second gasification stage is required to treat the solid residue. The carbon conversion takes place at 700–1,200 °C, where it reacts with oxygen to produce carbon monoxide.

3. **Supercritical method**

   Using supercritical fluid instead of organic solvent can help avoid pollution. In Canada, lignocellulosic biomass is the main source of available raw materials. Canada A1cell Company uses supercritical ethanol as solvent to extract hemicellulose and lignin of plants and separate them from cellulose. Five monosaccharides including glucose, galactose, mannose, xylose and arabinose were obtained by hydrolysis. The whole process basically produces no waste, does not pollute the environment, and also makes full use of energy.

In addition, some nonchemical and combined chemical separation methods can also be used to separate cellulose efficiently.

Green chemistry opens a new field of biomass chemistry, which is the study of chemical conversion from green plants, by which cellulose, hemicellulose and lignin are separated from effective and cheap methods, and then degraded to glucose monomers, alcohol and other organic chemical raw materials, in which way, the earth's huge green plants are transformed into human food, livestock feed, energy, and organic chemical raw materials. This discipline is not only to protect the ecological environment, but also an important backup for traditional technology that can not meet the needs of human survival and development. China uses 7% of the world's arable land to feed 22% of the world's population. The arable land is reduced at a rate of millions hectares per year. In addition to developing new ecological agriculture products, biomass is directly converted into usable chemical products. It will be more efficient. In the United States, many kinds of biomass, such as sugar and starch crops, are used for biomass conversion. The sugars isolated from biomass will become value-added products after further processing.

9.2 Properties and Analysis Methods of Main Components in Biomass

9.2.1 Physical and Chemical Properties of Cellulose

1. **Structure of cellulose**
   1) **Chemical structure**

      Cellulose is a macromolecular polysaccharide composed of D-glucose in the form of a β-1,4-glycosidic bond with a molecular weight of about 50,000 to 2,500,000, which is equivalent to about 300 to 15,000 glucose groups. The molecular formula can be written as $(C_6H_{10}O_5)_n$, where n is the degree of polymerization. The degree of polymerization of cellulose in nature is about 10,000. Figure 9.5 shows the structure of cellulose.

      Figure 9.5 shows that in addition to the two glucose residues at the head and tail of cellulose, the intermediate residues contain only three free hydroxyl groups: a primary hydroxyl group, and two secondary hydroxyl groups, their reaction activities being different. The primary hydroxyl group is not involved in the formation of intramolecular hydrogen bonds, but it can participate in the formation of adjacent intermolecular hydrogen bonds.

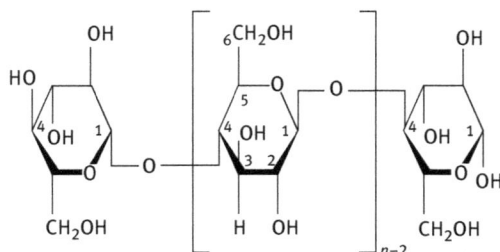

Figure 9.5: Structure of cellulose.

2) **Molecular Weight and Degree of Polymerization**

The molecular formula of cellulose can be simply expressed as $C_6H_{11}O_6$-$(C_6H_{10}O_5)$ $_n$-$C_6H_{11}O_5$, a base ring of cellulose with relative molecular weight of 162. The polymerization degree of cellulose is DP = n + 2, and the relative molecular weight of cellulose is M = DP × 162 + 18.

When DP is large, in the above formula 18 is negligible, so the relationship between the relative molecular mass M of the cellulose and the degree of polymerization of polymerization DP is M = DP x 162 + 18 or DP = M/162.

3) **Physical Structure**

The physical structure of cellulose refers to the relative arrangement of the structural units with different scales, which includes the chain structure and aggregation structure of the polymer.

The chain structure, also known as the primary structure, shows the geometric arrangement of atoms or groups in a molecular chain. There are two kinds of structures with different scales. The short-range structure is the first hierarchical structure, which refers to a single polymer within one or several structural units of the chemical structure and three-dimensional chemical structure. The remote structure is the second hierarchical structure, which refers to the size of a single polymer and the various shapes in the space. For example, it may be a straight chain, irregular or folded chain, spiral chain, etc.

Aggregation structure, also known as the secondary structure or the third hierarchical structure, refers to the internal structure of the polymer, which includes the crystal structure, amorphous structure, orientation structure and liquid crystal structure, and describes how the polymer aggregates are stacked between each molecule, i.e., whether it is intertwined with the group structure, the folding chain from the regular stack of crystals and so on.

The chain structure of polymers is the most important structure influencing various properties of the polymer, such as melting point, density, solubility, viscosity and adhesion. The aggregation structure is the main factor determining the performance of polymer products.

2. **Physical Properties of Cellulose**
   1) **Moisture Absorption and Desorption of Cellulose**
      The free hydroxyl groups of cellulose have strong affinity for polar solvents and solutions. Dried cellulose is placed in the atmosphere and it can absorb moisture from the air to a certain moisture content. The absorption of water or steam from the atmosphere by cellulose is called adsorption. The release of water or steam from cellulose due to a decrease in the partial pressure of steam in the atmosphere is referred to as desorption. The adsorption of water vapor from cellulose affects many of the important properties of cellulosic fibers. For example, as the amount of cellulose adsorbed changes in the fiber due to swelling or contraction, properties related to fiber strength and electrochemical properties will change. In addition, during the drying of the paper, desorption of water from cellulose takes place.
   2) **Swelling and Dissolution of Cellulose Fibers**
      The swelling of cellulose fiber can be divided into limited and unlimited expansion. There is a limit to the amount of cellulose absorbing swelling agent, and the degree of swelling is also limited. Infinite swelling means that the swelling agent can be swollen into the amorphous and crystalline regions of the cellulose, but does not form new swelling compounds, so the amount of swelling agent that enters into the amorphous and crystalline regions is not limited. Most of the cellulose swelling agents are polar, because the hydroxyl groups on the cellulose itself are polar. Usually, water or LiOH, NaOH, KOH, RbOH, CsOH aqueous solutions can be used as the cellulose swelling agent; phosphoric acid can also cause fiber swelling. In general, the greater the polarity of the liquid, the greater the degree of swelling.

      The dissolution of cellulose takes place in two steps: first in the swelling phase, where dissolution occurs when the cellulose swells indefinitely. The original X-ray pattern of the cellulose disappears and new X-ray image no longer appears. Cellulose can be dissolved in certain inorganic acids, bases, and salts. In general, the dissolution of cellulose uses a coordination compound of copper hydroxide with amine or ammonia, such as cuprammonium hydroxide solution or copper ethylenediamine solution. Cellulose can also be dissolved in non-aqueous solvents based on organic solvents.
   3) **Thermal Degradation of Cellulose**
      When the cellulose is heated, the degree of polymerization decreases. In most cases, the hydrolysis and degradation of cellulose occurs when the cellulose is thermally degraded. In severe cases, decomposition of cellulose occurs, and even carbonization or graphitization may occur. At the temperature of 25-150°C, the water sorption of cellulose physically begins to desorption; at 150-240°C some of the glucose groups in the cellulose structure begin to dehydrate; at 240-400°C, the glycosidic bonds in the cellulose structure begin to break, and some C-O bonds and C-C bonds also began to break and

produce some new products and low molecular weight volatile compounds; above 400° C, the remaining part of the cellulose structure undergoes aromatic cyclization and gradually forms a graphite structure.

3. **Chemical Properties of Cellulose**
   1) **Degradation Reactions of Cellulose**
      In a variety of environments, cellulose has the potential for degradation. There are several different types of degradation.
      (1) Acid hydrolysis degradation
         Cellulose can be dissolved in Schwitzer (copper hydroxide solution) reagent or concentrated sulfuric acid. Although cellulose is not easy to be acid hydrolyzed, dilute acid or cellulase can degrade cellulose to produce D-glucose, cellobiose and oligosaccharides. Cellulose in the acid or alkali catalysis are prone to hydrolysis reaction, but the reaction is different. In the acidic medium, the hydrolysis of the beta glycosidic bond is given by:

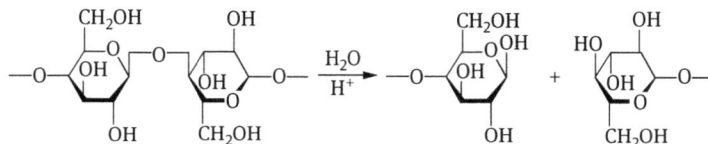

         When fully hydrolyzed, the final product is D-glucose. Under mild conditions, hydrolyzed cellulose is obtained after hydrolysis (cellulose with reduced degree of polymerization; when n falls below 200, it is in powder form).
         In the dilute acid and high-temperature hydrolysis, the monosaccharides can be further decomposed:

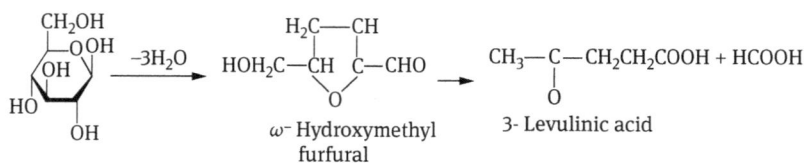

         The hydrolysis of cellulose at high temperature has the effect of autocatalysis.
      (2) Alkaline Degradation
         In alkaline medium, cellulose β-glycosidic bond is more stable, however, alkaline hydrolysis can be performed under high temperature, and the reaction is very complicated. The first is that the terminal group is ring-opened into an aldehyde form and converted into a ketone form under the action of a base, causing cellulose to remove the glucose group one by one from the terminal group, and undergoes a series of isomerization reactions.

(3) Oxidative Degradation

Cellulose is oxidized by air, oxygen, bleach, and the free hydroxyl groups at the C-2, C-3, and C-6 positions of the cellulose glucosyl ring, and at the reducing terminal C-1 position. According to different conditions, aldehyde group, ketone group or carboxyl group, and oxidized cellulose are correspondingly generated. The structure and properties of oxidized cellulose are different from those of the original cellulose. In most cases, the degree of polymerization of cellulose decreases with the oxidation of hydroxyl groups.

(4) Microbial Degradation

Cellulose is degraded by the action of microbial enzymes, resulting in degradation of cellulose polymerization degree. In the study of enzymatic hydrolysis of cellulose, we hope to find a low-cost and high-efficiency method.

2) Esterification and Etherification of Cellulose

Cellulose is a compound containing polyols. Organic and inorganic acids react to produce ester derivatives; although a strong acid such as nitric acid, sulfuric acid and phosphoric acid can directly react with cellulose, to generate inorganic esters, other strong acids such as perchloric acid and other halogen acids can directly esterify cellulose. Organic acids, acid anhydrides and acid chlorides act on cellulose to produce organic acid esters, of which only formic acid can directly esterify the cellulose with a relatively high degree of substitution of the ester; the degree of substitution of other organic acids is low, even at the temperature of its boiling point. However, the conversion of these organic acids to anhydrides can esterify cellulose and have a high degree of substitution. In addition, an inorganic acid or a salt such as magnesium perchlorate is generally used as a catalyst in the esterification reaction of an organic acid and cellulose.

The alcoholic hydroxyl groups of the cellulose can be etherified with an alkyl halide or other etherifying agent under basic conditions to form the corresponding cellulose ether. For example, under alkaline conditions, cellulose and dimethyl sulfate react to yield cellulose methyl ether, referred to as methyl cellulose. Methyl cellulose can continue to take part in methyl substitution reaction:

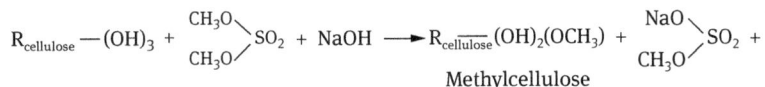

$$R_{cellulose}-(OH)_3 + \begin{array}{c} CH_3O \\ CH_3O \end{array}\!\!>\!\!SO_2 + NaOH \longrightarrow R_{\overline{cellulose}}(OH)_2(OCH_3) + \begin{array}{c} NaO \\ CH_3O \end{array}\!\!>\!\!SO_2 +$$

$$\text{Methylcellulose}$$

Another kind of cellulose ether commonly used in industry is carboxymethyl cellulose (short for CMC), which is obtained by reacting chloroacetic acid and alkali cellulose:

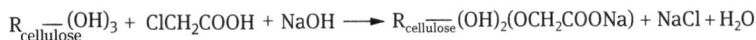

$$R_{\overline{cellulose}}\!-\!(OH)_3 + ClCH_2COOH + NaOH \longrightarrow R_{\overline{cellulose}}(OH)_2(OCH_2COONa) + NaCl + H_2O$$

3) **Chemical Modification of Cellulose**

Cellulose being natural polymer, there are some shortcomings in performance, with regard to chemical resistance, limited strength, etc. Cellulose
has been chemically modified to obtain new cellulose derivatives with better
performance, including resistance to heat and microbes, abrasion resistance,
acid resistance and improved wet strength, adhesion and absorption of dyes.
More graft copolymerization and cross-linking reactions were applied in the
chemical modification of cellulose. Chemical modification has a wide range,
such as: free radical or ionic graft copolymerization; under the action of heat,
light, radiation or cross-linking agent, covalent bonds are formed between
cellulose chains to produce gels or Insolubles form a cross-linking reaction of
the ester. This chemical modification achieves fire-resistant, heat-resistant,
microbe-resistant, abrasion-resistant, acid-resistant, and increases cellulose
wet strength, adhesion and dye absorption.

9.2.2 Physical and Chemical Properties of Hemicellulose

Hemicellulose is not a homogeneous glycan, but a group of complex glycans. The raw
materials are different, and the components of the complex glycans are also different. The
relative molecular mass of hemicellulose is not large, and the degree of polymerization is
usually around 200. The molecules are basically linear but with various short side chains.

1. **Chemical structure of Hemicellulose**

The hemicellulose is composed of D-glucose, D-mannose, D-xylose, L-arabinose,
D-glucuronic acid, 4-methoxy-D-galacturonic acid and D-glucuronic acid. There
is also a small amount of L-rhamnosyl, L-fucosyl and various neutral glycosyl
groups with methoxy and acetyl groups. These constituent units are generally
not homoglycans, i.e., composed of one structural unit, but are heterogeneous
polysaccharides composed of 2–4 kinds of structural units. The specific composition varies greatly with tree species.

The types and quantities of hemicellulose contained in coniferous wood,
hardwood and grass raw materials are different. The hemicellulose of coniferous
wood is dominated by polygalactose glucose, mannose and poly-arabinose-
4-methoxy-glucuronic acid xylose. Broad wood hemicellulose is mainly poly-
acetyl-4-methoxy-glucuronic acid xylose, accompanied by a small amount
of polyglycans mannose. The hemicellulose in gramineous plants is mainly
poly-arabinose 4-methoxyl glucuronic acid xylose.

2. **Physical Properties of Hemicellulose**

In general, the hemicellulose has a higher solubility than the natural state of
hemicellulose. Poly arabinose galactose is soluble in water. Polytetra polyglycine
glucuronic acid xylose is easy to dissolve in water, while hardwood polyglycolic
acid xylose is less soluble in water than softwood. When birchwood is extracted

by alkaline extraction, polyxylose containing more glucuronyl groups is easily extracted. Due to the random structure and branching of hemicellulose, and with many hydrophilic groups on the main chain and the side group, their moisture absorption is larger than that of cellulose. The swelling capacity of hemicellulose is large; the hemicellulose contains a large number of free hydroxyl groups, which can play a role in bonding when pressed.

Due to the random structure and branching of hemicellulose, and the main chain and pendant with many hydrophilic groups, its hygroscopicity is larger than cellulose. The swelling ability of hemicellulose is larger; hemicellulose contains a large number of free hydroxyl groups, which can play a binding role when it is under hot pressed.

3. **Chemical properties of Hemicellulose**

The chemical properties of hemicellulose and cellulose are similar under hydrolysis, esterification, etherification, graft copolymerization and cross-linking, pyrolysis and combustion reactions, as well as phosphide reactions. Furthermore, due to the random structure of hemicellulose and the presence of pendant groups, it is easier to carry out such reactions with hemicellulose than in the case of cellulose. Therefore, it is often regarded as a kind of polysaccharide that is most susceptible to external conditions and most likely to change and react in wood.

1) **Acid hydrolysis**

The hemicellulose bond is broken in the acid medium, leading to degradation, which is the same as the acid hydrolysis of cellulose. However, hemicellulose and cellulose are very different in the structure, For example, hemicellulose has many types of glycosyl radicals, such as pyridine, furan, α and β glycosides, and D and L type. The connections between glycosyl groups are also diverse, with $1 \rightarrow 2$, $1 \rightarrow 3$, $1 \rightarrow 4$, and $1 \rightarrow 6$ connections, so the reaction is more complex than cellulose.

The hydrolysis rate of methyl pyran arabinose with glycosides was observed to be the fastest when the hydrolysis was carried out with 1.5 mol/L hydrochloric acid at 75 °C. The sequence of hydrolysis rate of hemicelluloses with different structures is: methyl-D-galactopyranoside glycoside, methyl-D-pyrano xyloside glycoside, methyl-D-pyrano mannose glucoside. And the most stable are methyl-D-pyranoglucose glycosides.

In most cases, the β-D type of each glycoside is more hydrolyzable than the alpha-D type. In general, furan aldose glycosides are much more acidic than the corresponding pyran aldose glycosides. Glucuronic acid glycosides have a slower hydrolysis rate than the corresponding glucose glycosides because of the stable effect of the carboxyl groups on the glucoside bond.

2) **Alkaline Degradation**

Hemicellulose can degrade under alkali conditions, and alkaline degradation includes alkaline hydrolysis and peeling.

In the 5% NaOH solution, the hemicellulose bond can be broken by hydrolysis at 170 °C, which is called alkaline hydrolysis. In each pair of glycosides,

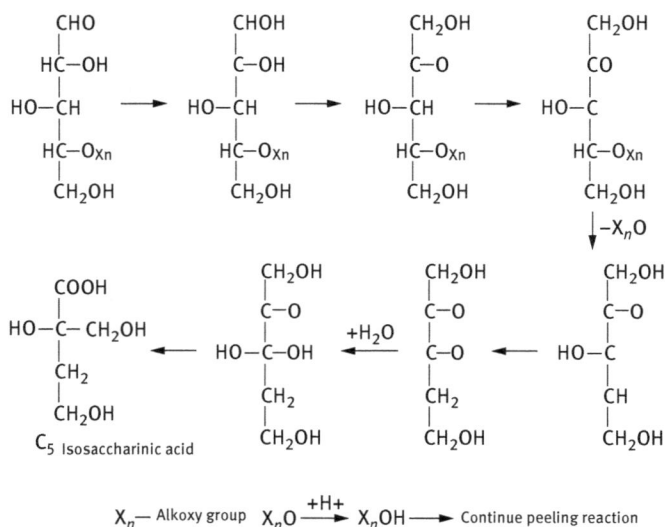

Figure 9.6: Hemicellulose reaction.

where the methoxylate of the glycoside is opposite to the hydroxyl group on the second carbon atom, the rate of alkaline hydrolysis is much higher than with the corresponding glycoside. The alkaline hydrolysis rate of furan-type glycosides is much higher than that of pyran-type glycosides. The basic hydrolysis rate of methyl-α-pyran glucuronic acid glycoside with methyl-β-pyran glucuronic acid glycoside is higher than that of furan-type glycosides.

Peeling reaction can occur under mild alkaline conditions. With cellulose, the peeling reaction of hemicellulose is also carried out by glycosylation of the reducing end groups of the glycans. However, since hemicellulose is a homogeneous polysaccharide composed of a plurality of saccharides, the reducing end groups of hemicellulose have various glycosyl groups, and are also branched, so the peeling reaction is more complicated. The peeling reaction of hemicellulose is described below with only 1→4 linked xylose as an example (see Figure 9.6).

As with cellulose, the alkaline peeling reaction of hemicellulose is also terminated to some extent, and the termination reaction is the same as that of cellulose, and the reducing terminal group is converted into a biocidal group. Since no aldehyde group is present on the terminal group, the peeling reaction can no longer occur and the degradation is thus terminated.

9.2.3 Physical and Chemical Properties of Lignin

Generally, the woody plants are extracted with water and phenyl alcohol, the left material after the hydrolysis with inorganic acid to remove cellulose and hemicellulose is called lignin. The lignin content of coniferous wood is 25–35% and hardwood is 20–25%, while gramineous plants generally contain 15–25% lignin. The content of lignin in plant raw materials varies greatly with the different morphological parts and with plant varieties.

1. **The Structure of lignin**
    Lignin has high carbon content (60–66%) and low hydrogen content (5–6.5%), showing that it is aromatic. Phenylpropane is as a basic structural unit of lignin, and is connected to each other via a ether bond and a carbon bond to form a polymer having a three-dimensional structure. The type, number, and connection of the structural units vary greatly with tree species. The structural elements of the three forms are shown in Figure 9.1. The three structural units contain hydroxyl groups, but their methoxy content is different. The three carbon atoms on the side chain of the benzene ring can be attached to different groups such as methoxy, hydroxy, carbonyl, double bonds and the like.

    The basic structural units of lignin in broad-leaved and coniferous wood are different. The broad-leaved wood lignin contains a large number of Guaiacyl-based and lilac-based structural units, and the coniferous wood lignin contains a large number of Guaiacyl-based structural units.

    There are two ways of connecting phenylpropane-based structural units: one is through an ether linkage and the other is a carbon-carbon bond, with the ether linkage being the main mode. In the lignin, about two-thirds to three-fourths of the phenylpropane structural units are connected through the ether bond, while only one-fourth to one-third of the phenylpropane structure unit is connected through the carbon-carbon bond. Some of the connected sites can occur between the phenolic hydroxyl groups of the benzene ring, and some between the three carbon atoms in the side chain, while others occur between the benzene ring and the side chain and so on.

2. **The physical properties of lignin**
    The physical properties of lignin are closely related to the source of lignin samples, such as plant species, organization and location, separation of samples and purification methods.

    1) **Solubility**
        The main solvent parameters determining the solubility of lignin are the hydrogen bonding capacity and cohesive energy density. Lignin is a kind of aggregate. There are many polar groups in the structure, especially more hydroxyl groups, resulting in strong intramolecular and intermolecular hydrogen bonds, so the log lignin is insoluble in any solvent.

        During separation of lignin by condensation or degradation, many physical properties, such as the solubility, will be changed; the soluble lignin has amorphous structure, while the insoluble lignin has a raw fiber structure. The presence of phenolic hydroxyl and carboxyl groups allows lignin to dissolve in concentrated, strong alkaline solution. Alkaline and lignin sulfonates are usually dissolved in dilute alkali, water, salt solutions and buffer solutions. Lignin is not soluble in all solvents.

2) **Thermal properties**

Lignin is an amorphous thermoplastic polymer. It is slightly brittle at room temperature, does not form a film in solution, and has only glass transition properties. Below the glass transition temperature (Tg, the thawing temperature of the segment movement), lignin is in a glass solid state; above the glass transition temperature, the molecular chain When exercise occurs, lignin softens and becomes viscous.

The glass transition temperature of separated lignin varies with tree species, separation method, molecular weight and water content. The softening temperature of dried lignin is around 127–129 °C, and the softening temperature decreases with the increase of water content of lignin sample. The softening point of the dry periodate is 193 °C, and when the water content is 27.1%, the softening point drops to 95 °C. Water plays the role of a plasticizer in lignin. The higher the molecular weight of lignin, the higher the softening point.

3. **Chemical properties of lignin**

The chemical properties of lignin include various chemical reactions of lignin, such as halogenation, nitration and oxidation reactions occurring on the benzene ring; reactions occurring on the side chain of benzyl alcohol, aryl ether and alkyl ether bonds; as well as the modification reaction and the color reaction.

1) **Halogenation reaction**

Lignin readily reacts with chlorine (available chlorine or chlorine) to produce chlorinated lignin. This method is used to achieve the purpose of paper bleaching. When the halogenation reaction is carried out at room temperature, the substitution reaction is carried out only on the aromatic ring, and the halogenation reaction may occur on the side chain as the temperature increases.

Bromine reacts less easily with lignin in the bromination reaction, but in acid medium the reaction is also easy to carry out. Commonly used bromination reagents are NaBr, NH_4Br, BrCl (chlorine and bromine reaction products) and so on. Early wood bromide treatment is based on the lignin-based bromination reaction, which can achieve the purpose of wood flame retardant.

2) **Sulfonation reaction**

In the sulphite pulping process, a sulfonation reaction occurs between the lignin and sulphite in the wood to form lignin sulfonate, which is dissolved in the sulphite waste liquid and makes the pulp wood Elements removed. Lignin sulfonation reaction occurs mainly in the side chain of lignin, due to the presence of active benzene on the side chain structure, as well as alkyl ether and aryl ether.

3) **Oxidation reaction**

Lignin is more easily oxidized than cellulose and hemicellulose, and can be oxidized in neutral, alkaline or acidic media. Wood drying process becomes darker, but also due to lignification oxidation of the formation of dark quinone-type structure. The reason why the surface properties of wood and its products deteriorate under sunlight and air is not only the photochemical degradation of lignin but also the lignin oxidation reaction.

In the strong conditions of the oxidation reaction, the lignin can undergo decomposition, and lead to the formation of a variety of low-grade aliphatic hydrocarbons such as carboxylic acid and other substances. In the presence of a base, oxygen in the air can oxidize lignin to yield humic acid.

4) **Reaction with formaldehyde**

Phenolic hydroxyl in lignin has the properties of general phenols, and one can introduce hydroxymethyl groups in the ortho position and para position of phenolic hydroxyl groups. Because of its large spatial barrier, the reaction is more difficult than with ordinary phenols. Under the catalysis of the base, most of the formaldehyde enters the aromatic ring, and a small part of the formaldehyde enters the active site on the side chain and is present in the free hydroxymethyl state. Under acidic conditions, formaldehyde is protonated and the hydroxymethylation reaction proceeds easily. Hydroxylation immediately follows a condensation reaction, which results in the formation of methylene cross-links between lignin molecules, the increase in molecular mass, and the occurrence of resination.

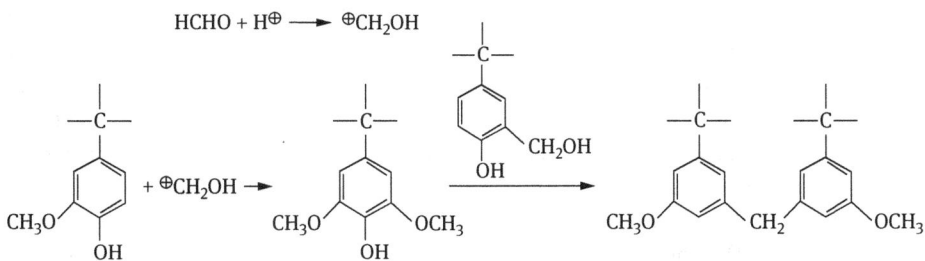

5) **Pyrolysis reaction**

The three main components of plants have the thermal stability of the order: hemicellulose <cellulose <lignin. Lignin decomposes at 300–350 °C under a severe pyrolysis reaction, by which time hemicellulose and cellulose have long been decomposed. The pyrolysis of lignin was not terminated until 400–450 °C. It is mainly converted to coke, so it is the bearer of flameless combustion of wood.

9.2.4 Biomass Solvent System and Law

1. Pretreatment solvent on lignocellulose raw material

The three major components of lignocellulosic biomass, including cellulose, hemicellulose and lignin, are intertwined to form a complex structure, which hinders the hydrolysis of cellulose. At the same time, because of the hydrogen bond, the cellulose chains form a more compact crystalline region, which is an important reason for the degradation and effective utilization of natural lignocellulose.

At present, the pretreatment methods of lignocellulose generally include four categories: physical, chemical, physical-chemical and biological methods. Physical methods mainly include grinding, as well as grinding and radiation; chemical methods use acid, alkali and organic solvents and other raw materials; physical-chemical method involves both physical and chemical processes, the typical treatment being neat blasting; while biological methods use lignin-degrading microorganisms or enzymes such as laccases produced by them to remove lignin to release its effect on cellulose.

Among them, the chemical method is mainly to cellulose, hemicellulose and lignin swelling, and destruction of its crystalline structure, so that to destroy its dense structure by degrading part of their components. Commonly used solvents include acids, bases, ammonia and organic solvents. Acid can be concentrated or dilute, the current dilute acid treatment process is more feasible. For example, about 1% of the cellulose materials with dilute acid treatment in 106–110 °C for several hours leads to hydrolysis of hemicellulose into monosaccharides and dissolution into the hydrolyzate while the lignin content remains unchanged. The average degree of polymerization of cellulose decreases, and the rate of enzymatic hydrolysis increases significantly in this process. The effect of alkali treatment is to weaken the hydrogen bonds between cellulose and hemicellulose and the ester bond between saponified hemicellulose and lignin. NaOH treatment can be used as a pretreatment method, because it has a strong role of delignification. In the ammonia treatment, the cellulose was immersed for 24–28 h with 10% mass fraction to remove most of the lignin in the raw material. Some organic solvents can dissolve lignin, the use of organic solvents or organic solvents without inorganic acid catalyst (HCl or H_2SO_4) mixed solution can also damage the lignocellulose raw materials within the lignin and hemicellulose in the connection between the bonds. Common organic solvents, such as methanol, ethanol, acetone, ethylene glycol and trivinyl alcohol, and organic acids, such as oxalic acid, acetyl salicylic acid and salicylic acid in the organic solvent treatment process, can be used as catalyst.

2. **Cellulose solvent**

Due to the strong intramolecular and intermolecular hydrogen bonds, as well as higher crystallinity, cellulose is difficult to dissolve and melt. The earliest cellulose solvents in use are viscose solution and copper ammonia solution; however, the traditional adhesive method, due to the addition of large amounts of harmful substances such as CS_2 in the production process, caused serious pollution. In order to solve this problem, many nontoxic, nonpolluting new solvent systems made of cellulose have been developed and studied.

The dissolution of cellulose is divided into direct physical dissolution (nonderivatizing solvent) and partial derivatization dissolution (the cellulose is

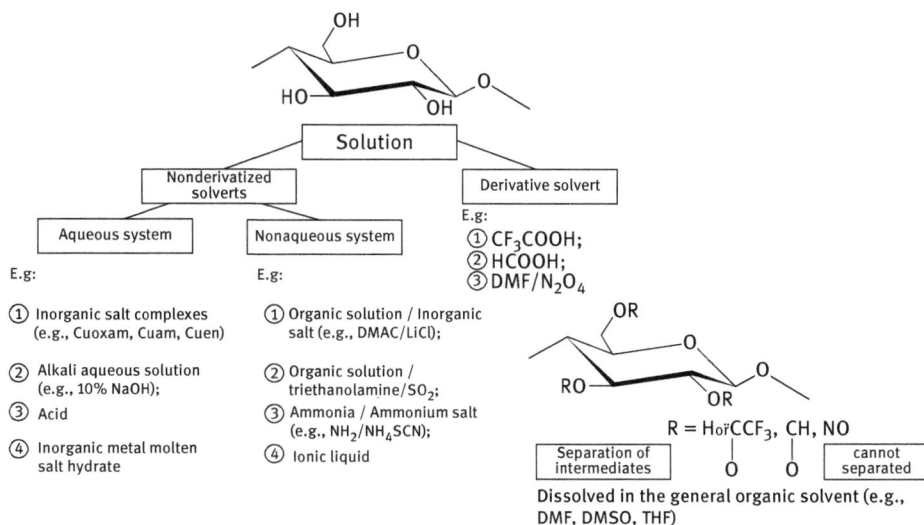

Figure 9.7: Classification of cellulose solvents.

dissolved in the derivatized solvent). The latter often introduces new functional groups in situ through the covalent bond to form fibrous intermediates. "Cellulose intermediates" are hydrolyzed, unstable cellulose species that can be separated from the derivatizing reagents. Soluble cellulose trifluoroacetate esters such as cellulose and trifluoroacetic acid/trifluoroacetic anhydride can be separated and re-dissolved in a typical organic solvent. Nonderivatized solvents include single-component and multi-component systems. Figure 9.7 is the classification of cellulose solvents, mainly divided into three categories, namely, derivatization system, nonderivatized aqueous phase system and nonaqueous system. Among them, the most commonly used systems are NaOH/CS2 (viscose method), copper ammonia solution, NMMO, DMAC/LiCl, DMSO/PF and so on.

9.2.5 Biomass Structure Analysis Method

1. Crystallinity analysis method

Cellulose is a homogeneous polycrystalline material. According to the X-ray diffraction results, it has been found that there are four crystal forms of cellulose, namely fiber home I, II, III and IV. The powder X-ray diffraction pattern of their crystalline morphology is shown in Figure 9.8. The diffraction angles corresponding to the respective planes of the cellulose crystals are shown in Table 9.1. Natural cellulose, including bacterial cellulose, seaweed and higher plants (such as cotton, lorillars and wood) belong to the cellulose of type I. The cellulose of type I molecular chains are stacked in parallel in the unit cells, as shown in

Table 9.1: The diffraction angles corresponding to the respective diffraction planes of the cellulose crystal variants.

Crystal form	Diffraction angle (2θ)/(degree)			
	110	110	020	012
Cellulose I	14.8	16.3	22.6	
Cellulose II	12.1	19.8	22.0	
Cellulose III$_I$	11.7	20.7	20.7	
Cellulose III$_{II}$	12.1	20.6	20.6	
Cellulose VI$_I$	15.6	15.6	22.2	
Cellulose VI$_{II}$	15.6	15.6	22.5	20.2

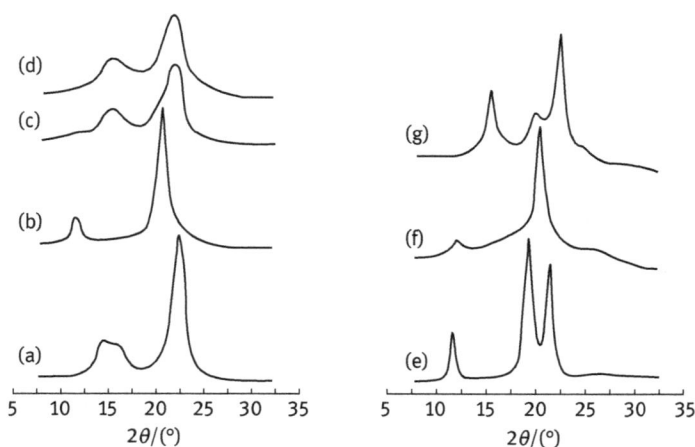

Figure 9.8: Cellulose X-ray diffraction pattern: (a) castor cellulose I; (b) castor-derived cellulose II; (c) cellulose III$_I$-derived cellulose IV$_I$; (d) cellulose prepared by liquid nitrogen IV$_I$; (e) cellulose II; (f) cellulose III$_{II}$; and (g) cellulose IV$_{II}$.

Table 9.1. According to the different sources of cellulose, their microfibril crystallinity (Xc), crystal size (Dhkt) and parallel size (d) are significantly different. Crystal structures of different types of cellulose can also be characterized using infrared spectroscopy, Raman spectroscopy, electron diffraction and solid-state cross-polarization/magic-angle sample-spinning nuclear magnetic resonance (CP/MAS ^{13}C NMR). Figure 9.9 shows the chemical shifts of the various cellulose crystals (CP/MAS ^{13}C NMR spectra, different crystalline cellulose C1, C4 and C6 are listed in Table 9.2). From the ^{13}C NMR spectrum and Table 9.2, it can be seen that the chemical shifts of C4 and C6 in different crystalline cellulose dextrose residues are significantly different, reflecting the difference in their crystal structure, which is due to the change in the chain conformation of different crystalline fibers. Crystalline piles are caused by differences in the effects of the glucopyranose units in C4 and C6. There are also significant differences in the CP/MAS ^{13}C NMR spectra of

Table 9.2: Chemical shift ranges for different crystalline cellulose C1, C4 and C6.

Crystal form	^{13}C chemical shift/ppm		
	C1	C4	C6
I	105.3–106.0	89.1–89.8	65.5–66.2
II	105.8–106.3	88.7–88.8	63.5–64.1
III$_I$	105.3–105.6	88.1–88.3	62.5–62.7
III$_{II}$	106.7–106.8	88.0	62.1–62.8
VI$_I$	105.6	83.6–83.4	63.3–63.8
VI$_{II}$	105.5	83.5–84.6	63.7

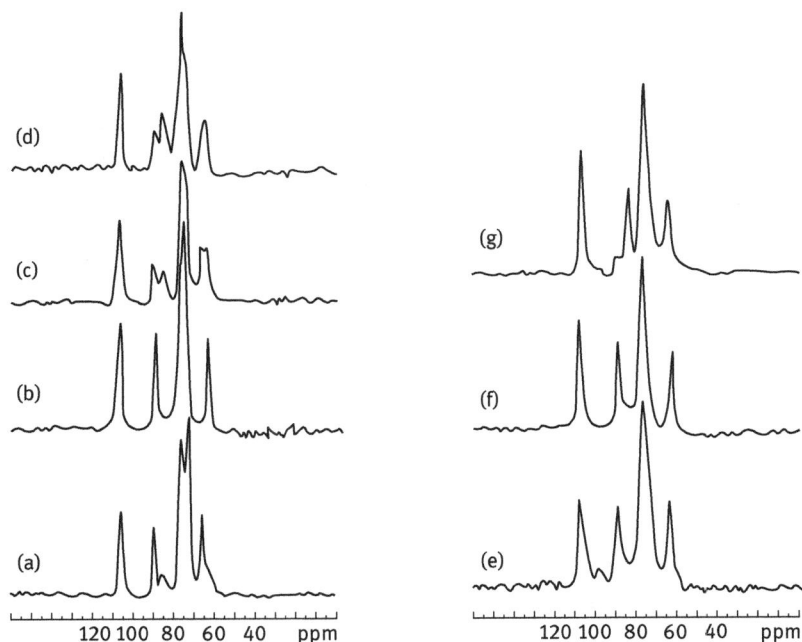

Figure 9.9: Solid CP/MAS ^{13}C NMR spectra of cellulose I cluster and cellulose II clusters. Cellulose I cluster: (a) Ramie cellulose I; (b) Ramie-derived cellulose III$_I$; (c) cellulose III$_I$-derived cellulose IV$_I$; (d) cellulose IV$_I$ prepared by liquid nitrogen at −33 °C; cellulose IIcluster; (e) cellulose II; (f) cellulose III$_{II}$; and (g) cellulose IV$_{II}$.

cellulose I. Figure 9.10 is a mixture of several different sources of cellulose I. From CP/MMAS ^{13}C NMR spectrum, it can be seen that the fine structure of cotton and linen cellulose and bacterial cellulose and Valonia cellulose on C1, C4 and C6 are different, indicating that these two types of cellulose are around 1,4-glycosidic bond, and that C5-C6 conformation, hydrogen bond or molecular stacking were significantly different. For ^{13}C NMR analysis of natural cellulose from different sources, it was found that cellulose I had two different crystal structures, namely

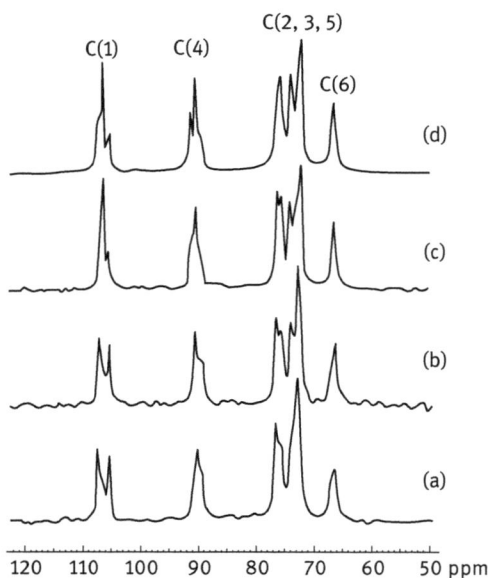

Figure 9.10: CP/MMAS ^{13}C NMR spectra of different natural cellulose: (a) cotton; (b) Ramie; (c) bacterial cellulose; and (d) Oak shell cellulose.

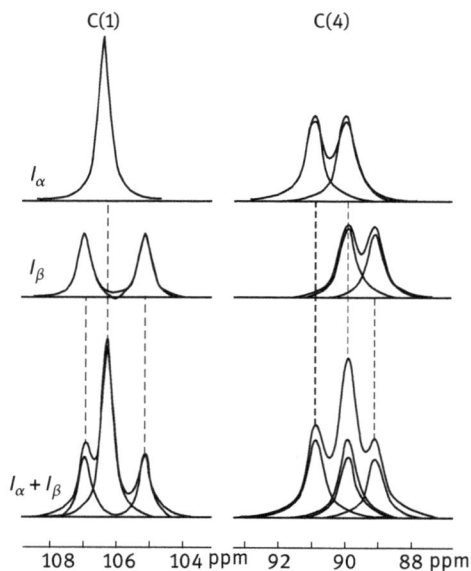

Figure 9.11: ^{13}C NMR spectra of C1 and C4 of cellulose Iα and Iβ.

cellulose Iα and Iβ. Figure 9.11 are the C1 and C4 ^{13}C NMR spectra of cellulose Iα and Iβ. Maximum difference can be observed for the chemical shift in C1, Iα is a single peak while Iβ is a bimodal. The bacterial cellulose mainly has Iα crystal structure, regenerated cellulose I is composed of almost pure Iβ crystals. After proper treatment, cellulose Iα can be converted to Iβ.

2. Surface morphology analysis method

1) Transmission electron microscopy

The structure of the transmission electron microscope (TEM) includes an illumination system, an imaging system, an observation and recording system. Figure 9.12 shows the optical path device and imaging principle of the TEM.

The samples of the TEM are thin slices. When the electron beam hits the sample, being easily permeable, some electrons are transmitted, called transmission electrons; other electrons come into contact with the nuclei and are scattered, and their movement direction and velocity change – these electrons are called scattering electrons. Transmission electron microscopy is the use of transmission electrons and partial scattering electron imaging, which shows different degrees of contrast, mainly amplitude contrast and phase contrast, with the amplitude contrast mainly including the scattering contrast and the diffraction contrast. The scattering contrast is the main cause of the amorphous formation contrast, and the diffraction contrast is the main contrast of the crystalline sample. High-resolution electron microscopy is used to determine phase contrast.

The morphology and structure of the amorphous and crystalline polymers were observed by transmission electron microscopy, including the morphology and size of the particles below the micron level, the molecular size and shape of the polymer molecules with large molecular weight, the size of nanoparticles and the shape of the peeling and intercalation. Figure 9.13(a) is a TEM photograph of carbon nanofibers.

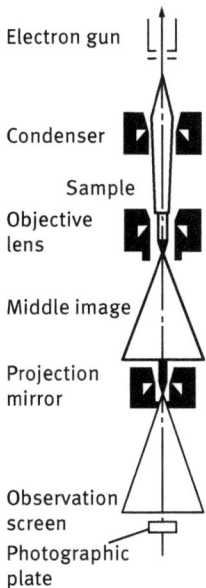

Electron gun
Condenser
Sample
Objective lens
Middle image
Projection mirror
Observation screen
Photographic plate

Figure 9.12: Schematic diagram of the optical path device of transmission electron microscopy.

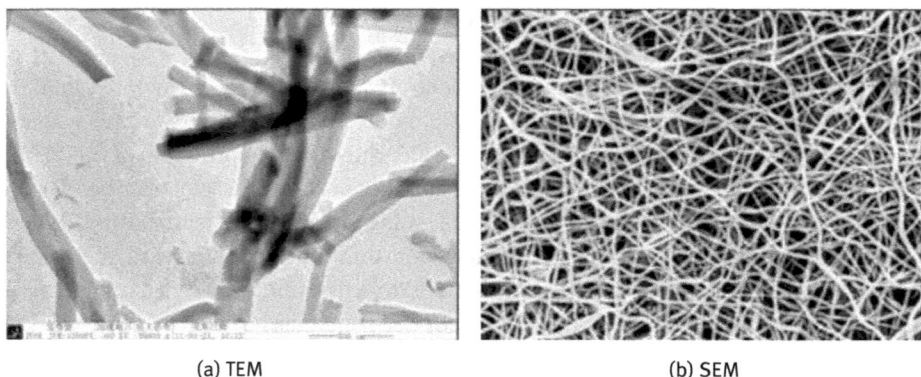

| (a) TEM | (b) SEM |

Figure 9.13: TEM (a) and SEM (b) photographs of carbon nanofibers.

2) Scanning electron microscopy

The basic device of the scanning electron microscope (SEM) is shown in Figure 9.14, which includes the electronic optics system, sample room, signal processing and display system and vacuum system.

There are two main types of scanning electron microscopy: ① In surface morphology contrast, the second power transmission, although the number depends on the surface of the sample ups and downs; sharp edges; small particles and the edge of the pit on the secondary electronic yield have a greater contribution. ② Atomic number contrast: The amount of secondary electron emission also depends on the atomic number of the element, and the element with the higher ordinal number produces more secondary electrons. Therefore, for the polymer material composed of elements such as C and H, it is necessary to sputter-layer the heavy metal (Au, Ag, Pt, etc.) conductive layer on the surface of the sample in order to increase the emission amount of the secondary electrons. The general sample preparation process of SEM samples involves: drying, mounting with sticky tapes, coating with gold and observation. Most of the natural polymer materials are composed of elements of low atomic number, such as C, H, O, N and other elements and the vast majority of insulating materials leads to the need to spray a layer of conductive layer in the sample surface, generally gold, platinum or carbon.

SEM can be used to observe the surface and internal structure and morphology of natural macromolecular materials, the surface morphology and size of microspheres and microfibers, the two phases of blending polymers and dislocations in materials. Figure 9.14 (1) consists of SEM photographs of carbon nanomaterials.

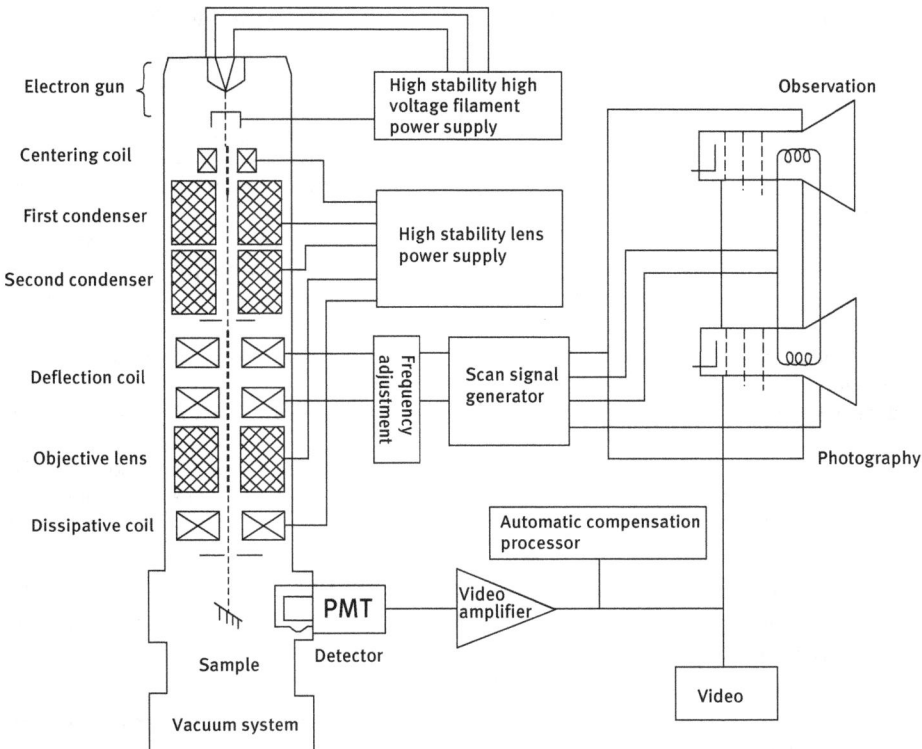

Figure 9.14: Schematic diagram of the instrument structure of a scanning electron microscope.

9.2.6 Biomass Composition Analysis Method

1. Determination of Molecular Weight and Degree of Polymerization

There are many ways to measure the molecular weight of cellulose, each having its own advantages and disadvantages and applicable molecular weight range. The statistical mean values of the molecular weights measured by various methods are not the same. Table 9.3 lists the various molecular weights. Mostly, the cellulose material to be tested is dissolved in a cellulose solvent and then measured using the resulting cellulose solution.

1) Boiling point rise and freezing point method

Since the vapor pressure of the cellulose solution is lower than that of pure solvent, the boiling point of the solution is higher than that of the pure solvent and the freezing point of the solution is lower than that of pure solvent.

It can be seen from thermodynamic reports that the boiling point elevation value ΔTb and the freezing point decrease value ΔTt of the solution are proportional to the concentration of the solution and inversely proportional

Table 9.3: Measurement significance and range of various molecular weight determination methods

Methods	Type of average molecular weight	Determination of relative molecular mass range	Type of methods
Vapor pressure dropping method	M_n	$2 \times 10^4 - 4 \times 10^4$	Relative law
Boiling point rising method	M_n	3×10^4	Relative law
Freezing point depression method	M_n	4×10^4	Relative law
Osmotic pressure method	M_n	$2 \times 10^4 - 5 \times 10^5$	Absolute law
Terminal group determination method	M_n	5×10^5	Absolute law
Light scattering method	M_w	$5 \times 10^3 - 5 \times 10^6$	Absolute law
Sedimentation balance method by supercentrifuge	M_w/M_n	$1 \times 10^4 - 1 \times 10^5$	Absolute law
Sedimentation rate method by supercentrifuge	Various averages	$1 \times 10^4 \sim 1 \times 10^7$	Absolute law
Viscosity method	M_η	$1 \times 10^3 - 1 \times 10^7$	Relative law
Gel permeation chromatography method	Various averages	$1 \times 10^3 \sim 5 \times 10^6$	Relative law

to the molecular weight of the solute. Therefore, the molecular weight of the cellulose can be determined by measuring the boiling point and freezing point of the solution.

2) **Vapor pressure drop method**

The solvent molecules are continuously subjected to irregular movement, and their movement depends on the temperature, since the movement of the molecules in the liquid phase into the gas phase results in liquid vapor pressure. If cellulose is added as a solute, the solute molecules can block the meta-regular movement of the solvent molecules, thereby lowering the vapor pressure. The molecular weight can be measured by measuring the vapor pressure drop after the cellulose is dissolved.

3) **Osmotic pressure method**

When the solvent pool and the solution cell are separated by a layer of semipermeable membrane it allows only the solvent molecules to pass through and does not allow the solute molecules to pass through, resulting in liquid column height difference. At this time, the solvent in the solution tank will also reverse osmosis into the solvent pool. When the liquid column height difference reaches a certain value, the number of solvent molecules returned by reverse osmosis per unit time is exactly equal to the unit time permeating the solvent pool into the solution tank. The number of solvent molecules reaches the balance of permeation. Meanwhile, the pressure generated by the liquid column height difference between the solution and the solvent pool is the osmotic pressure. This osmotic pressure produces different osmotic pressures due to the molecular weight of the cellulose polymer, so this method is used to determine the molecular weight of the fiber cord.

4) **Light scattering method**

When a beam passes through a medium, the phenomenon of varying light intensity can be observed in all directions outside the direction of the incident light, called scattering. Usually, the light intensity of the scattered light in the polymer solution is much larger than that of the pure solvent, and the scattered light intensity also increases with the increase of the molecular weight and the solution concentration of the cellulose, and is related to the particle size and shape of the solute.

5) **Ultracentrifugation method**

The molecules in the suspension will gradually sink in the gravitational field. From the speed of settling, the mass of suspended particles can be calculated. The mass of the suspended particles can be calculated from the rate of settling. However, the solution formed by the cellulose molecule is a polymer solution and must be allowed to settle in a very strong force field. Therefore, it is necessary to use the ultracentrifuge to produce a large centrifugal force to allow the cellulose molecules to settle.

6) **Viscosity method**

Viscosity refers to the internal friction when liquid flows. When the internal friction is large, the liquid has a larger viscosity and a slower flow. On the contrary, when the coefficient of viscosity is small, the flow is faster. The viscosity method is a method in which cellulose or its derivatives are dissolved in a solution, and then the viscosity of the solution is measured to calculate the molecular weight and degree of polymerization of the cellulose. The viscosity of the solution is related to the molecular weight of the cellulose, and also the structure, morphology and degree of expansion in the solvent.

2. **Degradation of the product after analysis**

The low-molecular-weight organic compounds of the polymer material after complete or almost complete degradation can be identified by infrared spectroscopy, gas chromatography, liquid chromatography and nuclear magnetic resonance. This section does not elaborate these methods.

9.3 Chemical Conversion Principle of the Key Components of Biomass

9.3.1 Chemical Conversion of Cellulose Components

The cellulose molecule has three active hydroxyl groups, which may undergo a series of hydroxyl-related chemical reactions such as esterification, etherification, graft copolymerization and cross-linking. Through these reactions, cellulose can be synthesized into a range of cellulose derivatives.

1. **Cellulose Acetate**

   Cellulose acetate, also known as cellulose acetate ester, is obtained from the acetoxylation of the fiber, hence it is also called acetyl cellulose. The three hydroxyl groups on the cellulose glucose residue are almost completely esterified to give triacetate or cellulose triacetate with an esterification degree of 3 and acetic acid content of 60.5–62.5%. Triacetate cellulose hydrolysis with esterification degree of 2.0–2.7, and acetic acid content of 48.8–58.8%, can be converted to cellulose diacetate.

   Cellulose acetate is an ester produced from acetic anhydride and hydroxyl groups in cellulose. The reaction is as follows:

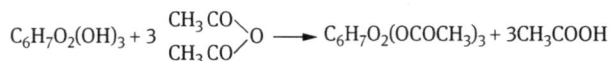

   $$C_6H_7O_2(OH)_3 + 3 \begin{array}{c} CH_3CO \\ CH_3CO \end{array}\!\!\!\!\!O \longrightarrow C_6H_7O_2(OCOCH_3)_3 + 3CH_3COOH$$

   Cellulose acetate is a white solid, with characteristics of flexibility, transparency, good gloss, high strength, good toughness, good melt flowability, easy molding, thermoplasticity, etc. It is the largest output of all kinds of cellulose derivatives currently produced in the industry. Important varieties are widely used in fibers, film, paint and many other fields.

2. **Hydroxyethylcellulose**

   Hydroxyethylcellulose is a nonionic soluble cellulose ether prepared by etherification of basic cellulose and ethylene oxide. Its formula is $[C_6H_7O_2(OH)2OCH_2CH_2OH]_n$ (where $n \geq 100$, represents the degree of polymerization of cellulose), and it is a white or light yellow, tasteless, nontoxic fibrous or powdered solid. The total reaction formula for etherification in a mixed solvent of base is as follows:

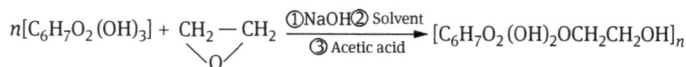

   $$n[C_6H_7O_2(OH)_3] + \begin{array}{c} CH_2 - CH_2 \\ \diagdown O \diagup \end{array} \xrightarrow[\textcircled{3} \text{ Acetic acid}]{\textcircled{1}\text{NaOH}\textcircled{2}\text{ Solvent}} [C_6H_7O_2(OH)_2OCH_2CH_2OH]_n$$

   The average molecular weight range of hydroxyethyl cellulose ranges from 6.8×10^4 at low viscosity to 8×10^6 or more at high viscosity, and it can be used as thickener, binder, stabilizer and film-forming material. At present, foreign countries mainly apply paints, oilfields, building materials, and polymerization reactions; domestically, they are mainly used in textiles, building materials, etc.

3. **Carboxymethylcellulose**

   Carboxymethylcellulose is a cellulose ether produced by the action of natural cellulose under the action of sodium hydroxide and chloroacetic acid. The important reaction is that of cellulose and sodium hydroxide aqueous solution to produce alkali cellulose; alkali cellulose and sodium chloroacetate undergo etherification reaction to form carboxymethyl cellulose. The main chemical reactions are:

   (1) Alkalization, where cellulose reacts with alkaline water to produce alkali cellulose:

   $$[C_6H_7O_2(OH)_3]_n + nNaOH \longrightarrow [C_6H_7O_2(OH)_2ONa]_n + nH_2O$$

(2) Etherification, where alkali cellulose and sodium chloroacetate undergo etherification reaction:

$$[C_6H_7O_2(OH)_2ONa]_n + nClCH_2COONa \longrightarrow [C_6H_7O_2(OH)_2OCH_2COONa]_n + nNaCl$$

Carboxymethyl cellulose is an anionic surfactant; it is a white or light yellow powder, nontoxic, odorless, tasteless, hygroscopic, insoluble in acid and methanol, ethanol, ether, acetone, chloroform and benzene and other organic solvents. It is easily soluble in water and colloidal in water. Its aqueous solution has functions of emulsification, thickening, film formation, bonding, moisture retention, colloid protection and suspension. The maximum theoretical value of substitution is 3, which determines its solubility and stability. When the degree of substitution is above 0.8, its acidity and salt tolerance are better.

Carboxymethyl cellulose is widely used in detergents; foodstuffs; toothpaste; textile printing and dyeing; papermaking; mining; medicine; oil drilling; manufacture of petroleum, building materials,, ceramics, electronic components, rubber, paint, pesticide, cosmetics, leather, and so on.

9.3.2 Chemical Conversion of Hemicellulose Components

The distribution of hemicellulose in plants varies according to plant species, maturity, cell type and morphological location. For example, the content of hemicellulose in coniferous wood is 15–20%, mainly galactose grape mannose, while the content of broadleaf and gramineous grass is 15–35%, mainly polyglucose. In addition, the light cells of the needle and the broad-leaved plants have more poly-xylose than the tube cells and fibroblasts. In the middle layer of the secondary wall of softwood cells, the content of xylan was the lowest, but it was higher in the secondary and outer layers of the secondary wall, whereas the distribution of galactomannan was the opposite. Therefore, the process of chemical conversion to hemicellulose is mainly divided into the conversion process of hexose and pentose.

1. **Chemical transformation of hexose**

    The hexose present in the pre-hydrolyzate can be used to produce alcohol (ethanol) and sorbitol (hexavirol). Glucose, mannose and galactose can be fermented to produce alcohol, This is the main direction of the current comprehensive utilization of hexose. This method can not only obtain alcohol, but also reduce the sugar content in lignosulfonate. The reaction is as follows:

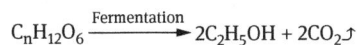

$$C_nH_{12}O_6 \xrightarrow{\text{Fermentation}} 2C_2H_5OH + 2CO_2\uparrow$$

The reaction produces CO_2, which can be made into dry ice.

Hexose can also be reduced to sorbitol; for example, glucose with nickel as a catalyst can undergo hydrogen reduction at 120–130 °C to yield sorbitol, the reaction is as follows:

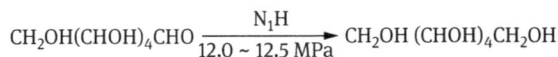

$$CH_2OH(CHOH)_4CHO \xrightarrow[12.0 \sim 12.5\ MPa]{N_1H} CH_2OH\ (CHOH)_4CH_2OH$$

Sorbitol is sweet and has a soft taste and can be used as a sweetener for confectionery. Its industrial application is very rich. For example, it can be used as a raw material for manufacturing and vitamin C, and a substitute for glycerol. It can also be used to regulate humidity in the production of coated paper, drown or break the cigarette in the production of cigarettes. In addtion, it can be used as Additives such as toothpaste, food, and cosmetics can also be used for the production of paints, surfactants, and plasticizers.

2. **Chemical conversion of pentose**

The reaction chamber was heated with a dilute acid under high pressure to distill off the furfural, and the reaction was as follows:

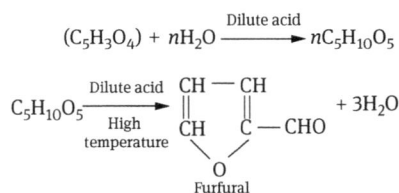

$$(C_5H_3O_4) + nH_2O \xrightarrow{\text{Dilute acid}} nC_5H_{10}O_5$$

$$C_5H_{10}O_5 \xrightarrow[\substack{\text{High} \\ \text{temperature}}]{\text{Dilute acid}} \underset{\text{Furfural}}{\begin{matrix} CH{-}CH \\ \| \quad \| \\ CH \quad C{-}CHO \\ \diagdown\ O\ \diagup \end{matrix}} + 3H_2O$$

Corn cob, sugar cane bagasse, cottonseed hull and waste wood are used as raw materials in industries in the hydrolysis of xylan, which after dehydration leads to the formation of furfural. Furfural can be used for the refining of lubricating oils, and can also be used for the production of solvents, furan resins, nylons, etc. It is an important chemical raw material.

In addition, xylan can be made into crystalline xylose or xylose syrup by hydrolysis. Xylose is used in the manufacture of candy, and in canned fruit and ice cream industry. The human body can only digest 15% to 20% xylose, while animals can digest 90% of xylose, animals require high-calorie food. Agricultural by-products, especially corn cob, are good raw materials for the production of xylose. With 0.1–0.25% dilute sulfuric acid, hydrolysis of corn cob can be used to obtain a high yield of xylose. The pretreatment of corn cob at 140 °C for 90 min can remove part of the ash, water-soluble sugar and protein, which is beneficial to improve the yield and purity of xylose. The yield of xylose can reach 15% and the purity can reach 94%. Hydrolysis with xylose produces better results than acid hydrolysis.

Xylose can be reduced to xylitol by hydrogenation, the reaction is as follows:

$$
\begin{array}{ccc}
\begin{array}{l}
\text{H}-\text{C}=\text{O} \\
\text{H}-\text{C}-\text{OH} \\
\text{HO}-\text{C}-\text{H} \\
\text{H}-\text{C}-\text{OH} \\
\text{CH}_2\text{OH}
\end{array}
&
\xrightarrow[\text{Aqueous 120 ~ 130 °C solution}]{\text{Na}_2\text{H } 9-10\text{MPa}}
&
\begin{array}{l}
\text{CH}_2\text{OH} \\
\text{H}-\text{C}-\text{OH} \\
\text{HO}-\text{C}-\text{H} \\
\text{H}-\text{C}-\text{OH} \\
\text{CH}_2\text{OH}
\end{array}
\end{array}
$$

Xylitol is odorless, white, heat-stable crystalline powder, and its sweetness and calorie contents are the same as sucrose. The metabolism of xylitol is not controlled by insulin and it is suitable for consumption by diabetic patients. Xylitol cannot be used by oral bacteria, so it does not cause dental caries and is suitable for applications in chewing gums. Xylitol can also be made into injections that can be used as anti-keto agents and metabolic correctors.

Xylose with a density of 1.2–1.4 g/cm³ nitric acid at 60–90 °C can undergo oxidation for 2–3 h to produce trihydroxy glutaric acid; the reaction is as follows:

$$
\begin{array}{ccccc}
\begin{array}{l}
\text{H}-\text{C}=\text{O} \\
\text{H}-\text{C}-\text{OH} \\
\text{HO}-\text{C}-\text{H} \\
\text{H}-\text{C}-\text{OH} \\
\text{CH}_2\text{OH}
\end{array}
&
\xrightarrow[60-90\,°C]{\text{HNO}_4}
&
\begin{array}{l}
\text{COOH} \\
\text{H}-\text{C}-\text{OH} \\
\text{HO}-\text{C}-\text{H} \\
\text{H}-\text{C}-\text{OH} \\
\text{CH}_2\text{OH}
\end{array}
&
\xrightarrow[60-90\,°C]{\text{HNO}_4}
&
\begin{array}{l}
\text{COOH} \\
\text{H}-\text{C}-\text{OH} \\
\text{HO}-\text{C}-\text{H} \\
\text{H}-\text{C}-\text{OH} \\
\text{COOH}
\end{array}
\end{array}
$$

Three hydroxy glutaric acid

Trihydroxy glutaric acid has a pleasant sour taste, so in the food industry it can replace citric acid; it can also be used in the preservation of plasma, and also as a stabilizing powder.

9.3.3 Chemical Conversion of Lignocellulosic Components

1. Lignin amine

Hydrogen atom in Lignin molecules are relatively active in aldehyde, ketone and sulfonic acid, with which Mannich reaction could take place. The Mannich reaction is a reaction in which the active hydrogen atom is substituted with an amine methyl group, and the amine compound is condensed with an aldehyde and a compound containing an active hydrogen atom, and the reaction is as follows:

$$
\text{>N}-\text{H}+-\overset{|}{\underset{\overset{||}{O}}{\text{C}}}-\text{H}+\text{Z}-\overset{|}{\underset{|}{\text{C}}}-\text{H} \longrightarrow \text{Z}-\overset{|}{\underset{|}{\text{C}}}-\text{CH}-\overset{|}{\text{N}} + \text{H}_2\text{O}
$$

When the lignin is subjected to the Mannich reaction, the ortho and para positions of the phenolic hydroxyl groups on the benzene ring and the hydrogen atoms at the α position of the carbonyl group on the side chain are more active and react easily with the aldehydes and amines to form lignin amines. Depending on the amine groups involved in the reaction, they can be divided into primary amine type lignin amine, secondary amine type lignin amine, tertiary amine type lignin amine, quaternary amine type lignin amine and polyamine type lignin amine. Take the primary amine type lignin amine as an example:

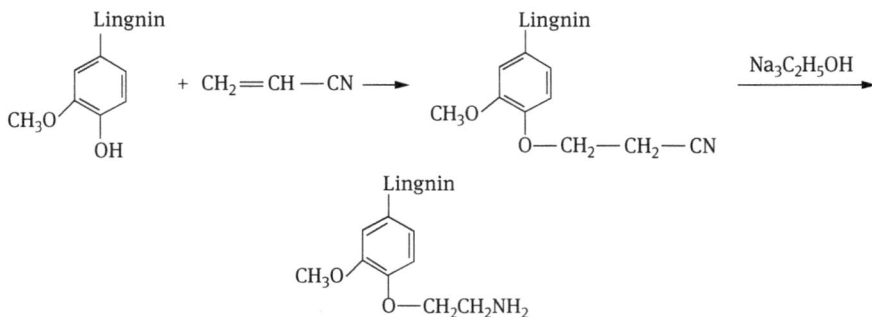

The strong surface activity of lignin amines makes it useful as an emulsifier for emulsions, flocculants and the like.

2. **Lignin alcohol ether**

In the lignin molecule, access to ethylene oxide or propylene oxide can change its surface activity, and greatly broaden its scope of application. The reaction principle is as follows:

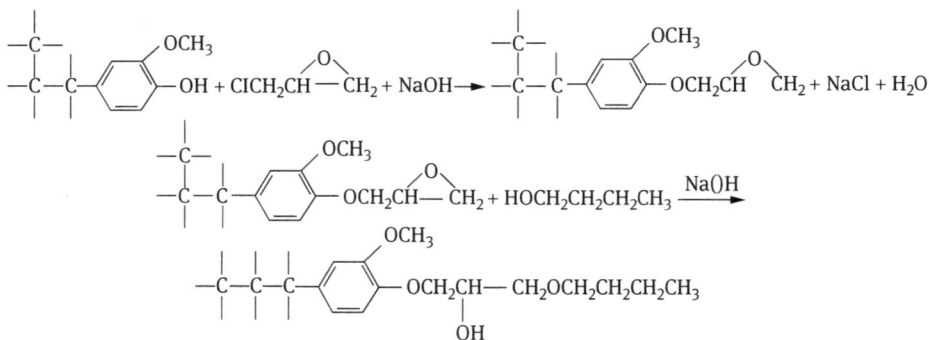

Lignin alcohol ether as a nonionic surfactant has a better ability to reduce the surface tension of aqueous solution, and has a good emulsifying power.

3. **Lignosulfonate**

Lignosulfonate is produced directly from the sulfite pulping process and can also be obtained by sulfonation of alkali lignin. Its structure is very complex, see Figure 9.15 for its partial structure.

Lignosulfonate has good water solubility, dispersibility, adhesion and surface activity.

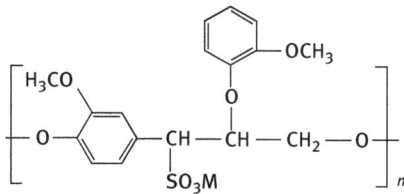

Figure 9.15: Schematic diagram of lignin sulfonate.

Anionic surfactant lignosulfonates are synthesized by sulfonation modification; cationic hydrophilic groups can also be synthesized on lignin to yield cationic surfactants. In addition, lignosulfonate can also be synthesized through oxidation, formaldehyde condensation, graft copolymerization and other chemical modification methods and used to prepare new surfactants, thereby enhancing product performance.

9.4 Principle and Technology of Clean Separation of Biomass Components

Lignocellulose is mainly composed of cellulose, hemicellulose and lignin, combined among the molecules by different binding forces. The binding force between molecules of cellulose and hemicellulose, cellulose and lignin is mainly hydrogen bond. Apart from hydrogen bonds, there still exists a chemical bond between hemicellulose and lignin. Efficient separation of three components is required before using lignocellulose. Therefore, separation and transformation are two key technical steps in biomass energy conversion and separation technology. That is to say, the full utilization of agricultural and forestry waste material relies on effective component separation.

9.4.1 Basic Principles of Separation of Components

At present, there are some disadvantages related to the separation technology for lignocellulose. (1) In the extraction of a target component, the structures of other components are severely damaged. The structure of cellulose, hemicellulose and lignin components obtained is incomplete, their chemical or biological reactivity is low and they cannot be effectively used; (2) serious environmental pollution is caused in the process of separation. For example: In order to get high purity cellulose, hemicellulose and lignin will be dissolved by using the solubility and catalysis action of acid which is aim to destroy the mutual connection between cellulose, lignin and hemicellulose in acid pretreatment process. In the process of dissolution of hemicellulose and lignin, the structure of the material will become loose, and the crystal structure of cellulose will be damaged, so that the efficiency of cellulose hydrolysis, enzymatic hydrolysis is improved [12, 13].

Recent research suggests, to solve the key problems related to the complexity of the cell wall structure in total component separation and for the development of new separation method, a detailed study of their ultrastructure at the molecular level is essential. In recent years, a new concept, biorefinery, which based on the understanding of the biomass of all components utilization and process economy, were put forward, even the utilization of lignin and hemicellulose in the process of cellulose separation is considered. In theory, a single chemical separation method can lead to changes in structure and properties because the components of lignocellulose are closely connected. However, low-strength processing technology, as well as one-step or two-step processing technology can be used to achieve an efficient and clean separation of components while maintaining the structure of hemicellulose and lignin [14, 15].

9.4.2 Component Separation Based on Steam Explosion

9.4.2.1 Steam Explosion
As a kind of physical and chemical method with less or no chemicals, no pollution to the environment, low energy consumption, steam explosion is a technology of pretreatment lignocellulose with rapid development, more effective and low cost in recent years. This method can fracture the biomass structure through the thermal physical and chemical processes (shearing force of hot steam, water blasting, and the hydrolysis of glycosidic linkage). That is softening of lignin under high pressure and high temperature, and then quickly decompress to fracture the cellulose crystals, to achieve the separation of lignin and cellulose. Blasting can make the connected bonds within and between molecules of molecular lattice of lignocellulose breaking, hemicellulose was dissolved, while a small amount of lignin was dissolved.

Steam explosion is a pretreatment method of biomass developed by W.H. Mason in 1928, in which the raw material is humidified to 0.69–4.83 MPa by 160–260 °C saturated steam, and the conditions are maintained for a few seconds to a few minutes, after which the mixture is suddenly depressurized to atmospheric pressure. In the process of steam explosion, high pressure steam infiltrated the fiber interior, then released from the closed hole in airflow. It would make the fiber partial mechanical fractured and the destruction of internal hydrogen bonding was intensified under high temperature and high pressure. That aims to release new free hydroxyl group, the ordered structure of cellulose is changed.

Different steam explosion temperatures and retention times are needed for different materials, in order to achieve efficient separation of hemicellulose, lignin and cellulose, and thus to achieve better conversion of cellulose by hydrolysis.

The method has the following advantages: it is energy saving – compared with mechanical crushing method, steam explosion can save 70% of the energy; no pollution is caused and the method achieves high enzymolysis efficiency; blasting does not need the addition of any catalyst, and the conversion of cellulose was found to be 90% after pretreatment; the method has wider applications – it can be used for a variety of plant biomass, pretreatment conditions are easy to adjust, and the inhibitor produced during steam explosion can be controlled by adjusting the conditions of steam explosion; hemicellulose, cellulose and lignin can be separated in stages (as water-soluble, alkali-soluble and alkali-insoluble components). Its disadvantage include: it will damage part of xylan in processing, and the aldehydes formed by five-carbon sugar degradation have an adverse effect on the subsequent biological treatment; the pretreatment materials need a large amount of water to remove the inhibitor and water-soluble hemicellulose, therefore reducing the total sugar yield; and the operation of steam explosion involves high-pressure equipment, higher investment costs [16–20].

9.4.2.2 Ammonia Fiber Explosion

Ammonia fiber explosion (AFEX) method is similar to steam explosion; in this method the lignocellulose are treated under high temperature and high pressure, then suddenly decompressed to fracture the cellulose crystals. In typical process of AFEX, the treatment temperature ranges from 90 to 95 °C, maintenance time is 20–30 min and PH less than 12. The hemicellulose and lignin can be partially dislodged by this method; the crystallinity of cellulose can be reduced, further increasing the contact between cellulase and cellulose; and it can cause the swelling of cellulose and the change of crystal structure, i.e., convert cellulose (I) into cellulose (II). The AFEX method is highly efficient for pretreatment of herbaceous crops and its agricultural waste, moderately efficient on hardwood, but not suitable for cork wood.

AFEX treatment is different from other pretreatment methods in that: After AFEX treatment, hemicellulose can not be effectively dissolved, so its components of biomass materials has little difference with untreatment. Cellulose and hemicellulose seldom or could not be degraded after ammonia fiber explosion treatment; it does not produce fermentation inhibitors, the material after the treatment did not need to neutralize before enzymatic hydrolysis; the expense of ammonia explosion equipment is lower than that of acid treatment – hydrolysate can directly ferment without treatment, and residual ammonium salts can be used as nutrient for microorganisms; ammonia destroys the crystalline structure of cellulose and increases hydrolysis rate of cellulose; ammonia fiber explosion equipment is basically the same as steam explosion equipment, in addition, it needs a recovery device of ammonia compression, so the investment cost is high.

9.4.2.3 Acid Gas Explosion

Acid gas explosion is used to add acidic gas in the process of steam explosion pretreatment for improving the efficiency of pure steam explosion, the primary saccharification rate and subsequent rate of hydrolysis of cellulose and hemicellulose. After explosion, change of the lignin structure is not very clear, but the original structure has obviously changed, lignin forms more aggregates. Reactive groups in alpha position, such as hydroxyl groups, are oxidized to carboxyl groups, or produce a methoxyl group, and the formation of the C–C bond leads to disappearance of alpha activity. Commonly used acid gases are CO_2, SO_2 and so on.

In the process, acid formed by partial acid gas was found to promote hydrolysis. The process significantly improves the degree of hydrolysis of hemicellulose, and there is no inhibitor of microbial fermentation produced in the process. Its cost is higher than that of steam explosion, but lower than the ammonia fiber explosion. Walsum [21] pretreated maize straw through carbon dioxide explosion, and found that xylose and furanose yields were significantly higher than that obtained with steam explosion after hydrolysis, the effect of treatment is associated with pressure.

The chamber of the steam explosion equipment is composed of a cylinder, the upper part is sealed by the cover, and the lower part is sealed by a static steam seal system, and the sealing system is operated by compressed air to control the opening and closing of the chamber. The cavity is connected with the steam generator through a valve, the pressure of the steam generator and the explosion chambers are respectively displayed by two pressure gauges, which is used for controlling pressure. In the biomass steam explosion pretreatment, first the cavity of the steam explosion equipment on the cover is opened, then the biomass added, cavity cover is closed, the steam is passed into the cavity to a set pressure and pressure is maintained for a set period of time, after which it is released. The steam works to push the material through the open lower cover to the collection room, and complete the blasting process.

9.4.3 Component Separation Process Based on Alkali Peroxide System

Alkali oxidation is a new method developed for pretreating lignocelluloses, it has the characteristics of requiring mild reaction condition, causing little pollution and producing better treatment effects. The study of alkali oxidation is mainly focused on the crop straw.

In this method the mixed solution of hydrogen peroxide and sodium hydroxide is used as pretreatment solution. Sodium hydroxide can destroy the ester bond and ether bond linkage between hemicelluloses and lignin, and cause the reduction of the bonding force between fibers. Therefore, it is more effective in dislodging the gum from raw materials and can partially degrade hemicelluloses and lignin, to destroy the lignin structure of raw materials. Swelling of cellulose will happen under alkaline conditions. The effect of hydrogen peroxide is mainly to produce peroxy radical for oxidizing lignin. Hydrogen peroxide is stable in acidic environment, but it is very unstable and easy to dissociate under alkaline conditions. Thus, in a mixture containing hemicelluloses and lignin, the basic conditions provided by sodium hydroxide can promote the dissociation of hydrogen peroxide; although the oxidation ability of peroxy radical ions is weaker than that of hydrogen peroxide, it has higher activity to its electrophilic center, and therefore, lignin can be more efficiently oxidized and dissolved. Moreover, hydrogen peroxide can oxidize the phenolic compounds formed by degradation of lignin, and further improve lignin oxidation, and this oxidative degradation can make the fiber cellulose further exposure to alkali environment, the moistening effect becomes more obvious, both have synergistic effects.

When using sodium hydroxide alone, a part of the lignin and a small amount of hemicellulose components can be removed without changing the composition of cellulose. While using hydrogen peroxide alone under normal temperature and pressure, only a small fraction of lignin is removed, but the cellulose and hemicellulose components do not change significantly. Under combined conditions, the degree of lignin degradation is higher than when used alone, and a small part of cellulose is removed, but the loss of hemicellulose components is minimum. Because hydrogen peroxide is a weak binary acid, ionization in aqueous solution is as follows: $H_2O_2 \rightarrow H^+ + HOO^-$, and HOO^- is a nucleophile, which can further initiate hydrogen peroxide forming free radicals: $H_2O_2 + HOO^- \rightarrow HOO \cdot + HO \cdot + OH^-$. If alkali is added to hydrogen peroxide, hydroxyl ions can neutralize hydrogen ions produced by ionization of hydrogen peroxide. This makes the concentration of HOO^- increase, thus further increasing the degradation ability. But the hydrogen ion produced by ionization of hydrogen peroxide neutralizes the hydroxyl ions of alkali, decreasing the solubility of hemicelluloses in alkali.

9.5 Green Process for Chemical Utilization of Biomass

With the development of economy, there is a growing need for a variety of chemicals; at the same time, there is also accelerated consumption of the one-off resources (e.g., coal and oil). Thus, the management of resources has become a common concern across the world. A huge amount of biomass resources, such as lignocellulose, can be constantly regenerated, and play an important role in sustaining energy and resources in future. So far, the degree of utilization of biomass resources has been very low. Only a small part of them, less than 1.0% of the total, is used as raw material in papermaking, feed and preparation of chemicals. If these resources can be converted into a wide range of basic chemicals, it will have a profound significance in solving two current problems of managing resources and energy.

Chemical conversion of biomass is the process by which biomass is degraded by chemical methods and then produces other chemicals. The main focus of the research on the green chemical conversion of lignocelluloses is the preparation of environmentally friendly chemicals through the green conversion processes [22].

9.5.1 Ethanol Produced from Biomass

Ethanol is an important organic chemical, mainly used as solvent, fuel and preservatives. The conventional method of preparing ethanol is via grain fermentation. In modern chemical industry, ethylene from petroleum cracking is generally used as raw material for preparation of ethanol by hydration. Since 1970s, the rapid development of biofuel industry characterized by usage of ethanol as fuel, especially the first generation of fuel ethanol with sugarcane and corn has been industrialized. The total global output of fuel ethanol is about 58 million 600 thousand tons in 2009, an increase of 12.7% over 2008, of which the United States accounted for 54.1%. By 2030, the output of biofuel will be expected to reach 120 million tons, and accounted for 5% of total amount of transportation fuel consumption. However, with the growth of demand for fuel ethanol, which is produced based on grain as raw mate rials, we will face the contradiction of "Competing food with mankind, and need more land to grow grain.", therefore, fuel ethanol from grain is partially controlled and large-scale production is constrained.

Fuel ethanol prepared from lignocellulose biomass is the second generation of biomass energy, often called bioethanol, by virtue of its properties of being clean, cheap, safe and environmentally friendly. Thus, it has become the most promising source of energy, and the focus of research and utilization of biomass in recent years. Using lignocellulose biomass to produce cellulosic ethanol can not only alleviate the shortage of energy, but also solve the problem related to raw materials for fuel ethanol production; furthermore, the usage of cellulosic ethanol can also reduce greenhouse gas emissions [23, 24].

The preparation method consists of first converting biomass to fermentable sugars, then using microbial agents to convert sugar into ethanol by fermentation. The fundamental bioprocess for preparing bioethanol can be divided into four parts: pretreatment, hydrolysis, fermentation and purification. At present, a variety of pretreatment methods have been developed with specific characteristics. The hydrolysis process is using acid or enzyme to hydrolyze the polymer, then makes it soluble monosaccharide. Enzyme hydrolysis, which has a high conversion rate, close to the theoretical value, is considered to be the most promising commercial hydrolysis method. The fermentation process is used for transforming hydrolyzates (pentose and hexose) to ethanol by microbial fermentation. Product purification is used to get pure ethanol through distillation, etc. Differences between bioethanol preparation process and the traditional ethanol fermentation lie in the fact that biomass hydrolysate often contains components which are harmful to microbial fermentation and the content of five-carbon sugar is relatively high. The removal of fermentation inhibitors and the use of five-carbon sugar is the key problem which needs to be addressed in studies on the development of bioethanol production [25].

Pretreatment is the key step determining the commercialization of ethanol from biomass, and is one of the most expensive steps in the whole process. It has a great influence on the size of raw materials in the process of enzymatic hydrolysis and fermentation. If the pretreatment achieves good separation, less enzyme is used in hydrolysis. The purpose of pretreatment is to destruct the internal structure of biomass which hinders saccharification and fermentation, to break the lignin protection on cellulose and crystal structure of cellulose, to provide full contact of raw material with the biological enzyme, in order to achieve good hydrolysis effects. The criteria for evaluating the effectiveness of pretreatment methods include the following: need to crush the raw materials before the pretreatment; its ability to retain the pentose structure in hemicelluloses; its ability to limit the production of fermentation inhibitors; energy consumption, etc. [26, 27].

Pretreatment methods commonly used for preparing bioethanol are of four kinds: physical method, chemical method, physical and chemical method and biological method. The physical method mainly refers to mechanical crushing method, which has shortcomings such as high energy consumption, high cost and low efficiency. The chemical method mainly refers to the use of acid, alkali and organic solvents as pretreatment catalyst to destroy the covalent bond between lignin and hemicelluloses, break down the crystal structure of cellulose and promote the dissolution of cellulose. The physical and chemical method mainly is of two kinds, steam explosion and ammonia fiber explosion, which were detailed Section 5.4. The biological method is the use of microorganisms and other bacteria to degrade lignin; these microbes can produce enzymes that selectively degrade lignin in the pretreatment step. Because the activity of lignin enzyme is low, and it needs a long period time, its development is slow [28].

9.5.2 Butanol and Acetone Production from Biomass

The microbial production of acetone and butanol relates to the discovery of fermentation. In 1861, Pasteur observed that butanol appears as a by-product during the fermentation of butyric acid from lactic acid or calcium lactate. In 1914, Dr. Weizmann successfully isolated and obtained *Clostridium acetobutylicum*, a bacterium which can ferment all kinds of starchy materials to yield butanol, acetone and ethanol in ratio 6:3:1. At the beginning of the twentieth century, the rapid developments in the automobile industry resulted in insufficient supply of natural rubber, this even promoted research on synthetic rubber. At that time the British discovered that acetone, when used as raw material, can yield rubber when reacted with isoprene on polymerization. Butyl alcohol, when used as raw material, was found to yield synthetic rubber through polymerization with 1, 3-butadiene. The increased demand for acetone during First World War stimulated the development of acetone and butanol via fermentation. In 1914, the British established the first fermentation factory for the synthesis of acetone. After the First World War, a process for synthesizing butyl acetate from butanol was developed by DuPont Company; butyl acetate can be used as an excellent solvent for nitrocellulose lacquers. In 1945, the United States began to use sugar honey as raw material for producing acetone and butanol, and develope their new uses. At the end of 1960s, with the development of petrochemical industry, acetone and butanol fermentation process was eliminated due to low cost advantage of the petrochemical industry. But the oil crisis and renewable biomass utilization made the acetone-butanol fermentation technology have been gained attention again [29]

Lignocellulose was hydrolyzed by acid catalysis, and then the hydrolyzate was used for microbial fermentation to yield butanol. Because the composition of monosaccharide in hydrolyzate of lignocellulose is complicated, it requires that the microbial are worse specific for all monosaccharides, and its utilization efficiency is high. The fermentative microorganism use monosaccharide produced by hydrolysis as carbon sources for acetone and butanol fermentation. Clostridium acetobutylicum can digest five-carbon sugar. This feature determines that acetone-butyl alcohol fermentation is more suitable for the combination with the cellulose hydrolysis technology, because the utilization efficiency of monosaccharide is higher

The traditional methods to produce acetone-butyl alcohol involved batch fermentation and distillation-extraction processes, which gave low yield, showed reduced energy efficiency and poor competitiveness. The main problem is that the energy consumption of separation of low concentration products becomes large. The fundamental way to enhance competitiveness of acetone-butanol fermentation is to enlarge the concentration of acetone-butanol in fermenting liquid and develop low energy consumption process. An improved production process of acetone-butanol fermentation mainly has extractive fermentation, gas stripping fermentation, pervaporation and fermentation of cheap raw materials.

The process of extractive fermentation involves the separation of fermentation product acetone-butanol from the fermentation broth using an extractant. The key to increasing the effectiveness of this method is to select the extractant with high separation factor and no toxicity to microorganisms. Studies show that enhanced butanol production is achieved when biodiesel was used as the extractant.

The stripping method allows an inactive gas to pass through the diluents, stripping the given component into its gas phase, so as to achieve the simultaneous separation of acetone and butanol.

Pervaporation is a new membrane separation method involving low energy consumption. This technology is applied to the separation of liquid mixtures, and has a bright future in separation of acetone-butyl alcohol for its characteristics of high separation efficiency and low energy consumption of pervaporation.

9.5.3 Polyols Production from Biomass

Biomass compounds called polyols refer to the polyhydroxy compounds from C_2 to C_6, including xylitol, mannitol, sorbitol, maltitol, glycerin and ethanediol. The raw materials for the traditional preparation of polyol were derived from petroleum and natural gas resources, but a renewable biomass can also be used as raw material to prepare polyol, making the research on biomass-based polyols more and more attractive.

Initially, polyol was used in food and pharmaceutical industries, but with the increasing emphasis on polyol and the development of industrial production technology, polyol has now been widely used to prepare polyurethane materials, paraffin, hydrogen, fuel and chemical intermediates and other fields, and becomes a energy platform compound. In 2004, the U.S. Department of energy ranked polyols such as glycerol and sorbitol as one of the most important among 12 'Building Block' molecules in the development process of biomass [30–33].

In 2006, Fukuoka et al. [34], has used Pt or Ru supported on γ-Al$_2$O$_3$ or γ-Al$_2$O$_3$ -SiO2) as the catalyst to achieve the conversion of cellulose in aqueous solution at 463 K. Six carbon alcohol yield of 30% was obtained under the action of Pt/Al$_2$O$_3$ bifunctional catalyst. The result shows that it has been greatly improved in product separation and catalyst recycling if using environmentally friendly solid acid to replace the traditional liquid acid.

Haichao Liu [35] developed a process to produce acid in situ using high-temperature water to catalyze the hydrolysis of cellulose, then combined the catalyst, Ru/C, to hydrogenate glucose to produce six-carbon polyols. Yields of 23.2% were achieved at 518 K in this reaction. The catalytic activity of Ru/C catalyst is better than that of Pt/Al$_2$O$_3$, because Ru results in better hydrogenation performance for C=O double bonds [36].

Zhang Tao et al further improved the rate of hydrolysis of cellulose, at 518 K, with activated carbon supported Ni-W$_2$C as the catalyst, resulting in efficient catalytic conversion of cellulose to ethanediol with a yield of more than 70%. After the reaction, the catalysts can be recycled and reused without loss of activity. W$_2$C is a Pt-like catalytic material, and has good catalytic activity in the process of C – C bond breaking. The process of converting cellulose into ethylene glycol is similar to that producing six carbon alcohol by cellulose hydrolysis. Firstly, the monosaccharides (glucose) were produced by cellulose hydrolysis. Then, monosaccharide were transformed into ethanediol under the catalytic effect of catalyst. This reaction has avoided the use of precious metal, and its efficiency is high, it is expected to achieve industrialization.

Except glycerol, which is obtained by plant or animal oil hydrolysis, polyols can be synthesized from epoxidized vegetable oil acid by ring-opening reactions. Several polyols with secondary hydroxyl groups with high functionality and hydroxyl value have been synthesized in industries using this method.

Zhang Long has developed a patented technology for the preparation of polyether polyols with starch as raw material, which has been used to replace the polyether polyol products synthesized from petroleum. Currently, a pilot-scale test for 10^3 tons of products is ongoing.

9.5.4 Levulinic Acid Produced from Biomass

In recent years, research of foreign scholars shows that another compound, levulinic acid (LA), was obtained after the further dehydration and lossing formic acid of glucose. It is widely used and likely become a new platform chemical [37].

It can be seen from the molecular structure of levulinic acid, it has a carboxyl group, and a ketone group, so it has good chemical reactivity, and can be used a catalyst in various reactions, such as esterification, oxid-reduction, substitution reaction and polymerization reaction, for synthesizing many useful compounds and new polymer materials. At the same time, the carbonyl group on the 4 position is a prechiral group, it can be transformed its chiral compounds by asymmetric reduction. Levulinic acid is also a biologically active molecule. In addition, it can dissolve in gasoline in any proportion and is used as an additive in automotive fuel, to improve the octane number and reduce exhaust pollution.

Production of levulinic acid by catalyzed hydrolysis of furfuryl alcohol is first invented in foreign countries. In recent years, this product began to be small-scale produced in domestic, its production cost is high and even have the serious pollution of acidic waste .

The US-based company Biofine used cellulosic biomass resources (including sawdust and waste paper) as raw materials to prepare levulinic acid by direct hydrolysis under high temperature and high pressure, achieving a conversion of 80 to 90%, and thus opened up a new direction for the production of this platform chemical.

In recent years, high-temperature and high-pressure hydrolysis processes without catalyst has attracted much attention. Research shows that it has advantages such as faster response, high selectivity for the target product, produces no pollution and involves no catalyst recovery and wastewater treatment.

In the near-critical region, compressed liquid water at a temperature between 250 and 350 °C has good solubility and acts as an effective acid-base catalyst. Hydrolysis in water in the near-critical region shows high selectivity for the target reaction. Lv Xiuyang [38] carried out a research showing that cellulose hydrolysis in near-critical water mainly yielded organic acids (e.g., formic acid, acetic acid and acetylpropionic acid), soluble polysaccharide, grape sugar, fructose, methyl-glyoxal, 5-hydroxymethyl furfural, etc, with a high content of acetylpropionic acid and formic acid and low contents of other ingredients.

9.5.5 Adipic Acid Production from Biomass

Adipic acid is the most important commercial aliphatic dicarboxylic acid. As it has two active methylene groups in alpha carbon atoms, it readily undergoes condensation with polyfunctional compounds and has wide applications. In 1933, W.H. Carothers first used adipic acid and hexanediamine to synthesize nylon 66. In 1935, the DuPont Company started to commercialize nylon 66. With the development of the petrochemical industry, the production of adipic acid starts to use cheaper petro chemical materials as raw, so that the production capacity is greatly increasing.

Frost [39] provides a clean route for producing adipic acid using biological methods. This route, firstly, involves transforming D-glucose into catechol under the catalysis of an enzyme; the catechol undergoes further conversion to cis, cis-hexadiene diacid, which finally yields adipic acid on hydrogenation. Du Pont Company developed a biocatalytic process at the beginning of 1990s, using coli bacillus to transform D-glucose into *cis, cis*-muconic acid, which was then hydrogenated to yield AA (adipic acid). Recently, this company has developed a new biological process, coded for enzymes by using a gene cluster, which isolated from aerobic denitrifying bacteria (Acinetobacter sp.) to obtain the synthase of AA by cyclohexanol conversion.

Niu [40] reported a process in which benzene and its derivatives are used to synthesize adipic acid using Escherichia coli as catalyst.

Bioanalysis involves the use of renewable substances as raw material for realizing green production; but it is expensive, and large-scale industrialized production has not been achieved yet.

9.5.6 Hydrogen Produced from Biomass

Hydrogen as an energy source has many advantages: it is clean, produces no pollution, can be easily stored and transported, and is widely used in chemical industry, metallurgy, aerospace, transportation and other fields. Among chemical industries, hydrogen is the most commonly used in the manufacture of ammonia; according to statistics, about 60% of the world's hydrogen is used in manufacturing synthetic ammonia, in China the proportion is even higher. In the smelting industry, hydrogen is widely used in the production process of desulfurization, hydrogenation and synthesis of chemical products. In the aerospace industry, hydrogen has become one of the most important fuels for spacecraft. In transportation industry, hydrogen fuel cells can be used to power vehicles, and involve zero emissions, high efficiency and other advantages. With the steadily growing applications of hydrogen, various countries have accelerated the development of hydrogen technology [41].

At present, the main methods for hydrogen production can be divided into two categories: one is to use fossil fuels (coal, natural gas, low carbon hydrocarbon or naphtha) as raw materials to obtain hydrogen, this method accounts for 95% of hydrogen production; the other is to produce hydrogen by electrolysis of water. The former is an energy-intensive process; and the primary energy used is still fossil energy. In the hydrogen production process, carbon in raw materials is converted into CO_2, and directly discharged into the atmosphere, leading to serious environmental problems. The latter requires large amounts of electric energy, producing hydrogen with high grade electric energy cannot meet the energy matching criterion, and the conversion efficiency of the net energy is low and the cost is higher. In the long run, hydrogen should be prepared by a way, which is economic, sustainable, nonstone materials (materials not belong to petrochemical raw materials) and without producing greenhouse gas. The ideal method is to produce hydrogen from biomass [42, 43].

9.5.6.1 Hydrogen Production by Pyrolysis

Pyrolysis refers to thermal decomposition of biomass without air, and its main products consist of tar, char and gaseous products. The traditional methods of pyrolysis often used low heating rate for the separation of char. In recent years, in order to obtain a higher yield of bio-oil, high heating rate has been widely used. Demirbas pointed out that when hydrogen is the target product, high temperature, high heating rate and long residence time should be used in pyrolysis. Steam reforming of the hydrocarbons in pyrolysis products is an important method of obtaining hydrogen. In addition, water-gas reaction can also lead to an increased hydrogen production.

Hydrocarbon steam reforming,

$$C_n H_m + H_2O \rightarrow CO + H_2$$

Water-gas shift reaction,

$$CO + H_2O \rightarrow CO_2 + H_2$$

The National Renewable Energy Laboratory (U.S. National Renewable Energy Laboratory NREL) proposed liquefaction of biomass by fast pyrolysis for hydrogen production, the method involves first conversion of biomass into bio-oil, then the residual oil is used to yield hydrogen by steam reforming or cracking.

9.5.6.2 Hydrogen Production by Gasification

Unlike pyrolysis, gasification is carried out in the presence of limited oxygen. In the gasification process, pyrolytic material and charred residue continue to react with air, steam, oxygen, carbon dioxide or hydrogen. Using this method not only increases the gas production, but also provides heat to the gasification reactor. Similarly, in order to obtain higher hydrogen production, it needs to do steam reforming and water-gas reaction respectively for the hydrocarbons and carbon monoxide in products.

The gasification medium for conversion of biomass includes air, oxygen and steam. Air was used as gasification medium in early studies, but the nitrogen in the air will reduce the content of hydrogen and combustible gas. Although the use of oxygen can effectively overcome the shortcomings brought by nitrogen, oxygen will lead to increasing cost and safety concerns. The use of steam is able to furthest ensure the steam reforming and water-gas reaction turning toward to the direction of hydrogen production. At present, steam gasification has become a development trend of biomass thermochemical hydrogen production. Based on this principle, the partial oxidation method of water vapor was invented. This method uses high-temperature steam as gasification medium to gasify biomass in order to obtain hydrogen-rich fuel [44].

9.5.6.3 Hydrogen Produced by Supercritical Water

Producing hydrogen from biomass in supercritical water involves the production of hydrogen-rich gas by catalytic splitting decomposition of biomass in supercritical water. Supercritical water refers to the water at the temperature of 647.2 K, at a pressure of above 22.1 MPa, and possesses both liquid solubility and allows gas diffusion. Even oil and organic solvents are soluble in it. All organic matter can be decomposed by oxidation in supercritical water. During catalytic gasification of biomass in supercritical water, the biomass gasification conversion can reach 100%, and the content of H_2 in the gas can exceed 50 vol%; besides, the process does not produce tar, charcoal and other by-products.

The General Atomic Co. of Santiago (General Atomics) conducted a research on the commercial feasibility of supercritical water gasification of biomass for hydrogen production, using the combustible portion of sewage, sludge, paper sludge, municipal solid as energy sources. The research shows that the method was beneficial for treating toxic and harmful pollutants with high water content [45].

9.5.6.4 Hydrogen Production by High-temperature Plasma

After the pyrolysis of biomass by arc plasma under nitrogen atmosphere, the main components in gas product are hydrogen and carbon monoxide, and tar. In plasma gasification, steam can be added to adjust the proportion of hydrogen and carbon monoxide, and for the preparation of other liquid fuel. There are many methods to generate plasma, such as using aggregation furnace, laser beam, flash tube, microwave plasma and arc plasma. An arc plasma is a typical thermal plasma, which is characterized by high temperature, reaching tens of thousands of degrees Celsius, and the plasma also contains a large number of charged ions, neutral ion and electron active species.

9.5.6.5 Hydrogen Production by Biomass Bio-technology

Under mild conditions, the biomass bio-hydrogen production technology is to transform the organic matter or water into hydrogen by the metabolism of the organism itself. According to the energy sources needed in the growth of hydrogen producing microorganisms, bio-hydrogen production technology can be divided into two categories: hydrogen production by photosynthetic microorganism and hydrogen production by fermentation [46–50].

Hydrogen production by photosynthetic microorganism involves direct conversion of solar energy into hydrogen, but the method has the following problems: it cannot be used to degrade macromolecular organisms; substrate utilization is limited; nitrogenase itself needs more energy, and solar energy conversion and utilization efficiency is low; stability of microbial metabolism is poor, and leads to low hydrogen yield and photobioreactors occupy a large area. These problems led to the difficulty of technology industrialization.

Compared with hydrogen production by photosynthetic microorganism, the technology of bio-hydrogen production by fermentation has certain advantages, mainly: the stability of this process is better than that involving photosynthetic microorganism; the process mainly uses organic substrate degradation to obtain energy, and process control is easy to realize; production capacity is larger; growth rate of microorganism during fermentation is faster; and ease of storage and transportation makes the fermentation of biomass more easily realizable for large-scale production; various substrates are available for bio-hydrogen production by fermentation, including glucose, sucrose, xylose, starch, cellulose, hemicellulose and lignin, and the hydrogen production efficiency of substrate is significantly higher than that of photosynthetic hydrogen production, so the overall cost of hydrogen production is low.

At present, the technology of bio-hydrogen production by fermentation is still in the laboratory research stage. The key problems related to this method are: ①High efficient producing hydrogen producing bacteria is needed: the performance and stability of bio fermentation hydrogen production strain restrict the further development of biomass fermentation technology. It is especially important to study the bacteria with superior performance, stable operation and strong adaptability to substrate. ②Stability of bacteria and continuity of fermentation process are still needed to be further studied: Most of the existing hydrogen producing microorganisms have limited utilization of substrates, which largely limits the industrialization of bio oxygen production. 3) The development of large-scale hydrogen production reactor in the biological fermentation process, the reactor plays a vital role in the process of comprehensive control. Most of the existing biological hydrogen producing reactor is transformed by the existingindustry fermentation reactor, the hydrogen producing efficiency of the reactor is generally not high, stability is poor, and the scale is mostly at the experimental stage. Therefore, it needs to enlarge and optimize the biological hydrogen producing reactor.

9.6 Green Chemical Conversion of Natural Oils and Fats

9.6.1 Profile

The main component of natural oil is fatty acid triglyceride (called triglyceride for short). From the point of view of molecular structure, triglyceride is made up of a glycerol molecule chemically combined with three fatty acid molecules. Triglyceride molecules includes two parts, i.e. glyceryl and fatty acyl, and its three acyl groups can be different fatty acyl groups. Fatty acids in vegetable oils are mainly saturated and unsaturated C_{16} and C_{18} fatty acids. The glycerol moiety in the fat molecule is constant, its molecular weight is 41, and accounts for 4% to 6% of the total oil content; while fatty acyl accounts for about 95% of the total oil content (its content varies with different kinds of oil), and influences the physical and chemical properties of triglyceride.

Therefore, fatty acid chemistry is an important field in oil chemistry. The most common vegetable oils in the world include soybean oil, rapeseed oil, peanut oil and sunflower oil, with China being a major producer of plant oil, including soybean oil, rapeseed oil, sunflower oil, which is extracted from herbaceous plants such as peanut, rapeseed, sesame, sunflower, canola, cotton, soybean, castor, etc., and woody oil plant such as tea-oil tree, tung tree, excoecaria sebifera, oil palm, Jatropha curcas and light skin tree.

Typical fatty acids include oleic acid, which contains a double bond ($C_{17}H_{33}COOH$); linoleic acid with two double bonds ($C_{17}H_{31}COOH$); linolenic acid with three double bonds ($C_{17}H_{29}COOH$) and stearic acid without unsaturated double bonds ($C_{17}H_{35}COOH$). The structure and type of fatty acids have decisive effects on their performance.

The development and application of plant oil chemicals have always been a hot research topic. Compared with petroleum products, they are characterized by low price and renewability. They have been widely used in food, detergents, cosmetics, drugs, petrochemical product substitutes, etc. [51, 52].

9.6.2 The Principle of Natural Fatty Acids and of Chemical Conversion

Fatty acid, an aliphatic carboxylic acid compound, is one of the hydrolysis products of fats and oils. Fatty acid has the properties of carboxylic acid and undergoes all chemical reactions of carboxylic acid. The chemical properties of oil mainly depends on the chemical properties of unsaturated fatty acid, with characteristic reaction site being the double bond, and characteristic reactions being addition, oxidation, isomerization and polymerization. The chemical reaction, which normally occurs at the long carbon chain double bonds of oils and fatty acids, has an increasingly important application in the industry.

Three acyl groups in animal and vegetable oils are usually derived from fatty acids with 12–22 carbon atoms. Plant fatty acid consists of saturated fatty acids, monoene acids, diene acids, triene acids, polyene acids, mono hydroxy fatty acids, mono hydroxy conjugated enyne acids and so on; most of them are unsaturated fatty acids, such as oleic acid (monoenoic acid with eighteen carbons) and linoleic acid (dienoic acid with eighteen carbons). The position of the double bond of unsaturated fatty acids is typically carbon 9 and 10 and linolenic acid has double bonds in carbon 12 and 13, and even in carbon15, 16, most of these double bonds are not conjugated, and their polymerization activity is low. Structural differences of most vegetable oils are only in the conjugated degree of saturation and unsaturated bonds, their chemical properties are similar, especially the chemical modification mechanism of linseed oil, soybean oil, rapeseed oil, corn oil etc. It is basically the same.

9.6.2.1 Hydrolysis of Fats and Oils
The reaction between oil and water to generate fatty acids and glycerol is called the hydrolysis of fats and oils.

The hydrolysis reaction is divided into three steps: first, the triglyceride loses an acyl group to generate diglyceride; then, the diglyceride loses an acyl group to generate monoglyceride, and, finally, the monoglyceride loses an acyl group to generate glycerol and fatty acids. Typically, the rate of hydrolysis in first step is slow; in the second step is very fast, and the third step is reduced. This is because when in the initial hydrolysis reaction, the solubility of water in oil is relatively low, and in the process of latter reaction, fatty acid has an inhibitory effect on the hydrolysis.

Because the hydrolysis reaction is reversible, the reaction is usually carried out under the conditions of high temperature, high pressure and in the presence of catalyst. The commonly used catalysts are inorganic acids, bases, sulfonate and metal oxides, and the lipase extracted from the plant.

9.6.2.2 Esterification

The process to generate esters via reaction of fatty acids and monobasic alcohol in the presence of acid catalyst is called esterification. Esterification and hydrolysis are inverse processes of each other.

The In the reaction, if the OH group of the carboxyl groups and H group of the hydroxyl groups were removed, this reaction is a inverse process. Its reaction rate is slow, it could be balanced by acid catalyst. Adding an excess of fatty acid or alcohol or even removing the generated water will speed up the reaction to completion. For the saturated fatty acids, their reaction rate is nearly the same; the closer the double bond of unsaturated fatty acid to carboxyl groups, the slower the reaction rate is. Esterification is important in industry and in the analysis of fats and oils.

Fatty acid methyl ester is mainly prepared by alcoholysis of methanol and oils and fats or esterification of fatty acid with methanol. Esterification of fatty acids with other polyols, such as ethylene glycol, propylene glycol, polyoxyethylene alcohol, pentaerythritol or sorbitol etc., can produce esters with different uses.

9.6.2.3 Ester Exchange Reaction

Interesterification includes alcoholysis of ester and alcohol, acidolysis of ester and acid, and exchange between ester.

The alcoholysis reaction is a reversible reaction, can be catalyzed by acid or alkaline. For example, triglyceride can be directly esterified with methanol by acid or alkaline catalyst, and fatty acid methyl ester is obtained. The commonly used base catalysts include sodium methanol, sodium hydroxide, potassium hydroxide, etc. Among them, sodium methoxide shows the best activity; for example, while using 0.1–0.5% sodium methoxide, an reaction is carried out at 20–60 °C for about 2 h, alcoholysis is complete achieved. So it can avoids the volatilization of lower fatty acid at high temperature and oxidation of unsaturated fatty acid. When the percentages of the free fatty acid contained in oils is higher than 2%, alkaline catalyst is not recommended, as the fatty acid will become stable carboxylate, and hinder the progress of reaction.

Alcoholysis normally yields a mixture of mono-, di- and triglycerides. Among them, monoglyceride accounts for 40–60% of products, and more than 95% purity can be obtained by molecular distillation. This is the main method for the preparation of food emulsifier of monoglyceride.

9.6.2.4 Reaction of Carboxyl Group of Fatty Acid

Carboxyl group of long-chain fatty acids can generate acyl halide, acid anhydride, salt, amide, peroxy acid. They can also be carried out the process of oxyalkylation, pyrolyzation etc, their derivatives have different applications in the industry.

(1) Salt: fatty acids can react with sodium hydroxide, potassium hydroxide, oxide and carbonate to produce soap, through saponification reaction.

(2) Acyl chloride: fatty acid reacts with reagents such as phosphorus trichloride or phosphorus pentachloride, thionyl chloride and so on to produce acyl chloride.

(3) Acid amide: amide is a fatty acid ammonium, generated by the reaction of fatty acid and ammonia (or amine) at high temperature, and then dehydrated. Long-chain fatty acid amide is widely used in surfactants. For example, ethanolamine or diethanol amine with long-chain fatty acid yields fatty acid monoethanolamide or diethanolamide, the product is mainly used as a bulk soap additive and high foam detergent.

(4) Anhydride: fatty acid anhydride can be generated by the dehydration of fatty acid, but the product yield is low. For example, on heating tetradecanoic acid for 12 min under the temperature of 343 °C, the yield was only about 30%.

(5) Peroxy acid: peroxy acid RCOOOH can be generated by reaction of fatty acid with hydrogen peroxide (30–98%) in acid catalyst (commonly used concentrated sulfuric acid). The reaction can tend to be completely under conditions of excess of fatty acids (C_1–C_5), high concentration of hydrogen peroxide (>70%), or removal of the water by azeotropic method.

(6) Oxyalkylation: the direct reaction speed of fatty acid and ethylene oxide is very slow; commonly, alkaline catalysts such as KOH, Na_2CO_3 and CH_3COONa are used to speed up the reaction. Its product contains fatty racid polyethylene glycol ester and fatty acid polyethylene glycol diester.

(7) Carboxyl reduction: the carboxyl group of fatty acid can be reduced to alcohol hydroxyl group under suitable condition to generate fatty alcohol. Fatty alcohol is an important raw material of fine chemical products, and the aliphatic alcohol derivatives obtained are widely used in surfactants, lubricants and for other industrial purposes.

9.6.2.5 Reaction of Alpha H on the Fatty Acid Carboxyl Group

Alpha halogenated acid: carboxyl alpha H can be replaced by halogen in the presence of a small amount of phosphorus, to generate alpha bromo-acid or alpha chloro-acid.

Alpha sulfonated fatty acids: saturated fatty acids react with SO_3 to generate mixed anhydride, and then molecular rearrangement were occurred, sulfonic acid group substituted for alpha -H, finally, alpha sulfonated fatty acids (alpha sulfonic acid for short) were produced

9.6.2.6 Reaction of the Double Bond on the Carbon Chain of Fatty Acid

The double bond on the carbon chain of the fatty acid is very active; it can be reacted with various reagents by addition, oxidation, sulfonation, isomerization and polymerization, to produce various industrially important chemicals.

9.6.3 Typical Products and Processes of Green Conversion of Natural Fatty Acids

9.6.3.1 Fatty Acid Methyl Esters (Biodiesel)

The fatty acid methyl ester or ethyl ester is obtained by chemical conversion of natural oil and fat, and is called biodiesel (mainly refers to fatty acid methyl ester). Biodiesel is one of the most attractive biofuels, being a high-quality diesel oil substitute and clean source of renewable energy. The research on application of biodiesel production technology is gaining more and more attention. Biodiesel is derived from natural plant oils, its properties are very similar to that of common diesel, and it can be used for diesel engine. The properties of biodiesel with regard to cloud point, flash point, cetane number, sulfur content, oxygen content and biodegradability and so on make it an even better fuel than ordinary diesel. For it has great strategic significance for promoting alternative energy, alleviating environmental pressure and even reducing air pollution etc. Thus biodiesel has become a kind of real 'Green Energy Sources'.

Compared with fatty acid and other raw materials, fine chemicals which prepared by FAME (Fatty Acid Methyl Ester) has the advantages of milder reaction conditions, the product performance is better and so on. The main products are fatty acid poly oxyethylene ester, fatty alcohol-polyoxyethylene ether, fatty alcohol-polyoxyethylene ether sulfate, fatty acid methyl ester alpha sulfonate, sucrose fatty acid ester, fatty acid mono-(di-)ethanol amide, fatty alcohol etc. So FAME is an important raw material for the production of fine chemicals with high added value and biodegradablity [53].

Biodiesel can be made from vegetable oil and animal fat by chemical methods, which include ester exchange method, direct mixed dilution method, microemulsion method and thermal cracking method.

Transesterification refers to the process of obtaining fatty acid methyl ester (or fatty acid ethyl ester) by transesterification of short-chain alcohols, methanol or ethanol with triglycerides in oils and fats in the presence of catalyst. The triglycerides are broken into three long-chain fatty acid methyl esters, thereby shortening the length of the chain, to reduce fuel oil viscosity and improve the fluidity and vaporization properties, to meet the requirements for use as fuel oil.

The catalysts in transesterification, which can be basic or acidic substances or biocatalysts, are classified as homogeneous and heterogeneous catalysts. The production process includes three steps: pretreatment, reaction and post treatment. Pretreatment includes the process of removing impurities from raw oil, steam cooking, deodorization and vacuum dehydration. Post treatment is the purification of crude biodiesel, and includes separation, distillation and other processes.

The transesterification reaction catalyzed by alkaline substances is an irreversible reaction. It can give high yield at low temperature; the reaction proceeds quickly and involves less alcohol consumption, and therefore has been successfully applied in industry; it is the most commonly used method at present to produce biodiesel.

In the preparation of biodiesel by transesterification, it is key to improve the conversion of raw materials and the purity of biodiesel. At the same time, some people adopt the method of steam distillation to refine biodiesel at low temperature, which is used for preparing biodiesel from rapeseed oil.

9.6.3.2 Aliphatic Epoxide

The epoxy grease is produced by the epoxidation of oil; it is a compound which contains epoxy three-membered ring structure, and includes epoxidized soybean oil and epoxidized rosin [54–55].

Epoxy grease is a widely used, nontoxic, tasteless PVC plasticizer and stabilizer; it has good stability to light and heat, and good compatibility, low volatility as well as less mobility. It can not only absorb the hydrogen chloride released in the decomposition of polyvinyl chloride resin, but also be compatible with polyvinyl chloride resin; thus it can be used in almost all PVC products. Unlike plasticizers, such as dibutyl phthalate, dioctyl phthalate and the mixed of dibutyl /dioctyl phthalate, they are not suitable for food packaging. But epoxidized soybean oil is recognized as nontoxic plasticizer, and national food administration as well as the US Food and Drug Administration (FDA) have approved the use of epoxidized soybean oil as packaging materials for food, medicines, toys, additives of household decoration materials and so on.

Vegetable oil is a kind of unsaturated fatty acid triglyceride, even there have unsaturated double bonds in its structure, epoxidation of vegetable oils belongs to the category of olefin epoxidation. At present, the preparation of epoxidized soybean oil is mainly based on the reaction of formic acid (or acetic acid) with hydrogen peroxide to produce peroxy formic(or acetic) acid, which then undergoes epoxidation with soybean oil under the action of catalyst; this is followed by alkali washing, water washing and vacuum distillation, to yield the final product. This method is not environment-friendly, and the products have poor color and low epoxy value. It is necessary to find a new method for the synthesis of epoxy compounds in accordance with the requirements of green chemistry.

The method of direct epoxidation is superior to the chlorine alcohol method in terms of technology, economy, environmental friendliness and other aspects, its invention shows strong vitality. Inspired by the direct oxidation method of ethylene, people are still trying to explore the technique of other olefins through direct oxidation by air or oxygen to produce epoxides until now. The method of molecular epoxidation has been studied for the past 20 years. The catalysts used mainly belong to the following two categories: (1) metal organic compounds (heteropoly acids) and (2) biomimetic catalyst (types of metalloporphyrin). Hydrogen peroxide is considered to be a 'green' oxidant, because its by-product generated in the reaction is only water, and its content of oxidation active species is high, compared with other species, the price is relatively low. Since 1980s, catalyst such as metal organic compounds (heteropoly compounds, methyl rhenium thrioxide), TS-1 molecular sieve, hydrotalcite, biomimetic catalysts (e.g., metal porphyrin complex) etc. have been appeared and applied. Epoxidation of olefins which used the safe concentration (27–50%) of hydrogen peroxide as oxidant aroused people's interest.

At present, one way to solve the problem of homogeneous catalyst separation and recovery is to achieve homogeneous catalysis. It has reported that the heteropoly compounds immobilized on carriers such as ion exchange resin, silicon dioxide and hydrotalcite were prepared to use for olefin epoxidation.

9.6.3.3 Fatty Acid Polybasic Alcohol

The fatty acid polybasic alcohol is a polyol nonionic surfactant, which has a structure of amphiphilic molecule, with a lipophilic chain and hydrophilic hydroxyl group in the molecule. The corresponding fatty acids in ester can be saturated fatty acid or unsaturated fatty acid, of which the number of carbon atoms ranges from 8 to 22, and the corresponding polyols can be sorbitol, glycerol, etc., or sugar and other carbohydrates. The tranditonal raw materials for preparing polyols are mostly from resources such as oil and natural gas. The study of the conversion of cellulose to prepare polybasic alcohol is key for efficient biomass utilization [56].

The synthetic pathway for obtaining polybasic alcohol is generally one of the following.
(1)   alcoholysis of oil and fat;
(2)   direct esterification of fatty acids with alcohols;
(3)   alcoholysis of methyl ester.
       Preparation of lauric acid monoglyceride is used for illustration:

Direct esterification of fatty acids and glycerol yields a mixture of monoglyceride, diglyceride and triglyceride.

Transesterification of glycerol and triglyceride generates monoglyceride.

Enzymatic method includes enzyme-catalyzed hydrolysis, esterification and transesterification.

Enzymatic hydrolysis was carried out by using triglyceride or diglyceride as raw material, with immobilized lipase as catalyst for localized hydrolysis, at a temperature of 40 °C. The product, being a mixture of monoglyceride and carboxylic acid, is easy to separate.

Hydroxyl protection. Three hydroxyl groups in the glycerol molecule, if they direct esterify with fatty acid, their opportunities of esterification are the same, and they will generate the mixture of monoglyceride and diglyceride and triglyceride. In order to obtain high content of monoglyceride product, some compounds were used to protect two hydroxyl groups of the glycerol, the remaining hydroxyl group were esterified with fatty acids. Then by hydrolysis under certain conditions to release the protection groups, monoglyceride were generated. This can even directly obtain high purity monoglycerides product. The protective agents include boric acid, ketones and aldehydes, and so on.

Ionic liquid is an environmentally friendly solvent and catalyst, which can be used for catalytic esterification. In view of its advantages of thermal stability and solubility, adjustable acidity, adjustable polarity and coordination ability, recyclability and so on, ionic liquids have great potential to enable efficient synthesis of monoglycerides.

Questions

1. Compare cellulose with hemicellulose, highlighting similarities and differences with regard to physical and chemical properties.
2. Discuss lignocellulose raw materials, lignin, cellulose and semi-fiber chemical structure.
3. Explain how the degree of polymerization of cellulose molecules can be determined.

References

[1] Wang Jun, editor. Biomass Chemicals. Beijing: Chemical Industry Press, 2008.
[2] Zhan Yixing, editor. Green Chemical Chemical Industry Episode 1. Changsha: Hunan University Press, 2001.
[3] Zhan Yixing, editor. Green Chemical Chemical Episode 2. Changsha: Hunan University Press, 2002.
[4] Guo Guolin. Step into the new world of chemistry. Shijiazhuang: Hebei Science and Technology Press, 2000.
[5] Chen Hongzhang, Wang Lan. Biomass Biochemical Conversion Technology. Beijing: Metallurgical Industry Press, 2012.
[6] Zhang Jianan, Andy Lau. Biomass Energy Utilization Technology. Beijing: Chemical Industry Press, 2009.
[7] Zhang Lina, editor. Natural Polymer Modified Materials and Applications. Beijing: Chemical Industry Press, 2005.
[8] Zhang Xiaoyang, Du Fengguang, Changchun and other editors. Cellulose Biomass Hydrolysis and Application. Zhengzhou: Zhengzhou University Press, 2012.
[9] Yang Shuhui, editor. Plant Cellulose Chemistry. Beijing: China Light Industry Publishing, 2001.

[10] Huang Xiaolei, Wu Yanliang. Biomass transformation method to produce fuels and chemicals potential. International Paper, 2011, 30(2): 45–48.

[11] Zuo Zhiyue, Jiang Jianchun, Xu Junming. Solvent liquefaction of fiber biomass and its application in polyurethane materials. Cellulose Science and Technology, 2010, 18(4): 55–64.

[12] Wang Qiang. Study on the clean separation technology of bagasse based on pulping. Guangzhou: Doctoral Dissertation of South China University of Technology, 2012.

[13] Panwar, N. L., Kaushik, S. C., Kothari, S. Role of renewable energy sources in environmental protection: A review. Renewable and Sustainable Energy Reviews, 2011, 15(3): 1513–1524.

[14] Wang Kun. Study on pretreatment, component separation and enzymatic saccharification of lignocellulosic biomass. Beijing: Doctoral Dissertation of Beijing Forestry University, 2011.

[15] Cheng Heli. Study on the separation and characterization of hemicellulose from corn stalk and the modification of sulfuric acid. Guangzhou: Doctoral Dissertation of South China University of Technology, 2011.

[16] Wang Xutao. Study on the technology and application of steam explosion pretreatment of biological fiber raw materials. Zhengzhou: Doctoral Dissertation of Henan Agricultural University, 2008.

[17] Sun Changzheng, Li Chunling, Xie Chenghua. The present situation and research of steam explosion machine for lignocellulose. Chemical Equipment Technology, 2010, 31(1): 54–57.

[18] Shao Ziqiang, Tian Yongsheng, Xu Kun, et al. The film forming performance of steam explosion after modification of cellulose fiber. Journal of Beijing Institute of Technology, 2004, 24(3): 272–275.

[19] Wang Xin. Steam explosion pretreatment technology and its research progress on bioconversion of cellulosic ethanol. Forest Products Chemistry and Industry, 2010, 30(4): 119–125.

[20] Wang Kun, Jiang Jianxin, Song Xianliang. Research progress of steam explosion pretreatment of lignocellulose and its biotransformation. Biomass Chemical Engineering, 2006, 40(6): 37–42.

[21] Van Walsum, G. P., Garciagil, M., Chen, S. F., et al. Effect of dissolved carbon dioxide on accumulation of organic acids in liquid hot water pretreated biomass hydrolyzates. Applied Biochemistry and Biotechnology, 2007, 137–140(1): 301.

[22] Ragauskas, A. J., Williams, C. K., Davison, B. H., Britovsek, G., Cairney, J., Eckert, C. A., Frederick, W. J., Hallett, J. P., Leak, D. J., Liotta, C. L., Mielenz, J. R., Murphy, R., Templer, R., Tschaplinski, T. The path forward for biofuels and biomaterials. Science, 2006, 311: 484–489.

[23] Zhu Shengdong, Wu Yuanxin, Yu Zi Niu et al. Research progress on the production of fuel ethanol from plant cellulose. Chemical and Biological Engineering, 2003, 20(5): 8–11.

[24] Lee, J. Biological conversion of lignocellulosic biomass to ethanol. Journal of Biotechnology, 1997, 56(1):1–24.

[25] Karimi, K., Taherzadeh, M. J. Conversion of rice straw to sugars by dilute-acid hydrolysis. Biomass and Bioenergy, 2006, 30(3):247–253.

[26] Yu Jun. The effect of different pretreatment on enzymatic hydrolysis of rice husk. Wuhan: Doctoral Dissertation of Huazhong Agricultural University, 2008.

[27] Wang Xiaojuan. A new technology for preparation of fuel ethanol from biomass. Dalian: Doctoral Dissertation of Dalian University of Technology, 2011.

[28] Chen Yuru et al. Research progress on pretreatment technology of plant cellulose raw materials. Chemical Engineering Progress, 1999, 18(4):24–26.

[29] Wang Yilei. Study on Fermentation of acetone butanol. Tianjin: Master degree thesis of Tianjin University, 2008.

[30] Fernando, S., Adhikari, S., Chandrapal, C., et al. Biorefineries: Current status, challenges, and future direction. Energy and Fuels, 2006, 20: 1727–1737.

[31] Wen Chunjun. Process for preparing biological polybasic alcohol by using rapeseed oil and its application in polyurethane foam. Nanjing: Master's thesis of Nanjing University of Science and Technology, 2011.

[32] Edgar, K. J., Buchanan, C. M., Debenham, J. S., Rundquist, P. A., Seiler, B. D., Shelton, M. C., Tindall, D. Advances in cellulose ester performance and application. Progress In Polymer Science, 2001, 26: 1605–1688.

[33] Chen Wenting. Study on preparation of polyurethane foam by vegetable oil polyol [D]. Nanjing: Master Thesis of Nanjing Forestry University, 2009.

[34] Dhepe, P. L., Fukuoka, A. Cellulose conversion under heterogeneous catalysis. ChemSusChem, 2008, 1: 969–975.

[35] Jia Yuqing, Sun Qianhui, Liu Haichao. Study on the selective hydrogenolysis of biomass based polyol on supported palladium catalyst. Annual meeting of Chinese Chemical Society. 2014.

[36] Zhou Likun, Wang Aiqin, et al. Selective production of 1,2-propylene glycol from Jerusalem artichoke tuber on Ni-W2CAC catalysts. Chemsuschem, 2012, 5(5): 932–938.

[37] Aman Chang. Analysis of biomass hydrolysis product of levulinic acid and formic acid and glucose and study on its separation method. Hangzhou: Master degree thesis of Zhejiang University, 2005.

[38] Lv Xiuyang, Sakoda Akiyoshi, Suzuki Motoyuki. Decomposition kinetics and product distribution of cellulose in subcritical water. Acta Chemical Sinica, 2001, 52(6): 556–559.

[39] Frost, J., Draths, K. Synthesis of adipic acid from biomass-derived carbon source. Biotechnology Advances, 1997, 15(1): 294–294(1).

[40] Niu Wei, Draths, K. M., Frost, J. W. Benzene-free synthesis of adipic acid. Biotechnology Progress, 2002, 18(2): 201–211.

[41] Li Tao. Basic research on hydrogen production by biomass fermentation. Zhengzhou: Doctoral Dissertation of Zhengzhou University, 2013.

[42] Lloyd, T. A., Wyman, C. E. Combined sugar yields for dilute sulfuric acid pretreatment of corn stover followed by enzymatic hydrolysis of the remaining solids. Bioresource Technology, 2005, 96(18): 1967–1977.

[43] Kadam, K. L., Wooley R. J., Aden A., et al. Softwood forest thinnings as a biomass source for ethanol production: A feasibility study for California. Biotechnology Progress, 2008, 16 (6): 947–957.

[44] Sanchez, O. J., Cardona, C. A. Trends in biotechnological production of fuel ethanol from different feedstocks. Bioresource Technology, 2008, 99(13): 5270–5295.

[45] Bicker, M., Kaiser, D., Ott, L., Vogel, H. Dehydration of D-fructose to hydroxymethylfurfural in sub- and supercritical fluids. The Journal of Supercritical Fluids, 2005, 36: 118–126.

[46] Chang, V. S., Holtzapple, M. T. Fundamental factors affecting biomass enzymatic reactivity. Applied Biochemistry and Biotechnology, 2000, 84(1): 5–37.

[47] Sun, Y., Cheng, J. Hydrolysis of lignocellulosic materials for ethanol production: A review. Bioresource Technology, 2002, 83(1): 1–11.

[48] Zhang Shaochun. Catalytic conversion of cellulose to prepare polyols and 5-hydroxymethyl furfural. Dalian: Master degree thesis of Dalian University of Technology, 2010.

[49] Shimokawa, T., Ishida, M., Yoshida, S., Nojiri, M. Effects of growth stage on enzymatic saccharification and simultaneous saccharification and fermentation of bamboo shoots for bioethanol production. Bioresource Technology, 2009, 100: 6651–6654.

[50] Adams, J. M., Gallagher, J. A., Donnison, I. S. Fermentation study on saccharina latissima for bioethanol production considering variable pre-treatments. Journal of Applied Phycology, 2009, 21: 569–574.

[51] Wang Xingguo, Jin Qingzhe, et al. Oil Chemistry. Beijing: Science Press, 2012.

[52] Chen Jie. Oil Chemistry. Beijing: Chemical Industry Press, 2004.

[53] Huang Hui. Study on deactivation mechanism of Cu/Zn catalyst for hydrogenation of fatty acid methyl ester to fatty alcohol. Shanghai: Doctoral Dissertation of East China University of Science and Technology, 2010.

[54] Li Xiaolei. Study on catalytic oxidation of vegetable oils by carbonyl compounds. Wuxi: Jiangnan University master's degree thesis, 2012.

[55] Ye Xia. Study on the preparation of mesoporous molecular sieves and their effects on the epoxidation of unsaturated fatty acids. Wuxi: Master Thesis of Jiangnan University, 2011.

[56] Chen Min. Study on green epoxidation of unsaturated organic compounds with long carbon chain. Wuxi: Master degree thesis of Jiangnan University, 2010.

10 Green chemistry in exploiting marine resources

The crisis of marine resource shortage will be more serious in the twenty-first century. In this regard, marine reserves have been receiving more and more attention owing to their huge potential to meet the increasing human needs among the countries. The application of chemical technology to exploit natural resources for economic growth has been accompanied by environmental pollution. Thus, green chemical technologies are required to ensure efficient, safe and sustainable use of natural, in particular marine, reserves.

10.1 The Reserves and Application of Marine Resources

10.1.1 Marine Resources

Marine resources can be divided into material resources and space resources, according to its nature or function. Marine material resources mainly include marine biological resources, marine mineral resources and seawater resources. The classification and utilization of marine material resources are shown in Figure 10.1.

Marine biological resources are also known as marine aquatic resources. It refers to the ocean economic animals and plants which can propagate themselves and constantly update. Marine biological resources include fish, mollusks, crustaceans, mammals, marine plants, marine microorganisms and viruses. More than 80% of the biological resources on earth are in the sea. A number of 20,278 kinds of marine organisms have been recorded, including 3,032 kinds of fish, 1,923 kinds of spiral shellfish, 734 kinds of crabs, 546 kinds of shrimps and 790 kinds of algae. The main economic species are more than 200 varieties. The biological marine resources are abundant, and the amounts of plants and animals in the sea are 1,000 times of those on land. The seafood can feed 30 billion people.

There are over 100 kinds of elements on the earth; of these, more than 80 kinds can be found in the ocean and more than 60 elements can be extracted. These elements exist in different forms in the sea. A total of 11 kinds of elements (chlorine, sodium, magnesium, potassium, sulfur, calcium, bromide, carbon boron, strontium and fluorine) account for more than 99.8% of the total dissolved substances in seawater, and 99% of bromide on the earth is in the ocean. There are 5.5×10^6 tons of gold and 5.5×10^7 tons of silver in seawater. The reserves of other elements, such as barium, sodium, zinc, molybdenum, lithium and calcium, are estimated to be billions of tons, and some of these are estimated to be even more than 100 billion tons. The total amount of manganese nodules at the bottom of the sea is estimated to reach up to 3,000 billion tons, with the Pacific Ocean containing the largest fraction, about 1.7×10^{12} tons, apart from 4.0×10^{11} tons of manganese, 1.64×10^{10} tons of nickel, 8.8×10^9

https://doi.org/10.1515/9783110479317-010

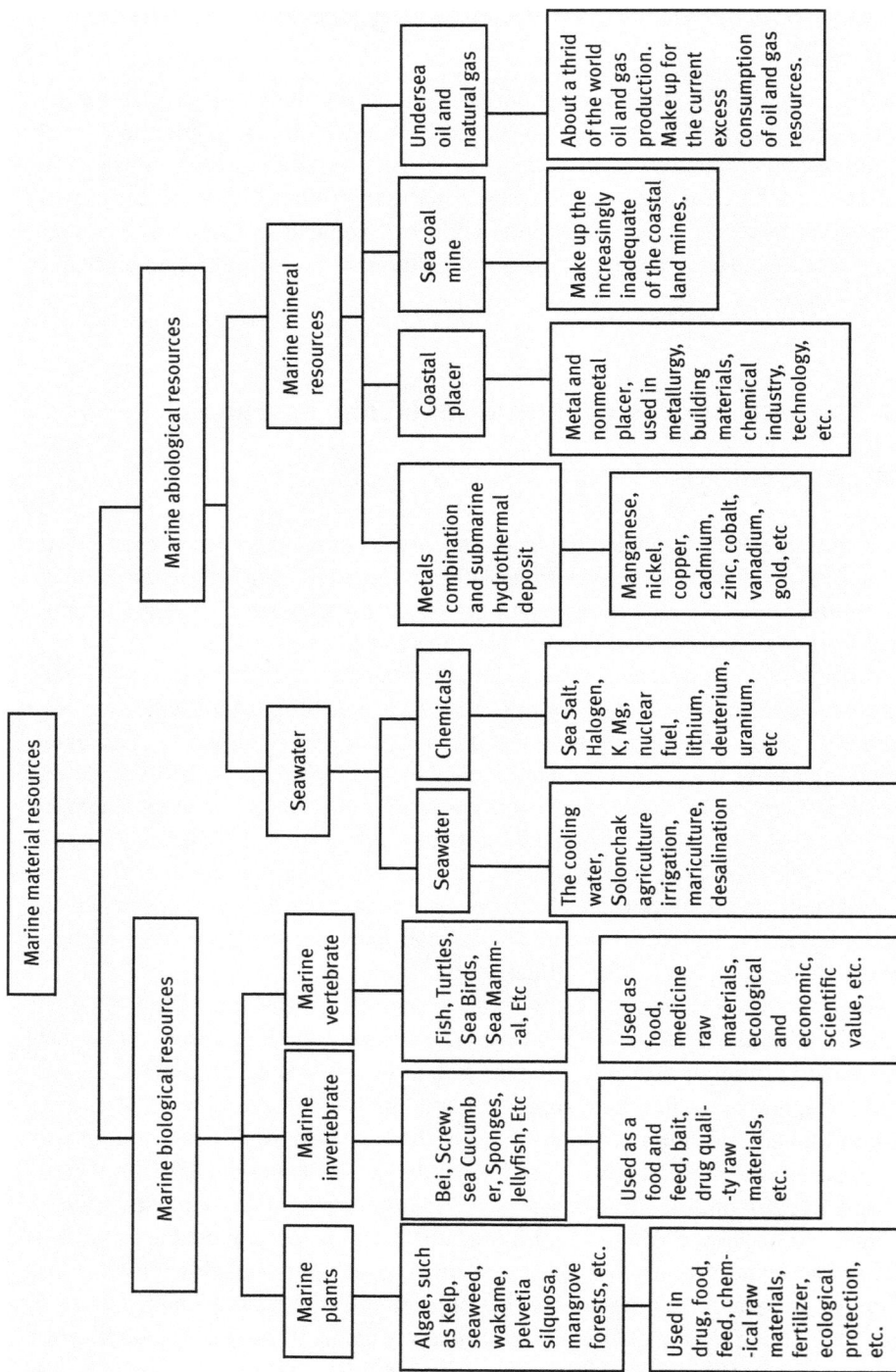

Figure 10.1: Classification and utilization of the marine material resources.

tons of copper and 5.8×10^9 tons of cobalt. These reserves are more than 400 times the manganese reserves, 1,000 times the nickel reserves, 88 times the copper reserves and 5,000 times the cobalt reserves on the land. According to the current annual consumption in the world, these minerals can meet human needs for thousands or even tens of thousands of years. More importantly, the nodules in the bottom of the sea are still growing. The manganese nodules at the bottom of Pacific Ocean grow at an annual rate of about 1.0×10^7 tons. The manganese produced by these nodules in one year is enough to satisfy the total world demand for a few years.

There are also plenty of reserves (uranium and heavy water) in seawater that can be used as nuclear fuel. The uranium reserves are estimated to be about 4.5×10^9 tons, which is 4,500 times that of the total reserves on the land. According to the energy it could release, uranium in seawater can afford the whole world at least 10,000 years. Heavy water is the main source of heavy hydrogen, which is used as a fuel in nuclear fusion reactors. The reserves of heavy water in the sea are about 2.0×10^{14} tons. The total quantity of heat generated by heavy hydrogen is equal to thousands times the heat from all the fossil fuels in the world. Seawater can be used as water for human consumption and irrigation after desalination.

In addition, oil and gas resources are also abundant in the ocean. The global proven oil reserves are 1.757×10^{11} tons and that of natural gas is $1.73 \times 10^{14} \, m^3$, while the global offshore oil resources are about 1.350×10^{11} tons and the proven reserves are about 3.80×10^{10} tons. On the other hand, the global marine natural gas resources are about $1.40 \times 10^{14} \, m^3$ and the proven reserves are about $4.0 \times 10^{13} \, m^3$. Therefore, the potential of marine reserves to meet future energy demands is enormous.

10.1.2 The Application of the Marine Resources

Humans have been using marine reserves since ancient times. About 18,000 years ago, sea cockle shells were found to be used for decorational purposes in Zhoukoudian Upper Cave, Beijing. Around 4000 bc, coastal residents began to "boil seawater for salt." Seawater salt had been commercially used in China 2,700 years ago. Using marine biological resources to obtain drugs also has a long history. There are records in the ancient Chinese medicine books about using crab, fish and algae as drugs. According to the Compendium of *Materia Medica*, there are about 1,900 kinds of drugs, more than 200 kinds of which can be derived from marine limnetic organisms.

In the view of the use of marine resources, marine aquatic products are the oldest and the most widely used reserves, including marine fish, invertebrates and algae. The world production of aquatic products increased from 1.20×10^6 tons in 1800 to 4.00×10^6 tons in 1900, 2.10×10^7 tons (the marine aquatic products were 1.88×10^7 tons) in 1938 and 7.08×10^7 tons (the marine aquatic products were 6.07×10^7 tons) in 1970. The total global output of marine aquatic products increased from 1.88×10^7 tons in 1938 to 6.458×10^7 tons in 1980, which increased by 2.4 times. The global marine fishery output was 28.3 billion dollars in 1978. At present, the ocean provides more than 9.0×10^7 tons of marine aquatic products to human being each year. It is estimated that the growth rate of marine fish products is as much as 6.0×10^8 tons per year. It can catch for $(2.0–3.0) \times 10^8$ tons marine fish could be caught every year under the premise of without resources destroying. It is 2–3 times of the current fishery production worldwide. It is estimated that the global demand for seafood will increase to 6–7 times the current yield by 2025, which is about 7.0×10^8 t/a.

Sea salt industry is another early industry using seawater resources. NaCl is one of the earliest minerals extracted from seawater. Sea salt is mentioned in the *Vinaya Pitaka*, a Buddhist scripture compiled in the mid-fifth century bc. The salt production of China is the highest in the world, reaching 3.177×10^7 tons in 2007, and the output value reached 5.5 billion yuan in 2009.

The marine resource chemistry focuses on chemicals extracted directly from the sea around 1930. The air blowout method for the extraction of bromine from seawater and the chemical precipitation method for the extraction of magnesium, respectively, from seawater were developed. Industries for preparation of bromine and magnesium from seawater were established, respectively. In 1935, the experiment of extracting potassium from seawater through the amide method was completed. Since 1952, the desalination technology has been widely used to remove salt from seawater. The bromine extracted from seawater accounts for about 70% of the global total bromine output, while the annual production capacity of magnesium from seawater is 2.57×10^6 tons, which accounts for 34% of the global total magnesium output.

Marine organisms, such as seaweeds, marine fungi, actinomycetes, microalgae, mangroves and bacteria are rich resources of products with pharmaceutical activities. The chemicals from marine organisms have greater diversity in structural and functional features owing to their environment having extreme variations in pressure, salinity, temperature and so forth. The efforts to extract drugs from the sea started in the late 1960s. However, the systematic investigation began in the mid-1970s. During the decade from 1977 to 1987, about 2,500 new metabolites were reported from a variety of marine organisms, clearly demonstrating the potential of marine environment as an excellent source of novel chemicals, not found in terrestrial sources. Until 2008, more than 25,000 compounds had been isolated from marine organisms, with hundreds of new compounds still being discovered every year. About 300 patents on bioactive marine natural products were issued between 1969 and 1999 [1]. Some of the marine drugs authorized by the Food and Drug Administration (FDA) so far include cytarabine (Ara-C), vidarabine (Ara-A), eribulin mesylate (E7389) from sponges, ziconotide from cone shell, omega-3-acid ethyl esters, omega-3-acid from fish and brentuximab vedotin from seaweed. Some drugs, such as marizomib from bacteria, plinabulin (NPI 2358) from fungi and plitedespin from ascidian *Aplidium albicans*, are in the stage of clinical trials.

10.2 Extraction and Preparation of Food Additives from Marine Resources

Besides being used as food, marine fauna and flora are also the main sources of many food additives, such as algae polysaccharide, cod liver oil, chitin, protein powder, calcium and trace elements. These substances have a wide range of applications in the food industry. The following section explains the characteristics and the green extraction processes of polysaccharides from seaweed and cod liver oil.

10.2.1 Algal Polysaccharide

Polysaccharides, the generic term for carbohydrates, are the main constituents in marine algae. The chemical structure of algal polysaccharide is very complex, and these molecules account for more than 50% of the dry weight of most algae. Generally, algal polysaccharide, also called seaweed gum, is water-soluble and possesses high viscosity. Seaweed gums mainly include alginates, agar and carrageenan [2], and have been used widely in food owing to their chemical properties such as high viscosity, emulsifying and film-forming property. Seaweed polysaccharides are used in milk, jelly, ice cream, spices and soft drinks as additives such as emulsifiers, thickeners, stabilizers and coagulants. Some polysaccharides with low surface tension can be used as natural foaming agent and other food additives. The preparation of algal polysaccharide mainly involves extraction and purification. The conditions of the process can be controlled to prepare algal polysaccharides with different purity. The detailed process is shown in Figure 10.2.

10.2.1.1 Extraction of Algae Polysaccharide

Solvent extraction is one of the most common methods of extracting algal polysaccharides and is mainly based on the principle of "like dissolves like." Generally, pure water, dilute acid and dilute alkali are used as the solvent to extract polysaccharides, depending on the polysaccharide structure. Mostly, algal polysaccharides, being water-soluble polar polymers, are extracted with hot water. Moreover, in order to achieve a high extraction rate, basic solution such as dilute NaOH solution can be used with acidic sugar residue. Furthermore, some polysaccharides show higher extraction rate in an acid solution. The equipment used in solvent extraction is simple and of low cost; however, the extraction rate is low and the extraction time is long. When acidic or basic solution is used to extract polysaccharides, the acid/base concentration must be strictly controlled to prevent the degradation of polysaccharides, decrease of the yield and deterioration of the quality.

In recent years, many new technologies have been used to extract polysaccharides from algae, such as microwave-assisted extraction (MAE), ultrasound-assisted

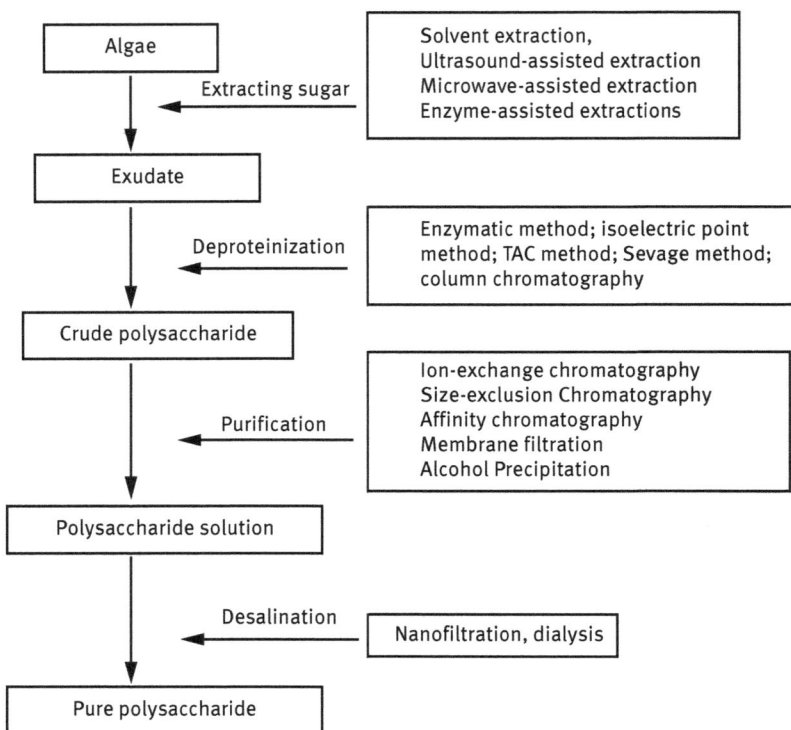

Figure 10.2: Seaweed polysaccharide extraction and purification process.

extraction (UAE), enzyme-assisted extraction (EAE), high-intensity pulsed electric field-assisted extraction and freezing and thawing methods [2]. Among these methods, UAE, MAE and EAE are more popular and practical.

(1) UAE: Seaweed polysaccharide is an intracellular polysaccharide. The cell wall of seaweed plants needs to be destroyed to release polysaccharides. In this range (the frequency of ultrasound is more than 20 kHz), sound waves migrate through the solvent, inducing pressure variations and cavities that grow and collapse. The sound waves are transformed into mechanical energy, which rupture the cell walls in an instant, and the contact between the solvent and algae is enhanced owing to the vibrations caused by the ultrasound. Then, the active ingredients are dissolved, released and diffused in the solvent. In addition, ultrasound can produce thermal effects, crushing, diffusion, emulsification, cohesion effects and a series of secondary effects, which further accelerate the release of active ingredients and enhance the extraction rate. The extraction efficiency of polysaccharides by UAE is higher than normal solvent extraction without ultrasound. Shi et al. [3] extracted polysaccharide from *Chlorella pyrenoidosa* by UAE, in water bath at 100°C for 4 h after passing 400-W ultrasound for 800 s, obtaining a yield of 44.8 g/kg.

(2) MAE: MAE is an energy-assisted extraction method. Due to the even heat field and high temperature created during the microwave irradiation, the temperature and the pressure of algal cells increase. The cell structure containing polysaccharide is decomposed quickly, and then the components within the cell are released. At the same time, the pressure generated by water vapor damages the cell wall and cell membrane, which promotes the spread of polysaccharides to the solvent. Microwave can also reduce the water in the cell wall and inside the cell, which leads to cracks, holes in the cell surface. The solvent diffuses inside the cell's interior. The ingredients within the cell are released and dissolved in the solvent, leading to an increased extraction rate. Rodriguez-Jasso et al. prepared polysaccharides from brown algae by MAE. The optimum extraction conditions were as follows: the pressure was 120 psi, the ratio of material to liquid was 1:25 and the extraction time was 1 min. Compared with the traditional solvent extraction method, MAE is more simple, faster, more efficient and more environmentally friendly [4].

(3) EAE: In the process of extracting polysaccharides, enzymes can be used to decompose the plant tissues and promote the release of substances from the tissues. In recent years, EAE has been extensively researched due to its high catalytic efficiency, high specificity and mild reactive conditions. Many enzymes have been used in polysaccharide extraction, such as protease enzyme, carbohydrase, amyloglucosidase, agarase, alcalase, carragenanase, cellucast, cellulose protamex, kojizyme, neutrase, termamyl, ultraflo, umamizyme and xylanase [5]. The method of EAE is ecofriendly, nontoxic and can be used in large scale. In the case of EAE of algal polysaccharide with cellucast, the effects of cellulase dosage, reaction temperature and reaction time on the extraction rate were investigated. The highest extraction rate of 7.41% was obtained for the conditions: cellulase dosage of 1.2%, extraction temperature of 45°C and enzymolysis time of 100 min, which is higher than that obtained through water extraction. However, the use of enzymes is limited due to their high price in industrial applications.

(4) Extraction in ionic liquid (IL) or supercritical liquid.
    An IL is a salt in the liquid state, sometimes it specifically refers to the salts whose melting point is below 100 °C. ILs are largely made of ions and short-lived ion pairs. IL has many unique characteristics, such as low vapor pressure, low combustibility, high thermal stability and high solvating ability for a range of polar and nonpolar compounds. IL is known as a green solvent and widely used in chemical reaction, extraction and separation. IL has also been used to extract algal polysaccharides. Few choline-based bio-ILs (choline formate, choline acetate, choline caproate, choline caprylate and choline laurate) have been employed for the selective precipitation of agarose from the hot seaweed extract of *Gracilaria dura* under ambient conditions [6]. Among the bio-ILs, choline laurate was found to be the most effective for the isolation of agarose

at lower usage level (4.0%, w/w) with the yield of (14.0, 84, 0.5)% w/w. Agarose obtained by this process had the properties required for molecular biological applications and gel electrophoresis. Moreover, the bio-ILs could be recycled and reused without affecting the yield and quality of biopolymer. The method provided a much "greener" and economical means for algal polysaccharides extraction.

Supercritical liquid is another popular green solvent in chemistry, which has also been used in the extraction of polysaccharides. Microwave-assisted hydrothermal experiments were performed on supercritical CO_2 ($SSCO_2$) deoiled samples. Experiments were carried out mostly using a Mars5 IP microwave extraction device at a set maximum microwave power of 600 W. Deoiled *Undaria pinnatifida* algae were treated under hydrothermal conditions heated by microwave radiation at a temperature range 110–200 °C and for a treatment time of 5–120 min to recover fucoidan or to degrade it into more valuable low-molecular weight products of around 5–30 kDa. Microwave heating can enable extraction and degradation of fucoidan faster than the conventional heating methods. $SSCO_2$ was found to simplify the purification process of crude polysaccharides. The purification of the crude polysaccharide to remove protein and other impurities is carried out through EtOH extraction and centrifugation [7] (shown in Figure 10.3).

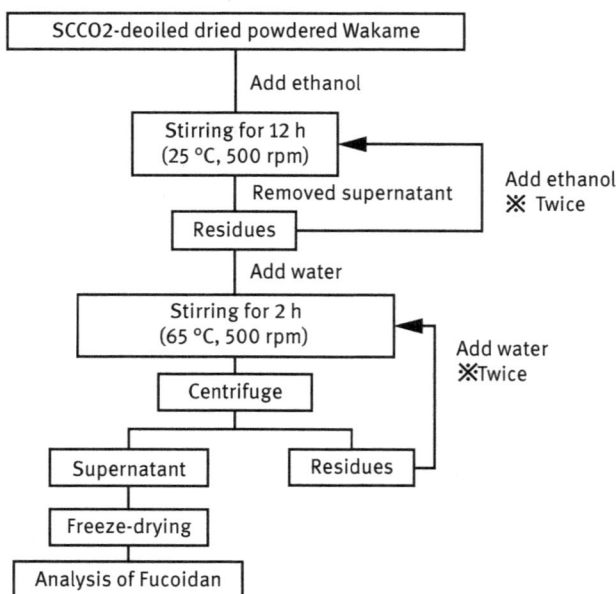

Figure 10.3: EtOH pretreatment procedures of deoiled *U. pinnatifida* for removal of proteins [7].

10.2.1.2 Purification of Algal Polysaccharide

There are many impurities in the crude polysaccharide obtained through solvent extraction, including protein, pigment, small molecules of sugar and inorganic salts. It is necessary to purify and remove these impurities one by one in order to obtain pure polysaccharides.

1. **Traditional methods**

   For removing protein from crude polysaccharide, there are many methods, including isoelectric point method, enzymatic method, organic solvent method (Sevage method also known as chloroform−n-butanol method, trifluorotrichloroethane method, trichloroacetic acid method) and column chromatography. In the isoelectric point method, enzymatic method and organic solvent method, the proteins are all transferred into insoluble state, which can be separated from the polysaccharide solution through centrifugation or sedimentation. In the Sevage method, the most efficient protein removal is achieved when a mixture of chloroform and pentanol or butanol (v:v is 4:1) is used as the solvent. Sevage method is the most popular and most widely used method for protein removing. However, there are many disadvantages in the Sevage method, including incomplete removal of protein, low polysaccharide yield, difficulty in solvent recovery and environmental effects. In the isoelectric point method and enzymatic method, the polysaccharide is easily lost, the yield is low and protein removal is not complete. Thus, the Sevage method and protease method have been combined, which reduces the use of organic solvents and makes the polysaccharide extraction process greener. Column chromatography and membrane filtration are newly developed technologies for polysaccharide purification.

2. **Column chromatography**

   Column chromatography uses the principle of chromatographic separation to separate the impurities from the polysaccharides. The high performance and resolution, high recovery rate of polysaccharide, together with large sample handling capacity and automation make column chromatography a desirable technique for purification of polysaccharides. Column chromatography can also be used to separate different types of polysaccharides or fractionate polysaccharides after deproteinization. Ion-exchange resins adsorb different types of polysaccharides owing to different affinity between the resin and the polysaccharide molecules. The adsorbed polysaccharides are washed via elution by stepwise or linear NaCl salt gradient. Diethylaminoethyl cellulose is the most popular resin for separating polysaccharides with different charges (positive, negative or neutral polysaccharide).

   Size-exclusion chromatography is used to fractionate polysaccharides. The polysaccharides are separated based on their size as they pass through a porous matrix of particles with chemical and physical stability. Dextran is the most popular chromatography packing material to fractionate polysaccharides. Many commercial columns are available in laboratory scale, such as single-column Sepharose CL-6B, PL aquagel-OH, Sephacryl S-300 or Superdex 200, or successively connected columns.

3. **Membrane filtration**

Membrane filtration is based on the separation of the polysaccharides according to molecular weights of interest. For polysaccharide separation and purification, ultrafiltration and nanofiltration are the common membrane filtration technologies. Membranes with different molecular weight cutoff are used to filtrate the polysaccharide to obtain fractions with different molecular weight. The separation membrane has an asymmetric microporous structure. The particles and macromolecules flow through the membrane surface during the separation process, while the small molecules pass through the pores in the membrane under pressure to achieve the separation effect. Ultrafiltration is mainly used for the purification of polysaccharides, while nanofiltration is used to remove inorganic salts. Membrane filtration has the advantages of high separation efficiency, continuous production, being environmentally benign, causing no harm to the polymer structure and activity and being easy to scale up. However, there are disadvantages limiting its commercial application, such as excessive membrane fouling, high energy inputs and frequent membrane cleaning or replacement.

There are many methods for removing the pigment in the polysaccharide solution, such as oxidation, ion-exchange resin, adsorption and metal complex method. Commercially, seaweed polysaccharide solution was processed with hydrogen peroxide. The pigment is oxidized to colorless products. The method involves low cost and is easy to operate. However, the pigment residues still remain in the solution, which requires further separation and purification.

10.2.2 Cod Liver Oil

Cod liver oil is a kind of liquid extracted from the liver of deep-sea fish. It is praised as liquid gold since it is a valuable source of nutrition, such as vitamin A, vitamin D, Docose Hexaenoie Acid (DHA), squalene and alkoxy glycerol. The naturally present vitamin A and vitamin D in cod liver oil are very helpful for the growth of infants and young children. There is a kind of biochemical substance called squalene in the liver of giant sharks, which are found in the deep sea. The drugs made from squalene have many significant clinical effects in treating cancer, hepatitis, heart disease, high blood pressure and other diseases. The alkoxy glycerol in the liver oil can effectively delay aging of various organs in the whole body and tissue cells. Alkoxy glycerol is also called the "king of antioxidants."

10.2.2.1 Extraction of Cod Liver Oil

In the eighteenth century, the process of extracting cod liver oil from the manual workshop was mechanized and many innovations and breakthroughs were made in this field. In the middle of the nineteenth century, the extraction process became

Cod liver oil →(Extraction)→ Thick cod--liver oil → Filter, removing impurity → Acid enzymatic degumming → Alkali refining acid → Decodorization → Refined oil

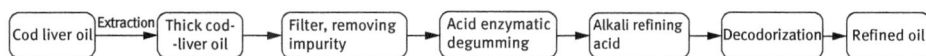

Figure 10.4: Extraction process of cod liver oil.

commercial. Cod liver oil preparation process now includes extraction and refining (mainly divided into four steps, including degumming, deacidification, decolorization and deodorization). A flowchart of the preparation process is shown in Figure 10.4.

The extraction process of cod liver oil has undergone four technical upgrading in the course of its development process. On the basis of the old squeezing method, several new technologies, such as cooking technology, solvent extraction, dilute alkaline hydrolysis and the EAE method, have been developed

Squeezing method: This method is the oldest physical method in the early stage. First, the fat combined with proteins is destroyed by physical pressure. Then, the crude cod liver oil is obtained after centrifugation. Most of the cod liver oil obtained by this method is a by-product of fish waste processing. The extraction efficiency is very low and the quality of the cod liver oil is poor. This method is not used anymore.

Cooking technology: The liver is stewed directly in a pot with two layers. The clear cod liver oil is obtained after precipitation for a period of time. This method can be divided into indirect steam refining and water vaporization. These two processes are basically similar. The difference is that the indirect steam refining method uses water vapor as the heating medium. The shortcomings of this method are low efficiency, long operation time and loss of most of the essence of fish liver.

Solvent extraction: The solvent extraction method involves extracting the oil with organic solvents (such as chloroform) based on the solubility of the gradients. The fish oil and fat-bound protein cannot be separated by this method. The oil extraction rate is low and there is organic solvent residue in the product, which will affect the quality of the fish oil and the method causes environment pollution.

Dilute alkali hydrolysis: Cooking the chopped fish liver together with dilute alkali solution (such as NaOH) is helpful to extract cod liver oil. The low concentrations of base can decompose the fish protein tissue and destruct the combination of protein and fish oil, which can accelerate the separation of cod liver oil. There is chemical residue and the process will cause environmental pollution by the traditional dilute alkali hydrolysis. The method has been modified by using other medium to help in extraction, such as KCl, ammonia and ammonium salt. The raw material and the by-products of the modified method, being traditional fertilizer, can be further utilized.

EAE: Due to the selective hydrolysis of protease to proteins, the combination of protein and fat is destroyed by protease. The stability of the emulsion is reduced and the oil is then released. Enzymatic reaction conditions are mild and the active ingredients in the liquid are not destroyed. The quality of oil can be improved by

EAE. The small molecules, peptides and amino acids in the enzyme solution can be fully used. To increase the extraction yield, EAE and solvent extraction have been combined. The catfish viscera was enzyme digested for 2h at 45°C to obtain crude oil, with the solid/liquid ratio of 1:6 and the amount of enzyme 1,400 U/g. The crude oil was extracted and filtrated. Then the mixture of n-hexane/isopropanol was used to extract the oil with the solid/liquid ratio of 1:5 for 25 min. The extraction rate of fish oil was 25.3%.

10.2.2.2 Purification of Cod Liver Oil

In the process of preparing fish oil, protein, pigment, phospholipids and other nonglycerol ester components remain in the crude oil. These substances will affect the stability of oil. Further processing, including degumming, deacidification, bleaching and deodorization, is required to refine and remove the nonglyceride components and other impurities. In the process of refining oil, the physical and chemical properties of the oil need to be measured in different stages.

Degumming: The aim of the degumming process is to remove glial impurities in the crude cod liver oil, such as phospholipids, proteins and viscous substances. The degumming methods include acid refining degumming, hydration degumming, membrane filtration degumming and enzymatic degumming. The phospholipid content in fish oil is low. Since acids can neutralize the charge of colloidal particles and precipitate the colloidal particles, the acid refining degumming method is commonly used for removing impurities. The glial impurities are removed after centrifugation. The heavy metals and pigments in fish oil can also be removed during degumming. Phosphoric acid, sulfuric acid and citric acid are the popular acids used in the degumming process. The oil obtained is called crude oil.

Deacidification: The aim of deacidification is to remove free fatty acids in the crude oil. Alkali refining deacidification is a frequently used method of deacidification. In the process, sodium hydroxide is added to the crude oil. The free fatty acids are converted into insoluble soap. The soap can absorb other impurities in the oil. The soapstock is removed by centrifugation. The degummed oil is pumped into the neutralization pot and heated to 30–40°C. The alkaline is added into the pot with stirring at 60 rev/min and heated up to 60°C and the stirring speed is reduced to 30 rev/min after a while. The soapstock is released after continuing the reaction for 5–8 h. Water with a volume of 30%–50% oil is added and heated to 40°C with stirring. The fish oil is neutralized by washing and centrifuged to remove the residual soapstock and water.

Decolorization: Crude oil is generally brown owing to the products resulting from the Maillard reaction between the carotene in fish oil and protein decomposition products and their oxides. In order to remove the pigment, adsorption method is often used. The common adsorbents used are activated clay and activated carbon. In order to avoid oil oxidation in the process of decolorization, vacuum is often used. The deacidified fish oil is pumped into a vacuum desiccator to be dehydrated under reduced pressure, and then heated to 100°C in a bleaching tank. Then a mixture of 0.5–1% of acidic clay and activated carbon (8:1) is added and stirred under reduced pressure. When the chroma of the product meet the requirement, the product is filtered.

Deodorization: The aim of deodorization is to remove the degradation products of proteins from the raw materials and the introduced pollutants the odorous ingredients produced by oil oxidation and deterioration, such as ketones, aldehydes, lower acids, peroxides. The oil deodorization methods include vacuum distillation deodorization, gas blowing method, polymerization and steam deodorization method. The odors cannot be removed completely at low temperature and with short process time. However, high temperature and long reaction times will lead to oil oxidation. Therefore, the products are often injected into a high-temperature steam under the high vacuum to remove the stink in cod liver oil.

10.3 Extraction and Synthesis of Drugs from Marine Resources

The development of modern marine drugs began in the 1960s, when the Japanese scholars started researching Hipodoxin. In the past few decades, more than 15,000 compounds have been separated from marine animals, plants and microbes. Cefotaxime sodium salt, conotoxin, cytarabine and arabinosyl adenosine are typical examples of drugs developed from marine precursor compounds. The analgesic drug Conotoxin (Ziconotide) is considered as the first drug of marine origin. The precursor compound of Ziconotide is the peptide toxin from conon (w-conotoxin). Ziconotide received the US FDA approval in 2004. Currently, ziconotide is prepared by synthetic methods.

The research on modern marine drugs is focused on marine bioactive components (i.e., marine natural products). The main marine bioactive components include lipid, amino acids, peptides, sugars, glycosides, terpenoids, carotenoids, steroids, nonpeptide nitrogen compounds, and so on. The pharmacological effects of marine active substances include antitumor, antibacterial and antiviral, anti-inflammatory, analgesic and antioxidative activities and they are mainly beneficial to the central nervous system and the cardiovascular and cerebrovascular system. There are several ways to develop marine drugs, including in vivo extraction of bioactive components from marine organisms, artificial synthesis and biological fermentation.

10.3.1 Extraction and Degradation of Chitin/Chitosan

10.3.1.1 Extraction of Chitin/Chitosan
Chitin (poly(1,4)-2-acetylamino-β-D-glucan) was first extracted from the crustacean shell by Odier in 1823. It has been confirmed that chitin can be extracted from

arthropods, fungi, butterflies, mosquitoes, flies, silkworm and other pupa shell. The content of chitin in crustaceans is about 20-30%, sometimes up to 58-85%. At present, commercial chitin is usually obtained from shrimp and crab shell. Their content in shrimp shell is about 20–25% and in crab shell is 15–18%. Chitosan is the deacetylated product of chitin, which is soluble in dilute acid. Low-molecular weight chitosan can even dissolve water. Therefore, chitosan is usually used as the raw material to develop chitin. The bioproperties of chitosan include its ability to decrease blood pressure, decrease blood sugar, decrease cholesterol, improve human immunity, promote cell production as well as antibacterial, antioxidation, anticancer activities. Many pharmaceutical products from chitosan have been commercialized, such as "Chitosan" for weight loss, chitosan hemorrhoids antibacterial gel, eye protection solution, cosmetics. Generally, the product soluble in 1% HAc is called chitosan, and that not soluble in 1% HAc is called chitin. Chitin contains about 40% 2-acetylglucosamine sugar residues. The molecular structure of chitin and chitosan is shown in Figure 10.5.

The extraction process of chitosan varies with the raw materials. In the industry, chitosan is extracted by treating shrimp and crab shell with acid and alkaline. The process mainly includes decalcification, deproteination, decolorization and deacetylation [8] (as shown in Figure 10.6). First, the shrimp or crab shells are washed with water and dried. Then the shells are soaked in 4–6% dilute hydrochloric acid at room temperature for 2 h to remove the calcium carbonate in the shell. After filtration, they are neutralized by washing, and the shells are placed in 10% NaOH solution and left to boil for 2 h to remove off the protein. Chitin is obtained after filtration, and neutralized by washing and drying. To obtain chitosan, chitin is placed in 45–50% NaOH solution and hydrolyzed at 100–100°C for 4 h or in 40% NaOH solution in an oven at 84 °C for 17 h. Chitosan is obtained after filtration, washing to neutral and drying. In the reaction process, the concentration of NaOH and reaction time should be controlled. When the sodium hydroxide concentration is less than 30%, no matter how high the reaction temperature and how long the reaction time are, the deacetylation degree (DD) can only reach about 50%. In the alkali solution with high concentration, hydrolyzed degradation of the chitin polymer chain occurs. Therefore, the reaction time must be strictly controlled. Due to the presence of astaxanthin in shrimp and crab shells, oxidants such as potassium permanganate and sodium bisulfite are used (to remove the color of the products). Or an organic solvent such as acetone is used to extract the pigment.

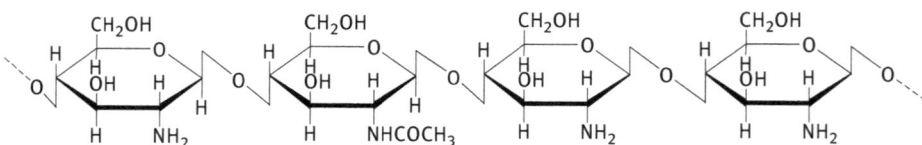

Figure 10.5: Molecular structure of chitin and chitosan.

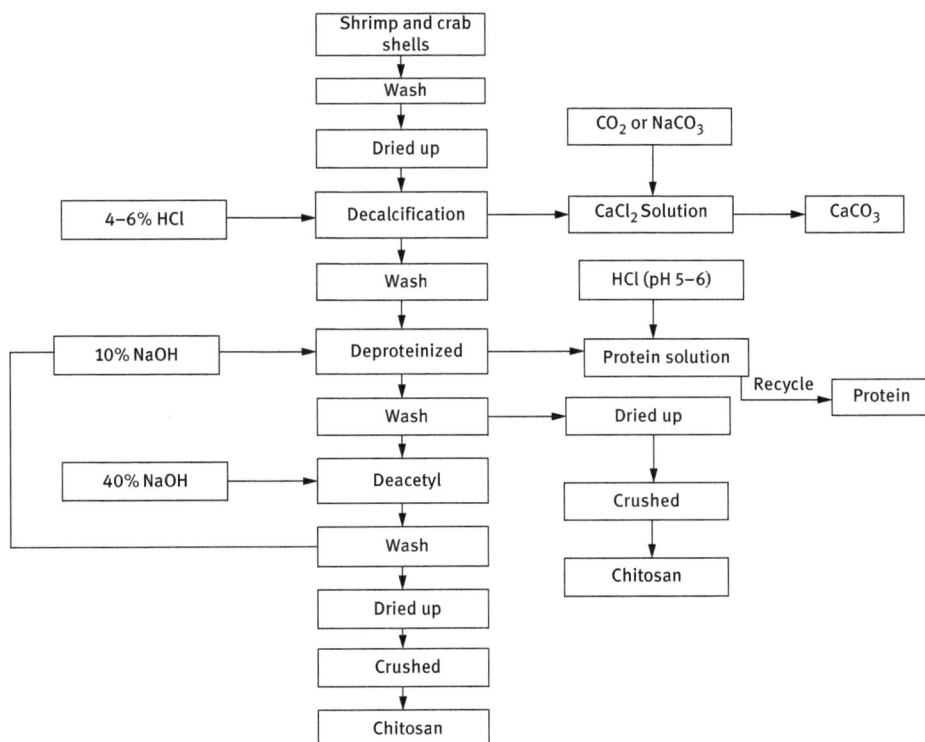

Figure 10.6: Extraction process of chitosan in shrimp and crab shell.

10.3.1.2 Degradation of Chitosan

The molecular weight of chitosan has a significant effect on its physical properties, chemical properties and bioactivity. Chitosan with molecular weight less than 10,000 Da has many good physiological activities. Chitosan with molecular weight of 1,500 and 3,000 has excellent hygroscopicity and moisture retention ability. Its degree of moisture absorption is stronger than that of hyaluronic acid, sodium lactate or glycerol. The bacteriostatic ability of chitosan gradually increases with the decrease of average molecular weight, and chitosan with the molecular weight of 1,500 has the best antibacterial ability. Chitosan with degree of polymerization (DP) 6–8 has good anticancer activity. Therefore, chitosan and its degradation has become the focus of attention of researchers around the world since 1950S.

At present, the method of degradation of chitosan / chitin can be divided into chemical degradation (acid degradation and oxidation degradation), physical degradation (radiation degradation, ultrasonic and microwave-assisted degradation, mechanical polishing degradation, high-pressure homogeneous degradation) and enzyme degradation (specific enzyme degradation, nonspecific enzyme degradation, complex enzyme degradation). Chitosan can be degraded to different degrees according to the degradation conditions. Chitosan with molecular weight of 1,000–100,000 Da, even oligosaccharides of glucosamine with the DP of 2–10, can be obtained using these methods.

1. **Chemical degradation**
   1) **Acid degradation**

      The glycosidic linkage of chitosan is not stable in acidic environments and can be hydrolyzed, which causes the chitosan polymer chain to break and then form a number of fragments with different molecular weights. When the degree of hydrolysis is high, chitosan can even be degraded to monosaccharides. The acids used in chitosan degradation are usually strong inorganic acids (such as hydrochloric acid, sulfuric acid, hydrofluoric acid) and other weak acids such as acetic acid, nitrous acid, phosphoric acid. The cost involved in acid degradation is low and the process is simple. However, the molecular weight of the product is not easy to control. The molecular weight distribution of low-molecular weight chitosan is broader, which usually affects its properties.

   2) **Oxidative degradation**

      Oxidative degradation is the process of oxidation between chitosan and the oxidant, which leads to the chitosan polymer chain breakage. The popular oxidants used in chitosan degradation are H_2O_2, nitrite, peracetic acid, ozone, hypohalite and potassium persulfate [9–11]. H_2O_2 is the most widely used oxidant since no toxic chemical by-products are formed in the reaction and the process is easy to handle. However, there are many disadvantages: long reaction time is required, the reaction temperature needs to be strictly controlled, random degradation, the molecular weight distribution of the product is quite wide, browning reactions occur in the late stage of degradation. The disadvantages decrease the quality of the product. Oxidative degradation has also been carried out homogeneously in IL 1-(4-sulfonic acid) butyl-3-methylimidazolium hydrogen sulfate ([MIMBS]HSO$_4$). At the optimum conditions, aqueous solution of 8% [MIMBS]HSO$_4$, $n(H_2O_2):n$(chitosan) =5:1, reaction temperature equal to 80°C and reaction time 3 h, the yield of the regenerated chitosan is 74%, and the viscosity-average molecular weight is 21.2×10^3 [12].

      Acid hydrolysis and oxidation degradation involve the use of chemical reagents, which inevitably causes waste pollution. However, they are still the most commonly used methods in industry due to low cost and mature technology.

2. **Physical degradation method**

   In physical degradation, a physical electric field, high-energy electromagnetic waves, ultrasonic wave or microwave, are used to supply energy to break the polymer chain of chitosan. At present, the most researched and mature methods are radiation, ultrasonic and microwave degradation.

In radiation degradation, the polymer chains are irradiated and break. Radiation degradation is a kind of random degradation. The molecular weight of the obtained fractures distribute randomly. Irradiation by γ-ray produced by radionuclide Co-60 is the most common method used for chitosan radiation degradation. Ultrasonic degradation involves degrading chitosan by ultrasonic wave through destroying the glycosidic bond by the mechanical force generated by ultrasound, which leads to the breaking of chitosan molecular chain. During microwave degradation, microwave is used to induce particle movement or rotation, which causes friction between molecules and produces heat.The heat breaks the molecules into free radicals and then causes polymer chain break to produce low-molecular weight fractures [13].

Radiation degradation involves no additives. The cost of radiation degradation is low and the process is easy to control. There is no pollution generated during the process and the quality of the product is high. However, it requires a special ionizing radiation equipment, and the high radiation intensity is not safe to the operator. Radiation is also likely to cause some cross-linking and disproportionation reaction, leading to changes in the product structure. Ultrasonic degradation and microwave radiation degradation have the advantages of simple operation, short reaction time, simple posttreatment process, and of causing less environmental pollution [14].

3. **Enzymatic degradation**
Enzymatic degradation of chitosan involves cutting off the chitosan polymer chain through interaction between the enzyme and chitosan to yield chitosan with lower molecular weight. The enzymes that can degrade chitosan can be divided into specific and nonspecific enzymes according to their selectivity. Chitosanase is a chitosan depolymerization enzyme that can specifically degrade chitosan without degrading colloidal chitin. In addition, other enzymes such as chitinase, lysozyme, glucase, protease, esterase, cellulase, hemicellulase and pectinase can also degrade chitosan. However, these are nonspecific enzymes, which can degrade various substrates. In recent years, complex enzymes formed by different enzymes and immobilized enzymes are also being used to improve the degradation efficiency and selectivity of chitosan degradation. The postprocessing method is also simplified [15].

There are no by-products formed during enzymatic degradation. The degradation conditions are mild and the relative molecular weight distribution of the degraded products is easy to control. The bioactivity of the obtained chitosan is high and the product does not need to be desalted. However, large-scale production of the enzyme is difficult, which hinders the commercialization of enzymatic degradation.

4. **Combined degradation method**
Acid hydrolysis, oxidation and enzymatic degradation all have their own advantages and also many unavoidable disadvantages. Therefore, combining

several degradation methods together to improve the degradation efficiency has been investigated [16]. Saccharifying enzyme/H_2O_2 two-step degradation of chitosan involves the following degradation conditions: mass ratio of enzyme to chitosan of 0.008:1, temperature of 62°C, degradation time of 33 h, 12.6% H_2O_2 concentration; in this reaction chitosan with molecular weight of 2.5×10^5 Da could be degraded into chitosan with molecular weight of 470 ~ 1102 Da with a yield of 83.3%. The method has the advantages of being simple, producing no pollution, involving easy separation and purification steps, having mild reaction condition, yielding less by-products and having a high degree of degradation. Chitosan with DD of 70% and average molecular weight (Mw) 90.502×10^3 in dilute lactic acid solution containing H_2O_2 (1%) was effectively degraded by irradiation with gamma 60-Co radiation (1.3302 kGy/h) at doses in the range 4–16 kGy. There was particularly strong synergy between H_2O_2 and radiation for degradation at the lower radiation doses. Radiation scission yields (Gs) were found to be 2.202 and 0.202 µmol/J for 5% chitosan with and without 1% H_2O_2, respectively [17].

10.3.2 Total Synthesis of Marine Drugs

Up to now, a lot of marine active substances with good biological activities have been extracted from coral, sponge, conch, sea rabbit and other marine organisms. However, due to the very low content of active substances in marine organisms, and the difficulties involved in extraction and separation, direct separation and extraction of marine bioactive components has not been able to yield enough products to meet actual industrial and clinical scientific research requirements. Moreover, some substances have low activity and even high toxicity. Therefore, study and development of methods for artificial synthesis or chemical structure modification, using the natural marine bioactive substances as the precursor compounds, to obtain compounds with high activity and low toxicity have become research hotspots.

As the structure of the natural active material is usually very complex, the process of synthesis is tedious, cumbersome and involves too many steps. Therefore, the development of green chemical reaction methods to improve the atomic economy and reaction efficiency is important to make advances in this area of research. For example, the active substance Naamidine A isolated from sponge has excellent biological antagonistic activity. The minimum inhibitory rate of 85% in the nude mice transplanted tumor of human skin squamous cell carcinoma (A431) is achieved at the maximum tolerated concentration of 25 mg/kg. Naamidine A was first isolated from sponges by Carmely and Kashman in 1987, and the researchers began to explore the total synthesis of naamidine A. In 2000, Otha achieved the total synthesis of this compound for the first time, but the total yield of the 13-step reaction is only 1.2%. In 2006, Aberle et al. obtained a greener naamidine A synthesis method, which uses a relatively cheap commercial chemical Boc-Tyr(Bzl)-OH(1) as the starting material. *p*-tert-Butyloxycarbonyl (Boc)-protected amino acid was subjected to methylation reaction to give the intermediate (2) with a yield of 96%. There was no methanol ester by-product in the reaction. The methylation reaction is important for the subsequent formation of imidazole rings to ensure stereoselectivity, followed by the formation of Weinreb amide (3, yield 86%). The Weinreb amide is treated with Grignard reagent to give the protected *R*-aminoketone (4, yield 62%). Then, (4) is deprotected with 4 mol/L HCl in ether and reacts with cyanamide to yield trisubstituted 2-aminoimidazole, which is further catalyzed to remove benzyl to get naamine A (5, two-step process with yield 86%). 1-Methylparabanic acid (6) reacts with *N,O*-isotrimethylsilylacetamide to yield silylated methylglyoxide (7). Subsequently, (7) and (5) further react together to yield naamidine A in one step. This synthesis process is simpler than the old process, and the yield is increased from 1.2% to 35%. The synthesis scheme is shown in Figure 10.7.

10.3.3 Extraction of Active Substances from Microbial Secondary Metabolites

Since penicillin was discovered and successfully used in the treatment of infectious diseases, microbial metabolites have become an important source of new drugs. Marine microorganisms have unique living conditions, which affect their metabolic pathways. Secondly, the metabolites are often rich in variety and have novel structures and unique activities. Therefore, it has become a hot topic in the field of natural medicine chemistry to search for bioactive products among marine microorganisms. Especially, the study of bioactive products of marine-derived fungi has received a lot of attention.

Most of the secondary metabolites of marine microorganisms secrete out from the cell. The preparation of secondary metabolites mainly involves two processes: biological fermentation and subsequent fermentation. In the process of fermentation, the fermentation conditions (including pH, temperature, time, oxygen and shaking speed) should be optimized to obtain the highest content of active substances in the fermentation broth. The separation and purification process of fermentation broth involves the separation and purification of the active ingredients by centrifugation, extraction, precipitation, chromatography, membrane separation and other methods [19].

The marine fungus *Aspergillus fumigatus* H1-04 was processed with shaking fermentation, and followed by extraction [20]. An active extract of 3.2 g was obtained from 10 L of the fermented product through extraction with chloroform. Seven kinds of alkaloids with anticancer activity were separated and purified by silica gel column chromatography and prepared using liquid chromatography. The specific operation is

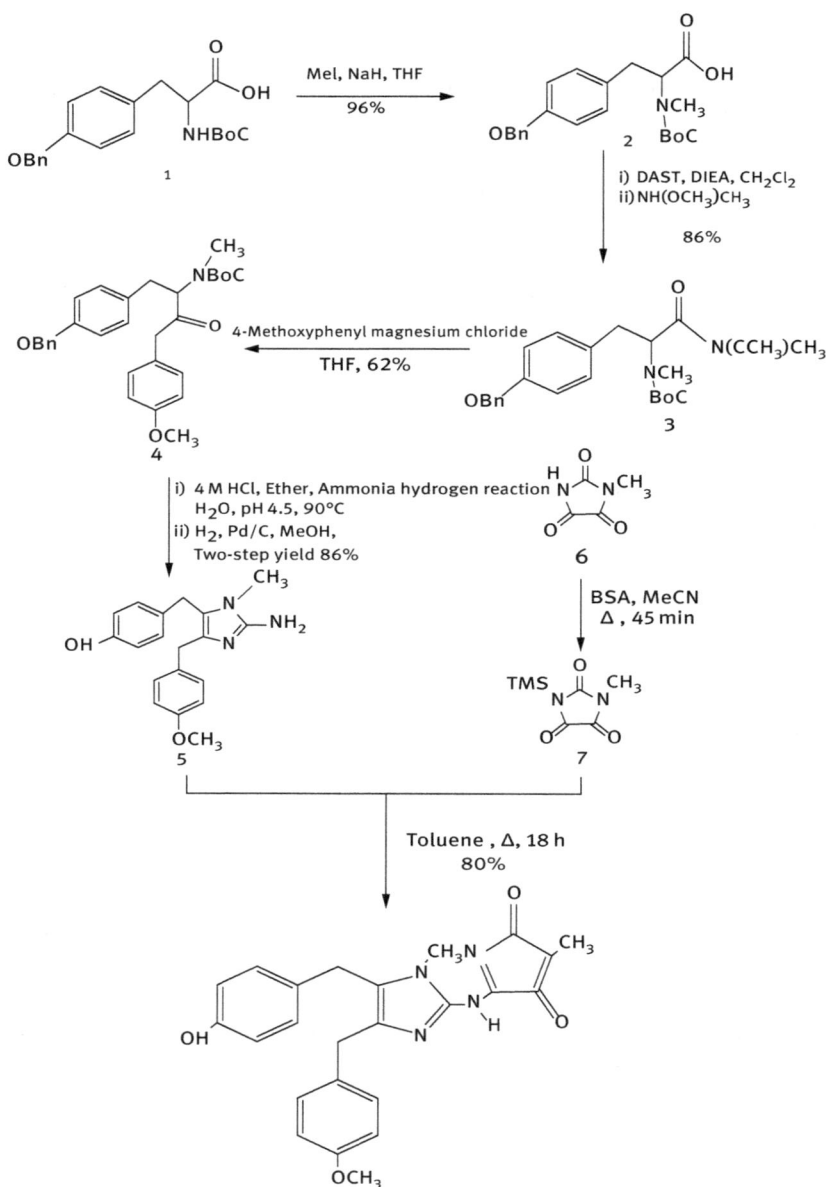

Figure 10.7: Synthesis scheme of naamidine A [18].

as follows: The H1-04 fungi is cultured in the medium of old seawater and potato juice agar (PDA). The fungi seed culture medium is obtained after several repeated subculture, slant culture, and liquid culture on PDA medium. The seed culture solution is inoculated into 100 500-mL Erlenmeyer flasks containing 100 mL of liquid medium.

The final volume fraction of the seed culture solution in the flask is 5%. The fungi are fermented at 28°C with a stirring speed of 120 rev/min for 6 days, obtaining 10-L fermentation broth containing the active products. The active components are obtained as the chloroform extract after subsequent centrifugation, extraction and fractional elution. Chloroform extract weighing 3.2 g is dissolved in an appropriate amount of the mixture of chloroform/methanol and 15 g of silica gel H is added to the sample and put into the column by dry column-packing method. Five alkaloids, tryptoquivaline and pseurotin A are obtained by gradient elution. One of the alkaloids (−)-(1R,4R)-1,4-(2, 3)-indolmethane-1-methy-l 2, 4-dihydro-1H-pyrazino [2, 1-b]quinazoline-3,6-dione was isolated from nature for the first time.

10.4 Extraction Rare Elements from Ocean

Ocean exploration is getting more and more attention since the 1960s. In recent years, the shortage of many important elements made it a global hotpot to search substitutes from the ocean. The work mainly focuses on extracting rare elements from the marine mineral, seawater and marine organisms. The tuberculosis mineral reserves of multiple metals (TMM) are abundant in the sea. There are about 3.0×10^{12} tons of TMM at a depth of 3,500–6,000 m, which contains dozens of elements, including Mn, Fe, Ni, Co and Cu. The storage of Mn could afford the world 18,000 years, and that of Ni 25,000 years

Seawater not only hosts thousands of life forms and abundant mineral sources, but also has a huge chemical reserves. There are more than 80 elements in seawater in the form of simple ions or coordinated ions. The elements become solid, in the form of salt, when the seawater is condensed and crystallized. Moreover, some elements can be extracted from marine organisms, such as iodine from seaweeds. Up to now, a dozen of elements have been extracted from seawater. The content and form of the elements are shown in Table 10.1.

Table 10.1: Some elements have been extracted from seawater.

Element	Existent form	Con. (µg/L)	Element	Existent form	Con. (µg/L)
H	H_2O	1.1×10^8	Ca	Ca^{2+}	4.1×10^5
O	H_2O, O_2	8.8×10^8	Br	Br^-	6.7×10^4
Cl	Cl^-	1.9×10^7	Sr	Sr^{2+}	8.0×10^4
Na	Na^+	1.1×10^7	I	IO_3^-, I^-	6.0×10^1
Mg	Mg^{2+}	1.3×10^6	Ba	Ba^{2+}	2.0×10^1
S	$SO_4^{2-}, NaSO_4^-$	9.1×10^8	Au	$AuCl_2^-$	4.0×10^{-3}
K	K^+	3.8×10^5	U	$UO_2(CO_3)_2^{4-}$	3.2×10^0

10.4.1 Potassium (K)

In the 1930s, the Norwegian scientist Jill had applied the first patent for the extraction of potassium from seawater by dipicrylamine method, following which a lot of scientists in the world have devoted themselves to develop technologies to recover potassium from seawater. The technologies are mainly focused on developing new extracting agents and optimizing extracting processes. The development of new extracting agent with high efficiency and low cost is a real bottleneck in the industrialization of extraction of potassium from seawater. The extraction methods include chemical precipitation, solvent extraction, membrane concentration and ion-exchange concentration. Chemical precipitation is the oldest method, in which proper precipitation agents are used to precipitate potassium according to the solubility differences of different potassium salts. Potassium is precipitated from seawater as insoluble salts. The precipitation agents include dipicrylamine, sulfate, phosphate, fluorosilicate and perchlorate. During solvent extraction, potassium is concentrated and separated since the partition coefficient of potassium is different in extracting agent and seawater. Some extracting agents can precipitate potassium selectively. Generally, the price of the extracting agent is high, and the extraction efficiency is low for extracting potassium from seawater due to the extremely low concentration of potassium in seawater.

Currently, membrane concentration and ion-exchange concentration are the most popular methods for extraction of potassium. Membrane concentration is a new separation technology in chemical engineering. It does not yield satisfactory results in the extraction of potassium from seawater owing to low capacity and selectivity, and is still in exploratory stage [21]. The ion-exchange method is more practical, and has been used in industries. Resin, inorganic ion exchanger, ion-sieve membrane and natural zeolite have been used to absorb potassium [22, 23].

Among all ion exchangers, zeolite is more popular and practical. Potassium is selectively absorbed, concentrated and then deabsorbed to obtain a solution with high concentration of potassium. In 1972, a technology for extracting potassium from seawater to prepare KCl with natural zeolite was developed. In 1975, a 10 ton scale technology was successfully finished and in 1983, a 1,000 t factory was built with natural zeolite as the concentration agent to extract potassium from seawater. The process is shown in Figure 10.8.

A technology named new continuous ionic exchange (ISEP) was developed for extracting potassium from seawater in 2007, where zeolite was used as the absorbent. Compared with fixed bed, the ISEP system has the following advantages: the concentration of K^+ of product and the recovery rate of K^+ are more constant, the utilization rate of zeolite is higher, and continuous manipulation is easier to carry out [24]. The technology solves the problem of potassium concentration and separation effectively and economically. Both the pilot scale and commercial tests have been successfully finished. It decreased the cost and made the extraction of potassium from seawater commercializable. The ISEP scheme is shown in Figure 10.9. There are 20 columns

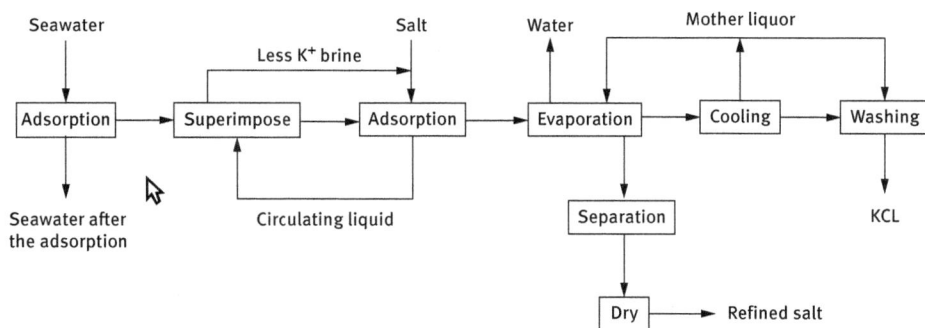

Figure 10.8: Extraction process of KCl from seawater with a natural zeolite as the enrichment agent.

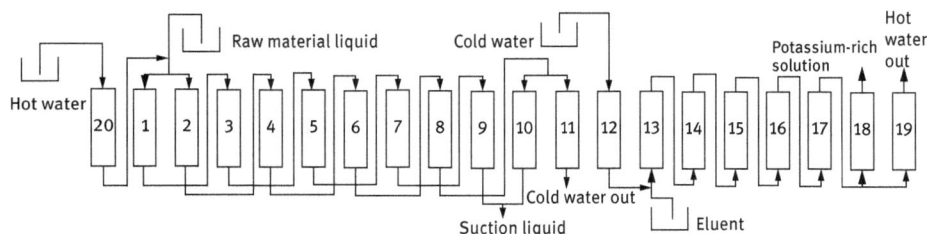

Figure 10.9: Technological process of the continuous ionic exchange system of extracting potassium from seawater [24].

(totally 7 kg of modified zeolite was uploaded) in the equipment. There are four sections, including the absorption section (column 1–10), cold water washing section (column 11–12), elution section (column 13–18) and hot water washing section (column 19–20). The raw seawater enters the adsorption section from top and comes out at the bottom after adsorption. Then it flows into the elution section from the bottom and comes out at the top as a potassium-rich solution. The adsorption, elution and washing steps can be carried out at the same time, which makes sure the operation is continuous and the feed and regenerator flow continuously. The components and the concentration of the product are constant. The concentration of K^+ in the enrichment solution can reach 63.07 g/L, which is 166 times more than that in seawater. The recovery rate of K^+ is 90.0%. The technology has many advantages: compact arrangement of the equipment; simple tube system; fewer auxiliary grooves, tanks and pumps and low operation cost.

10.4.2 Extraction of Bromine

Ninety-nine percent of bromine in the earth exists in ocean. That is why bromine is called "marine element." The concentration of bromine in seawater is so high that the total amount of bromine in seawater is up to 9.5×10^5 tons. Extraction of bromine from ocean has a history of more than half a century and research in this area has made considerable progress. Besides the earliest aniline process, many methods have been developed, including ion-exchange resin adsorption, gas membrane process, emulsion liquid membrane, agitated bulb membrane absorption and high-gravity air stripping technology.

Air blowout method is the most popular technology for bromine preparation and has been widely used to extract bromine from seawater. Bromine blowing out process is shown in Figure 10.10. The air blowout method involves the use of chlorine to oxidize Br in seawater to get Br_2. Then, Br_2 is blown out by air and becomes liquid bromine. Sulfur dioxide (SO_2) is used to absorb the Br_2 and obtain mist droplets containing hydrogen bromide (HBr) and sulfuric acid (H_2SO_4). The mist droplets are then captured in a glass fiber bed. A primary acid liquor (PAL) product stream that is rich in bromine (40% w/w) is then obtained, from which product bromine is easily recovered. In this process, the stripping rate is low. The stripping rate can only reach 75~85% even under the most suitable conditions. The technology also has high energy-consumption and narrow temperature range of application.

A modified air blowout method named high-gravity air stripping technology method (Figure 10.11) has been developed. In this process, a rotating packed bed is used to strip bromine out from the solution. Air is sent by the centrifugal fan to the bottom of the rotating bed, passing through the packing layer in the axial direction. The oxidizing liquid is introduced into the rotor cavity from the liquid inlet pipe. The oxidizing liquid is dispersed and broken into a large and constantly changing surface under high-speed rotation. Under the circumstances of high dispersion, high turbulence, strong mixing and rapid updating of the interface, the oxidizing

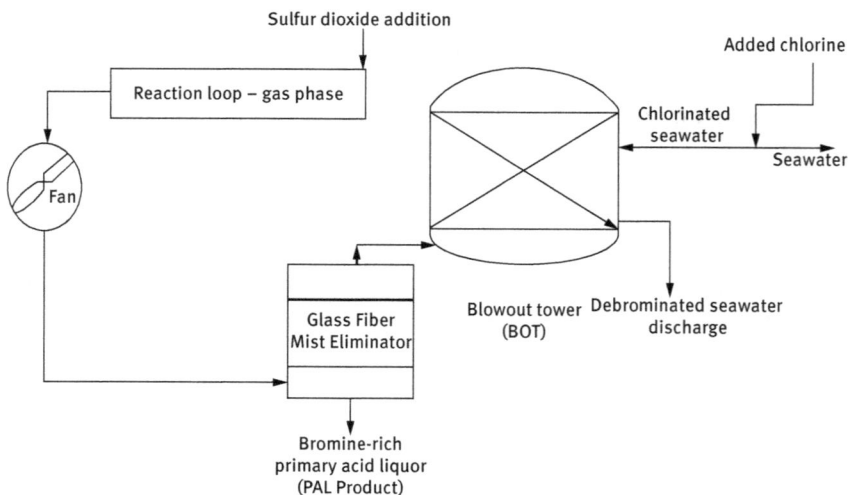

Figure 10.10: Bromine blowing out process [25].

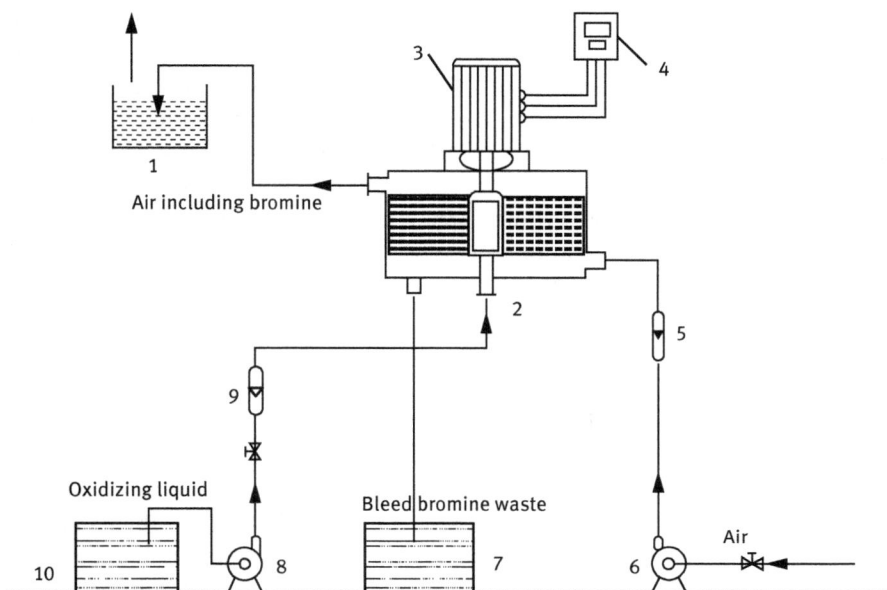

1 -Alkaline absorption tank; 2 -Rotating packed bed; 3 -Motor; 4 -Frequency converter;
5,9 -Rotor flow meters; 6 -Centrifugal Fan; 7 -Waste tank; 8 -Corrosion resistant pump; 10 -Storage tank

Figure 10.11: Process scheme of stripping free bromine in rotating packed bed [26].

liquid contacts with air reversely with a very high velocity in the curved channel.
The mass transfer process between the oxidizing liquid and air is great enhanced
and the free liquid bromine is blown out efficiently. The results show that under
the conditions of temperature 20–25°C, gas–liquid volume ratio of 120, pH 3.5 and
a high gravitational factor of 84.67, the single-stage strip rate of the oxidized liquid
with the total bromine concentration of 250 mg/L can reach 88% or more, and the
triple-stage strip rate can reach 93%. The strip rate is 10% higher than that of the
traditional tower equipment (being 75–85%). Under the same operation conditions,
the single-stage strip rate of the oxidation liquid with a total bromine concentration
of 2,000 mg/L is 94.5%. The stripping effect is much better and the energy consumption is reduced. This method has the advantages of high stripping rate, lower
pressure drop, small occupation area and low energy consumption, which make it
suitable for application.

10.4.3 Extraction of Lithium

Lithium is the lightest metal in nature. It is recognized as the energy metal which
promoted the world, which has broad application prospects in the field of chemical

power supply, new alloy materials and nuclear fusion power generation. As lithium ore resources in land are not sufficient to meet human needs at present, researchers have begun to explore technologies for extraction of lithium from seawater. One of the earliest lithium extraction methods is the solvent extraction method, which is suitable for aqueous lithium solution with high concentration. The concentration of lithium in seawater is very low, only 0.17 mg/L, which needs to be concentrated. Thus, the process is time-consuming and laborious, and not suitable for large-scale industrial application. Therefore, adsorption method is considered the most promising method to extract lithium from seawater with low lithium concentration [27].

In the adsorption process, there are some key processes like the development of adsorbents that are selective, recyclable and of relatively low cost. The adsorbents can be divided into organic and inorganic adsorbents. Organic adsorbents are usually organic ion-exchange resins. Inorganic ion-exchange adsorbent has high selectivity to lithium. Some inorganic ion-exchange adsorbents with ion-sieve effect particularly have excellent adsorption property. Inorganic ion-exchange adsorbents mainly include amorphous hydroxide, layered polyvalent metal acid salt, antimonates and ion-sieve adsorbent [28]. The representative materials and adsorption principles of the inorganic ion-exchange adsorbents are shown in Table 10.2.

As far as stability, selectivity and cost are concerned, the ionic sieve oxides, such as ion-sieve adsorbent, spinel manganese oxide (RMn_2O_4), have been widely researched. RMn_2O_4 has a spatial structure suitable for the migration of Li ions. In the spinel manganese oxides, the oxygen atoms are cubic close packed. Manganese atoms (Mn) are alternately located in the interstitial positions of the close packed oxygen atoms. The manganese oxide skeleton forms a three-dimensional network of tetrahedron (Figure 10.12) coplanar with octahedron, which is favorable for the diffusion of lithium ion (Li^+) and highly selective for Li^+ [29]. Lithium manganese oxide ($Li_{1.33}Mn_{1.67}O_4$) was synthesized by Japan Institute of Marine Resources and Environment. The highest adsorption capacity for lithium is 25.5 mg/g (powders without PVC) and 18 mg/g (powders with PVC). The maximum adsorption capacity of $Li_{1.6}Mn_{1.6}O_4$ to lithium was 40 mg/g after the addition of PVC [30].

Table 10.2: Adsorbents and absorption principle of lithium absorption.

Absorbents	Representative examples	Adsorption principle
Amorphous hydroxide	Aluminum oxides and hydrous oxides	Hydroxyl cation
Layered polyvalent metal acid salt	Arsenate and phosphate	Interlayer interaction
Antimony salt	$LiSbO_3$	Intermolecular interaction
Ion sieve	Spinel-type titanium oxide, Li-Mn oxide	Ion exchange

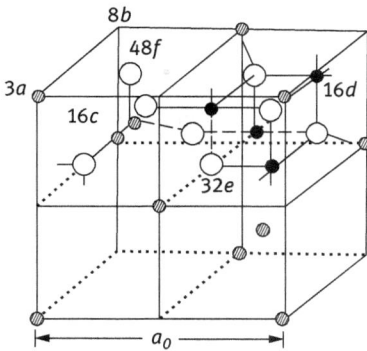

Figure 10.12: Spatial structure of spinel manganese oxide.

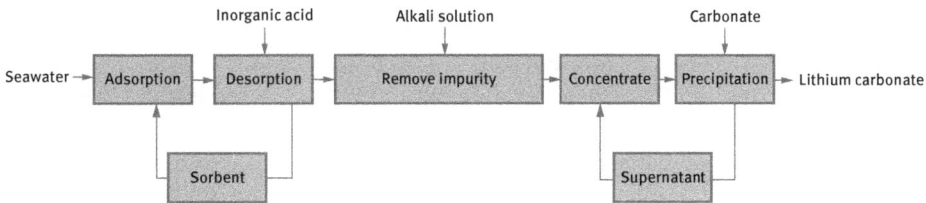

Figure 10.13: Seawater extraction process diagram.

Before and after lithium manganese oxides were eluted with suitable eluent, they were referred to as lithium ion-sieve precursor and lithium ion-sieve, respectively. Lithium ion-sieve still kept the original spinel structure, which made it a lithium-selective absorbent [4]. The ratios of both lithium extraction and manganese loss are important indicators to evaluate the effect of elution. The higher the lithium extraction ratio and the lower the manganese loss ratio, the better the elution effect of eluent. At present, many inorganic acids such as sulfuric acid, hydrochloric acid and nitric acid, salts such as ammonium persulfate, sodium persulfate, peroxydisulfate and potassium persulfate are main eluents for the precursor [31].

The facilities are important to extract lithium from seawater through adsorption. In a Japanese patent, granular adsorbent is loaded in a pressurized water tank of ship. The seawater enters the tank with adsorbent from the excess at the bottom of the ship. The excess has a check valve. Water passes through the absorbent bed to reach its upper part. Then, the seawater is pumped out the hull. The ship sails slowly in the outer sea, which makes the adsorbents in adsorbent bed tank fully in contact with the seawater. The ship returns after about 20 days. After landing, the adsorbent is pumped into the desorption tank on the land by a sand pump and immersed into 15% hydrochloric acid solution. The desorbed liquid is concentrated, processed and separated to obtain lithium. After desorption, the adsorbents are pumped back to the ship and recycled. The detailed process is shown in Figure 10.13.

10.4.4 Extraction of Uranium

Uranium is an important nuclear fuel. With the rapid development of atomic energy industry, the demand for uranium is increasing. However, the total amount of uranium on land is only 1 million tons, but the whole sea contains 4.5 billion tons of uranium, about 1000 times as much as is available from all known terrestrial ores. If a commercial uranium extraction technology can be developed, the ocean will likely become an almost inexhaustible uranium repository. However, developing a cost-effective uranium extraction process is a huge challenge due to the uranium concentration in seawater being very low, only 3.3 µg/L [32].

From the 1950s, Germany, Italy, Japan, the United Kingdom and the United States have carried out research on uranium extraction from seawater. The methods of adsorption, chemical precipitation, biological treatment, membrane treatment, flotation, superconducting magnetic separation, etc. have been investigated. Up to now, no country has successfully developed a commercially viable technology for the extraction of uranium from seawater. Due to uranium being present in seawater as the UO_2^{2+} cation and various carbonate and hydroxyl species, the adsorption method is recognized as one of the most efficient methods, mainly through chelation of UO_2^{2+}. A good absorbent of uranium requires many properties, including high selectivity to UO_2^{2+} cation, high adsorption capacity, fast adsorption speed, good chemical stability, high mechanical stability, ease of recovery and low cost. There are three technical steps in uranium extraction, including absorbent preparation, operation in sea, and uranium elution and purification. According to the report of Japan nuclear power bureau, in a commercial operation of uranium extraction with fiber absorbent in the production scale of 1,200 t/a, the absorbent costs more than 69%, operation in sea costs 29%, and uranium recovery and purification costs 2% [33].

The adsorbents commonly used in uranium adsorption are inorganic adsorbents (alkaline earth metal or transition metal oxide or salts, such as lead compounds, manganese dioxide, basic zinc carbonate, hydrated titanium oxide, metal sulfide); small organic ligands (such as phosphonic acids, amino phosphonic acids, amidoximes,1, 10-phenanthroline-2,9-dicarboxylic acid, and ligands with acid or amine groups); polymeric fiber sorbents [34], such as polyacrylonitrile, polypropylene, poly(1-vinyl)imidazole and polystyrene and the their copolymer and derivatives, and organic/inorganic hybrid adsorbents (such as polyacrylamide/attapulgite clay nanocomposites, polypropylene montmorillonite nanocomposites, polymethacrylic acid/ montmorillonite nanocomposites, polyacrylonitrile/montmorillonite nanocomposites and poly(amidoxime)-reduced graphene oxide). In general, inorganic adsorbent material has the advantages of fast adsorption speed, easy recovery and elution, and the disadvantage of poor selectivity to uranium adsorption. Organic adsorption materials have better selectivity for uranium, while there are many problems, such as slow adsorption rate and low adsorption capacity. Organic/inorganic hybrid materials have the advantages of high mechanical strength, high uranium adsorption capacity, and low interference by the coexistence ions. The hybrid material is more ideal for uranium extraction.

American Oak Ridge National Laboratory and Hill have developed a polyethylene composite fiber adsorbent (HiCap) which has high surface area and high adsorption capacity. HiCap is has remarkable uranium adsorption capacity, adsorption rate and selectivity. HiCap has a uranium adsorption capacity of 146 g/kg in the mixed solution whose uranium concentration is 6 ppm at 20°C, that is 7 times the capacity of the most advanced adsorption material in the world. HiCap has an adsorption capacity of 3.94 gU/kg in seawater, which is more than five times that of the other best absorbents. In terms of selectivity, it is seven times more selective than the current best adsorbent. In June 2012, HiCap was awarded as one of the most important technological innovation of the year by the United States "R & D magazine" and won the "100 R & D Award" [35].

In 2013, Lin Wenbin, who is a professor of chemistry at the University of North Carolina in the United States, reported a metal organic skeleton material (MOFs) that was prepared using the amino-TPDC or TPDC bridging ligands containing orthogonal phosphoryl urea groups (TPDC is p,p_0-terphenyldicarboxylic acid). The MOFs exhibited near quantitative removal of uranium from water and artificial seawater, surpassing that of amidoxime polymers by at least fourfold. MOFs were shown to be highly efficient in sorbing uranyl ions, with saturation sorption capacities as high as 217 mg U/g sorbent, which is equivalent to binding one uranyl ion for every two sorbent groups. The ligand structure and SEM image of MOF are shown in Figure 10.14. The production cost of uranium from seawater in Japan is about US$ 560/lb uranium (US$ 1,230/kg uranium). US production costs are US$ 300/lb uranium (US$ 661/kg uranium). The research and development of materials by Lin Wenbin reduced the production cost of uranium to $ 150/lb (330 US dollars/ kg of uranium), which is very close to the highest spot price ($ 137/lb) in the world's uranium market in the past 25 years. It has been shown that the MOF material that has great commercial application prospects. In 2015, the group of Lin reported a

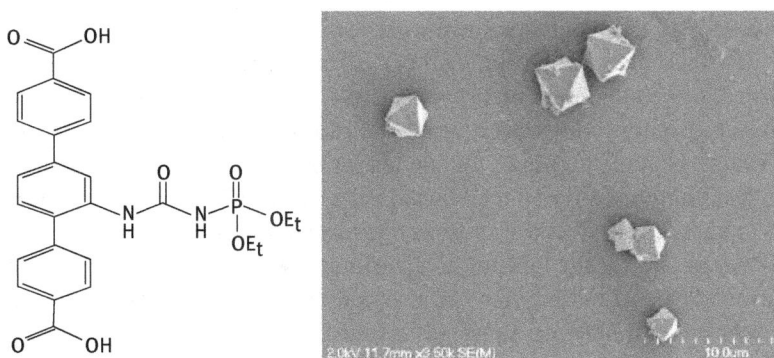

Figure 10.14: The ligand structure (a) and scanning electron microscopy (b) of MOF [37].

novel bifunctional chelator, (Z)-2-[2-(N'-hydroxycarbamimidoyl)phenoxy]benzoic acid which has a ultrahigh uranium uptake capacity. In artificial seawater (pH 8.2), an exceptional uranium uptake of 553 mg of uranium/g of sorbent with a theoretical saturation capacity of 710 mg/g was obtained by fitting isotherm data with the Langmuir–Freundlich model [36].

Ion-imprinting is a good method to prepare absorbent with good ion selectivity of uranium over other ions, especially vanadium in seawater. Surface ion-imprinted polypropylene nonwoven fabric is prepared by copolymerization of 4-vinylbenzyl chloride and 1-vinylimidazole in the presence of uranyltricarbonate complex. The sorption capacity was 133.3 mg/g within 15 h at pH 8.0 and 298.15 K. The imprinted fabric shows excellent selectivity toward uranium (K_d = 201) over vanadium (K_d = 98.8) and the other coexisting ions (K_d< 10) in seawater. It also exhibits good salt-resistant stability and can be regenerated efficiently after five cycles. The obtained imprinted fabric has potential for extracting uranium from seawater [38].

Material for extracting uranium from seawater can be packed on different adsorption devices to ensure that the adsorbent makes good contact with seawater. Currently, there are several types of devices, including adsorber, bed, biological device and membrane. In the type of adsorber, the adsorbent is put into the fishnet bag or compiled into a rope, coupled with the floating body, and then placed in the sea. The natural seawater is allowed to flush the adsorbent net bag to achieve the purpose of uranium adsorption. In the type of bed, two dams are built in the sea to form a pool, in which the absorbent is placed. The adsorbent is placed in the pool. Taking advantage of tide fluctuation, seawater passes through the absorption bed in the dam, scouring the adsorption bed for uranium adsorption. To achieve commercial acceptance, the adsorbent material and the adsorption methods need to be researched to improve the adsorption efficiency and reduce production costs.

10.5 Desalination

Water is the source of all life, while the freshwater resources on land are very scarce and unevenly distributed in the world. The oceans account for 71% of the total area of the earth. The total volume of seawater is $1.37 \times 10^{20}\,m^3$. If there is a suitable "desalination technology" to convert seawater into freshwater which can meet human needs, water resource problems will no longer be an issue to human beings. The history of desalination can be traced back to the AD third century. The sailors used a sponge to absorb the water vapor from the seawater and then squeezed the condensed freshwater out for the journey at that time. It was in the late eighteenth century that seawater desalination was realized commercially. The earliest desalination plant was built in 1881 on the Mediterranean island of Malta. Seawater desalination technology has been used in 155 countries around the world to solve water shortage problems of more than 100 million people. In 2016, the global water production by desalination was projected to exceed 38 billion m^3 per year. Although the desalination technologies are mature enough to be a reliable source for freshwater from the sea, a significant amount of research and development has been carried out in order to constantly improve the technologies and reduce the cost of desalination.

Seawater desalination is an industrial technology involving the use of chemical or physical methods to remove the salt contained in seawater in order to obtain water with low salt concentration. According to the separation process, seawater desalination mainly includes distillation, membrane separation and freezing method. Distillation is the process of heating seawater to evaporate water and then condensing the steam to obtain freshwater. It can be divided into multistage flash, low-temperature multieffect distillation and gas distillation [39]. The membrane method involves using external energy or chemical potential difference as the driving force to separate salt and water with a natural or synthetic polymer film or polymer composite membrane (such as cellulose acetate, polyamide composite, nanotube- or aquaporin-based membranes) [40]. Based on the resources of driving power, it can be divided into electrodialysis, reverse osmosis and so on. Low-temperature multieffect distillation technology has developed rapidly in recent years due to its high energy efficiency. The scale of the devices is growing and the cost is increasingly reduced. Reverse osmosis desalination technology is undergoing rapid development, while the project cost and operating costs continue to decline. Both thermal distillation and membrane technology have a wide range of applications. The investment and performance of major desalination technologies are shown in Table 10.3.

Low-temperature multieffect distillation desalination technology involves desalination at a maximum evaporating temperature of 70°C. A series of horizontal tubes falling film evaporator collect serially. The evaporation temperature is lower than that of the previous effect. The secondary steam generated from last effect is used as the heating steam of next effect. The secondary steam is reused, resulting in more distilled water than the initial steam volume. The key techniques involve

Table 10.3: Comparison of investment and performance of major desalination methods.

Desalination method	Low-temperature multieffect distillation	Reverse osmosis	Multistage flash	Steam distillation
Major equipment investment	Low	Lower	Slightly higher	Slightly higher
Water intake and pretreatment investment	High	Low	Slightly higher	Low
Operating costs	Lower	Lower	Lower	Slightly higher
Equipment life (a)	Long	Long	Long	Long
Water quality (salinity mg/L)	500	5	5	5
Technical maturity	Mature	Mature	Mature	Mature

controlling the maximum operating temperature below 70°C, slowing down and preventing equipment corrosion and scaling problems. In addition, the lower operating temperature makes it possible to use other low-cost materials, such as aluminum alloy heat-transfer tubes and carbon steel shell with special anticorrosion coating to reduce the cost of seawater desalination equipment. The process of low-temperature multieffect desalination is shown in Figure 10.15. Low-temperature multieffect distillation of desalination technology has been basically mature. The focuses of its research are the application of new materials and new technology, scale up in single unit, etc. in aim to reduce equipment costs and operating costs and improve competitiveness.

The concept of seawater reverse osmosis desalination (SWRO) was first proposed in 1930. In 1953, the research results of Prof. C. E. Reid of the University of Florida proved that freshwater could be obtained from seawater through reverse osmosis on cellulose acetate membrane. In the past two decades, reverse osmosis technology has developed rapidly. Reverse osmosis technology is to use a film to separate two kinds of water with different salt concentration. The pressure is higher than the osmotic pressure difference, which forces the water to diffuse from the high concentration side to the low concentration side. The detailed process of SWRO is shown in Figure 10.16. SWRO mainly involves the use of seawater preheating system and reverse osmosis desalination system. The seawater preheating system is usually used in winter at low temperature. First, the raw water is preheated with water from the desalination system and the concentrated water. Then the raw seawater is further heated with the recycled hot water from the boiler to meet the minimum temperature requirements for reverse osmosis systems. Reverse osmosis desalination system mainly comprises a high-pressure pump, booster pump, pressure exchange energy recovery device and reverse osmosis membrane group. About 40% of the raw seawater is pressurized through high-pressure pump and about 60% through the energy recovery device and booster pump. The two seawater are mixed and enter the reverse osmosis membrane group. The freshwater passes the reverse osmosis membrane (product water) and flows to the product pool outdoor and then is transported to the end user by variable-frequency water pump. The concentrated raw seawater (concentrated brine) is discharged to the saltern to prepare salt.

Reverse osmosis membrane group is the core of seawater desalination. Therefore, in order to adapt to the industrial needs and reduce membrane costs, new membrane materials are being constantly developed. The earliest commercial membrane is cellulose acetate asymmetric membrane developed by Reid. At present, the reverse osmosis membrane and its components are quite mature. The commercial membranes are mainly aromatic polyamide-based films, whose desalination rate is higher than 99.3%. The water flux and antipollution and antioxidation capabilities are continuingly increasing and the price is stable with a slight decline.

In the SWRO process, different effects can be achieved by varying the number and combination of membrane modules. The current process combinations are mainly

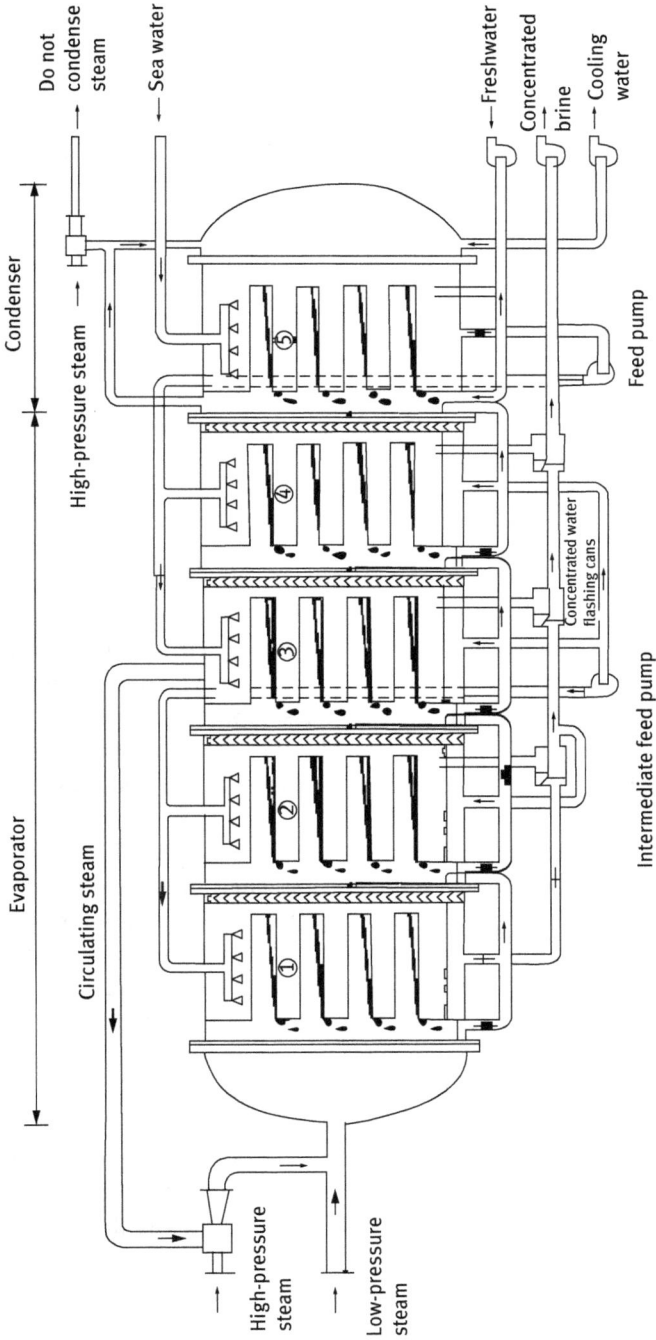

Figure 10.15: Process flowchart of low-temperature multieffect distillation seawater desalination.

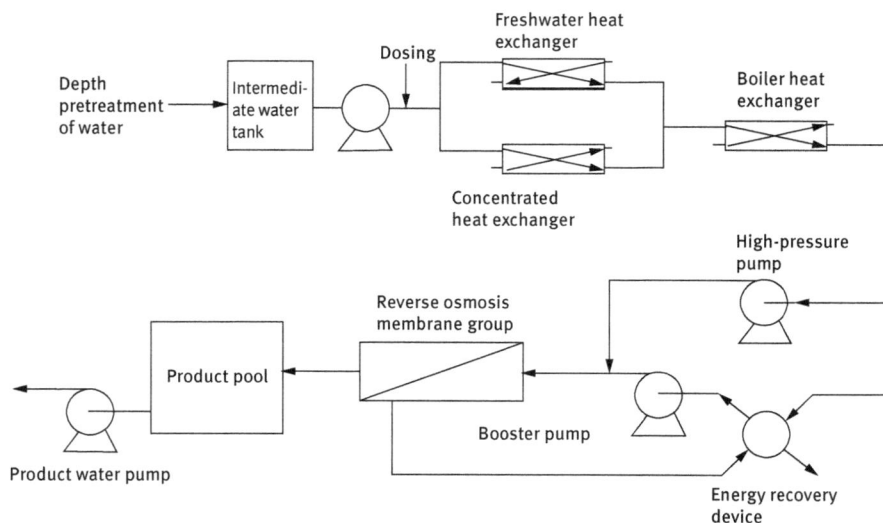

Figure 10.16: Process flow of reverse osmosis seawater desalination.

single level, parallel, cutoff level and product level. The single level is the simplest combination, having only one membrane module with suitable capacity. Parallel means that multiple membrane modules are connected in parallel to increase the yield, while the system desalination rate and the recovery rate do not change. The cutoff level is also known as multistage or series, in which the concentrated saltwater from the first level is used as the feed water for the second level to improve the recovery rate. Using the product water from the first level as the feed water for the second level can improve the desalination rate.

The twenty-first century is the century of exploiting ocean. To obtain the expected chemicals from ocean efficiently, economically, safely and environmental-benignly, more attentions should be paid to new extraction and separation methods. Extraction with the assistance of new energy (including ultrasound, microwave, etc.)in greener solvents(such as IL and water) is potential in exploiting marine resources. Separation processes, such as adsorption, membrane separation, chromatography, are prospective to obtain pure products. Enzyme methods are also a good choice to exploit marine resources with high efficiency under mild conditions. New synthesis methodologies such as combinatorial chemistry are required to obtain new drugs from marine bioresources at increased product scale and decreased cost. Structure design with computational chemistry using natural chemicals separated from bioorganisms and microbial secondary metabolites as the lead compound is also a shortcut to obtain new drugs with high pharmaceutical activities. Moreover, long-term international cooperation needs to be established to preserve and utilize marine reserves.

Questions

1. Why is it important to develop green chemistry for exploiting marine resources?
2. What kinds of resources exist in ocean?
3. Please introduce the green extraction methods of algal polysaccharides and their advantages.
4. Please illustrate the preparation methods of pharmaceuticals from marine resources.
5. What kinds of methods can be used to extract potassium from seawater? What is the mechanism?
6. What is the main difficulty in extracting lithium from seawater? What extraction methods of lithium are there currently?
7. What kinds of absorbents can be used to obtain uranium from seawater?
8. Why do we need to desalinate seawater? Please introduce the mechanism of operation of the green desalination methods.

References

[1] Sithranga Boopathy, N., Kathiresan, K. Anticancer drugs from Marine Flora: An overview. Journal of Oncology 2010, 2010 (1687–8450): 214186.
[2] Garcia-Vaquero, M., Rajauria, G., O'Doherty, J. V., Sweeney, T. Polysaccharides from macroalgae: Recent advances, innovative technologies and challenges in extraction and purification. Food Research International 2016.
[3] Ying Shi, Jianchun Sheng, Fangmei Yang, Qiuhui Hu. Purification and identification of polysaccharide derived from Chlorella pyrenoidosa. Food Chemistry 2007, 103(1): 101–105.
[4] Rodriguez-Jasso, R. M., Mussatto, S. I., Pastrana, L., Aguilar, Cristóbal N., Teixeira, José A. Microwave-assisted extraction of sulfated polysaccharides (fucoidan) from brown seaweed. Carbohydrate Polymers 2011, 86(3): 1137–1144.
[5] Nanna Rhein-Knudsen, Marcel Tutor Ale, Anne S. Meyer. Seaweed hydrocolloid production: An update on enzyme assisted extraction and modification technologies. Marine Drugs 2015, 13(6): 3340–3359.
[6] Sharma, M., Chaudhary, J. P., Mondal, D., Meena, R., Prasad, K.. A green and sustainable approach to utilize bio-ionic liquids for the selective precipitation of high purity agarose from an agarophyte extract. Green Chemistry 2015, 17(5): 2867–2873.
[7] Armando T. Quitain, Takahisa Kai, Mitsuru Sasaki, Motonobu Goto. Microwave–hydrothermal extraction and degradation of Fucoidan from supercritical carbon dioxide deoiled undaria pinnatifida. Industrial & Engineering Chemistry Research 2013, 52(23): 7940–7946.
[8] F.A. Al Sagheer, M.A. Al-Sughayer, S. Muslim, M.Z. Elsabee. Extraction and characterization of chitin and chitosan from marine sources in Arabian Gulf. Carbohydrate Polymers 2009, 77(2): 410–419.
[9] Shih-Chang Hsu, Trong-Ming Don, Wen-Yen Chiu. Free radical degradation of chitosan with potassium persulfate. Polymer Degradation & Stability. 2002, 75(1): 73–83.
[10] Fernandes, A. L. P., Morais, W. A., Santos, A. I. B., et al. The influence of oxidative degradation on the preparation of chitosan nanoparticles. Colloid and Polymer Science 2005, 284(1):1–9.

[11] Xueqiong Yin, Xiaoli Zhang, Qiang Lin, Yuhong Feng, Wenxia Yu, Qi Zhang. Metal-coordinating controlled oxidative degradation of chitosan and antioxidant activity of chitosan-metal complex. Arkivoc 2004, 37(3): 66–78.

[12] ChangchunWu, Hongjun Zang, Daqing Li, Mingchuan Zhang, Jianchun Yu, Bowen Cheng. Oxidative degradation of chitosan by H_2O_2 in acidic ionic liquid aqueous solutions. Polymer Materials Science & Engineering 2014, 30(4): 75–79.

[13] Naeem, M. El-Sawy, Hassan A. Abd El-Rehim, Ahmed M. Elbarbary, El-Sayed A. Hegazy. Radiation-induced degradation of chitosan for possible use as a growth promoter in agricultural purposes. Carbohydrate Polymers 2010, 79(3): 555–562.

[14] RenataCzechowska-Biskup, BozenaRokita, Salah Lotfy, PiotrUlanski, Janusz M. Rosiak. Degradation of chitosan and starch by 360-kHz ultrasound. Carbohydrate Polymers 2005, 60(2): 175–184.

[15] Gooday, G. W. Physiology of microbial degradation of chitin and chitosan. Biodegradation 1990, 1(2): 177–190.

[16] Carlos, W., Michael, H., Marmadou, B. D., Warren, H., Jessica, A. J.. Comparison of enzymatic and accelerated oxidative degradation methods to evaluate chitosan. Frontiers in Bioengineering & Biotechnology 2016, 4.

[17] Nguyen QuocHien, Dang Van Phu, Nguyen Ngoc Duy, Nguyen Thi Kim Lan. Degradation of chitosan in solution by gamma irradiation in the presence of hydrogen peroxide. Carbohydrate Polymers 2012, 87(1): 935–938.

[18] Aberle, N. S., Lessene, G., Watson, K. G. A concise total synthesis of Naamidine A. Organic letters 2006, 8(3): 419–421.

[19] Daniela Giordano, Daniela Coppola, Roberta Russo, et al. Marine Microbial Secondary Metabolites: Pathways, Evolution and Physiological Roles. In: Robert K. Poole, editor, Advances in Microbial Physiology, Oxford, Academic Press, 2015, 357–428.

[20] XiaoxianHan, XiaoyanXu, Chengbin Cui, QianqunGu. Alkaloidal compounds produced by a marine-derived fungus, Aspergillusfumigatus H1-04, and their antitumor activities. Chinese Journal of Medicinal Chemistry 2007, 17(4): 232–237.

[21] Li Pan, Ao-Bo Zhang, Jie Sun, Ying Ye, Xue-Gang Chen, Mei-Sheng Xia. Application of ocean manganese nodules for the adsorption of potassium ions from seawater. Minerals Engineering 2013, 49(8): 121–127.

[22] Ying Yao, Yushan Zhang, Laibo Ma, Huifeng Zhang, Ying Wang. Study on exchange behavior of potassium ion with different kinds of molecular sieves. Journal of Salt & Chemical Industry 2015, 44(4): 18–22.

[23] Junsheng Yuan, Yingying Zhao, Qinghui Li, ZhiyongJi, XiaofuGuo. Preparation of potassium ionic sieve membrane and its application on extracting potash from seawater. Separation and Purification Technology 2012, 99: 55–60.

[24] Junsheng Yuan, Shu'e Yang, Huining Deng. The technology of continuous ionic exchange and its application in extracting potassium from seawater. Journal of Salt and Chemical Industry. 36(3): 27–30.

[25] El-Hamouz, A. M., Mann, R. Chemical Reaction Engineering Analysis of the Blowout Process for Bromine Manufacture from Seawater.

[26] Youzhi Liu, Linna Zhang, Li Yu, Weizhou Jiao, Xiangdan Song, Jiangze Han. Study on high gravity air stripping technology in the extraction of bromine from brine. Xiandai Huagong 2009, 29(8):78–81.

[27] Lee, J. C. Current status in the recovery of metal value from seawater. Journal of the Korean Society for Geosystem Engineers 2005, 42(5): 513.

[28] Harvianto, G. R., Kim, S. H., Ju, C. S. Solvent extraction and stripping of lithium ion from aqueous solution and its application to seawater. Rare Metals 2016 , 35 (12) :948–953.

[29] Özgür, C. Preparation and characterization of $LiMn_2O_4$ ion-sieve with high Li^+ adsorption rate by ultrasonic spray pyrolysis. Solid State Ionics 2010, 181(31–32): 1425–1428.

[30] Chitrakar, R., Kanoh, H., Miyai, Y., Ooi, K. Recovery of Lithium from seawater using Manganese Oxide adsorbent ($H_{1.6}Mn_{1.6}O_4$) derived from $Li_{1.6}Mn_{1.6}O_4$. Industrial & Engineering Chemistry Research 2001, 40(9): 2054–2058.

[31] ZhiyongJi, Mengyao Zhao, Yingying Zhao, Jie Liu, Jingling Peng, Junsheng Yuan. Lithium extraction process on spinel-type $LiMn_2O_4$ and characterization based on the hydrolysis of sodium persulfate. Solid State Ionics 2017, 301: 116–124.

[32] Macfarlane, A. M., Miller, M. Nuclear energy and uranium resources. Elements 2007, 171(66): 125–216.

[33] Tamada, M. Cost Estimation of uranium from seawater with system of braid type adsorbent. Journal of the Atomic Energy Society of Japan 2006, 5(4): 385–363.

[34] Janke, C. J., Dai, S., Oyola, Y. Fiber-Based Adsorbent having High Adsorption Capacities for Recovering Dissolved Metals and Methods Thereof: US, 8722757B2[P]. 2014-05-13.

[35] Material could economically extract uranium from water. Oak Ridge National Laboratory. Engineer (Online Edition), 2012-8-22.

[36] Piechowicz, M., Abney, C. W., Zhou, X., Thacker, N. C., Li, Z., Lin, W. Design, synthesis, and characterization of a bifunctional chelator with ultrahigh capacity for uranium uptake from seawater simulant. Industrial & Engineering Chemistry Research 2015, 55(15): 27–43.

[37] Carboni, M., Abney, C. W., Liu, S., Lin, W. Highly porous and stable metal-organic frameworks for uranium extraction. Chemical Science 2013, 4(6): 2396–2402.

[38] Zhang, L., Yang, S., Qian, J., Hua, D. Surface Ion-Imprinted Polypropylene nonwoven fabric for potential uranium seawater extraction with high selectivity over vanadium. Industrial & Engineering Chemistry Research 2017, 56(7): 1860–1867.

[39] Khawaji, A. D., Kutubkhanah, I. K., Wie, J.-M. Advances in seawater desalination technologies. Science Direct 2008, 221(1): 47–69.

[40] Elimelech, M., Phillip, W. A. The future of seawater desalination: Energy, technology, and the environment. Science 2011, 333(6043): 712–717.

11 The greening of the energy industry

11.1 Clean Utilization Technology of fossil fuel

11.1.1 Impact of Energy Consumption on the Environment

Fossil fuels (mainly include oil, natural gas and coal) are the main energies used worldwide. The exploitation, processing, transportation and consumption of fossil fuels have a great impact on the environment mainly as follows:

1. **Environmental pollution**

   Pollutants that are caused by fossil fuel combustion mainly include CO, CO_2, SO_2, H_2S, NO_x, CH_3SH, as well as flying ash, all kinds of trace metal elements, radioactive particles, heavy metals such as Hg, Cd, Pb, Zn. SO_2 and NO_2 produced by the fossil fuel combustion (coal combustion mainly) are the main reasons for acid rain, which can affect human health and destroy the ecosystem in all kinds of ways, therefore, becoming one of the major environmental problems in the world.

2. **Greenhouse effect**

   In addition to nitrogen and oxygen, there are also some trace gases in the atmosphere. Some trace gases like moisture and CO_2 are transparent to solar short-wave radiation while they can easily absorb the long-wave radiation. By absorbing the long-wave radiation released by the earth's surface, they preserve some of the ground radiation heat in the atmosphere like a greenhouse, which is known as the greenhouse effect. Since the amount of CO_2 is larger than other gasses in the atmosphere, it becomes the most important greenhouse gas.

   The rising temperature brought about by the greenhouse effect will not only cause and exacerbate the spread of infectious diseases, but also result in higher morbidity and mortality rates of heat-related diseases such as heart disease and hypertension. The increasing amount of CO_2 will also lead to adjustment of the global atmosphere circulation and the extension of the climate zone polar regions. Middle latitude regions including the northern part of China will be drier, owing to the decrease of precipitation and the increase of evaporation. Climate warming will cause a rise in the sea level, which will directly flood coastal lowland areas with dense population and developed industry and agriculture, therefore threatening hundreds of millions of people's lives. In order to avoid the threat of global warming, the Third Conference of the Parties to the United Nations Framework Convention on Climate Change was held in Kyoto, Japan, in December 1997, and the "Kyoto Protocol" was passed aiming at limiting greenhouse gas emissions from developed countries to curb global warming. On February 16, 2005, "Kyoto Protocol" came into effect. This is the first action in human history to limit greenhouse gas in the form of regulation.

https://doi.org/10.1515/9783110479317-011

11.1.2 Clean Combustion and Efficient Utilization Technology of Coal

Coal is the most abundant fossil fuel resource in the world, accounting for 92% of global fossil fuel reserves. It is an energy source with the second largest demand, only after oil. Coal contains a large amount of carbon and a small amount of hydrogen (only 5%). In addition, it also contains a little nitrogen, sulfur, oxygen and some inorganic minerals. After the combustion of coal dust, SO_2, NO_x, CO, C_xH_y, CO_2 and other gases are emitted to the atmospheric environment and cause serious pollution and damage.

China is in lack of oil, inadequate in gas while rich in coal. In the next 30–50 years, the coal-centered energy structure will not have any fundamental change. Due to underdeveloped technology and low utilization rate, coal combustion has resulted in serious environmental pollution, which has been a main factor restricting the economy development. In the mid-1910s, the United States witnessed the rise of clean coal technology, a general term referring to all new technologies that can reduce pollution and improve utilization efficiency in the process of coal mining, processing, combustion and pollution control. The clean coal combustion technology is now the most advanced combustion technology in the world, for it can solve environmental problems and energy-saving problems in a good way. Therefore, it is the key strategy for China to develop and promote this new technology to transform the current energy structure, so that China's coal-dominated energy production system can be transformed to a sustainable mode which saves energy and protect the environment. As a main measure to change China's energy structure, the clean coal combustion technology has been highly valued by the country.

1. **Clean combustion technologies of coal**

    Clean combustion technologies of coal mainly include purification technology before combustion, clean combustion technology during combustion and purification technology after combustion.

    1) **Purification technology before combustion**

        Purification technology before combustions mainly includes coal washing and preparation, briquette processing and coal–water slurry technology.

        (1) Coal washing and preparation.

            Coal washing and preparation is a processing technology which, according to the physical and chemical differences between coal and impurities (waste rock), separates coal from impurities using physical, chemical or microbial preparation methods, and then produces coal into well-distributed coal products with different functions. Based on different coal preparation methods, it can be classified into physical coal preparation, physical and chemical coal preparation, chemical coal preparation and microbial coal preparation.

            Physical coal preparation is conducted based on different physical properties (size, density, hardness, magnetism and electrical property) between coal and impurities. Main methods for physical coal preparation include ① gravity coal preparation, including jigging coal preparation, heavy media coal preparation, chute coal preparation, shaking coal preparation and wind coal preparation; ② electromagnetic coal preparation, namely, sorting coal by different electromagnetic properties between coal and impurities. However, this method is not used in the actual production.

Physical and chemical coal preparation, also known as floating coal preparation (referred to as flotation), is based on the differences in physical and chemical properties of the mineral surface. Currently, there are many kinds of flotation equipment in use, mainly including mechanical agitated flotation and nonmechanical stirring flotation.

Chemical coal preparation is a process of removing impurities and harmful components by enriching the useful components in coal with the help of chemical reaction. At present, a common chemical way used in library is desulfurization which, classified by types of chemicals used and different reaction principles, include alkali treatment, oxidation and solvent extraction.

Microbial coal preparation mainly uses the metabolites of some autotrophs and heterotrophs to dissolve sulfur directly or indirectly, in order to achieve the purpose of desulfurization.

Physical coal preparation and physical and chemical coal preparation are two commonly used technologies, which can effectively remove inorganic sulfur (pyritic sulfur) in coal. Chemical and microbial coal preparation can also remove organic sulfur. At present, coal preparation methods commonly used in industrial production are jigging, heavy medium and flotation. Moreover, dry coal preparation also has also developed fast in recent years.

Generally speaking, coal preparation plants contain the following processes shown in Figure 11.1.

① Raw coal preparation: including raw coal acceptance, storage, crushing and screening.

② Raw coal sorting: mainly including jigging-flotation combined process; heavy medium-flotation combined process; jigging-heavy medium-flotation combined process; lump coal heavy medium-fine coal heavy medium splitter combined process; and single jigging and single heavy medium processes.

③ Product dehydration: including dehydration of lump coal and fine coal, dehydration of fine flotation coal and dehydration of coal slime.

④ Product drying: drying coal using thermal energy, usually adopted by cold regions.

⑤ Slurry water processing.

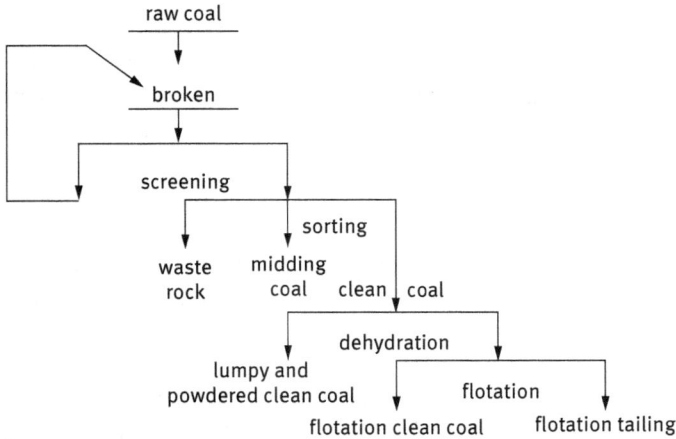

Figure 11.1: Processes of coal preparation.

Functions of coal washing and preparation:

① Improve the quality of coal and decrease pollutants emission caused by coal combustion.

Coal washing and preparation can remove 50–110% of ash, 30–40% of total sulfur (or 60–110% of inorganic sulfur). Coal washing and preparation can effectively reduce the emission of dust, SO_2 and NO_x. Washing 100 million tons of thermal coal can generally reduce 600,000–700,000 tons of SO_2 and 1.6×10^7 tons of waste rock.

② Improve utilization efficiency of coal and save energy

The improvement in coal quality can significantly improve the efficiency of coal utilization. Some studies have shown that if the ash content of cocking coal is reduced by 1%, the consumption of iron-making coke will be reduced by 2.66%, the usage factor of iron-making blast furnace improved by 3.99%; and the anthracite produced by synthesis ammonia washing saved by 20%. If the ash content of coal used for power generation is increased by 1%, heat energy will be decreased by 200–360 J/g, the standard coal consumption per kilowatt increased by 2–5 g; the thermal efficiency of separation coal for industrial boiler sand furnace burning improved by 3–11%.

③ Optimize the product structure and improve product competitiveness

The development of coal washing and preparation is conducive to the conversion of coal products from single structure and poor quality to multiple structures and high quality, therefore optimizing the products. China has a large number of coal consumers whose demand for the types and quality of coal is getting higher and higher. In some cities, the sulfur content of coal is required to less than 0.5%, ash content less than 10%. Thus, the development of coal washing is the key to meeting consumers' needs.

④ Save transport capacity

Since China's coal-producing areas are far away from coal-consuming areas that are also economically developed areas, the average distance for coal transportation is pretty long, i.e., about 600 km. Coal washing can help remove large amounts of impurities. Every 100 Mt coal washed can save 9,600 Mt·km transport capacity.

(2) Briquette processing

Briquette is a massive fuel and raw material with certain physical and chemical properties (cold strength, heat strength, thermal stability, water resistance, etc.). It is formed by one or several types of coal mixed according to their own characteristics with a certain percentage of adhesive, sulfur fixative and leavening agent. Briquette technology is good at saving energy and protecting environment. The sulfur fixing agent of briquette is mainly made by calces, limestone, dolomite and calcium carbide slag, and its main sulfur fixing content is CaO, which can effectively reduce the emission of SO_2 during coal combustion. The total reaction of lime as a desulfurization agent is $CaO+SO_2+2H_2O=CaSO_3 \cdot 2H_2O$. In China, there are two types of briquettes: industrial briquette and civil briquette, and their products are as follows: (i) industrial briquette includes briquette for industrial boiler, briquette for stream locomotive, briquette for coal-gas (fertilizer gas included), briquette for industrial furnace and briquette for coking; (ii) civil briquette includes honeycomb briquette (ordinary honeycomb briquette, aviation briquette, barbecue square carbon) and egg-shaped briquette (for civil cooking and heating, hotpot, hand warmer and barbecue).

(3) Coal–water slurry technology [1, 2].

Coal–water slurry technology is a new technology that appeared in the 1970s during the worldwide oil crisis. It aims at replacing oil by coal. The main feature of this technology is that it adds coal, water and certain amount of additives into the mill, and turns them into a flowable coal-based fluid fuel similar to oil after grinding.

The washed coal is pulverized into <6 mm coal particles and sent into a ball mill where the particles become grinding slurry mixed with water and dispersants. Then the slurry is pumped to a filter to remove the coarse particles and impurities into a mixing tank. The rest coal slurry is mixed with stabilizers to adjust its viscosity, and then reserved in a storage tank.

Coal–water slurry has good fluidity and stability, so it can be stored and transported like oil products. What's more, due to its nonflammable and nonpolluting characteristics, it is a relatively economical and practical clean fuel which can substitute oil. Since the coal–water slurry is prepared with washed coal, its ash content and sulfur content are low (less than 10% dry ash and less than 0.5% sulfur). During the combustion process, the water in the slurry reduces the center temperature of combustion flame and inhibits the generation amount of nitrogen oxides. Besides, once the coal is sent to the mill, the coal–water slurry can be transported by pipeline and tanker. Therefore, it will not cause any environmental pollution in this process, and can effectively protect the environment.

As a substitute fuel for oil, the water–coal slurry can replace heavy oil and crude oil for the burning of boiler and various furnaces. It mainly shows the following advantages:

Advantages of coal–water slurry technology:

① The combustion effect is good. The viscosity of coal–water slurry is lower than that of heavy oil, so it is easy to be adjusted, and the minimum load can be adjusted to 40%. It can be combusted in the boiler instead of heavy oil, with its combustion efficiency reaching to 96–99%, equivalent to that of the boiler efficiency, which is about 90%. Coal–water slurry is convenient to adjust, stable and reliable to operate.

② The environmental effect is obvious. The temperature of coal–water slurry combustion is between 1,200 and 1,300 °C, which is 100–150 °C lower than that of fuel oil and powdered coal, and the coal–water slurry has low sulfur content and pulverized lime content. Therefore, the concentration of SO_2 and NO_x emission is relatively low. Besides, desulfurizer can be added in the process of coal–water slurry preparation, and the desulfurization rate can be up to 40%. Finally, coal–water slurry produces less dust and noise. The ash after combustion can be utilized as a cement admixture, causing no secondary pollution.

③ The process shows many advantages. First, the application of wet ball mill in the process of grinding is quite safe because of its low temperature at only 50–60 °C; what's more, coal–water slurry has low ash content. Therefore, compared with fire coal, it causes less wear to the heating surface of boilers and costs less maintenance fee; furthermore, it does not need coal preparation system or coal preparation field, and the area of its ash discharge field accounts for only one-fourth that of fire coal.

④ The cost of coal–water slurry is lower than that of powdered coal. Water–coal slurry as a substitute of oil can make full use of the original equipment with simplified processes and low investment. What's more, the cost of water–coal slurry is only one-third to one-half that of powdered coal, and its conversion time is only one-third that of powdered coal. Hence, it shows significant economic effect to replace oil with water–coal slurry.

2) **Clean combustion technology during combustion**

Clean combustion technology during combustion mainly includes fluidized bed combustion technology and advanced burner technology. Fluidized bed, also known as ebullated bed, contains bubble bed and circulating bed. The low combustion temperature can decrease the emission of nitrogen oxide. The emission of sulfur dioxide can also be reduced if lime is added to the coal. Through the comprehensive use of slag, the amount of low-quality coal can also be decreased. Advanced burner technology refers to the technology which reduces the emission of sulfur dioxide and nitrogen oxides by improving the structure of boiler and furnace as well as the combustion technique.

3) **Purification technology after combustion**

Purification and treatment technology after combustion mainly refers to smoke and dust removal technology as well as desulfurization and denitrification technology. There are many kinds of smoke and dust removal technology, among which the electrostatic cleaner has the highest efficiency of up to 99%. Therefore, it is adopted by most power plants. Desulfurization includes dry and wet desulfurization. The former makes the slurry lime spray to react with sulfur dioxide in the gas to generate highly dry calcium sulfate particles, and then collects them using an integrator; the latter leaches dust using lime water, and then generates sulfurous acid for emission. Both of these two methods have a desulfurization rate reaching to 90%.

The real development of flue gas desulphurization technology began around 1970 when the United States and Japan first installed flue gas desulfurization devices, and began to implement the emission control standards for SO_2, which dramatically promote the development of flue gas desulphurization technology. At that time, flue gas desulfurization technology mainly adopted wet washing method. At present, there are basically three methods widely applied to medium and large capacity unit, including:

① Wet limestone fuel gas desulfurization (WLFGD). This is the most widely used and most reliable desulfurization process. Since WLFGD has been developed for a long time and has been the most mature flue gas desulfurization process, it is the dominant desulfurization process in power plants abroad.

② Lime spray drying process. It began to be used for desulfurization in power plants in the middle and late 1970s.

③ Limestone Injection into the furnace and activation of calcium.

2. **Efficient use of coal**

Efficient use of coal is a way that uses the coal after purification treatment as a fuel or raw material according to the terminal needs, in order to realize its value. Efficient use of coal includes efficient combustion and efficient conversion. Efficient combustion uses coal as a fuel, and it also includes two methods. The first method converts the chemical energy of coal into heat energy for direct use; while the second method coal first converts chemical energy into heat energy, then into electric energy for use. Efficient conversion uses coal as a raw material and converts it to gas, liquid and solid fuels or chemicals and carbon materials with special purpose. Efficient use of coal mainly contains the following four types of technology:

1) **Coal gasification technology**

Coal gasification refers to a process when a series of chemical reactions occur between the organic matter in the coal and the gasification agent (such as steam, air or oxygen) in certain equipment at certain temperature and pressure, and then the solid coal is converted into combustible gas containing CO, H_2 and CH_4 as well as noncombustible gas containing CO_2, N_2 and so on. After gasification, the coal contains no smoke, sulfur or dust, which can greatly reduce environmental pollution. The reactions during gasification are shown as follows:

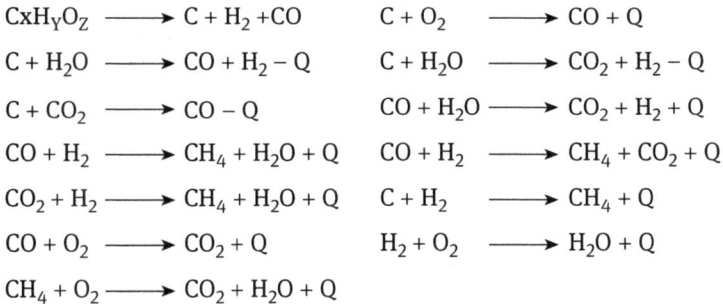

$$CxH_YO_Z \longrightarrow C + H_2 + CO \qquad C + O_2 \longrightarrow CO + Q$$
$$C + H_2O \longrightarrow CO + H_2 - Q \qquad C + H_2O \longrightarrow CO_2 + H_2 - Q$$
$$C + CO_2 \longrightarrow CO - Q \qquad CO + H_2O \longrightarrow CO_2 + H_2 + Q$$
$$CO + H_2 \longrightarrow CH_4 + H_2O + Q \qquad CO + H_2 \longrightarrow CH_4 + CO_2 + Q$$
$$CO_2 + H_2 \longrightarrow CH_4 + H_2O + Q \qquad C + H_2 \longrightarrow CH_4 + Q$$
$$CO + O_2 \longrightarrow CO_2 + Q \qquad H_2 + O_2 \longrightarrow H_2O + Q$$
$$CH_4 + O_2 \longrightarrow CO_2 + H_2O + Q$$

The coal gasification requires three conditions, namely, gasification furnace, gasification agent and heat supply, which are all indispensable. Reactions occurring in the gasification include coal pyrolysis, gasification and combustion reaction. Coal pyrolysis refers to the process when the solid coal is converted to gas, solid and liquid products. Coal gasification and combustion reaction include two types of reaction: heterogeneous gas–solid reaction and homogeneous gas-phase reaction.

Different gasification processes need different properties of raw materials. Therefore, the selection of coal gasification processes greatly depends on the properties and effects of coal for gasification. Properties of coal for gasification mainly include coal reactivity, adhesion, clinkering property, thermal stability, mechanical strength, particle composition as well as content of moisture, ash and sulfur.

The coal gasification process can be classified according to the pressure, the gasification agent, the heating method and others In general, it is classified based on the way coal materials contact with the gasification agent in the gasification furnace, including:

(1) Fixed bed gasification: In the process of gasification, coal is added from the top of the gasification furnace, and gasification agent is added from the bottom. Then the coal and the gasification agent countercurrent contact with each other. Compared with the rate of rise of the gas, the rate of descent of the coal is quite slow and can even be regarded as immobilized. That's why the process is called fixed bed gasification. However, in fact, the coal is moving downward at a very slow speed, so it would be better called moving bed gasification.

(2) Fluidized bed gasification: The raw materials for fluidized bed gasification are small coal particles whose size is 0–10 mm. These particles spread and suspend in the upward airflow, and then have the gasification reaction in the boiling state. In this way, temperature within the coal can remain the same. What's more, this method is easy to control and it can improve the gasification efficiency.

(3) Entrained flow bed gasification: It is a kind of cocurrent gasification. The coal particles with a size of under 100 μm are brought into the gasification furnace by a gasification agent; or the coal powder can be made into coal–water slurry and then pumped into the gasification furnace. Then the coal has combustion and gasification reactions with the gasification agent at a temperature higher than its ash melting point. At last, the ash is discharged out of gasification furnace in the form of liquid.

(4) Bath bed gasification: In this process, coal powder and gasification agent are sprayed at a high speed into a molten pool with high temperature and stability. In this way, a part of the kinetic energy is converted to the molten pool, in order to make the liquid melt rotate in a spiral way and get gasified. Currently, this process has stopped being developed.

2) **Coal liquefaction technology**

Coal liquefaction technology is an advanced clean coal technology that converts solid coal into liquid fuel, chemical raw materials and products. It can be classified into direct coal liquefaction technology and indirect liquefaction technology.

Direct coal liquefaction technology is a process that makes the solid coal to react with hydrogen at high temperature and high pressure. After degradation and hydrogen, it turns coal into liquid oil, so the technology is also known as hydrogenation liquefaction. Generally, a ton of anhydrous ash-free coal can be converted into over half a ton of liquefied oil. Direct coal liquefaction oil can produce clean and good-quality gasoline, diesel and aviation fuel. The process is shown in Figure 11.2.

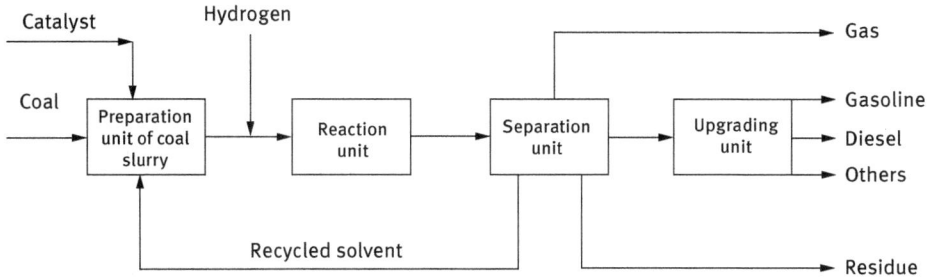

Figure 11.2: Direct coal liquefaction process.

First, the coal is pulverized into powder which becomes coal slurry with the liquefied heavy oil (circulating solvent) generated by itself. Then hydrogen is added into the coal slurry at a high temperature (450 °C) and a high pressure (20–30 MPa) to convert coal into gasoline, diesel and other oil products. A ton of anhydrous ash-free coal can produce 500–600 kg of oil.

The indirect liquefaction technology of coal first gasifies the coal, and then produces fuel, chemical raw materials and products. The process is shown in Figure 11.3.

The indirect coal liquefaction process first gasifies all coal into syngas (hydrogen and carbon monoxide), and then synthesize the syngas into gasoline using the catalyst. About 5–7 tons of coal can produce 1 ton of oil.

Advantages of indirect coal liquefaction are listed as follows:
(1) It can be applied to more types of coal than direct liquefaction technology.
(2) Gasoline can be produced if the existing chemical fertilizer factories have gasification furnace.
(3) The reaction pressure is 3 MPa, which is lower than that of direct coal liquefaction. The reaction temperature is 550 °C, higher than that of direct coal liquefaction.

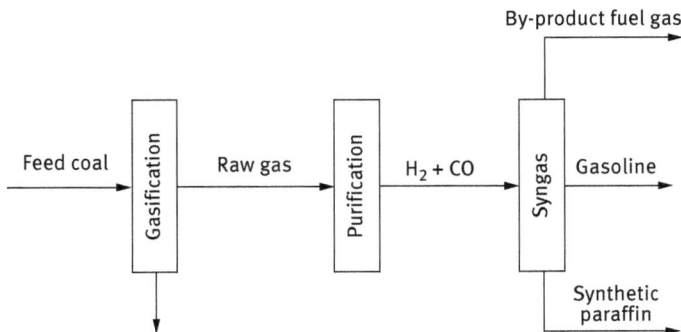

Figure 11.3: Indirect coal liquefaction process.

(4) The amount of oil produced is less than that of direct liquefaction. Five to seven tons of coal can produce 1 ton of oil, which costs much more than direct liquefaction.

3) Coal gasification combined cycle power generation technology

Coal-fired combined cycle power generation technology has a variety of methods. It can be classified into integrated gasification and partial gasification. The former means that after coal gasification and desulfurization, we can get clean coal gas which can be used for gas turbine. The desulfurization rate in integrated gasification combined cycle power plant can be up to 99%, while the NO_x emission is only 15–30% to that in the conventional power plants. Therefore, complete gasification is good for the environment. The latter includes partial gasification of differential bed combined cycle and coal partial gasification (coal can directly be combusted in the circulating fluidized bed boiler, and the high-temperature gas drives the gas turbine combined cycle). Different schemes have their own advantages, difficulties and complexities.

Coal partial gasification combined cycle power generation is a kind of clean coal power generation technology, and it was first proposed by the British Coal Research Institute in the late 1970s. Compared with the integrated coal gasification combined cycle, this generation system has slightly lower generating efficiency, but it has simpler system, as well as lower power consumption, technical requirements and investment. What's more, it can be easily matched with subcritical and supercritical steam turbine power generation systems. Thus, it is a competitive clean coal power generation technology. The main applications of this technology are listed as follows:

(1) Second-generation PFBC system of Foster Wheeler [3]

The system process is shown in Figure 11.4: coal, desulfurization agent and air are put into the pressurized fluidized bed gasifier to produce coal gas and semicoke with low calorific value through gasification. The coal gas can be sent into the precombustion chamber after two stages of dust removal and purification. The high-temperature flue gas generated will be sent into the gas turbine. Then semicoke is put into the pressurized fluidized bed combustion boiler, and the high-temperature flue gas generated after the combustion is sent into the precombustion chamber, where it is mixed with the high-temperature flue gas generated after coal combustion. The combined gas then enters into the gas turbine and works there. The rest heat of the gas turbine is used for generating high-temperature steam. The system is expected to have a thermal efficiency of up to 45% if it adopts a gas turbine whose initial temperature is 1,260 °C.

(2) APFBC technology of the United Kingdom

Compared with the scheme of Foster Wheeler, all the heat released by the combustion of semicoke in atmospheric circulating fluidized bed is

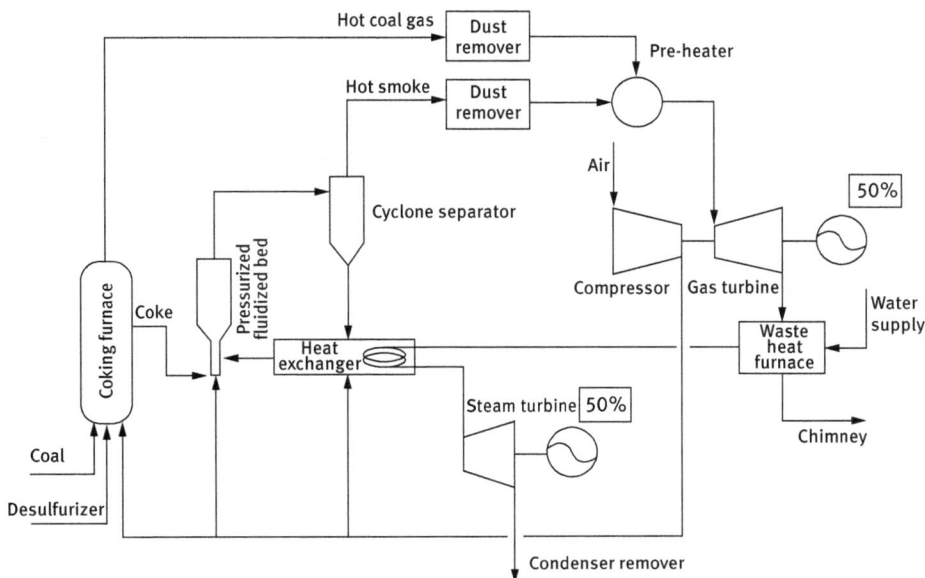

Figure 11.4: Second-generation PFBC system process of Foster Wheeler.

used for steam generation; at the same time, the gas turbine exhaust also provides heat to the steam system. The system process is shown in Figure 11.5. Its efficiency can be up to 46–48%.

G. Lozza et al. proposed a combined cycle scheme of partial gasification process and atmospheric fluidized bed combustion (AFBC), as shown in Figure 11.6. The feature of this scheme is that the oxygen required by semicoke during AFBC is provided by the turbine exhaust (the high-temperature flue gas after combustion of coal gas needs to be cooled to the entrance temperature required by gas turbine, and the precombustion chamber is mixed with a lot of cooling air, so the oxygen content in the exhaust gas machine is very high). The temperature of AFBC boiler is 1,170 °C, and the heat of exhaust gas is used for heating water and producing steam in the waste heat boiler, the thermal efficiency reaching 45–47%.

4) **Coal-fired magnetic fluid power generation technology**
   Coal-fired magnetic fluid power generation technology, also known as plasma power generation technology, is a typical application of magnetic fluid power generation. When the high-temperature (more than $2.6 \times 10^6\,°C$) plasma gas produced through coal combustion enters into a strong magnetic field at a high speed, electrons in the gas are affected by the magnetic force and therefore flow toward the electrode along the direction that is perpendicular to the magnetic line. Then the electrons produce direct current which is inverted to alternating current (AC) and sent to the AC network.

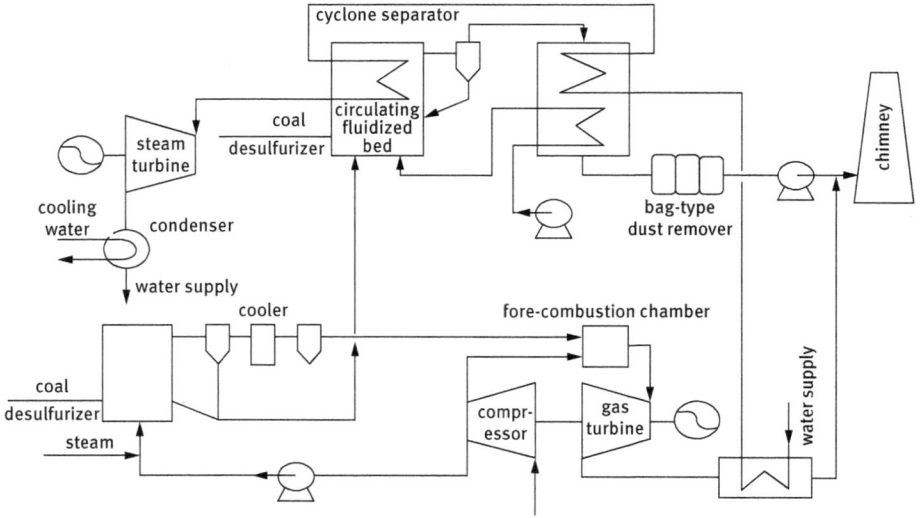

Figure 11.5: APFBC technology process.(3) PGFBC-CC power generation scheme proposed by G. Lozza et al.

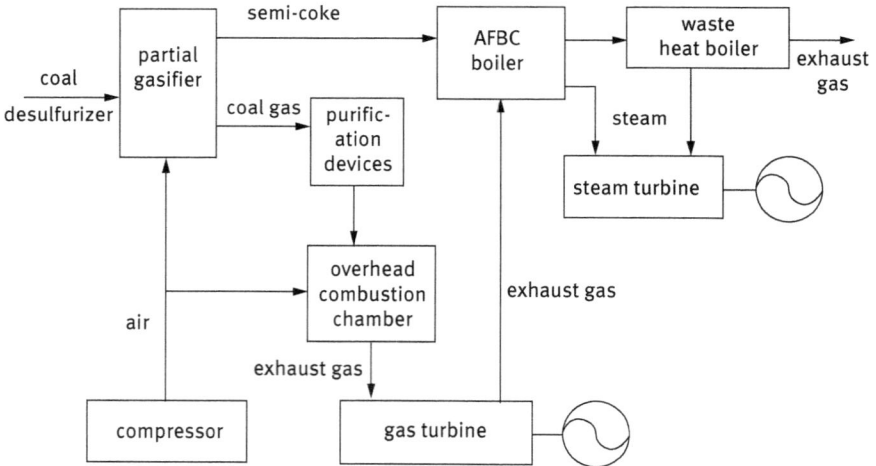

Figure 11.6: PGFBC-CC power generation process.

The efficiency of magnetic fluid power is only about 20%, but due to its high exhaust temperature, the exhaust gas discharged from the magnetic fluid can also be sent to the general boiler to combust and generate steam. The steam can drive a turbine to generate power and form combined cycle power generation with high efficiency. The total thermal efficiency can be up to 50–60%, the highest among all the high-efficiency power generators that are under development at present. Similarly, it can be desulfurized effectively to control the production of NO_x. Therefore, it is also a low-pollution coal gasification combined cycle power generation technology.

As to the magnet fluid power generation technology, high-temperature ceramics not only decides whether it can work normally at 2,000–3,000K magnetic fluid temperature, but also determines the lifetime of the channel which is the key to whether the coal-fired magnetic fluid power system can work normally. Currently, the highest tolerable temperature of high-temperature ceramics has reached 3,090k.

There has been the model project of coal-fired open-loop magnetic fluid power generation. It is expected to be locally commercialized in 2010, which will make a significant contribution to saving energy, reducing CO_2 emission and realizing green production in the power industry. In terms of nonequilibrium ionization closed-loop magnetic fluid power generation, due to its low working temperature, it is suitable for 100–300 MW medium-sized units, and has a huge potential in the development of power generation industry with coal as fuel. What's more, there is also closed-loop liquid metal magnetic fluid power generation. It is suitable for small-sized power devices and has bright prospect. But it is still in the stage of basic research.

11.2 Research and Development of Biomass Energy

Biomass energy is the energy converted by green plants from solar energy through chlorophyll and stored inside biomass. Biomass energy includes all kinds of energy in the natural world, such as biogas, biodiesel, forestry processing wastes, crop straw, organic waste and other wild plants. These kinds of energy are converted from substances such as all kinds of plants, human and animal wastes, and urban and rural organic wastes. Biomass energy comes from solar energy, so it is an environmental-friendly and renewable green energy. Despite the advantages of low pollution, wide distribution, easy for combustion and low ash content, it also has disadvantages such as scattered distribution, low energy density, low calorific value and low thermal efficiency.

Each year, plants on the earth can fix 2×10^{11} tons of carbon through photosynthesis, with 3×10^{11} tons of energy, and the amount of energy that can be developed is equivalent to 10 times the world energy consumption each year. The biomass can be found all over the world, with enormous reserves. Only plants on the earth have an annual capacity that is 20 times the mineral consumption, or 160 times the food energy produced by the worldwide population at present. Although the yield of biomass in different countries has a big difference, every country does have some types of biomass. Biomass energy is the source of heat energy and provides the basic fuel for mankind. Biomass resources in China are rather abundant. Theoretical biomass energy is equivalent to 5 billion tons of standard coal, about four times the total energy consumption currently in China [4].

The development of green energy has become an important means for industrialized countries to save energy and reduce pollution. At least 14 industrialized countries have made great achievements on the development of green energy, some of which have greatly relieved national energy shortage and improved the environment. The development of biomass energy includes: use of alcohol plants as green oil, use of methanol plants for power generation, use of oil-producing plants and algae, development of algae and bacteria in the sea, promotion of firewood and "energy forest," use of biogas digesters, development and utilization of human biological power, research of bacterial mining technology and so on.

11.2.1 Utilization Status of Biomass Energy at Home and Abroad

1. Utilization status of biomass energy abroad

Biomass energy has become an important task for the research and development of renewable energy in the world. Many countries have developed corresponding research and development plans, such as Japan's new Sunshine Plan, India's green energy projects, the US energy farms and Brazil's alcohol energy plan. In Germany, biomass is used to mix with coal for power generation and gas production. The United Kingdom has set a target that 10% of the country's electricity needs to come from biomass within a decade. The EU put forward in the white paper in 1998 that by 2010, biomass energy utilization should account for 12% of the total energy consumption, more than double the 5.6% in 1998. France has set a goal to triple the production of biomass fuels within 2 years, bringing energy crop acreage to 1 million hectares and eventually becoming Europe's largest producer of biomass fuels. Some scientists predict that by 2050, the biomass energy will provide 60% of the world's electricity and 40% of the liquid fuel (vegetable oil, alcohol), and reduce the CO_2 emission greatly. It may become a major sustainable energy source in the future. At present, the United States has become a world leader in the field of biogas power generation, with a total capacity of 340 MW, a total installed capacity of 104 MW, a single capacity reaching 10–25 MW, and the biomass energy utilization accounting for about 4% of the total primary energy consumption. The Staten Garbage Disposal Station in New York has got an investment of 2×10^{11}. The waste is disposed of by wet treatment, and the biogas is recovered for power generation and fertilizer production. The

University of Western Kentucky has developed a new type of biomass air gasification technology for the production of gas with high calorific value and low
tar content. This technology has high carbon conversion rate and gasification
efficiency. The United States National Renewable Energy Laboratory conducted
research on high-pressure coal gasification combined with biomass fluidized
bed, and obtained satisfactory results. They also analyzed and evaluated all
kinds of biomass utilization technologies. University of Western Ontario, Canada,
developed a direct ultra-short biomass liquefaction technology with a cost of
only 50 Canadian dollars/ton for large-scale industrial production, which is a
major breakthrough in biomass liquefaction technology. Europe is very active
in developing and utilizing biomass energy. The German government subsidies for biogas power network to encourage farmers to use biogas technology,
and gives special funds for the development and research of biogas technology.
According to the calculation of the German Biogas Association, at current technical level in Germany, the annual consumption of biogas power is 6×10^{10} kW,
which accounts for 11% of the total power consumption. University of Tübingen
in Germany developed low-temperature splitting devices to dispose city garbage,
and the feeding flow reached 2 ton/h. University of Twente in Holland developed the rotating cone process. Austria has successfully implemented the plan
of establishing a region whose power supply comes from the combustion of wood
energy. Currently there have been more than 90 regional heating stations with a
capacity of 1,000–2,000 kW and an annual supply of heat up to 1,000 MJ. Sweden
and Denmark are now engaged in the implementation of biomass cogeneration
plan so that the biomass energy can provide high-grade electrical energy while
meeting the heating requirements. Brazil is the world's largest producer and consumer of ethanol made from sugarcane. Therefore, biomass, especially the waste
of sugarcane or bagasse, plays a decisive role in Brazil's energy structure. The
total ethanol production in Brazil reached 1.75×10^{10} L in 2006, accounting for
38% of the worldwide total output. What's more, 44% of Brazil's transport fuel is
ethanol. It appears that the cross-fusion of physical and chemical conversion and
biological technology will be the trend for the development of biomass application technology.

2. **Utilization status of biomass energy utilization at home**
In China, biomass energy is the fourth largest energy resource only after coal,
oil and natural gas, accounting for 20% of the total energy consumption. China
attaches great importance to biomass energy utilization. Biomass energy utilization has been listed as a key scientific and technical project in four consecutive
"Five-Year Plans." Great progress has been made on the research and development of new technology for biomass energy utilization.

In recent years, significant progress has been made by domestic scientific research institutes in biomass gasification. Guangzhou Institute of Energy
Research, Chinese Academy of Sciences, has made a series of progress on

gasification power generation by circulating fluidized bed, and has built and operated several sets of gasification power systems; Institute of Forest Products and Chemical Industry of Chinese Academy of Forestry has made achievements on biomass fluidization gasification technology, and oxygen-rich gasification by cone circulating fluidized bed; Xi'an Jiaotong University has been committed to basic research of supercritical catalytic biomass gasification; University of Science and Technology of China has conducted research on biomass plasma gasification and biomass gasification synthesis; Shandong University has studied technology of fixed bed gasification. At present, the gasification technology has been put into use, especially the concentrated biomass gasification technology and the small-scale biomass gasification technology. Due to small investment, these technologies are suitable for decentralized use in rural areas, with good economic and social benefits.

Since the mid-1980s, China started the research on fuel for biomass solidification. By introducing and learning from foreign advanced models, various types of biomass briquetting machines suitable for China's national conditions have been developed, used for producing rod-shaped, massive or granular biomass briquette. Henan Agricultural University has developed the HPB2III bidirectional hydraulic-drive straw extrusion molding machine, and actively explored to launch it in the market. Overall, there is a long way ahead of China to catch up with the international advanced level in the field of biomass solidification.

Studies on biomass pyrolysis and liquefaction technology in China are still in the initial stage. In 1997, Dong Liangjie of Shenyang Agricultural University adopted the Kissinger and Dzawai methods to verify the dynamic parameters, after which he studied the pyrolysis reaction of sawdust and its components; University of Science and Technology of China developed an electric fast-fluidized bed for biomass pyrolysis liquefaction, which can be used for the liquefaction of various solid biomass. China is one of the countries with great utilization of biogas in the world. China's biogas technology has entered the commercial application stage. Technology of large-scale biogas projects for sewage treatment has also entered the stage of commercial demonstration and preliminary promotion. Currently, 1.7×10^7 biogas digesters have been built in rural areas, and more than 2,400 large- and medium-sized biogas projects have also have been built, contributing to an annual biogas output of more than $11 \times 10^9 \, \text{m}^3$.

11.2.2 Biomass Energy Utilization Technology

1. Biomass pyrolysis synthesis technology [5, 6]

It refers to a thermochemical reaction process in which the biomass is decomposed into biocoke, bio-oil and combustible gas in the conditions of complete oxygen deficiency in the reactor or only limited oxygen and no catalyst. After the biomass pyrolysis, 80% to 90% of its energy is converted into high-grade fuel.

Solid and liquid fuel produced by agricultural and forestry waste through biomass pyrolysis have no smoke while combusted. The sulfur content in the exhaust gas is low, and combustion residue is little, which reduces environmental pollution. After the sludge produced by city garbage and wastewater is pyrolyzed, its volume is greatly reduced, and its odor, chemical pollutants and pathogens are all removed. In this way, environment pollution is transformed into energy. Different raw materials and processes for pyrolysis also cause different in the production ratio and calorific value among 3 kinds of products, namely, biocoke, bio-oil and fuel gas.

Main products made by biomass pyrolysis mainly include gas, liquid, solid product (products ratio varies with different process conditions), as shown in Figure 11.7.

Based on the heating rate, pyrolysis can be divided into low-temperature slow pyrolysis and fast pyrolysis. The temperature for the first one is below 400 °C, and its main product is coke (30%); the second one is developed abroad. It is carried out at a temperature of 500 °Cand a heating rate of up to 1,000 deg/s, and the liquid fuel oil is produced after transient cracking. Fast pyrolysis is a promising biomass application technology, with the liquid fuel oil rate being up to more than 70%, and a calorific value of 1.7×10^4 kJ/kg. It is hard to control the conditions of fast pyrolysis, which affects the yield. Bio-oil is a liquid product that has high oxygen content and the low hydrogen–carbon ratio. Its unique properties cause instability, especially its thermal instability. Therefore, it can be used as fuel only after catalytic hydrogenation and catalytic cracking. Fast pyrolysis technology has been developed rapidly since 1980s when it was first put forward. Now there have been various kinds of process. The fluidized bed reactor created by University of Waterloo in Canada, rotating cone reactor invented by University of Twente in Holland, and free-fall reactor made in Swiss all increase the yield of liquid products to the maximum. China started to conduct research on fast pyrolysis since the "10th Five-Year Plan." Currently, this technology is still in the research and experiment stage.

2. **Biomass gasification technology [7]**

The research of biomass gasification technology developed very fast all over the world. This technology mainly includes pyrolysis and gasification, as well as anaerobic fermentation.

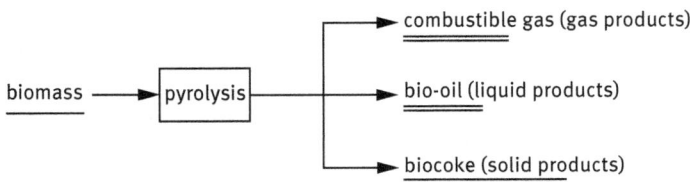

Figure 11.7: Main products made by biomass pyrolysis.

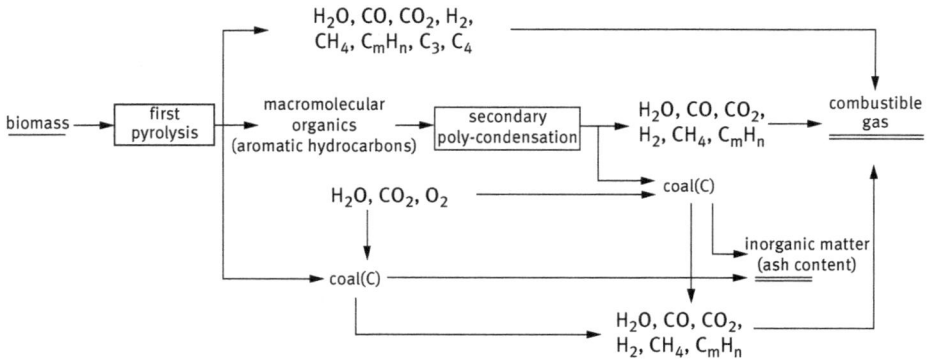

Figure 11.8: Biomass pyrolysis and gasification.

1) **Pyrolysis and gasification technology**

   Countries abroad use different kinds of biomass as raw materials, and generally adopt the combustion pressure gasification technology to drive the gas turbine. There are also generator gas methanation, fluidized bed gasifier and fixed bed gasification furnace. Gasification technology in the United States, Japan, Canada, Sweden and other countries have been mature enough to produce water gas in a large scale. Figure 11.8 shows the process of biomass pyrolysis and gasification.

2) **Anaerobic fermentation technology for biogas production**

   Under anaerobic conditions, organics can be fermented by microorganism to produce a combustible gas: biogas, also known as biogenic gas. The main composition of biogas is methane which accounting for about 60%. The calorific value per cubic meter of biogas is equivalent to 1 kg of coal heat.

   Anaerobic fermentation is a relatively mature technology for biogas production, and there is no energy consumption in this process. In fact, it is believed that the fossil energy on earth was formed by biomass in anaerobic conditions, so anaerobic fermentation is the most promising and sustainable way of producing biogas. Before the 1980s, developing countries mainly developed biogas technology, using crop straw and animal manure as raw materials for biogas production; while developed countries mainly used livestock manure and high-concentration organic wastewater as raw materials. After the 1980s, large-scale biogas projects began to emerge, and entered the stage of industrialization and commercialization.

   Currently, developed countries including Japan, Denmark, Holland, Germany, France and the United States are all disposing animal wastes by the anaerobic method. IC Netherlands has made the gas production rate of beer wastewater treatment reach 10 $m^3/(m^3 \cdot d)$; the United States, Britain, Italy and other developed countries mainly use biogas technology for waste treatment; Britain turns wastes into raw materials to generate biogas power that reached 111 MW. In the next decade, another 150 million pounds will be invested to build waste biogas power plants.

In the southern part of China, biogas digesters have been quite popular among rural households, and a number of large-scale biogas projects have been established. After over 10 years of research and development, anaerobic fermentation technology has made some progress. For example, the gas production rate of the anaerobic fermentation in the swine manure has reached $2.2\ m^3/(m^3 \bullet d)$, and a considerable number of design, construction and equipment manufacturing enterprises as well as management service enterprises have been involved in the biogas project construction. However, the gas production rate of biogas digesters in rural households is not high. In addition, biogas can never become a commodity by the natural fermentation in households. We must promote industrialized production, improve the fermentation rate of biogas and solve climate restrictions and other problems.

3. **Biomass liquefaction technology**

Liquefaction refers to the process when biomass is chemically converted into liquid products. It mainly includes direct liquefaction and indirect liquefaction. Direct liquefaction is to put biomass in high-voltage equipment for reaction with suitable catalyst in certain conditions. The liquefied oil produced can be used as automotive fuel or chemical products if processed further. Indirect liquefaction is to gasify biomass first and then make it into liquid products through further catalytic reaction. This kind of technology is one of the hotspots of biomass research. The oxygen content in the biomass is high, which is favorable to the formation of synthesis gas $(CO + H_2)$. There is no CO_2, CH_4 or other impurities in the plasma gasification gas, which greatly reduces the gas refining cost and provides favorable conditions for the preparation of syngas. Although China has studied Fischer Tropsch synthesis for many years, it has not been industrialized yet. The research and development of catalysts and their reactor systems are the key to larger production in the future. In particular, in terms of the characteristics of biomass synthesis gas (such as gas composition and tar), it is necessary to study the reaction mechanism, transform the existing technologies and catalysts, improve the quality and economic efficiency of the product, in order to realize the industrialization.

Alcohols are hydrocarbons that contain oxygen. Methanol and ethanol are two kinds of commonly-used alcohol. Methanol can be obtained by distillation of lignocelluloses, or by the catalytic reaction between carbon monoxide and hydrogen. Raw materials for methanol production are cheap, but the equipment costs a lot. Ethanol can be obtained by the synthesis of acetylene and ethylene, but the energy consumption is too high. Generally, more than 60% of the ethanol production cost is used for raw materials. Therefore, cheap raw materials are very important for reducing the cost of ethanol.

Biodiesel is a kind of fatty acid ester whose raw materials are vegetable oils and animal fats. It is made through the interesterification reaction with alcohols (methanol and ethanol). Biodiesel can be used as diesel engine fuel instead of diesel, or be mixed with diesel. This kind of environmental-friendly fuel can alleviate the constantly exhausted fossil energy, so it is imminent to be developed.

4. **Biochemical conversion technology**

Ethanol preparation from biomass: pure alcohol or the mixture of gasoline and alcohol can both be used as disposable fuel. Materials for the preparation of liquid alcohol can be classified into three types: the first contains sugar, such as sugarcane; the second contains cellulose, such as straw, husk, wood and its machining remainder; the third contains starch, such as sweet potato, corn and wheat. Different components of biomass require different liquefaction methods, but the technologies have all been mature.

1) **Biomass hydrolysis technology**

The main raw materials for the ethanol preparation from biomass are sugar, starch and lignocelluloses. The process is described as follows: crush the biomass first; convert starch, cellulose or hemicelluloses into polysaccharide through chemical hydrolysis (usually sulfuric acid) or catalytic enzymes; then convert sugar into ethanol through ferments, after which products can be obtained with relatively low volume fraction of ethanol (5–15%); distill water and remove other impurities; and condense the concentrated ethanol (95% ethanol can be obtained by one-step distillation process) into liquid ethanol. Transformation of lignocellulosic biomass is rather complicated and its processing cost is high. The cellulose can only be converted into sugar through hydrolysis of several kinds of acids, and then ethanol can be obtained after fermentation. This hydrolysis technology has high energy consumption, high pollution and high cost; therefore, it lacks economic competitiveness. Now the method of catalytic enzymatic hydrolysis is being developed, but due to the high cost of enzymes, the method is still in the research stage.

2) **Anaerobic fermentation technology**

Anaerobic fermentation refers to the decomposition of biomass by bacterial action in the absence of oxygen. Put the organic wastewater (pharmaceutical wastewater, human and animal manure) into an anaerobic fermentation tank (reactor and digester), and hydrolyze the complex organics by anaerobic fermentation bacteria. After fermentation, products such as organic acid, alcohol, H_2 and CO_2 can be obtained. Then metabolize the organic acid and alcohols into acetic acid and hydrogen by hydrogen-producing acetogens, and composite the acetic acid, H_2, and CO_2 that have been produced into CH_4 by CH_4 producing bacteria. Finally, a gas mixture of CH_4 (55–65% volume fraction) and CO_2 (30–40% volume fraction) can be obtained. This method can be used for the hydrogen production from many obligate anaerobic and facultative anaerobic microorganisms, such as *Clostridium butyricum*, *Escherichia coli*, *Enterobacter aerogenes* and *Azotobacter nodulans*. Under the action of nitride enzyme or hydrogenase, their substrates can be decomposed to produce hydrogen. Since oxygen can inhibit the synthesis and activity of hydrogen-production microbial catalyst, the anaerobic fermentation method for hydrogen production should be carried out in the absence of oxygen. Due to the high specificity of transformation bacteria, different kinds of bacteria decompose different substrates. Therefore, in order to completely decompose substrates and produce a large amount of hydrogen, and to achieve the complete decomposition of substrate and to produce a large amount of hydrogen, co-cultivation of different kinds of bacteria is needed. The output of hydrogen production from anaerobic fermentation is low, the energy conversion rate being only about 33%. In order to improve the yield of hydrogen, it is necessary to not only breed oxygen-resistant bacteria, but also develop advanced technology.

3) **Biomass technology**
The photosynthetic microbial hydrogen production mainly relies on photosynthetic bacteria and algae which can decompose the substrates of biomass through photosynthesis to produce hydrogen. In 1949, Gest et al. first reported that the photosynthetic bacterium *Rhodospirillum rubrum* can produce molecular hydrogen using organic matter as a hydrogen donor in anaerobic light. Since then, a series of related research has been conducted. The present studies show that microorganisms related to the photosynthetic microbial hydrogen production mainly are concentrated in more than 20 strains of seven genera, including *Rhodopseudomonas*, *Rhodospirillum*, *Clostridium*, red sulfur bacteria, *Ectothiorhodospira* butyric acid, bacillus and rhodomicrobium. The mechanism of this method is generally believed as follows: the photons energy is captured and sent to the photosynthetic reaction center for charge separation, which results in the formation of high-energy electrons and a proton gradient. Adenosine triphosphate (ATP) can therefore be obtained. After the charge separation, high-energy electrons can produce iron ferrite reduction protein (Fdred) which, together with ATP, can be used for produce hydrogen by the azotase.

11.2.3 Biodiesel

1. **Definition of biodiesel**
Biodiesel is a kind of fatty acid methyl ester which is made through the ester exchange between methanol and different kinds of raw materials, including oil crops (soybean, cottonseed and rapeseed), oil fruits (oil palm and *Pistacia chinensis* bunge), oil aquatic plants (engineered microalgae), animal fats and waste food oil. It is a kind of clean, renewable biofuel and an excellent substitute for diesel. The molecule of biodiesel is composed of about 15 carbon chains.

Similarly, the molecule of vegetable oil generally consists of 14–18 carbon chains, which is why it is called biodiesel.

2. **Features of biodiesel**

Compared with general diesel oil, biodiesel has the following features:

(1) It is conducive to environmental protection. The content of sulfur is low in biodiesel, which lowers the emission of sulfur dioxide and sulfur by about 30% (70% with catalyst). Aromatic hydrocarbon that is harmful to the environment is not contained in biodiesel, so the waste gas of biodiesel causes less damage to human body and the environment. Tests showed that compared with ordinary diesel, biodiesel can reduce 90% of the air toxicity, 94% of the cancer rate. Due to the high content of oxygen in biodiesel, there is less smoke during the combustion. The emission of carbon monoxide is 10% (95% with catalyst) less than that of diesel. At last, biodiesel also has high biodegradability.

(2) It can start the engine at low temperature. The cold filter point can be up to −20 °C without additive.

(3) It has good lubricating property, which can reduce the wear rate of the fuel injection pump, engine cylinder and rod, and make them more durable.

(4) It has good safety performance. Due to the high flash point, biodiesel does not belong to dangerous goods. Therefore, its transportation, storage and use are pretty safe.

(5) It has good fuel performance. The high ratio of cetane makes it easier to combust than diesel. The combustion residue is slightly acidic, which can prolong the service time of the catalyst and engine oil.

(6) It is a renewable energy. Different from oil, the supply of biodiesel can last forever through the efforts of agricultural and biological scientists.

(7) It can be directly added for use, with no need to change the diesel engine. What's more, there is no need to add other refueling equipment, storage equipment or special technical training.

(8) It can be mixed with fossil diesel in a certain proportion, which can reduce fuel consumption, improve power performance and reduce exhaust pollution.

The excellent performances of biodiesel make the engine emission index not only meet the current European III emission standard, but also can ease the global warming problem caused by carbon dioxide emission.

Although biodiesel shows many advantages, it has some disadvantages as well:

(1) High price is the biggest obstacle to the widespread use of biodiesel. The main material of biodiesel is soybean, so its price is changing with the yield of soybean. The United States is developing other raw materials. As the types and quantity of raw materials increase, the price of biodiesel will decrease.

(2) The viscosity of biodiesel is higher than that of conventional diesel, so it is not suitable for low-temperature regions, which limits its application in

Canada, America and most European countries. In these regions, biodiesel is mainly used as an additive, namely, mixed in a proportion of 5–10% with conventional diesel.

(3) Biodiesel also has storage problem. The use of pure biodiesel in the old engine can easily damage sealing waxes and gaskets, so the pure biodiesel is better to be stored in a clean oil tank. Such problems can be avoided if 20% biodiesel (B20) is added in the diesel, and the performance of the engine using B20 is exactly the same as the performance of No. 2 fossil diesel.

3. **Production method of biodiesel**

The biodiesel production methods include chemical method, biological enzymatic method and engineered microalgae method. The chemical method is the earliest one, but its cost is quite high; the biological enzymatic method has received an increasing attention, but it has to figure out how to reduce the toxicity component of the enzyme; the engineered microalgae method takes the oil-rich engineered algae as a raw material. It is believed to have a great prospect.

1) **Chemical method**

The production of biodiesel by chemical method is to adopt bio oil with methanol or ethanol and other alcohols, and use sodium hydroxide (accounting for 1% of the oil weight) or sodium methoxide (sodium methoxide) as a contact agent. Under the action of acid or alkaline catalyst, at a high temperature (230–250 °C), the transesterification reaction will take place, which generates corresponding fatty acid methyl or ethyl ester. After washing and drying process, biodiesel can be obtained. Methanol or ethanol can be recycled in the production. Similar to general oil production equipment, the production equipment of this method generates about 10% of by-product glycerol.

However, the chemical method has the following disadvantages: high reaction temperature and complicated process; excessive use of methanol in the reaction process; need of corresponding alcohol recovery device; complex process and high energy consumption; oil and free fatty acids in the raw water seriously affecting the quality and productivity of biodiesel; complex product purification; esterification products difficult to recycle; byproducts of the reaction difficult to remove, large amount of wastewater generated by the use of alkali catalyst; and second pollution caused by alkali (acid) solution.

The cost problem of chemical method also cannot be ignored, for the catalyst is alkalescent which requires crude oil as the raw material. For example, the cost of unrefined soybean oil and rapeseed oil accounts for 75% of the total cost, so the key to improving the practical use of biodiesel is to use cheap raw materials and improve the conversion rate. The United States has been studying plants with high oil content through genetic engineering method. Japan has adopted industrial waste oil and frying oil as raw materials. Europeans cultivate oil-rich crops on the lands that are not suitable for growing grain.

2) Biological enzymatic synthesis method

In order to solve the above problems, studies on biological enzymatic synthesis began to be made, namely, to prepare fatty acid methyl ester and ethyl ester using animal oil and low carbon alcohol in the process of lipase transesterification. The enzymatic synthesis of biodiesel has advantages such as mild conditions, low dosage of alcohol and little environment harm. In 2001, Japan adopted immobilized *Rhizopus oryzae* cells to produce biodiesel, with a conversion rate of about 80%. The microbial cells can be used for 430 consecutive hours.

On June 4, 2005, *China Environmental News* reported that Tsinghua University successfully passed the pilot scale test of new enzymatic synthesis technology for biodiesel production, the biodiesel yield reaching above 90%. The pilot product specifications met the United States and Germany biodiesel standards, and also China's No. 0 superior diesel standards. The contrast test of the pilot products showed that compared with the commercially available fossil diesel, if the diesel oil was mixed with 20% biodiesel, carbon monoxide, engine exhaust emissions of hydrocarbons, the concentration of main toxic components of smoke could be decreased significantly while the engine dynamic characteristics basically remained unchanged.

Thanks to its advantages such as mild reaction conditions, low dosage of alcohol and no pollutant emissions, the biological enzymatic method wins growing attention. However, it also has some problems: lipase is effective on esterification or transesterification of long-chain fatty alcohols, but the conversion rate of short-chain fatty alcohol (such as methanol or ethanol) is low, generally 40–60%. What's more, methanol and ethanol have a certain toxicity which may inactivate the enzyme. Furthermore, the by-products, glycerol and water, are hard to recover, which not only affects the formation of product, but also have toxic effects on the enzyme. Finally, short-chain fatty alcohol and glycerol both affect the reaction activity and stability of enzyme, which may greatly shorten the life of immobilized enzyme. These problems are the main factors that restrict the industrialized production of biodiesel by enzymatic synthesis method.

3) Engineered microalgae method

The engineered microalgae method opens up a new approach for biodiesel production. Through modern biotechnology, the US National Renewable Energy Laboratory (NREL) has created the engineered microalgae, a type of diatom Cyclotella. In laboratory conditions, the content of lipid in the engineered microalgae can be increased to more than 60%, and more than 40% in outdoor production; while in natural state, the lipid content in microalgae is generally 5–20%. The increase of lipid content in the engineered microalgae is mainly because of the high expression of acetyl-CoA carboxylase (ACC) gene in microalgae cells, which plays an important role in the control of lipid accumulation. At present, studies are being made to find appropriate molecular carriers so that the ACC gene can be fully expressed in bacteria, yeast and plants, and the modified ACC gene can be brought into microalgae for higher expression. The engineering microalgae method for biodiesel production has important economic and ecological significance: microalgae has high productivity, and the use of seawater as a natural medium can save agricultural resources; the productivity is several times higher than that of terrestrial plants; the biodiesel produced by this method contains no sulfur and emit no toxic gases; after being discharged into the environment, it can be degraded by microorganism without polluting the environment. In a word, the development of oil-rich microalgae or engineered microalgae is a major trend in biodiesel production.

4. **Biodiesel standards**
   1) **Current biodiesel standard**
   Many countries in the world have developed biodiesel standard to ensure both the quality of the diesel and user demands. The international standards for biodiesel are ISO 14214A and the ASTM International Standard ASTM D 6751 which is adopted by the United States and is legally recognized by the US Environmental Protection Agency in 1996 under Section 211 (b) of the Clean Air Act. Another widely accepted one is the German DIN system of biodiesel standards, which is by far the most detailed system of biodiesel standards. This standard system has different DIN standards for different manufacturing materials: DIN E 51606 biodiesel standard of RME (rapeseed methyl ester) and PME (vegetable methyl ester) for biodiesel whose raw materials are rapeseed and pure vegetable seeds; DIN V 51606 biodiesel standard of FME (fat methyl ester) for biodiesel whose raw material is vegetable oil mixed with animal fat. The European Union also enacted the EN14241 biodiesel fuel standard in November 2003. In addition, Austria, Australia, the Czech Republic, France, Italy, Sweden and other countries also developed biodiesel fuel specifications.
   2) **DIN V 51606 biodiesel standard in Germany**
   The biodiesel standards mainly evaluate the overall process of production, the removal of glycerol, catalyst and alcohol, and the absence of free fatty acids. Assessment indicators of biodiesel production include specific gravity, dynamic viscosity, flash point, sulfur content, residue, cetane number, ash, water, total impurities, triglycerides and free glycerol. Biodiesel standards are driving biodiesel adoption and legalization in the automotive industry in many countries. Meanwhile, as a large number of countries have recognized biodiesel as a new renewable energy, they are also promoting its globalized application.

   Since the current biodiesel supplied in the market is mainly a mixture of biodiesel and fossil diesel oil, there is also standard for the oil mixture. For example, a blend of 5% biodiesel and 95% conventional diesel is required to meet the EN590 (EN590: 2000) standard promulgated in 2000. Mixed oil that can meet this standard can be applied safely to all diesel engines. Although there is no need of stabilizer in the mixed oil, some foreigner experts suggested that it is necessary to add in the EN590:2000 standard that the biodiesel in the mixed oil must comply with the EN 14214 standard.

3) **National biodiesel standard in China**

China's first national biodiesel standard drafted by the Petrochemical Research Institute and other units has been approved by the State Administration of Quality Supervision since May 1, 2007, onward. This is the China's second standard for alternative biomass energy after the implementation of national standard for ethanol gasoline. National biodiesel standard is a relatively high standard that can give investors a clear reference to the development of biodiesel industry. The standard mainly involves detailed specifications on the composition, content, greasy property and cetane number of B100 biodiesel (biodiesel content is 100%), a total of 17 technical requirements. Currently China's diesel is mainly used for the agricultural power machinery as well as highway, waterway and railway transport machinery, similar to the United States. Therefore, China's biodiesel standards and technology requirements are made according to the ASTM D6751 – 03a Standard Specification for Biodiesel Fuel (B100) Blend Stock for Distillate Fuels made by the United States. What's more, the standards are also based on inspections and tests on biodiesel products of different raw materials in recent years.

In addition, China is currently formulating the B5 and B10 national standards for biodiesel diesel fuel, which are expected to be released in late 2007 or early 2008. Two petrochemical industry standards, Determination of Free Glycerol and Total Glycerol in Biodiesel and Determination of Oxidation Stability, have been completed and entered the approval procedure. Two analysis method standards, Determination of Biodiesel Content in Diesel Fuel and Determination of Ester Content in Biodiesel are also being developed.

5. **Application status of biodiesel**

At present, many countries in the world are vigorously developing biodiesel technology and actively promoting its industrialization process, especially in Europe, the United States and some other developed countries. In recent years, the governments are promoting the biodiesel industry by strong supports in tax policy and financial subsidies. As a result, the price of biodiesel is almost the same with that of oil. The EU is the region in the world with the fastest growing biodiesel industry, with a production capacity of 2.25 million tons in 2004 and a planned output of 8–10 million tons in 2010, accounting for 5.75% of the diesel market and 20% by 2020. The United States, Canada, Japan, South Korea and other countries have also increased efforts to develop biodiesel.

(1) The United States. The rise of biodiesel in the United States is due to the enactment of the Air Pollution Act of 1991 and the Decree on the Reduction of Sulfur

in Diesel Fuel and Emissions. The Energy Policy Act of 1992 defines a long-term goal: by the year 2000, oil-free alternative fuels will account for 10% of engine fuel and 30% by 2010. Biodiesel produced in the United States has been extensively tested with a large number of public transport vehicles, and tests show that biodiesel costs 2.5 times as much as fossil diesel.

(2)   Europe. The development of the biodiesel industry in Europe has benefited from two aspects. On the one hand, the Common Agricultural Policy has proposed a policy of idle land for other purposes (not for grain production), while providing stable financial subsidies as an encouragement. If the land is used to produce raw materials of biodiesel, the highest subsidies can be given. As the demand for industrial oilseeds (for production of biodiesel) increased, the idle land for the cultivation of oilseed crops in 1995–1996 had increased by 50% to 900,000 ha. According to the current production rate of industrial oilseeds, it may reach or exceed the equivalent of 1 million tons of soybean meal in the next few years. On the other hand, European countries have a high fuel tax, typically 50% or more of the retail price of diesel fuels. Most European governments believe that using grain to make alternative fuels does not improve the economy, in fact only increasing the pressure on agricultural budgets. Nevertheless, in February 1994, the European Parliament was still exempt from biodiesel tax policy of 90%. Legislative support, tax exemptions, and subsidies for the production of oilseeds have enabled biodiesel in many European countries to compete in price with diesel fuel. Since 1995, biodiesel production capacity in Western Europe has exceeded 1.1 million tons per year, most of the biodiesel produced through ester exchange. This will produce more than 110,000 tons of glycerol by-products per year, causing an excess of glycerol in the market. Germany is studying the use of cold rape seed for producing biodiesel, in order to avoid the problem of excess glycerol.

(3)   Canada. In Canada, the addition of biodiesel to diesel is called green diesel, and exhaust and engine tests show that green diesel has the same performance of conventional diesel fuel with commercial nitrates. Biodiesel and green diesel are not yet commercially available in Canada, but green diesel testing is in progress.

(4)   China. Compared with foreign countries, development of biodiesel in China is still in the primary research stage. It shows such defects as small scale, low level of technology, inadequate supply of raw materials, different quality standards, and nonstandard market circulation. What's more, the application of biodiesel is mainly carried out in provide enterprises and the level of comprehensive utilization and deep processing is expected to be improved.

At present, domestic enterprises with an annual output of millions of tons include Hainan Zhenghe Biological Energy Company, Sichuan Gushan Oil Chemical Co., Ltd., Fujian Zhuoyue New Energy Development Company and Xi'an Lantian

Biological Engineering Company. In addition, Sinopec Research Institute of Petroleum & Chemical and Shijiazhuang Refining & Chemical Co., Ltd. are building a biodiesel production plant with an output of 50,000 tons per year using a variety of raw materials. PetroChina is also planning to set up a biodiesel project in Sichuan with an output of 100,000 tons each year.

In recent years, by making full use of their own advantages and foreign technical supports, enterprises and local governments in China have actively carried out international cooperation on biodiesel development and production, and achieved preliminary results. For example, Sichuan University studied the gene of high-quality woody oil crops, in order to control the oil content and carbon chain length in biodiesel production. They also established a gene pool and a breeding base, and set up planting standards. What's more, southwestern regions such as Sichuan and Guizhou have actively carried out exchanges with relevant institutions of Southeast Asia, Europe and the United States, and have cooperated with them in the fields of planting base and technology research.

6. **Prospects**

Although biodiesel technology is already quite advanced, it is still not competitive economically. With improvements in the emission reduction device and fossil fuel combustion efficiency, the advantages of biodiesel will become less prominent. At the same time, biodiesel is also facing the fierce competition from other alternative fuels such as electricity, propane and natural gas. Nevertheless, biodiesel has legally entered the diesel market. In terms of refining efficiency and fuel economy, the position of diesel in the market should be maintained. But biodiesel possesses a certain market advantages if we aim at reducing dependence on imported gas, recovering the rural economy and protecting environment. In future, studies will continue to be made on the application of biodiesel in the maritime and mining industries. In addition to motor vehicles, equipment and facilities are also another potential market for biodiesel. It is also a subject of future research to reduce the production cost of biodiesel, improve fuel performance and further analyze the impact of biodiesel development on agriculture. In addition, we should seek for methods to deal with the by-products of biodiesel to improve resource utilization, increase revenue and reduce the production cost. Biodiesel still needs to overcome some difficulties in economy, rules and regulations, but it is believed to have a bright future.

11.3 Development and Utilization of Clean Energy

11.3.1 Solar Energy

Solar energy is the energy loaded by the sun. It is generally measured by the amount of sun radiation reaching the ground (including direct sun radiation and scattered radiation from the key). Therefore, it is an inexhaustible, environmental-friendly renewable energy. Solar power generation mainly includes three types of technologies: the first one converts solar energy into heat for thermal power generation; the second one directly converts solar energy into electricity, known as photovoltaic power generation technology; and the third one splits water into hydrogen and oxygen, and generates electricity using hydrogen energy [9].

1. **Solar thermal power generation technology**

   Concentrated solar thermal power generation (referred to as solar thermal power generation) is a technology which uses concentrating solar collectors to convert solar radiation into thermal energy, and generates power through thermodynamic cycle. Since the 1980s, the United States, Israel, Germany, Italy, Russia, Australia, Spanish and other countries have established different forms of demonstration devices, which effectively promoted the solar thermal power generation technology. According to the studies of the US Sunlab joint laboratory, by 2020, the cost of solar thermal power generation will be about 5 cents/kWh; therefore, as an alternative to conventional energy, it may become the most economic method for high-power generation. At present, there are three types of solar thermal power generation systems in the world, including solar power towers, parabolic troughs and solar dish systems. Since the early 1990s, parabolic troughs have been applied in the market; while the other two types are still in the commercial demonstration stage with huge potential for application.

2. **Photovoltaic cell power generation technology [8]**

   The interaction between the light quantum of sunlight and certain material can produce electric potential, and further convert the energy of light into electrical energy. Such energy conversion can be completed by a photovoltaic element called solar cell, also known as photovoltaic cell. In 1954, Bell Labs invented the first solar cell, but due to its low efficiency and high cost, it did not have too much commercial value. With the development of space technology, solar cells have become an irreplaceable part of space vehicles. In March 1951, the solar cell was first installed in Vanguard 1, which was launched by the United States. In May 1951, the third satellite launched by the Soviet Union was also installed with solar cells. In 1969, along with the American landing on the moon, the development of solar cells reached its first peak period. Since then, almost all man-made objects launched into space have been installed with solar cells. In the early 1970s, the Middle East War and oil embargo disrupted the oil supply of many industrial countries, which led to an energy crisis. It was then when people began to realize that traditional energy sources could not be relied upon for a long time. This is especially true in recent years as fossil fuel resources are reduced, and environmental pollution problems are increasing. Hence, the application of solar cells has been put on the agenda of many national governments.

   In recent years, the photovoltaic cell industry has developed rapidly. According to the 2004 data analysis, among a variety of solar cells, silicon-based solar cells accounted for 98% of the total output. Crystalline silicon solar cells accounted for

84.6% of total output, and polysilicon solar cells accounted for 56% of the total [9]. In 2005, the world's total output of solar cells were 1,656 MW, of which Japan was still the first with an output of 762 MW, accounting for 46% of world amount; Europe was with 464 MW, 28% of the total; the United States with 156 MW, 9% of total output; and other places with 274 MW, accounting for 17% of total output. In 2005, the world's photovoltaic installed capacity was 1,460 MW, with a 34% increase over 2004. Among all the countries, Germany had the largest installed capacity of up to 1,137 MW, accounting for 57% of the world's total installed capacity, a 53% growth compared with 2004. Europe had 920 MW, 63% of the world's total capacity; Japan had 292 MW, 20% of the world's total capacity, with an increase of 14%; the United States had 102 MW, 7% of the world's total capacity; and other countries and regions had 146 MW, 10% of the total capacity. According to statistics, from 2005 to 2010, the world average annual growth rate of solar cells was 25%.

1) **Advantages of solar power generation**
   (1) Solar energy is inexhaustible. The amount of solar energy on the earth is 6,000 times larger than that of human consumption. As long as a 160.9 km × 160.9 km giant photovoltaic power plant is established in the sunny southwest desert of the United States, it can generate enough power to meet the electricity needs of the whole country. Solar power generation is safe and reliable, free from energy crisis or the impact of unstable fuel market.
   (2) Solar energy is available everywhere, which can supply power with no need for long-distance transmission, and hence avoid losses such as transmission lines.
   (3) Solar energy does not need fuel, and its operating costs are low.
   (4) Solar power generation has no moving parts, so it is durable, easy for maintenance and especially suitable for unattended use.
   (5) Solar power does not produce any waste, pollution, noise or other public nuisance, and has no adverse effects on the environment. Therefore, it is an ideal clean energy.
   (6) The construction period of solar power system is short. The modular installation is convenient and flexible, as it can both be used for a small-scale solar power calculator of a few milliwatts, or a photovoltaic power plant as large as tens of megawatts. What's more, the capacity of solar cells can be increased or reduced according to load, which can avoid waste.

2) **Disadvantages of solar power generation**
   (1) Its application is intermittent. The power generation is correlated with climate conditions. For example, it generates no or little electricity in the evening or rainy days, which cannot meet the needs for power consumption. Therefore, it needs to be equipped with energy storage devices, and optimization design is required according to different locations.

(2) It has low energy density. In standard test conditions, the intensity of solar radiation received by the ground is 1 kW/m². Therefore, a large-scale use of solar energy requires a large area.

(3) Currently, its cost is still high, almost two to ten times that of the conventional power generation. The large initial investment also affects its popularization and application.

3. **Conversion of solar energy into chemical energy**

The photoelectrochemical process is the electrochemical process under the action of light, namely, the process when molecules, ions and solids absorb sunlight and excite electrons for charge transfer. Photoelectrochemical reactions are carried out at the interface of two conductive phases with different types of (electron and ion) conductance. Like the electrochemical reaction, the photoelectrochemical reaction system is accompanied by the flow of current. It usually involves studies of different kinds of conversion between solar energy and chemical energy in the electrochemical system. The most common one is the process when solar energy is converted into chemical (or electrical) energy through the photoelectrochemical reaction, and its inverse process when the chemical (or electrical) energy is converted into solar energy (e.g., electrogenerated chemiluminescence). As the research object of photoelectrochemistry, semiconductor is significantly different from the metal. Its energy band structure is the top of the valence band (Evb) that is fully charged with electrons and the bottom the conduction band (Ecb) that is unfilled or semifilled with electrons are separated by a band gap (Eg). Due to the band gap, the interaction of electronic states between the valence band and the conduction band is weak. Thus, after being excited by the light, the valence electrons of the semiconductor enter the conduction band and leave holes in the valence band. These valence electrons have a long lifetime (until recombination), allowing them sufficient time to participate in the electrochemical reaction at the interface of the electrode/electrolyte. It is such photoelectrochemical reactions caused by photogenerated electrons and photogenerated holes on the semiconductor electrode that will become the theoretical basis of solar photoelectric conversion, photochemical conversion and storage.

It is the most attractive way for hydrogen production to split water for hydrogen decomposition directly using solar energy. Since 1972 when Japanese scientists Fujishima and Honda reported the phenomenon of photochemical water splitting at a TiO electrode in the magazine *Nature*, photoelectrochemical and photocatalytic water splitting for hydrogen production has become a globally focused issue. At present, this technology is still in the research stage. Scientists are aiming to split water for hydrogen production by using visible light as a catalyst.

11.3.2 Wind Energy

Wind energy is the energy carried by the wind, and its amount depends on wind speed and air density. As a converted form of solar energy, wind energy is also an important natural energy. Currently, wind energy is mainly used for power generation. Among all the new energy projects, wind power generation is a project with mature technology, the greatest development conditions and good prospects for development. Wind power generation has gone through the development process from independent system to grid-connected system. Now in developed countries, wind energy is mainly developed for the construction of large-scale wind farms.

The principle of wind power generation is using wind to drive the wind turbine blade, and increasing the rotation speed of the blade through a booster machine to promote the generator for power generation. At present, large-scale wind turbine generators are generally horizontal axis wind generators, which consist of wind wheel, gear box, generator, yaw device, control system, tower and other components. The wind wheel can convert wind energy into mechanical energy. It is formed by two to three blades with great aerodynamic performance installed on a wheel. When the wind wheel is rotating at low speed, it can be speeded up by a gearbox, and then the power can be transmitted to the generator. The above components are installed on the plane of the engine room which is lifted by the tall tower. Due to the constant change of the wind direction, a windward device must be installed to utilize the wind energy effectively. After it measures the wind direction according to a wind direction sensor, the controller drives the small gear to rotate, so that the engine room is always turned to the wind.

In the past 20 years, wind power technology has made continuous breakthroughs and showed increasing benefits in economies of scale. According to the NREL, the cost of wind power has fallen by more than 90% from 1980 to 2005, much faster than other renewable sources. Germany is the leading country in the use of renewable energy. According to the statistics of its environmental, natural and nuclear security departments, the cost of wind power is significantly lower than other forms of renewable energy, close to that of traditional hydropower. The World Wind Energy Association estimates that by 2020, the worldwide wind power capacity is expected to reach 1.231 billion kW, and the annual generating capacity will be equivalent to 12% of the world's electricity demand. Therefore, for those countries which aim at building a resource-saving society, wind power generation is no longer a supplementary energy, but an emerging energy industry with the greatest prospects for commercial development.

China has large wind energy reserves with wide distribution. According to the data provided by the National Meteorological Bureau, China's available wind resources at the height of 50 m are 500 million kW, and its offshore wind resources are also over 500 million kW, far more than the available water resources of 371.1 million kW. Among different kinds of renewable energy, wind power generation is the cheapest with the most mature technology. If rationally utilized, it is expected to become the third largest power source, second only to thermal power and hydropower. At present, wind power generation is booming in the Yangtze River Delta and Pearl River Delta. The world's largest wind power plant will also be completed in 2007 in Jiangsu. Some experts predict that a worldwide upsurge of wind power generation has appeared. Wind power is expected to provide 10% of world's electricity demand by 2020 [9].

11.3.3 Geothermal Energy

Coming from the deep earth, geothermal energy is originated in the decay of melting magma and radioactive materials of the earth. Geothermal (water) resource is a very valuable renewable mineral resource, because it integrates heat, minerals and water together, and can be used in many ways. More importantly, geothermal energy shows the characteristics of clean energy and are good for human health. Therefore, it becomes increasingly popular and develops fast.

1. **Classification of geothermal resources**

   According to the mode of occurrence, geothermal resources can be classified into hydrothermal, hot-dry-rock and ground-pressure geothermal resources. According to technical and economic conditions, they can be divided into economical geothermal resources in depths shallower than 2 km, and subeconomic geothermal resources in depth of 2–5 km. According to the cause of formation, they can be classified into modern volcano type, magma type, fault type, rifted basin type and depressed basin type. According to the temperature, there are high-temperature and low-temperature geothermal resources. High-temperature ($>150\ ^\circ$C) resources are mainly distributed in the edge of the major plates, such as plate collision zones, plate cracking zones and modern rift zones. Middle- and low-temperature ($<150\ ^\circ$C) geothermal resources are distributed in the active fault belts within plates, rifted valleys and depressed basin areas. Internationally, there are three kinds of assessment for geothermal resources. The first is called "(available) resource base" which refers to the total heat accumulated 5km under the surface of the earth. This part of the heat can both be used theoretically and technically. The second is called "resources," referring to resources from the "resource base" that will have economic value in 40–50 years. The third is called "recoverable resources" which refers to resources from the "resource base" that will have economic value in 10–20 years. According to the estimate of Pomorini, the "resource base" of global geothermal energy is 1.4×10^{11} billion joules/year.

2. **Distribution and reserves of geothermal resources in the world**

   The global distribution of geothermal resources is unbalanced. Areas with obvious geothermal gradient of more than 30 $^\circ$C per kilometer can be defined as areas of geothermal anomaly. They are mainly distributed in the oceanic spreading ridges with growing or rifted plates and the subduction zones with colliding or decaying plates. There are four global geothermal zones:

(1) Geothermal zone on the Pacific Rim: On the collision boundary between the world's largest Pacific plate, the American, Eurasian and Indian plates. Many world famous geothermal fields, such as the Geiseles, Hase, and Roosevelt of the United States; Cerro and Prieto of Mexican; Macao of China Taiwan; and Matsukawa of Japan, are distributed in this zone.

(2) Mediterranean-Himalayan geothermal zone: On the collision boundary between the Eurasian plate, African plates and Indian plates. Italy's Larderello geothermal field, the first geothermal power station in the world, is located in the geothermal zone. China's Yangbajing geothermal field in Tibet and Tengchong geothermal field in Yunnan are also situated here.

(3) Geothermal zone on the mid-Atlantic ridge: On the crack of the Atlantic plate. Some geothermal fields in Iceland, such as Clafera, Namafia and the Azores, are located in this geothermal zone.

(4) Geothermal zone along the Red Sea-the Gulf of Aden-the East African Rift Valley: Geothermal fields in Djibouti, Ethiopia and Kenya are located in this zone. In addition to the crustal high-heat flow areas in the plate boundary, high-heat flow areas can also be formed near the plate boundary under a certain geological conditions. For example, in China's Jiaodong Peninsula and Liaodong Peninsula, North China Plain and Southeast coastal areas, the value of heat flow is greater than the mainland average value of 1.46 heat units, up to 1.7–2.0 heat flow units. The interior of the earth contains tremendous energy that is unimaginable. It is estimated that only 10 km within the outermost layer of the crust has an amount of energy of $1,254 \times 10^{24}$ joules, equivalent to 2,000 times the total heat produced by coal in the world. The total amount of geothermal energy is equivalent to 170 million times the total reserves of coal, which illustrates that the potential of geothermal resources is 100 times that of hydropower. In the world, the total energy amount of recoverable geothermal resources is 5×10^5 billion joules/year. Although it only accounts for a small portion among the "resource base" (whose total energy amount is 140×10^9 billion joules/year), the amount is still impressive, more than the annual global primary energy consumption (about 400 trillion joules/year).

3. **Utilization of geothermal energy [10]**

Although geothermal resources have been used for bathing for thousands of years, large-scale use of geothermal energy for power generation, heating, industry and agriculture only began in the twentieth century. In the 1970s when the world's oil crisis appeared, many countries were seeking for alternative energy, hence bringing about a boom for the development of new energy and renewable energy. However, as a member of the new energy family, geothermal energy plays a minimal role in the whole energy structure similar to the solar energy, wind energy and biological energy, except in a few countries. Compared with the solar energy, wind energy, and tidal energy, it is easy to see that geothermal energy is still the most realistic new energy that should be vigorously explored. In the 1990s, with the growing awareness of global environmental protection, there was a new upsurge in the direct use of geothermal energy in China, especially in the cold three-north (northeast China, north China and northwest China) areas of high latitude where geothermal heating (heating and living water) was greatly developed. The development not only reduced the emission of many harmful substances, but also obtained significant economic benefits. For example, in 1990, the geothermal heating was applied to a nationwide area of only 1.9 million m^2, but the number increased to 11 million m^2 in 2000. At present, some provinces such as Heilongjiang, Liaoning, Ningxia, Shandong, Hebei, Henan as well as several big cities including Beijing, Tianjin and Xi'an are actively working on the construction and promotion of demonstration projects using a variety of heating forms (including heat pump). In the west of China, Yunnan, Tibet, Xinjiang, Sichuan, Shaanxi and other provinces are working to develop geothermal tourism resources, in order to create new growth points for the local tourism industry. Coastal provinces in the southeast are also vigorously developing geothermal tourism (health care, convalescence) as well as planting and aquaculture of premium varieties. Meanwhile, they are also applying geothermal energy to refrigeration and drying projects. To sum up, the direct utilization of geothermal is developing throughout the country toward a large-scale, healthy and industrialized way. By 1999, the installed capacity of China's geothermal equipment for direct use had reached 22,112 MWt, with an annual energy output of 10,531 GWh, ranking the first in the world among 511 countries with direct use of geothermal energy.

Located in the world's two major geothermal zones, China is endowed with abundant geothermal resources that are mainly distributed in Yunnan, Tibet, western Sichuan (Himalayan geothermal zone or Yunnan and Tibet geothermal zone) and Taiwan. China ranks the first in the utilization of geothermal resources in the world. Geothermal energy is mainly applied for direct use and geothermal power generation. The geothermal fluid at different temperatures can be used as follows: 200–400 °C for direct power generation and comprehensive utilization; 150–200 °C for double-cycle power generation, refrigeration, industrial drying and industrial thermal processing; 100–150 °C for double-cycle power generation, heating, cooling, industrial drying, dehydration processing, recycling salt and canned food; 50–100 °C for heating, greenhouse, domestic hot water and industrial drying; 20–50 °C for bathing, aquaculture, raising livestock, soil warming and dehydration processing.

11.3.4 Ocean Energy

The vast ocean is not only an enormous solar collector, but also a thermal storage warehouse with very large heat capacity. The sun constantly supplies great heat to the earth. Except for the reflection and absorption of the atmosphere, there are about

110 trillion kilowatts of the solar radiation energy that can finally arrive at the earth. As for oceans that account for 71% of the total area of the earth, they can receive up to 64 trillion kilowatts of solar radiation energy. Except for part of the energy that has been consumed because of radiation on the sea surface and evaporation of seawater, a considerable part of solar energy is absorbed and stored by the ocean in different forms of conversion, therefore constituting a huge resource of marine energy [11, 12].

1. **Classification and characteristics**

   Ocean energy usually refers to the renewable natural energy in the ocean, mainly including tidal energy, wave energy, temperature difference energy, ocean current energy and salt difference energy. Tidal energy comes from the motion of planets in the solar system, while others come from the solar radiation. Ocean energy can be classified into mechanical, thermal and chemical energy by forms of storage. Tidal, current and wave energy is mechanical energy; seawater temperature difference energy is the thermal energy; and salt difference energy is chemical energy.

   Oceans have the following characteristics:

   1) **Different conversion methods and technology maturity for different kinds of ocean energy**

      Tidal power generation is similar to hydroelectric power generation. Wave power generation has various forms of devices, including mechanical, pneumatic, hydraulic and hydraulic ones. Temperature difference power generation uses the principle of thermodynamic cycle, and salt energy conversion mainly utilizes semipermeable membrane technology. The maturity of various technologies is also different. Tidal energy utilization technology has been applied in practice. Wave and current energy is nearly used in practice. The utilization of temperature difference energy is still in the experimental stage, while the salt difference energy is still in the conceptual test stage.

   2) **Large reserves, low energy density**

      Among various kinds of ocean energy, only the salt difference energy has high energy density, and the rest all have low density. For example, the great tide difference of tidal energy is 13–15 m, and usually 3–6 m in areas with relatively abundant tidal energy. The high flow rate of tidal energy is about 5 m/s, and usually 2–3 m/s. The height of large wave is 11–12 m, usually 1–3 m. The great temperature difference between ocean surface and deep seawater is 24 °C, usually about 20 °C with large reserves. According to IIASA estimates, each year 1.045 TW year/year (year/year, 1 TW year = 1,076 Mtce) of ocean energy in the world can be developed.

   3) **Energy distribution changing with space and time**

      Temperature difference energy is concentrated in low-latitude deep ocean waters. Tidal energy is distributed in coastal waters. Current energy is mainly distributed in the west side of the two oceans in the northern hemisphere. Wave energy is concentrated in the middle latitudes (30° to 40°) of the two oceans in the northern hemisphere and the Antarctic storm belt (40° to 50° S).

Salt difference energy mainly exists in coastal areas near the mouths of rivers. In terms of time, only the temperature difference energy and current energy are stable, while other kinds of energy all have significant changes. At present, accurate forecasts can be made on tidal and current changes.

4) **Harsh development environment, large investment and comprehensive benefits**

Due to dynamic effects such as wind, waves and currents, as well as problems of seawater corrosion, aquatic attachments and low energy density, the development of ocean energy requires enormous conversion devices with great corrosion resistance, complex construction technology, large investment and high costs. On the other hand, ocean power generation does not need to take up lands or relocate population. It can also bring about comprehensive utilization benefits in reclamation, breeding, marine chemical industry and tourism.

2. **Technical description and resource evaluation**

1) **Tidal energy**

Tidal energy is the potential energy generated by the daily rise and fall of the ocean water level at the coast. Tidal fluctuations are mainly caused by the gravity of the Moon (and, to a lesser extent, the Sun), acting on the ocean that rotates with the Earth. The energy of the tide is proportional to the square of the tidal range and the area of the reservoir. As long as the average tidal range is more than 3 m, there is development value. Tidal energy can be used for power generation. First, build a reservoir in the gulf. The seawater will pour into the reservoir at high tide, and go out of the reservoir when the tide falls. Then the difference between the high tide and the low tide can promote the rotation of the turbine to generate power. The construction of tidal power generating units should take into consideration many factors, such as variable power, low head, large flow and seawater corrosion, so it is far more complex than the construction of conventional hydropower station. The tidal power station can be classified into single-story and one-direction one, single-story and two-direction one and double-story two-direction one.

The theoretical amount of tidal energy in the world is 3 TW year/year, and the amount of exploitable resources is 0.44 TW year/year. In China, there are 424 tidal power dam sites, with a total installed capacity of 21.11 GW, of which 40.9% and 47.4% are distributed in Zhejiang and Fujian, respectively.

2) **Wave energy**

Wave energy is the transport and potential energy of ocean surface waves. It is directly proportional to the square of the wave height, the wave motion period, and the width of the wave surface. Wave energy is the most unstable energy in the ocean. Waves caused by typhoon can generate great power with the power density of up to thousands of watts per wave surface, while the average annual wave power is only 20–40 kW/m in the Europe North Sea with rich wave energy. In China, the average annual wave power density is 2–71 kW/m. Wave energy is mainly applied for power generation. In addition, it can also be used for pumping, heating, desalination and hydrogen production. There are many kinds of wave energy utilization devices, and basic principles used include oscillating movement of an object generated by waves, change in wave pressure, and conversion of wave energy to water potential taking advantage of the climbing waves.

At present, wave power generation technology has been used in practice. There are three types of devices: oscillating water column devices, pendulum devices and polywave reservoirs.

The amount of world's wave energy and ocean current energy that can be developed amounts to 0.05 TW year/year. In China, the theoretical amount of wave energy is 12.9 GW, of which one-third is in Taiwan; and the theoretical amount of ocean current energy is 14.0 GW, of which one-half exists in Zhejiang. In addition, coastal areas in Liaoning, Shandong, Fujian and Taiwan also have rich current energy resources. China is one of the regions with the highest power density of ocean current energy in the world. This is especially true in Jintang, Guishan and Xishaomen Waterway of Zhoushan Islands where the average power density is 20 kW/m^2, quite favorable for development.

3) **Ocean current energy**

Ocean current energy is the kinetic energy of seawater flow, mainly produced by the relatively stable currents in submarine channels and straits as well as the regular currents generated by tides. The energy of currents is directly proportional to the square of flow rate and the quantity of currents. Compared with waves, the change of ocean current energy is more stable and regular. For a waterway with a maximum flow rate of 2 m/s or more, its currents have development value. Ocean current energy is mainly used for power generation, and its principle is similar to that of wind power. Since the density of seawater is about 1,000 times that of air, and the device must be placed underwater, there are a series of technical problems in ocean current power generation, such as installation and maintenance, power transmission, corrosion protection, load and safety in marine environment, etc. Current power generating devices can be installed on the sea floor, or at the bottom of a floating body.

4) **Temperature difference energy**

Temperature difference energy is the thermal energy produced by the water temperature difference between the surface water and deep seawater (about 1,000 m). In tropical or subtropical waters, the temperature difference is above 20 °C. In addition to generating electricity, the utilization devices of ocean temperature difference energy can also obtain freshwater and deep seawater (for air conditioning, etc.), and can also be combined with the mining system in the deep-sea mining system. With the help of these devices, we can establish independent living space in the sea, supported by seawater desalination plant, marine mining, marine city and marine pasture. The conversion of temperature difference energy can be divided into two types: open circulation and closed circulation.

The theoretical resource amount of ocean temperature difference energy can reach 22 TW year/year, and the amount of resources that are developed is 1 TW year/year. China can develop about 150 GW of energy sources, and 99% of them are in the South China Sea.

5) **Salt difference energy**

Salt difference energy is the energy produced by the chemical potential different between the seawater and freshwater or between two kinds of seawater with different salt concentration. It mainly exists at the junction of river and sea. Among all ocean energy sources, salt difference energy has the greatest density. The energy density of the chemical potential difference between seawater (3.5% of salinity) and river water is equivalent to a 240 m waterhead. If a semipermeable membrane (through which water can pass while salt cannot) is placed between two kinds of seawater with different salinities, there will be a pressure gradient that forces the water to permeate from the side with lower salinity to the side with higher salinity, until the salinity of both sides is equal. This process converts the chemical potential difference between seawater of different salinity into water potential energy, and then uses the turbine to generate electricity. The theoretical estimation of the world's ocean salt difference energy is of the order of 10 TW, and the estimated value in China is 110 GW. The energy is mainly distributed in the marine outfall of major rivers. This energy also exists in some inland salt lakes in Qinghai and other places in China.

3. **Prospects for ocean energy utilization**

In the past 20 years, driven by the crisis of fossil fuel energy and the pressure of environmental change, the marine energy industry, which is one of the major renewable energy sources, has made great progress. With the support of high technology, the application technology of marine energy is becoming mature, showing a great prospect for the full use of ocean energy in the next century. The major marine countries all attach great importance to ocean energy, and invest a lot for its development and utilization. The United States has taken promoting the development of renewable energy as a cornerstone of national energy policy, with particular emphasis on the research of marine power generation technology. Experts pointed out that the twenty-first century is the century of the ocean. After 2020, the global utilization ratio of ocean energy will be hundreds of times the current ratio. Therefore, the ocean is also called the "energy source" of the future.

11.4 Renewable Energy and Sustainable Development

11.4.1 Renewable Energy

Energy has three basic characteristics, namely, mutual substitution for use, transmission and transformation, quality differences. The characteristics of renewable energy can be reflected in the following aspects [13]:
(1) Rich: It means that the source and quantity of renewable energy are both endless.
(2) Clean: It means that unlike conventional energy sources, renewable energy does not produce any pollutants during the production and utilization.
(3) Local: It means that renewable energy is local energy, which is convenient for use, but also subject to the characteristics of the region's natural environment. Therefore, the development and utilization of renewable energy must be in harmony with the natural environment of the region.
(4) Economical: It means that although construction and operating costs of renewable energy in the early stage are higher than that of conventional energy, its costs on environmental control and environmental protection are much lower. What's more, its medium- and long-term economic benefits are quite competitive.

With the development of world's economy and society, oil and gas resources have been continuously consumed, which not only brings about growing environmental problems but also provides an unprecedented opportunity to the development of renewable energy and new energy. It is estimated that in the next 10–20 years, renewable energy and new energy will lead an important role in the world's energy consumption structure.

Renewable energy and new energy developed so fast in the United States and some European developed countries that it has become an important part of the energy structure. In China, the development of renewable energy and new energy began a little later, but it has remained an average annual growth rate of more than 25% in recent years. From the perspective of technical and market potential, wind energy, solar energy, hydrogen and fuel cells will be the key fields for the development of renewable energy and new energy.

Due to the energy shortage, the renewable energy industrialization has become a major initiative for China to meet the needs of national energy development and seek for sustainable development. In order to speed up the process of renewable energy and new energy industrialization, China should strengthen industry supervision and provide policy supports. Relevant departments should also strengthen support for projects. Enterprises should combine production with research, focus on independent technological innovation and accelerate the process from research and development to demonstration, then to the industrialization, in order to develop renewable energy and new energy by leaps and bounds.

11.4.2 Research of Sustainable Energy Strategy

Energy industry is not only a basic industry of the national economy but also a technology-intensive industry. The energy system can effectively support China's economic, environmental and social sustainable development; at the same time it can also achieve its own sustainable development in a safe, economic, efficient, clean, low-carbon and fair way. The characteristics of modern energy technology can be defined as "safe, efficient, and low carbon," which is also a main direction for the future development of energy technology. Energy sustainable development is the prerequisite and important content of China's sustainable development. In the "12th Five-Year Plan" of China's energy, the overall idea has been clearly put forward as "the current energy system transiting to the modern system of sustainable development" [14].

Sustainable development refers to the development that can continuously meet the needs of the present and future generations for material, energy and information in production and life. The key lies in satisfying the needs of contemporary people but not harming the production and life of future generations. The issue of sustainable development has always been one of the most central issues in the world in the twenty-first century. It is directly related to the continuation of human civilization and becomes an indispensable basic element for direct participation in the country's highest decision making. To better reflect the requirements of sustainable development, we must stick to the fair and reasonable principles between generations; namely, the development of contemporary people cannot sacrifice the development of future generations. The principles of common prosperity, complementarity and equality should be implemented among different regions, in order to narrow the gap between people of the same generation. We should create external conditions suitable for the "natural-social-economic" support system, and constantly optimize the organization and operation mechanism of the system.

The two basic themes of sustainable development are man and nature, man and man. Its three most obvious connotations are development level, coordination degree and sustainability. On the one hand, sustainable development becomes a national or global strategic choice; and on the other hand, it becomes a standard for measuring whether the regional development is healthy or not.

China's sustainable energy strategy should consider at least two aspects. The first is how to ensure the reasonable and sustainable energy supply and efficient use of energy, and the second is how to solve environmental problems related to energy processes.

According to China's energy supply and demand, the construction of energy development strategy should be based on the principle of "improving efficiency, protecting the environment, ensuring supply, and promoting sustainable development."

1. **Put priority on energy efficiency and promote sustainable development**
   We should improve the development and utilization efficiency of energy, and put it in the first place in the energy development strategy. According to statistics, the utilization efficiency of China's comprehensive energy is about 33%, 10 percentage

points lower than that in developed countries; while the energy consumption per unit of output value is more than twice the world average. To fundamentally solve the energy problems in China, we cannot rely solely on energy construction; rather, we should save energy, improve energy efficiency, change the mode of economic growth, take a new road to industrialization and choose a development mode which is centered on resource conservation, quality and efficiency, and led by technology. In addition, we should vigorously adjust the industrial structure, product structure, technical structure and organizational structure, rely on technological innovation, institutional innovation and management innovation, and create an energy-saving mode of production and consumption in the whole country, in order to develop energy-saving economy and build an energy-saving society.

2. **Accelerate the development of natural gas and new energy, and strive to transform the energy structure [15]**

For a long time, the energy structure in China has always been based on coal, which caused low efficiency and environmental pollution. In recent years, due to changes in the demand structure and total amount of terminal energy, as well as the environmental requirements led by some central cities, it has become an important trend for energy development to optimize the energy structure. However, with the rapid increase of oil import as the sharp rise in international oil price, the security of energy supply has drawn wide attention. The development and utilization of natural gas requires large-scale infrastructure for long-distance transportation, which will increase the costs and price of natural gas. Therefore, people have intensely discussed and worried about future energy costs. There are still uncertainties about the optimization of energy structure.

In nowadays and the decades to come, oil and gas will continue to be two major sources of energy worldwide. In particular, the development of natural gas is still in the ascendant. The use of natural gas has a good environmental effect. What's more, the energy technology based on natural gas is also and will be the most energy-efficient technology. During the period of "10th Five-Year Plan," the "West-to-East Gas Transmission Project" is of great significance. Once the natural gas pipeline network is completed, the development of natural gas will be promoted, and the actual cost of natural gas will be significantly reduced. In the development and utilization of new and renewable energies, renewable energies including hydropower, solar energy, wind energy, geothermal energy, ocean energy and bioenergy have already accounted for about 22% of the world's energy consumption. China has rapidly become a great power in the global new energy industry, and has achieved breakthrough in the technologies, markets and service system of renewable energy. However, China still needs to learn the technology and experience from developed countries, and explore new energy storage technology, improve the economic efficiency, in order to lay a foundation for the comprehensive development and utilization of new energy. The development of natural gas requires national support and coordination.

3. **Combine economic and environmental benefits by implementing the clean utilization of coal [16, 17]**
    Optimizing the energy structure does not conflict with making full and rational use of China's coal resources. In the process of energy structure optimization, coal will undoubtedly withdraw from some areas, but its position as an important type of energy in China will still remain. At resent, use of technology and methods for coal in China are still very far away from the target for sustainable development in society and economy. In the energy strategy of sustainable development, the use of coal should first tackle problems related to environmental pollution.

    China has long been relying on the development and utilization of mineral energy, especially coal energy, which had caused great damages to the environment. With the environmental pollution getting more and more serious, the air pollution expanding from indoor to urban areas, and the acid rain pollution getting aggravated, more than one third of China's crops and forests are affected by the acid rain. Other problems also include damages to the ecosystem, soil erosion, desertification and grassland degradation, biodiversity reduction and the increase in respiratory diseases, which cause a serious threat to people's health. The development and utilization of new energy and renewable energy will effectively mitigate ecological and environmental damages and guarantee people's health. Therefore, it can achieve a win-win result between energy consumption and environmental protection, and realize the sustainable development of energy and environmental protection.

4. **Pay attention to environmental protection and implement energy-sustainable development**
    Environmental protection is a basis for sustainable development, also a fundamental driving force to promote the development of energy technology.

    At present, in developed countries, the requirement for environmental protection has become an important factor in determining energy structure and costs. In the future, the environmental protection will also become an important factor for deciding the energy structure in China. The clean energy structure will also play a significant role in promoting energy efficiency. In order to realize the strategy for sustainable development of energy, the energy development should take full consideration of environmental protection in all aspects. Energy infrastructure is a hug project with a long utilization period. Once the energy system is built, it not only costs a lot for reconstruction but also requires a long period. Therefore, we should take into account the long-term effect of environmental protection in energy construction.

    The development and utilization of renewable energy is conducive to developing circular economy, and further achieving sustainable development with economic, social and environmental benefits. First of all, the development and utilization of renewable energy can effectively protect the environment and reduce production costs. Through resource recycling, pollutants can be reduced and even eliminated; utilization rate of resources can be enhanced; and the ecological environment can be preserved. What's more, by promoting renewable energy, the field of employment can be expanded, and the new renewable energy industry can be developed.

In the "11th Five-Year Plan," China has put priority on major technology development in key areas such as energy and resources. It has formulated a long-term energy development plan, which not only attached great importance to the optimization of energy structure as well as the development of nuclear power and renewable energy, but also emphasized the energy-saving and efficiency-based principles. The "12th Five-Year Plan on National Energy Science and Technology (2011–2015)" (referred to as "the Plan") analyzed the development situation of energy technology. It pointed out that we should plan for the research, development and application of new technology by transforming the method of energy development and enhancing independent innovation. We should break through the constraints of limited energy and resources using unlimited scientific and technological strength. By promoting the development, transformation and utilization of energy and resources, and making full use of renewable energy technologies, we can innovate on energy production and utilization. According to the similarity and correlation of technologies in the energy production and supply chain, the plan divided four key technical fields: exploration and mining technology, processing and transformation technology, power generation, transmission and distribution technology, and new energy technology. In the planning and practices of key technical areas, it also adhered to the principle of "efficiency as a priority." Since January 1, 2006, China has formally implemented the "Renewable Energy Law" in which "renewable energy" is defined as nonfossil energy sources such as hydropower, wind, solar, geothermal, ocean energy and biomass. The significance of this law is to promote the development and utilization of renewable energy, and solve a series of problems in this process, such as difficulties in power generation, high Internet price as well as policy and market barriers [18].

In order to promote the sustainable and stable development of renewable energy and related industries, while implementing the "Renewable Energy Law" that has been formulated and enacted, it is also necessary to establish a series of laws and regulations with actual content and effects to promote the development and utilization of renewable energy and environmental protection.

(1) Establish an economic incentive system for the development and utilization of renewable energy. In order to strengthen policy support for the development of renewable energy and related industries, we should provide support to relative aspects including technical policy, industrial policy, tax policy, credit policy, personnel policy and market access. In addition, we should formulate necessary rules and regulations on economic incentive to encourage the development of renewable energy technologies and industries from the aspects of finance, taxation and credit.

(2) Set standards and quality management system of renewable energy technologies and products, and ensure the legal effect of these standards.
(3) Establish fund for the development of renewable energy technology to encourage scientific and technological innovation, and promote the development of renewable energy industries.
(4) Strengthen the legal system concerning the development and utilization of renewable energy and environmental protection. Establish the green energy marking system.

By establishing and improving a legal responsibility system for the green mark certification and management, we can promote the development and utilization of renewable energy as well as environmental protection, and further continues the sustainable development.

Questions

1. What is renewable energy? Which are its main features?
2. Please describes the formation of the greenhouse effect and its harm.
3. How many coal processing technologies are there? What is the significance of their application?
4. Please make a brief description of the coal gasification technology.
5. What are the advantages and disadvantages of biomass energy? What is the significance of research on it?
6. How many ways of biomass energy utilization are there?
7. Please name several forms for clean energy development and utilization, and give examples.
8. What is the classification of geothermal resources? What are the forms for geothermal resources utilization?
9. How to promote sustainable development from the aspect of energy use?

References

[1] Sun Cheng-gong, Xie Ya-xiong, Li Bao-qing, et al. Effects of structural characteristics of dispersant molecular on rheological properties of coal slurry. Journal of Fuel Chemistry and Technology, 1997, 25(3): 213–217.
[2] Cheng Jing-yan, Liu Ling, Liu Yan-hua. Study on a new type of stabilizer for coal water slurry. Coal Processing and Comprehensive Utilization, 1999 (3): 12–16.
[3] Robertson, A., Bonk, D. Effects of pressure on the second-generation P FBC-CC power plant. Engineering for Gas Turbines and Power, Transaction of ASME, 1994, 116(4): 345–351.
[4] Zhou Shan-yuan. New energy in the 21st century: Biomass energy. Jiangxi Energy, 2001 (4): 34–37.
[5] Wu Chuang-zhi, Ma Long-long. Modern Utilization of Biomass Energy Technology. Beijing: Chemical Industry Press, 2005.

[6]   Ma Long-long, Wu Chuang-zhi, Sun Li. Biomass Gasification Technology and Its Application. Beijing: Chemical Industry Press, 2003.
[7]   Liu Quan-gen. Development and utilization of world ocean energy. Energy Engineering, 1999 (2): 5–9.
[8]   Feng Zhi-ming. Production of Resources Science. Beijing: Science Press, 2004.
[9]   Yao Fei, Zhao Zhong-hua, Yu Zhu-feng. Analysis on the status of clean energy in China. Group Economics Research, 2007 (3): 69.
[10] Feng Xiao-mei, Zou Yu, Sun Zong-yu. Application of geothermal cascade utilization technology in energy-saving and land-saving residential heating project. China Housing Facilities, 2005 (6): 121–122.
[11] Wang Ya-min, He Dan. Sustainable development for marine resources. WTO Economic Guide, 2013 (11): 58–59.
[12] Mi Tie, Tang Ru-jiang. Comparison of biomass gasification technology and its gasification and power generation technology. New Energy and Technology, 2004 (5): 35–37.
[13] Yang Jin-huan, Ge Liang, Tan Bei-yue, et al. Application of solar photovoltaic power generation. Shanghai Electric Power, 2006 (4): 355–356.
[14] Lin Bo-qiang. Study on energy strategy adjustment and energy policy optimization in China. Power System & Clean Energy, 2012 (1): 1–3.
[15] Zhao Zheng, Zhang Liang-liang. How to realize the transformation of China's Energy Strategy in the New Situation. Economic Survey, 2013, 3: 011.
[16] Zhou Hong-chun, Wu Ping. Chinese energy strategy in the background of low Carbon. Theoretical Reference, 2013 (1): 15–18.
[17] Luo Zuo-xian. Coal chemical industry should become an important part in the energy strategy. Sinopec, 2013 (6): 9–11.
[18] Hui Feng. China's renewable energy development forum was held in Beijing. Renewable Energy, 2006 (1): 25.

12 Circular economy and eco-industrial parks

12.1 The Theoretical Basis of Eco-industry

12.1.1 Concept and Connotation of Eco-industry

Eco-industry refers to an integrated industrial production system with multiple levels, structures and purposes. It is established and developed with the application of modern science and technology based on principles of ecology and ecological economics. It turns industrial waste into raw materials, therefore achieving circular production and intensive management. Eco-industry differs from traditional industry mainly in that eco-industry concerns about ecology in the process of industrial production, and it regards the optimization of ecological environment as a standard for the quality of industrial development. Compared with traditional industrial production, eco-industry has the following characteristics:

(1) Instead of pursuing profit only, eco-industry strives for the unity of economy and ecology in industrial production and its resource utilization. In addition, it changes the mode of industrial production and management from uneconomically external one to the unity of both external and internal efficiency.

(2) In terms of industrial design, eco-industry attaches great importance to turning waste into resources, products, heat and energy. By doing so, it builds an industrial system of multi-level closing cycle with no waste or pollution.

(3) Eco-industry calls for bringing ecological protection into the decisive factors for industrial production and management. It pays attention to the study of environmental measures for industry and requires the production and management of modern industry to be strictly in line with the principles of eco-economy. Based on these principles, it plans, organizes and manages the production and living in the industrial parks.

(4) Eco-industry is an industrial production mode with low input, low consumption, high yield, high quality and high efficiency. It coordinates the ecological development and the economic growth [1].

12.1.2 Dual Nature of Traditional Industry

Industry has a dual nature: one is the social attribute, the other is the natural attribute. The former refers to the attribute of value increment during industrial production process, that is, gaining economic benefits by processing raw materials into products and then realizing the value of products. The natural attribute of industry is that the raw materials and energy needed during industrial production process are all from natural ecological environment and the "three wastes" generated in the

https://doi.org/10.1515/9783110479317-012

industrial production usually return to natural ecological environment. Industrial production relies on natural ecological environment. In the past, the social attribute of industry (namely the attribute of economic increment) was valued more. Therefore, people pursued only economic benefits while ignoring the fundamental natural attribute of industry, seeing nature as a supplier with inexhaustible resources and a trash to discharge the industrial "three wastes." As human beings were blindly intoxicated by the huge material achievements brought by the industrial revolution, it never dawned on them that this was also devastating. Although this industrial production method, which did not account for the environmental cost, helped human beings accumulate material wealth and develop by leaps and bounds, it would also become a weapon to stop human progress and destroy human social civilization. More specifically, the existence of industry is determined by the natural attribute of industry. Without raw materials, industrial production will be making bricks without straw. Eco-industry awakens human beings and leads them to reconsider the nature of industry and treat dialectically its two attributes. The natural attribute of industry is the prerequisite for the social attribute, and only the combination of these two attributes can guarantee a sustainable development of the industrial economy [2].

12.1.3 Industrial Eco-economic System

Like the eco-economic system in agriculture, there is also a similar system in industry, also called the industrial eco-economic system. It is the major environment which the eco-industry takes roots in and relies on. Unlike other types of eco-economic system, the industrial eco-economic system is a complex system composed of the socio-economic system and the ecosystem. In general, it consists of four fundamental parts as follows:
(1) Producer, including mental and physical labors as well as industrial microbiology;
(2) Technologies, including production equipment, tools and science and technology;
(3) Raw materials, including minerals, agricultural products, forest products, livestock products, fishery products and other raw materials;
(4) Natural conditions, including light, heat, water, air, land and other natural resources.

The socio-economic system continuously inputs social resources such as labor force, science and technology and demand information to the industrial eco-economic system.

Meanwhile, the ecosystem continuously supplies mineral products and biological products. After the initial processing, secondary processing and multiple processing by producers, a variety of products are produced to meet consumer demand. The ecosystem also supplies mineral products and biological products, which are processed by the producers for the first time, second time and multiple times, and then produces various kinds of products to meet human demand for consumption.

12.1.4 Theoretical Basis of Eco-industry

1. **Principle of chain-wise mutual restriction between resources**
   There are various resources existing in the industrial ecosystem. They are interdependent and mutually restrained through an ecological process similar to the biological food chain. For example, coal is used to generate power while the waste residue and fly ash are used as raw materials to build road and manufacture cement; grain is the raw material to make wine while vinasse can be used to generate biogas. In this way, the traditional production mode of "raw materials–products–waste" is changed into a sustainable production mode of "raw materials–products–waste–raw materials." The ecological process helps to extend the resource processing chain, maximize the development and utilization of resources, increase the resource value, protect the ecological environment and make full use of industrial products in the whole process [3].

2. **Multi-level utilization of energy and material recycling principle**
   In the industrial eco-economic system, there is a processing chain, also called an industrial ecological chain which is similar to the food chain in the natural ecosystem. It is not only an energy conversion chain but also a material delivery one. From the perspective of economic value, it is a value-added chain. In the industrial eco-economic system, there is also an ongoing material cycle with material and energy flowing along the industrial ecological chain level by level. Raw materials, energy, "three wastes" and various environmental factors form a three-dimensional structure where energy and resources are recycled and used to their maximum, realizing the regenerative multiplication of wasted resources. For example, a sugar refinery uses sugar cane as a raw material to produce all kinds of sugar, also output waste such as bagasse, sugarcane residue and remnant honey. Through multi-level processing and utilization of waste, bagasse is used for producing fiber panels, decorative panels and particle board; sugarcane residue for sugar cane wax; and remnant honey for alcohol. Matter and energy can be fully used in this processing chain, which not only improves the utilization of resources and energy, but also effectively enhances the purification rate and conversion rate of waste. Furthermore, it reduces industrial production cost, and achieves value increment and good eco-economic benefits.

3. **Principle of sustainable development**
   Industrial production is an economic activity by which human beings realize value of resources. It is aimed to increase the output and economic income, and improve human welfare. Since the industrial revolution, industrial production has experienced a stage when economic growth was the sole purpose, and then

a dilemma of economic benefit and ecological benefit. Nowadays, it enters a new period when there was a strong call for taking both economic and ecological benefits into account. People have completely been aware that the industrial production mode regardless of ecological damages is undoubtedly a "self-destruction." Although it seems that contemporary people need to increase their welfare through the rapid growth of industrial economy, when the long-term interests of future generations are taken into account, mankind must choose a sustainable development mode which the ecosystem can endure, that is, a mode which "can meet the present generation's needs without compromising the ability of future generations to meet their own needs for development." When it comes to industry, this principle means that as the output and income of industrial economy is increased, resources should also be rationally allocated, industrial waste be reasonably developed and utilized, and the industrial structure be properly adjusted. Green technology and green management can be used to balance the relationship between industrial ecology and industrial economy, in order to establish a coordinated pattern for industrial economic development.

12.2 Circular Economy

12.2.1 Background of Circular Economy

1. **Circular economy is a revolution to the traditional linear economy**
   Circular economy is the abbreviation of closing materials cycle economy. Since the 1990s, at the call of implementing sustainable development strategy, scholars and governments have gradually reached a consensus that the root cause of the increasingly serious environmental problems is the linear economic model formed after industrialization movement and characterized by high exploitation, low utilization and high emission (so-called two highs and one low). Therefore, they have proposed that in the future, people should establish an economy characterized by closing materials cycle (i.e., circular economy) to achieve both environmental and economic benefits as required by the sustainable development. In other words, people should promote economic growth while preventing resources and environment from degradation [4].

   From the viewpoint of material flows and forms, the traditional industrial economy is a unidirectional linear economy of "resources–products–emission." In the linear economy, people extract materials and energy from the earth, and then discharge a large number of pollutants and wastes into air, water, soil and vegetation, which are regarded as "gutter holes" or "rubbish bins" of the earth. It is through such continuous activities at the expense of natural resource consumption that linear economy achieves quantity-oriented economic growth. In contrast, circular economy advocates developing economy in harmony with the earth. It requires economic activities to be organized as a feedback process of "resources–products–renewable resources," so that all materials and energy shall be reasonably and sustainably used in this ongoing economic cycle. As a result, the impact of economic activities on the natural environment can be reduced to minimum.

Circular economy is an eco-economy in nature. It suggests that the economic activities of human society shall be guided by principles of ecology rather than mechanistic laws. The fundamental difference between circular economy and linear economy is that linear economy is merely a system with multiple material flows which do not interact with each other. Since the material flowing into and out of the system is far greater than the internal interactive material flows, the economic activities will be characterized by "high exploitation, low utilization and high emission." In contrast, circular economy suggests that materials shall be exchanged within the system in an interconnected way, so that materials and energy flowing into the system can be made full use of, and the economic activities can be changed to something with "low exploitation, high utilization and low emission." An ideal circular economy system usually includes four types of actors: resource miners, processors (manufacturers), consumers and waste handlers. Different types of actors have feedback-type and network-like interconnections, so their materials flowing to one another within the system are far greater than the flows in and out of the system. Circular economy can provide an overall idea to optimize the relationship between various components of human economic system, and a strategic paradigm to turn the traditional economy to an economy with sustainable development. Therefore, it can fundamentally eliminate the long-standing conflict between environment and development.

2. **The emergence and development of circular economy**

The idea of circular economy dates back to the era when the ideological trend of environmental protection began. In the 1960s, the American economist Boulding put forward his theory of "Spaceship Earth" which can be seen as an early representative of circular economy. Since the early stage of environmental movement, Boulding had keenly realized that we must explore the root cause of environmental problems in the economic process. According to him, the earth was like a well-provisioned ship sailing through the space (the Apollo Project was being implemented then), and it survived on its own limited resources. If human beings kept exploiting its resources and damaging the environment for economic activities, the earth will be destroyed just like a spaceship once it cannot bear the consumption anymore. Therefore, the spaceship economy requires a new circular economy to replace the old linear economy. Even today, Boulding's theory of "Spaceship Earth" is still quite advanced, for it means that the economic activities of human society should change from the mechanistic laws characterized by linearity to principles of ecology characterized by feedback.

However, in the 1970s when the international community started organizing environmental improvement campaigns, the idea of circular economy was still more like an advanced concept of forerunners, and the majority did not actively develop economy following this idea. At that time, the biggest concern of the whole world was how to control pollutants to reduce their impact, namely the so-called end-of-pipe treatment for environmental protection. In the 1980s, people started to pay attention to a resource-oriented approach to deal with waste, and there was an alteration in both their thoughts and policies. However, most countries still lacked insights and initiatives toward the fundamental question: whether the generation of pollution was reasonable and whether it should be prevented at the beginning of production and consumption? To sum up, in the 1970s–1980s, environmental movements mainly focused on the ecological consequences of economic activities instead of analyzing the economic mechanism. It was not until the 1990s, especially the years when the sustainable development became a world trend, did the idea of "prevention at the source and control in the whole process" replace the end-of-pipe treatment and become the mainstream of national policy on environment and development. A systemic circular economy strategy was finally likely to be formed. The important prerequisite for the ideological leap in the 1990s was that people realized the limits of end-of-pipe treatment as follows:

(1) It is a passive approach to treat pollution after the problem occurred, so it cannot fundamentally avoid pollution.
(2) Its cost will increase as the pollutants decrease, so it offsets the benefits of economic growth to a certain extent.
(3) It forms environmental markets which generate false and vicious economic benefits.
(4) It tends to strengthen rather than weaken the existing technology systems, which may sacrifice real technology innovations.
(5) It keeps enterprises content with complying with environmental legislation rather than investing on production modes which will generate less pollution.
(6) Instead of providing a comprehensive insight, it causes misunderstanding between environment and development, as well as between various fields concerning environmental governance.
(7) It prevents developing countries from directly accessing to more modern economic modes and makes them rely more on developed countries in the aspect of environmental governance.

3. **Three support for circular economy**

Under the above background, circular economy began to thrive from the 1990s onward. People presented a series of terms such as "zero-emission factory," "product lifecycle" and "design for environment" which embodied the concept of circular economy. Three key ideas of closing materials cycle were also formed at the three important levels of economic activities.

(1) Concept and practice of ecological efficiency

In 1992, the World Business Council for Sustainable Development (WBCSD) brought up the new concept of ecological efficiency in the report *Changing Pace* in Rio conference. In its essence, this idea requires that the materials and energy are recycled to minimize pollution. The WBCSD pointed out that enterprises' caring about ecological efficiency should have the following practices:

(1) Reduce the amount of materials used for production and services;
(2) Reduce the amount of energy used for production and services;
(3) Reduce the emission of toxic substances;
(4) Strengthen material recycling capacity
(5) Maximize the sustainable use of renewable resources;
(6) Improve product durability;
(7) Improve the service intensity of products and services.

The WBCSD is an organization consisting of 120 world renowned enterprises, with members coming from more than 33 countries and over 20 major industrial sectors. With shared idea of ecological efficiency, these members have effectively promoted the practice of circular economy at the enterprise level.

(2) Concept and practice of industrial ecology

It was believed that the concept of industrial ecology was related to Robert A. Froseh and Nicholas Gallopoulos who served in the research department of General Motors Corporation [5]. In 1989, they published an article in *Scientific American*, titled "Strategies for Manufacturing," in which they put forward a new concept of eco-industrial park (EIP). They called for building input/output relationship of waste between enterprises, which means organizing circulation of materials and energy at the level of enterprise symbiosis. From 1993 onward, countries began to construct eco-industrial parks. In the United States, the President's Commission on Sustainable Development (PCSD) set up a task force of eco-industrial parks. By 1997, there had been about 15 eco-industrial parks spread across the United States. In addition to Kalundborg in Denmark, similar plans had been promoted in many other countries, including Canada (Halifax), the Netherlands (Rotterdam) and Austria (Graz).

(3) Reuse and recycle of domestic waste received more attention

Since the 1990s, with Germany as a leader, the focus of domestic waste treatment in developed countries has shifted from harmless disposal to resource reduction. This actually requires materials and energy to be recycled within a broader community and during/after consumption. In 1991, in accordance with the idea of circular economy, Germany passed *German Packaging Ordinance* which requires that manufacturers and retailers should first control the generation of packages and second recycle them to significantly reduce the amount of packaging waste to be buried and burned. In 1996, Germany enacted *Closed Substance Cycle and Waste Management Act* which is more systematic and extends the idea of closing materials cycle from packaging to all domestic waste. Since the 1990s, Germany's philosophy of dealing with domestic waste has exerted a great impact on the world. Member countries of the EU, the United States, Japan, Australia, Canada and other countries had successively reformulated regulations on national waste management according to the ideas of closing materials cycle and avoiding the generation of waste. In 1995, the Worldwatch Institute published an important article titled "Creating a Sustainable Materials Economy" in *State of the World*. The article stated that in the twenty-first century, the world economy should be based on the reuse and recycle of renewable resources.

12.2.2 Basic Principles of Circular Economy

Circular economy is an optimized mode for the international community to promote the sustainable development strategy. What is more, it is also a new economic development mode which realizes the recycle of materials and energy in the process of economic development in accordance with the laws of ecology and economics as well as the "3R" principle (i.e., reduce, reuse and recycle). In the closed-loop recycle, it shows the features of low pollution and low emission, even zero emission. It integrates clean production, comprehensive utilization of resources, ecological design and sustainable consumption. Hence, it is an eco-economy.

As a new mode of production, circular economy is also a new paradigm of technological economy when the ecological environment becomes a restriction factor for economic growth and a good ecological environment becomes a public asset. It is a new economic form which is built on the basis that the human living conditions and welfare are equal, and it is intended to maximize the welfare of all members in the society. The closing materials cycle of "resource consumption–products–renewable resources" and the reduction, reuse and recycle of resources are merely representations of the technological economic paradigm. In nature, the paradigm is to adjust the relations of production and to pursue sustainable development.

Circular economy is a way of thinking which adapts to the economic reorganization and transformation brought by sustainable development strategy. In order to achieve the "triple win" result among economy, environment and society, circular economy suggests that the economical operation should follow the "3R" principle.

Circular economy is an advanced economic formation characterized by low investment, high utilization and low emission. As interpreted by the United Nations Environment Programme (UNEP), sustainable consumption means "the use of services and related products, which respond to basic needs and bring a better quality of life while minimizing the use of natural resources and toxic materials as well as emissions of waste and pollutants over the life cycle of the service or product so as not to jeopardize the needs of future generations." In the opinion of Wu Jisong, a renowned scholar in China, "circular economy is a concept in the comprehensive system of human, natural resources as well as science and technology. It involves the whole process of resource input, enterprise production, product consumption and waste. Through achieving the reduction, reuse and recycle of waste, it continuously improves the efficiency of resource utilization, and changes the traditional economic development which relies on resource consumption to a recycling economic development which relies on eco-type resources."

In terms of the connotation, circular economy and sustainable consumption are the same in nature. Both are based on the premise of not compromising the survival condition of future generations, both observe the basic guideline of reducing resource consumption and negative effects on the environment during the economic development and both aim at improving people's quality of life.

In terms of the principles, sustainable consumption is the intrinsic requirement for developing circular economy. This is because the implementation of each principle (i.e., reduce, reuse and recycle) of circular economy is based on sustainable consumption. Therefore, the development of circular economy must take sustainable consumption as a premise.

(1) The principle of reduce requires us to reduce not only the amount of materials entering into the production process, but also the amount of materials entering into the consumption process. In the field of consumption, this principle requires for economical and moderate consumption. Economical consumption means to reduce consumption of energy, water, electricity, materials, land, etc.; moderate consumption means that the consumption shall not exceed either the production level of consumer goods or the capacity of resource and environment. It is a state of consumption which adapts to the national situation and strength, level of productivity development and natural resources endowment. In the view of the relationship between production and demand, reduction in the field of consumption is more important for developing circular economy. This is because consumption is the driving force and the ultimate goal of production; in other words, production is oriented to consumers' demand. If the demand could not be reduced, it is impossible to reduce investment on material products from the source.

(2) The principle of reuse asks for extending the use of products and services, and reducing the generation of waste during production and consumption, in order to prevent resources and goods from becoming trash too soon. It is not only an objective requirement on the production behavior, but also an objective requirement on the consumer behavior. From the perspective of consumption, this principle requires that people use what they buy many times and in as many ways as possible, and reduce the use of disposable products. By reuse, people can prevent the goods from becoming trash too early and extend the time limits for the use of products and services.

(3) The principle of recycle means that the products can become available resources again after being used. Its purpose is to reduce the final treatment capacity of waste and to relieve the stress of harmless treatment of waste. This principle requires consumers to change their consumption behaviors and habits, asking them to sort and recycle waste and achieve the recycling of resources.

12.2.3 Typical Examples of Circular Economy

Many countries including Denmark, the United States, Germany and Japan have conducted eco-industrial practices since long before, and they have gained a wealth of experience. In China, the eco-industrial parks have also rapidly developed, thanks to the promotion and supports of relevant state departments. According to incomplete statistics, there are more than 10 eco-industrial parks of various types under construction.

1. **The mode of Kalundborg EID**
   Kalundborg EID in Denmark enjoys a high reputation in the world of international environmental protection. As the first EID and currently the most successful one in the world, it is considered as a "holy land" of circular economy. Kalundborg is a small industrial city in Denmark with only 20,000 residents. Located in the beach of the North Sea, about 100 km away from the west of Copenhagen, it is a tourist attraction with beautiful scenery.

   Kalundborg eco-industrial system (as shown in Figure 12.1) is gradually formed on a commercial basis. It is a spontaneous process within which all enterprises gain benefits by using the "waste" from each other. After many years of continuous development and optimization, at present the system has become a compound ecosystem consisting of power plants, oil refineries, biotechnology products factories, plastic board factories, sulfuric acid plants, cement plants, farming, animal husbandry, horticulture and heating system of Kalundborg town. Each unit (enterprise) in the system makes use of each other's waste heat, purified waste water and gas, and other by-products such as sulfur and calcium sulfide to minimize waste generated in the whole town, thus to reduce the cost of production by mutual cooperation and to obtain direct economic benefits.

   The success of Kalundborg eco-industrial system is based on the existing trust relationships and full exchange of information among partners. This cooperation mode is not subjected to government interference; rather, exchanges and trade among the factories rely on private negotiation and consultation. Some of these partnerships are based on economic interests, while others are based on shared infrastructure. Of course, in some cases, constraints of the environmental management system also stimulate the reuse of waste and ultimately contribute to the possibility of cooperation among the parties. Through cooperation among enterprises on by-products, raw materials and so

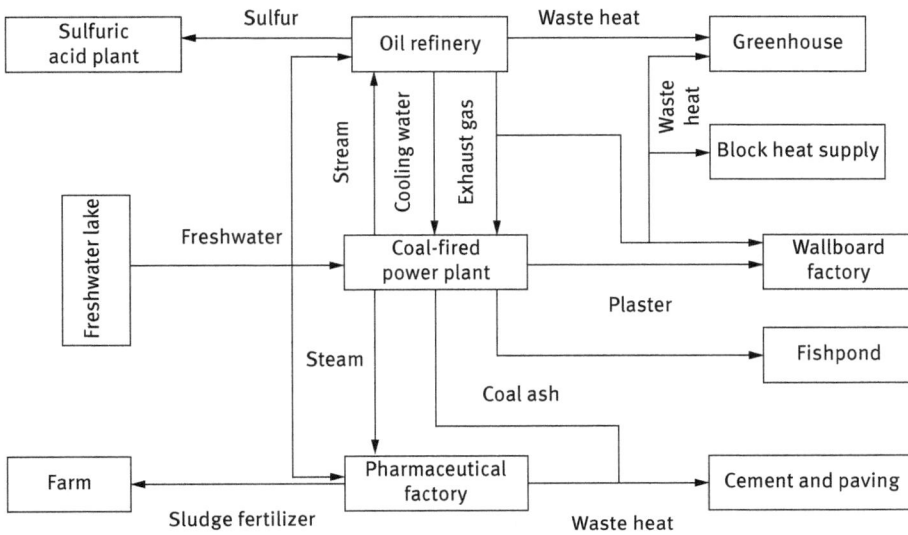

Figure 12.1: Kalundborg eco-industrial system.

on, Kalundborg has gained significant environmental and economic benefits, and established a virtuous circle for economic development and resources and environment.

2. **Circular economy of domestic chlor-alkali industry: mode of fine chemical industry**

Based on chlor-alkali installations, we should construct large-scale chlorine products, and then produce downstream products which consume hydrogen and caustic soda [6, 7]. At present, Sinopec Nanjing Chemical Industry Company Ltd and Yangnong Chemical Corporation are two companies with relatively ideal implementation of this model.

Relying on the 5×10^4 t/a diaphragm caustic soda unit, the chemical plant of Sinopec Nanjing Chemical Industry Company Ltd built an 8×10^4 t/a chlorobenzene device and a 1.2×10^5 t/a Nitrochlorobenzene device (these two sets are world-class production facilities). It built a 4×10^4 t/a aniline device using hydrogen to composite *N*-Phenyl-*p*-phenylenediamine by aniline and *p*-nitrochlorobenzene, and then produce *p*-phenylene diamine series rubber antioxidant using *N*-Phenyl-*p*-phenylenediamine as a raw material. It used aniline to produce cyclohexylamine, and then used aniline and cyclohexylamine as raw materials to produce a series of accelerator and antioxidant. It used alkali and synthetic hydrochloric acid to produce dihydroxybenzene, which is the main raw material of rubber adhesives. It used nitrochlorobenzene to produce 3, 3′-Dichloro aniline and *o*-Phenylenediamine, *p*-Phenylenediamine, *p*-phenol and 2, 4-DINITRO-chlorobenzene, etc. The products chain of the chemical plant is shown in Figure 12.2.

It can be seen that hydrogen, chlorine, caustic soda and downstream products are everywhere in the chain. Many intermediate products interact with each other and generate various fine chemicals, mainly rubber additives. The whole product chain consumes almost all chlorine, hydrogen and part of caustic soda, and the production of aniline, 3, 3'-two chlorine joint aniline, N-Phenyl-p-phenylenediamine and o-Phenylenediamine uses a clean process with hydrogen-added catalytic. What is more, rubber antioxidants RD, 4010NA, 4020 and promotion agents CBS, NS, DZ are all mainstream and ecological friendly rubber chemicals, which are in line with the international development trend of rubber auxiliaries. The company is testing and modifying the adiabatic nitration of p-nitrochlorobenzene and nitrobenzene devices. It applies resin adsorption technology to the wastewater treatment of chlorinated benzene and p-nitrochlorobenzene to recycle resources and reduce pollution. In other words, the company is running well in the mode of circular economy and gains remarkable economic benefits [8–10].

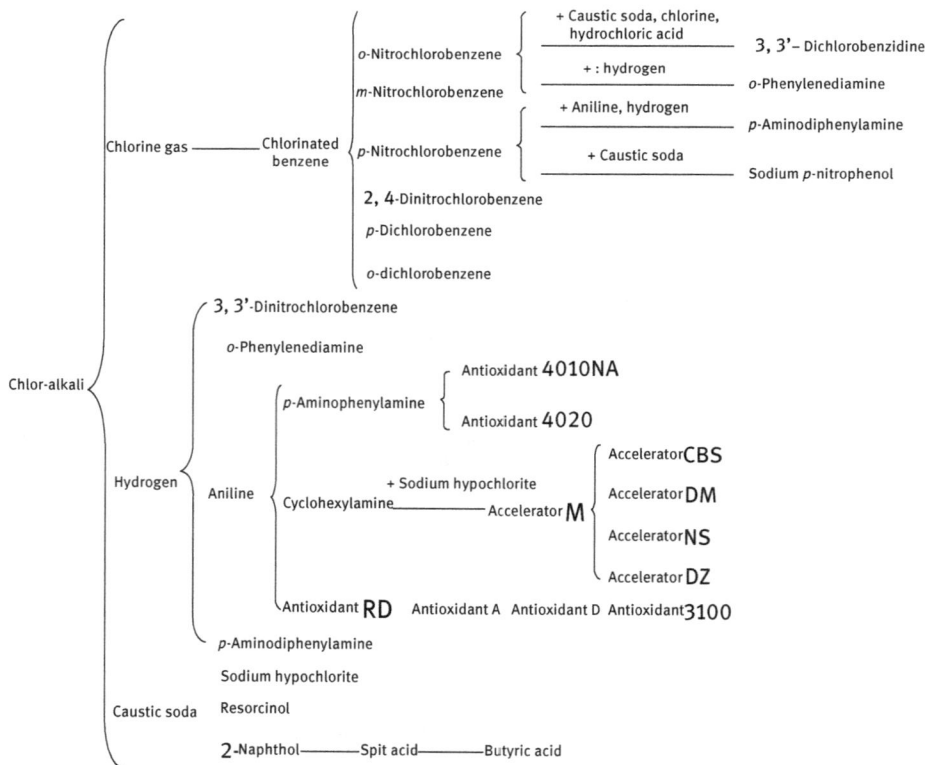

Figure 12.2: Products chain of the chemical plant of Sinopec Nanjing Chemical Industry Company Ltd.

3. **Development mode of Xinjiang Tianye**

Xinjiang Tianye Co., Ltd (referred to as Xinjiang Tianye) is an important chlor-alkali enterprise in the West. It is the first batch of pilot companies of circular economy set by the National Development and Reform Commission in 2005.

Xinjiang Tianye is a state-owned enterprise group whose core business is chlor-alkali chemical industry and plastics processing. Now it has a production capacity of 200,000 kW thermal power, 2.5×10^5 t/a calcium carbide, 2.6×10^5 t/a PVC resin, 2.3×10^5 t/a ion-exchange membrane caustic soda, 5 million acres of plastic water-saving equipment, 8×10^4 t/a tomato sauce, 1.6×10^4 t/a citric acid and a construction company with grade II construction qualification. The company achieves coordinated development of environment, economy and ecological benefits through the implementation of industrial circulation combined with circular utilization of resources including "coal-power-chemical-plastics processing-building materials." Figure 12.3 shows the company's circular economy industrial chain. On the one hand, the integration of coal and electricity has dominated the market because it reduces 30% cost of polyvinyl chloride resin, a major raw material for the

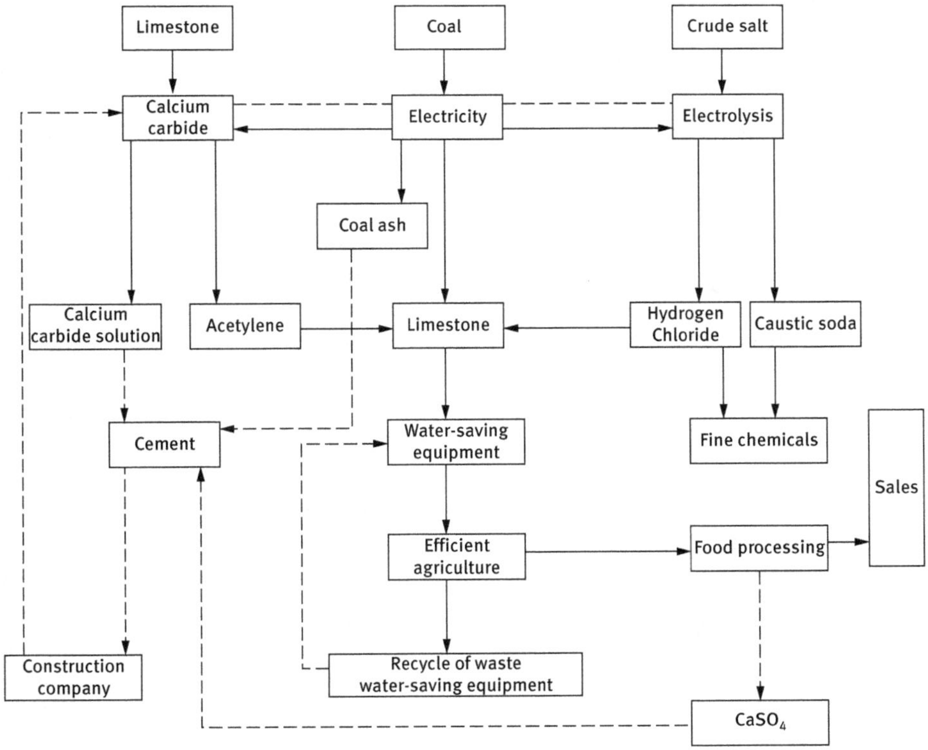

Figure 12.3: Circular economy industrial chain of Xinjiang Tianye Company.

water-saving drip irrigation system pipes. On the other hand, the newly developed technology of "disposable recyclable drip irrigation tape" increases the recycling rate of wasted drip irrigation tape by more than 80%, which not only solved the environmental problem, but also significantly reduced the cost of raw materials for drip irrigation tape by 50%. Currently, the one-time fund of Tianye drip irrigation devices is only 1/5 the investment of foreign products and 1/2 the investment of similar products in China.

Drip irrigation water-saving technology promotes the development of high-efficiency agriculture, which further provides a large number of high-quality raw materials for Xinjiang Tianye to develop the food processing industry. The food industry of the company has a capacity to process 4×10^5 t/a tomatoes and 3×10^4 t/a corn. The operation of drip irrigation system improves the competitiveness of the food industry of the company in the market. By-products (corn dregs and tomato peel) of the food industry are all supplied to the feed business, while waste calcium sulfate and carbide slag generated in the production of PVC and ash mix are all used to produce cement. To sum up, the circulating combination of various industries keeps Xinjiang Tianye in a positive and harmonious development track.

During the development of circular economy, Xinjiang Tianye focused on three aspects: the industrial cycle combination, circular utilization of resources and clean production. In terms of improving technology and reducing emissions, the company conformed to the state industrial policy by applying the calcium carbide furnace with internal combustion to the project of 2.5×10^5 t/a calcium carbide, supported by advanced dry bag dust control technology. Its investment in environmental protection facilities accounted for 30% of the total investment. The waste heat of calcium carbide furnace gas is used for drying the coke. Cooling water is recycled to achieve "zero emission." The recycled coke power is used as a fuel for the power plant. The recycled lime powder is used in the cement plant and citric acid plant as a substitution of lime. In this way, Xinjiang Tianye makes full use of waste and turns waste into treasure, therefore forming a circular economy industrial chain of "coal-power-chemical-plastics processing-building materials." It has achieved an annual economic income of about 130 million, accounting for more than 50% of total corporate profits, and has gained good environmental and economic benefits.

As a main pollution source and big environmental risk, calcium carbide slag is generated during production of PVC in calcium carbide processing devices. In the industrial chain of "coal-power-chemical-plastics processing-building materials" in Xinjiang Tianye, the primary calcium resource is, respectively, used as a carrier and a main raw material in the production of PVC and cement. Cooperating with a company specialized in cement production in Xinjiang, it built a production line of 3.5×10^5 t/a calcium carbide residue to make high-grade cement, with an annual use of various industrial waste of 3.7×10^5 t/a. This cement production line will consume calcium carbide slag generated by the 2×10^5 t/a PVC, iron slag generated by the 4×10^4 t/a sulfuric acid, calcium sulfate waste residue generated by the citric acid factory and lime dust generated by the calcium carbide plant. It can also consume a lot of fly ash generated by the power plant. Cement is directly provided to the Xinjiang Tianye Construction Company and other enterprises in bulk transport ways. The company will also further strengthen two core industries, that is, the chemical industry and the water-saving equipment industry, to build a "coal-power integration" project of 1×10^6 t/a PVC driven by the chain of circular economy, and to extend the water-saving irrigation area to more than 30 million acres, so that it can make greater contributions to the establishment of ecological environment and the development of China's PVC industry and water-saving industry.

12.2.4 Implementing Measures for Circular Economy

In the 1960s, countries such as Germany and the United States had entered the post-industrial era when they took a lead in the high-efficiency use of resources and energy. The development of circular economy in these countries originated from an environmental movement at that time: Waste Recycling Management Movement. Later, it was called "garbage economy" since it extended and transferred to the field of production and consumption.

Since the reform and opening up, China has maintained a relatively fast economic growth, and achieved fruitful results in promoting conservation and comprehensive utilization of resource as well as implementing clean production. However, due to China's large population, its per capita resource share only accounts for 1/3 of the world average. What is more, since China has long adopted the extensive mode of economic development with high material consumption, high energy consumption and high pollution, the environmental degradation and shortage of resources became more prominent. Meanwhile, it also faces problems such as incomplete and unsound regulations, policies, systems and mechanisms as well as poor technologies.

Therefore, it is of great significance to learn from the successful practices of circular economy in developed countries to promote circular economy in China. It should be carried out in the following three aspects [11].

1. **Strengthen legislation to promote circular economy**
   Both the development of circular economy and the establishment of a harmonious society require a sound legislative guarantee. By learning from the laws and regulations concerning circular economy in other countries, China can establish and improve its own laws and regulations according to the national conditions.

   Germany first established a number of related laws and regulations in individual areas, and then introduced the overall legislation of the circular economy. After continuous practice and amendments, the legal system of circular economy with more rigorous provisions and sound structure has been formed. All walks of life, from the production areas to the consumption field, from a single individual to the whole society, are all involved in this system. These comprehensive laws and regulations provide a strong guarantee for the development of circular economy.

Japan boasts its most comprehensive legislation of circular economy among the developed countries. It applies bottom-up legislation, namely, taking the "Basic Law for Promoting the Creation of a Recycling-Oriented Society" as a framework law under whose direction laws and regulations on circular economy in various fields are established. There are three levels of legislation in Japan: at the primary level is the fundamental law; at the second level are comprehensive laws, for example, "Law for the Promotion of Effective Utilization of Resources"; and at the third level are regulations specified for different industries and products, for example, "Law for Promotion of Sorted Collection and Recycling of Containers and Packaging." These laws and regulations reflect "three factors and one goal," that is, the reduction of waste, the reuse of old items, the reuse of resources and, ultimately, the establishment of a recycling-oriented society.

In accordance with the principle of "assigning responsibility to those who created pollution to clear it up, to those who damaged environment to restore it and to those who gained benefit to pay for it," China improved the tax system of ecological environment and resources, and perfected the compensation mechanism of ecological environment. At present, China has formulated and improved 9 environmental protection laws, 12 natural resource management and ecological conservation laws, 3 laws on disaster prevention and mitigation, more than 100 volumes of administrative rules and regulations on environment, resources and disasters, and local environmental regulations amounting to over 1,500 pieces. With the promotion of circular economy, the legal system will continue to be improved, which will put more stringent legal constraints on enterprises.

2. **Improve national industrial policy and guide the development of circular economy**

Policy instruments such as fiscal policy, tax policy and credit policy should be applied to strictly control bank credit funds for industries with high input, high energy consumption, high emission and low efficiency, and to increase funding support for the development of circular economy. When evaluating credit rating and project risks of enterprises, banks should pay more attention to environmental indicators such as waste emissions and frequency of energy use, thus to put more pressure on enterprises and force them to improve their behavior by adopting resource-saving and environment-friendly production modes.

Enterprises are profit-oriented organizations. Therefore, the development of circular economy must respect the economic laws, so that it becomes a conscious pursuit of enterprises. Governments should formulate various economic and industrial policies to guide enterprises to develop circular economy. In general, these policies mainly include tax incentives, that is, appropriately reducing sales tax for enterprises which purchase renewable resources, pollution control equipment and waste recycling equipment within the term of service; government procurement policy, namely, government intervening purchase behaviors at all levels to guarantee a big share of recycled product in government procurement; charging policy, that is, charging the sewage treatment fee on household water-consumption; taxation policy, which means levying a tax on fresh materials to reduce the use of raw materials and encourage recycling products.

3. **Attach importance to publicity and raise public awareness**
The implementation of circular economy not only requires government advocacy and self-discipline of enterprises, but also supports from the public. Otherwise, the circular economy cannot go deep into all aspects of the society.

Developed countries attach great importance to the use of mass media and other means to strengthen publicity of circular economy. Since 1997, the United States has declared December 15 as "America Recycles Day." Japan also declared October as "3Rs (Reduce, Reuse and Recycle) Promotion Month." In general, the publicity activities and education programs in developed countries present the following characteristics: (1) focusing on the foundation, that is, spreading the idea of circular economy to all levels of school education to influence students, and thus to parents and the whole society; (2) focusing on targeted publicity, that is, applying various forms of writing promotional materials to people at different levels; (3) focusing on interest, that is, making the promotional materials entertaining and suitable for all ages; (4) focusing on durability, that is, using various types of mass media including TV, websites, advertising shirts, calendar cards, bus and even litter bins, making it easier for people to watch and remember.

During the development of circular economy in developed countries, various associations have played highly active roles and become indispensable driving forces, especially the intermediary organizations engaged in waste recycling, such as the DSD recycling system which covers the whole Germany, the top five packaging recycling organizations in Sweden and the carpet recycling organization in California. Some organizations also assist the government to work on formulating laws and industry standards. For example, the "Anti-waste" organization in California once assisted the state legislature to enact the California Beverage Container Recycling and Litter Reduction Act which charged a container fee of 2.5 cent for each beer or soft drink bottle. In Japan, 41 industrial associations spontaneously developed environmental standards and objectives of the industry. Some intermediary organizations also conduct research on circular economy, establish information network and provide counseling and information services concerning circular economy.

In China, the circular economy mode still needs to go through a long process. Circular economy development is an effective carrier to promote the scientific outlook on development. It is also the only way to resolve the contradiction between economic growth and material consumption, and to overcome problems such as the shortage of material resources and environmental damages caused by large and extensive consumption. It is necessary for China to develop circular economy, constantly increase the proportion of recycled materials in the national economy and gradually reduce the use of non-renewable materials. Only by doing so, can China coordinate economic development with population, resources and environment, and find a civilized road to development featuring higher productivity, a well-to-do life and sound ecosystem while meeting the needs of the whole society for materials.

In summary, along with the rapid development of industrialization and urbanization in the twenty-first century, as well as the growing population, China must establish its own circular economy. According to *The Twelfth Five-Year Plan for National Economic and Social Development of the People's Republic of China*, it aims to double its GDP by the year 2010 on the basis of 2000, and the economy over the next 10 years still needs to maintain a momentum of rapid growth. Apparently, if it continues the traditional "three highs" mode of development to drive the economy, the sustainability of its socio-economic development will be shaken. In other words, the existing resources and energy in China can hardly meet the requirement of the "three highs" mode for rapid economic growth in a decade. Therefore, the traditional economy should be transformed by using high-tech and green technology while circular economy and new economy should be developed to lead China's economy and society to a sustainable road [12–15].

12.3 Eco-industrial parks

The concept of EID was first put forward by Professor Ernest Lowe from Indigo Development. It is a new form of industrial organization based on cleaner production requirements, the concept of circular economy and industrial ecology principles. In the park, all member units obtain greater environmental, economic and social benefits through the joint management of environmental and economic matters. In order to reduce production cost, enterprises within the park are allowed to discharge waste which may become raw materials for other enterprises, therefore preventing pollution and increasing resource utilization. The EID is the third-generation industrial park following the industrial park and the high-tech park. Under the guidance of the eco-industrial theory and the idea of circular economy, it creates a virtuous cycle for production development, resource utilization and environmental protection. It is an efficient, stable, balanced and integrated ecosystem which enables sustainable development and maximizes people's enthusiasm and creativity. It is also an upgraded version of high-tech industrial park, representing new industrial characteristics and reflecting requirements of the sustainable development strategy.

The process of constructing eco-industrial parks is in fact a process of urbanization. First, rural population living in those areas will lose their lands and become urban population. With the economic development of eco-industrial parks, these people have to adopt the urban lifestyle quickly. Furthermore, the construction of infrastructure in eco-industrial parks follows high standards of urban construction. Lands in eco-industrial parks have been irreversibly converted to urbanized areas. Finally, centering on the secondary and tertiary industries, the economic structure of eco-industrial parks shows typical characteristics of urban economy. Since an EID is an industry cluster, it will attract external population and cause rapid expansion of urbanization in terms of its population size. Therefore, in both quantity and quality, the process of constructing eco-industrial parks is an irreversible process of urbanization [16].

12.3.1 Development at Home and Abroad

1. **Development abroad**

   Many countries are conducting practices of eco-industrial parks according to their own national circumstances [17]. Currently, there are more than 60 projects of eco-industrial parks under construction or in planning around the world, most of which are in the West. The Kalundborg eco-industrial parks in Denmark is a typical industrial ecosystem with high efficiency and harmony. Besides, eco-industrial parks are thriving in many European countries, including Austria, Sweden, Ireland, Netherlands, France, Finland, the United Kingdom and Italy.

   (1) **The United States**

   In 1994, the Environmental Protection Agency (EPA) and the President's Commission on Sustainable Development (PCSD) in the United States specified four communities (e.g., Baltimore in Maryland, Cape Charles in Virginia, Brownsville in Texas and Chattanooga in Tennessee) as demonstration sites of eco-industrial parks. By 2005, there were already 20 eco-industrial parks under construction or in planning. In the United States, projects in the eco-industrial parks involved development of biological energy, waste treatment, clean industry, recycling of solid and liquid waste and many other industries. According to statistics, 20 city governments in the United States have been working with large companies on the establishment of eco-industrial parks since 1993, and two parks have basically been completed.

   (2) **Canada**

   The project of Burnside EID in Halifax, Canada, started in 1993. It adopted a cooperation method of production, study and research. College of Environmental Sciences in Dalhousie University is responsible for the maintenance and management of the ecological efficiency center within the park. The local government and enterprises in the park provide financial support. With the help of university scientific research, the project keeps optimizing material flow and energy flow, and promoting the by-product exchange and other cooperation among enterprises. At present, a relatively thorough industrial symbiotic has already been formed. Since 1995, eco-industrial parks have also been developed in the Portland industrial area of Toronto, Canada. This industrial area has brought together various manufacturing and service industries with potentials to exchange waste and energy. According to a recent study on the possibility of the integration of symbiosis and energy recycling, 9 of 40 industrial parks in Canada were considered to have a strong possibility to be developed into eco-industrial parks. The core industries involved included steam production, paper, packaging, chemistry (styrene/PVC), biological fuels, power, steel, oil refining and cement.

(3) **Asian countries**

Japan is one of the earliest countries to construct artificial ecosystem and is famous for its construction of ecological towns. Meanwhile, Japan is pushing forward the practice of eco-industrial parks such as the Kokubo EID and the Fujisawa EID. In Indonesia, India, Thailand and other Asian countries, construction or reconstruction of eco-industrial parks is being carried out, funded by German Organization for Technical Cooperation (GTZ). Indonesia established its EID in the suburb of Jakarta, and is currently studying the possibility of building material exchange network. India is building an EID in the industrial area of Naroda, and the park will be specialized in sugar industry which is similar to GuiTang Group in China. Thailand pays even more attention to this project. Led by the Industrial Estate Authority of Thailand, it is committed to reconstruct all the 29 industrial parks in the country into eco-industrial parks. In Philippines, the project of eco-industrial parks is funded by the United Nations Development Programme. It will first reconstruct five industrial parks which will later form an eco-industrial network. These eco-industrial parks will then collaborate with each other on by-product exchanges in regional areas. What is more, the possibility of building regional resource recovery system and enterprise incubator will be assessed.

2. **Domestic development**

In China, the concept of eco-industrial parks has already been accepted, and a number of productive practices have been carried out. Since 1999, China has been working on pilot projects on the construction of eco-industrial parks, and established the first state-level eco-industrial parks, Guangxi Guigang Eco-Industrial (sugar) Park. According to statistics, by the end of the "11th Five-Year Plan," there were more than 40 state-level eco-industrial parks under construction, among which 10 of them had passed the state acceptance inspection. In addition, with the help of United Nations Environment Programme, planning and reconstruction of eco-industrial parks have been conducted in the development zones of Dalian, Suzhou, Yantai, Tianjin and other cities. In general, China's eco-industrial parks are still in the pilot stage. Major eco-industrial parks in China are shown in Table 12.1 [18–21].

12.3.2 Principles and Contents of EID Planning

The EID is a new form of industrial organization based on the concept of circular economy. EID planning aims to establish an eco-industrial network by which

Table 12.1: Major eco-industrial parks in China.

	Region	Type	Core Enterprise	Major Industry	Relevant Industry
Guigang National EID	Western (Guangxi Zhuang Autonomous Region)	Reconstructed	Guitang Group	Sugar	Planting, Paper and and Alcohol Fuel
Shihezi National EID	Western (Xinjiang Uygur Autonomous Region)	Reconstructed	Tianhong Papermaking Group	Paper-making	Livestock, Livestock Product, Processing and Ecological Tourism
Baotou National EID	Western (Inner Mongolia Autonomous Region)	Reconstructed	Baotou Aluminium Group	Metallurgy, Machinery, Building Materials, Electricity and Rare Earth	None
Huangxing National EID	Central (Hunan Province)	Newly-Built	Broad Air Conditioning	Electronic Information, New Materials, Biological Pharmacy and Environmental Protection	None
Nanhai National EID	Eastern (Guangdong Province)	Newly-Built	None	Environmental Protection	Resource Recycling
Lubei National EID	Eastern (Shandong Province)	Reconstructed	Lubei Chemical	Chemical and Paper	None

by-products of a factory or enterprise can be used as inputs or raw materials of another factory. Through measures like the waste exchange, recycle and reuse, clean production and ecological management, the environment can be protected, and a sustainable development in the industrial park can be achieved. The EID is the third-generation mode of industrial parks in China, following the example of the economic and technological development zone and the high-tech development zone. Using the theory of industrial ecology, it seeks to achieve sustainable development of environment and economy by finding the connection among enterprises, linking industries and establishing the balance among enterprises. It is a higher level of the aforesaid two types of development zones.

1. **Six key principles of EID planning**
    (1) Principle of natural ecology: eco-industrial parks should be combined with the regional natural ecosystem and maintain as many ecological functions as possible. In the existing industrial parks, structure adjustment and technical renovation should be conducted in accordance with the requirements of sustainable development, so that the efficiency of resource use can be greatly improved, and pollution and pressure on the environment can be reduced. Site selection of newly built parks should take full consideration of the ecological environment capacity in that region, in order to minimize the impact of the parks on local landscape and hydrological context, the regional ecosystem and the global environment.
    (2) Principle of ecological efficiency: The concept of clean production shall be reflected in the layout, infrastructure, construction and industrial processes in the parks. More specifically, through clean production, the resource consumption and waste generation of each unit can be minimized; through the exchange of by-products among units in the park, the total material, water and energy consumption can be deceased; and through substitute materials and process innovation, the use of toxic and hazardous substances can be reduced.
    (3) Principle of coordination: Positive factors favorable for park construction should be taken full use of, while negative factors should be mitigated. Enterprises, markets, government and communities should coordinate with each other to increase the vitality and competitiveness of eco-industrial parks.
    (4) Principle of regional development: Eco-industrial parks should be developed along with the community and local economic characteristics, and the construction of the parks should be combined with regional ecological improvement. The planning of eco-industrial parks should be included in the local social and economic development planning and coordinated with the regional environmental protection plan.
    (5) Principle of high technology and high efficiency: Modern ecological technology and technologies for energy saving, water saving, recycling and information should be applied massively, and international standard for production process management and environmental management should be adopted.
    (6) Principle of attaching equal importance to software and hardware: The hardware refers to the construction planning of specific projects (industrial facilities, infrastructure and service facilities). The software refers to the construction of environmental management system and information support system as well as the formulation of preferential policies, and it provides a guarantee for the healthy and sustainable development of the eco-industrial parks.

On the premise of taking these principles into account, other technical measures should also be adopted. For example, we should encourage as many enterprises as possible to collaborate with each other and to recycle or exchange materials.

In addition, we should intentionally attract enterprises that can make good use of waste to enter into the parks, and also encourage "decomposers" because they are needed in an ecosystem. These "decomposers" include enterprises engaged in purchasing, selling or secondhand trading, and enterprises working on repair and maintenance. We should collect information of enterprises, provide necessary information for the enterprises on the premise of confidence and analyze the feedback information. What is more, a sound and perfect support system is crucial for the construction of eco-industrial parks. The support system mainly includes information management center, waste exchange center, environmental assessment center, educational planning center and application and research center.

2. **Contents of EID planning**

   Contents of EID planning cover a wide range, which include site selection, land use, landscape design, infrastructure, single facility and shared support services.

   (1) Energy: Efficient energy use is the main strategy for cutting costs and lessening the burden on the environment. Within an EID, individual enterprises not only seek to maximize the efficient use of electricity, steam and hot water, but also want to achieve the "energy flow" among each other. For example, the steam may be transmitted from a factory to home users who live in the same area. What is more, in many areas, renewable energy such as wind and solar power can be used for the infrastructure in the eco-industrial parks.

   (2) Material flow: The waste is seen as potential raw materials or by-products that can be used by the member units of the EID or sold to other units abroad. All the members and the whole community should optimize the use of all materials and reduce the use of toxic substances. Infrastructure in the EID should provide members with intermediate transfer functions, inventory sites and treatment facilities for common poison. Therefore, the site of an EID should be selected close to resources recycling companies, with a pattern of resource recycling, reuse and reprocessing around the park.

   (3) Water flow: Like energy, water can be recycled after pretreatment. Most of the water needed should flow in the infrastructure within the EID, which helps to improve the efficiency of water use. In addition, when designing the EID, we should consider constructing facilities to recollect rain and to use it.

   (4) Management and support services: Compared with the traditional industrial parks, an EID requires a more complex management and support system which can manage and support the exchange of by-products between units and help them adapt to changes (such as moving out of producers or consumers) in the industrial ecosystem. It should have a remote communication system which is connected with the sites for regional by-product exchange. It should also include training centers, buffet restaurants, day care centers, general supply and ordering offices, transportation and logistics offices, etc. With these shared services, enterprises in the park can cut down expenses. At last, the park should adopt a mode which is maintainable and easy to regroup.

(5) Land use and landscape design: Based on the basic principle of landscape management and design, we should carefully design the land use, construction, infrastructure, visual effects, environmental quality, afforestation, soil, hydrology, landscaping, lighting, traffic and environment.

At present, China's projects of eco-industrial parks are in the ascendant. However, these projects are either aggregations of enterprises that produce green products, enterprise communities that are mainly engaged in environmental protection or enterprise alliances that mainly exchange by-products. They are not real eco-industrial parks. Therefore, in the future, during the planning and development of eco-industrial parks, we must strictly follow principles of industrial ecology, circular economy, sustainable planning and construction, and combine them with the characteristics of ecosystem in the local area. What is more, we should also learn and draw lessons from the recent experience of other countries and regions, in order to bring environmental, economic and social benefits to each member (including enterprises, government and the community) in the park.

12.3.3 Construction of Eco-industrial parks

1. **Construction goal of eco-industrial parks**

   The construction of eco-industrial parks should be carried out under the guidance of circular economy ideas, and it should mainly reflect a unity of economy, society and environmental benefits with the economic development as a premise and the environmental protection as an important concern. Therefore, its goal is to gain economic benefits, social benefits and environmental benefits. Economic benefits involve annual growth rate of GDP and GDP per capita, etc. Social benefits refer to per capita housing area, Engel's coefficient of residents, average life expectancy and contribution rate of scientific and technological progress to the economy. Environmental benefits include the condition of air, water and noise, wastewater treatment rate, household waste treatment (innocuity) rate, drinking water quality, gasification rate, forestation rate and ISO14000 System Certification rate of industrial enterprises, green coverage rate, green space per capita, water consumption per unit of GDP, energy consumption per unit of GDP, resource consumption per unit of GDP, rate of multipurpose utilization, recycling rate, etc.

2. **Construction methods: new construction and reconstruction**

   There are two main construction methods of eco-industrial parks, that is, "Greenfield" and "Brownfield" in the mainstream of industrial ecology theory. Scholars of Greenfield suggest building completely new eco-industrial parks on idle lands or by removal of traditional industrial parks under the guidance of the government, namely new construction. Scholars of Brownfield suggest reconstructing the existing potential industrial parks which have material and energy-cycling bases, since new construction contains too much uncertainty and low feasibility. In other words, there are two methods to construct eco-industrial parks: to build new ones and to reconstruct the existing ones.

It is unrealistic for developing countries to build new eco-industrial parks, for the governments do not have enough financial resources to support new EID projects. It is also impossible to convince enterprises to take risk in investment and move to the newly built eco-industrial parks. In addition, there is no mature theory, practice or technology of constructing eco-industrial parks yet. Therefore, many scholars advocate combining traditional industrial parks with the theory of industrial ecosystem and building an industrial symbiotic system based on the existing industrial parks. In this way, traditional industrial parks can be transformed into eco-industrial parks and the mode of production can be changed from "throughput" to "roundput." Compared with new construction, this step-by-step approach is more practical and effective. Therefore, it is a more feasible way to construct eco-industrial parks by reconstructing the existing industrial parks.

3. **Construction of eco-industrial parks must adopt a progressive approach**
   The construction of eco-industrial parks should follow a progressive and autonomous approach [22]. Driven by interests during the reconstruction, more and more enterprises will autonomously participate in the construction of eco-industrial parks. It is impossible to construct eco-industrial parks all at once. Instead, we should first construct an industrial symbiotic system and then gradually introduce cooperation among more enterprises on material and energy recycling. At last, the system can be expanded and a sound EID will be finally established.

   According to the experience in constructing eco-industrial parks of various European countries, the breakthrough in the initial stage of the construction is some economic projects with less investment but quicker returns, like pollution control projects, rather than the material and energy recycling network. Just like the separation system for rainwater and sewage, which is commonly used in China's architectural design, eco-industrial parks can also separate different types of industrial wastewater, industrial wastewater of the same kind being transmitted to different devices of the wastewater treatment center through a single pipe system. In order to stimulate the industrial symbiosis system, governments should focus on those projects with less investment, higher return and lower risk in the early stage of development of eco-industrial parks and build public facilities as well as simple material and energy recycling network for enterprises. Through these projects with small investment and short construction period, enterprises will greatly save their costs on production and sewage treatment while benefits brought by these projects will draw their interest and attract them to make further investment. Experience has shown that the construction of eco-industrial parks will be a long and gradual process. When these prophase projects are completed and accepted by a large number of enterprises within and outside the system, governments can start to construct material and energy recycling projects which not only have long-term effects but also require huge investment funds. Finally, the construction of eco-industrial parks can be completed.

12.3.4 Examples of Eco-industrial parks

1. **Guangxi Guigang Eco-industrial parks**

Established in 1954, Guitang Group is the largest state-owned sugar enterprise in China, with over 3,800 employees and $14,700 \times 10^4$ m² planting area of sugarcane. During 1998–1999 when the sugar industry in China suffered from a total loss of 2.2 billion yuan, Guitang was still thriving. The reason is that Guitang (Group) has gone on a development path of circular economy, forming a primary EID with two main chains: sugarcane → sugar refining → producing alcohol with waste molasses → producing compound fertilizer with alcohol effluent; and sugarcane → sugar refining → producing pulp and paper with bagasse. Nothing is seen as waste in its logistics; rather, anything can be used as a resource. Therefore, the company can achieve full resource sharing.

Guigang eco-industrial system is composed of six subsystems, including sugarcane field, sugar refining, alcohol, papermaking, combined heat and power generation, and comprehensive environmental treatment. Various systems have their own products and they are connected with each other through mutual exchange of intermediate products and waste. Hence, these subsystems together form a relatively comprehensive ecosystem, which combines industry and plantation, and form an efficient, safe and stable industrial ecological park of sugar refining. To further improve the eco-industrial system, Guitang (Group) is currently working on six major systems and 12 projects with a total investment of 3.6 billion yuan. Many projects have already been in operation or under construction, and they cost over half of the total investment.

2. **Changsha Huangxing High-Tech Zone**

Changsha Huangxing Industrial Park in Hunan Province is the first multi-industry eco-industrial demonstration park in China. The EID positions high-tech industry as its dominant industry, including electronic information industry, new material industry, bio-pharmaceutical industry and environmental protection industry. Among these four types, electronic information industry is at the core position, new material industry and bio-pharmaceutical industry are attached great importance to, and environmental protection industry and other characteristic industries are moderately developed.

The construction planning of the park preliminarily determined 33 enterprises within the park and over 10 virtual enterprises outside the park, aiming at building 12 major industrial eco-chains and gradually enriching the industrial ecological network. The four different industries in the park can form an independent industrial eco-community, respectively. At the same time, they are also connected with each

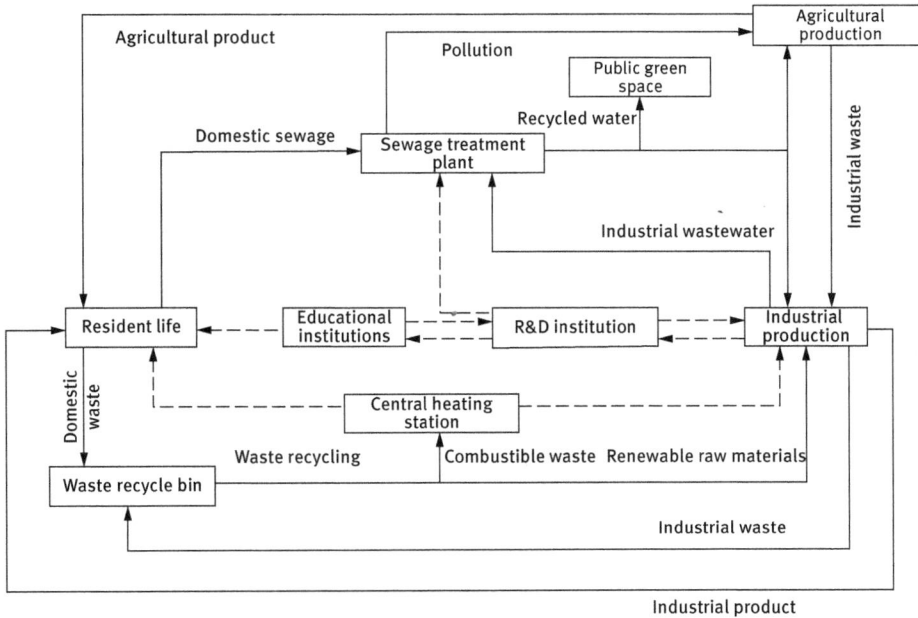

Figure 12.4: Symbiotic system of Changsha Huangxing Eco-Industrial Demonstration Park.

other through material, energy and information flows, therefore forming a network of circular economy with multiple material and energy links, as is shown in Figure 12.4.

Huangxing EID mainly relies on the electronic industrial eco-chain dominated by Broad Air Conditioning and its supporting industries, the new material industrial eco-chain dominated by antibiotic ceramics and its supporting industries, the biological product industrial eco-chain dominated by the deep processing of multiple agricultural products and the environmental protection industrial chain dominated by environmental protection equipment and environment-friendly building materials. On this basis, the EID builds an eco-industrial network with inter-coupled eco-chains, as is shown in Figure 12.5. What is more, along with industries like agricultural plantation, cultivation and ecological tourism outside the park, the construction of the park can also form a larger industrial ecosystem which would contribute to the benign development of regional economy.

3. **Shandong Lubei National EID**

Located at the Yellow River Delta, Lubei National EID is close to the Bohai Sea and Huanghua Port, commanding an area of 400 km². The main bodies of the Park are Lubei Enterprise Group General Company and Lubei Chemical Co., Ltd with 52 member enterprises covering 10 industries including chemical engineering, building material, light industry and electric power.

The Park implements the strategy of green civilization. Through technology integration and innovation, it has built three green eco-industrial chains,

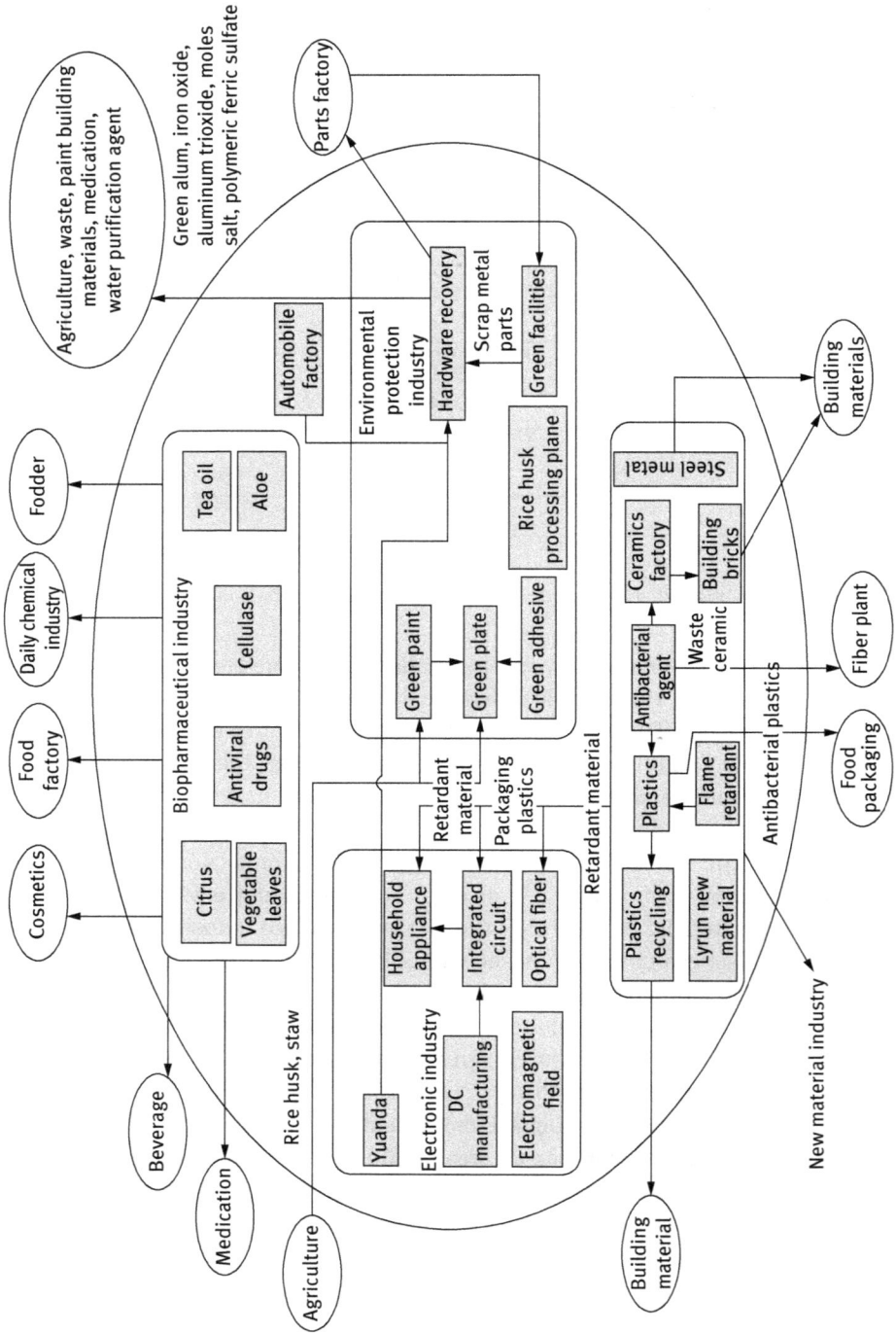

Figure 12.5: Overall planning of industrial ecological network of Changsha Huangxing Eco-Industrial Demonstration Park.

namely, the integrated production of ammonium phosphate–sulfuric acid–cement, the "multiple use" of seawater, and the integrated production of salt, alkali and thermoelectricity. An eco-industrial system is therefore established, which resolves the contradiction between industrial development and environmental protection, thus achieves the organic combination of scientific and technological innovation, industrial development, comprehensive utilization of resources, and environmental protection, and becomes a model of circular economic development using the ecological science and technology.

Thanks to the new road to industrialization and circular economy, Lubei EID obtains a new strategy and broad space for development. Currently, the Park is adopting the ecological science and technology industry as its support system, and 10 major projects as its main bodies, in order to implement a large-scale industrial upgrade [23].

(1) Develop innovative equipment for the integrated production of ammonium phosphate, sulfuric acid and cement, and based on this, develop fertilizer products which are environment-friendly and are required in featured agriculture. At the same time, transfer technology to domestic enterprises, export technology to countries like the United States and Russia, and promote the overall export of large-scale homemade equipment by using monopolistic technology.

(2) Build the largest ecological power supply base in China. Rely on local transportation, regional and ecological advantages, invest 20 billion yuan to build 2×300 MW generator sets and 5×600 MW generator sets. Together with the existing generating capacity, the generating scale will reach to 4 million kW by 2010.

(3) Build the largest high-quality salt export base in China. Cooperate with Japan's Mitsui & Co. and double the scale of the one-million-ton salt field, with an annual crude salt output of 500,000 tons.

(4) Build a large-scale chlor-alkali industrial base. Expand the industrial chains of "multiple use" of seawater as well as chlor-alkali, bromine and ocean chemical engineering, give full play to the advantages of resource, technology, energy and port, expand the chlorine-alkali project and enhance the production capacity of ionic membrane caustic soda to 500,000 tons per year.

(5) Integrate petrochemical engineering, chlor-alkali chemical engineering and salt chemical engineering, build the project of CPP oil-to-ethylene, extend the organic chemical industry chain related to chlorine consuming and bromine consuming projects and develop synthetic resin and high-tech fine chemical products.

(6) Develop the coal chemical industry base. Taking the geographical advantage of the coal terminal at Huanghua Port, expand ammonia plant to an annual production capacity of 300,000 tons, conduct research on coal gasification and coal coking technology, and develop clean fuels.

(7)   Work on large-scale titanium dioxide projects. Relying on the large-scale sulfuric acid and thermoelectric engineering, expand the scale of the titanium dioxide equipment from the existing 15,000 tons to 100,000 tons. The waste sulfuric acid generated in the production will be used to produce ammonium phosphate, and the in-house thermal power plant can satisfy the demand of titanium dioxide for steam, therefore guaranteeing the good quality of titanium dioxide as well as environmental protection, energy efficiency and cleanliness

(8)   Establish the Bridge Company. Increase investment and build the railway and highway combined Lugang Bridge leading to Huanghua Port.

(9)   Build two million mu papermaking forest and the "forest-paper integration" project with a capacity of 500,000 tons of paper pulp.

(10)   Make strategies for regional ecological economy relying on Jieshi Mountain and Huanghua Port. Accomplish the construction and development planning for the water conservation, tourism, transportation and culture of one mountain (Jieshi Mountain), one port (Huanghua Port), one island (Wangzi Island), one bridge (Lugang Bridge leading to Huanghua Port), two roads (highway and railway), two seas (South China Sea and North Sea), two lakes (freshwater lake) and three rivers (Majia River, Dehuixin River, Zhangwei New River), so as to mobilize the development of regional circular economy.

By the year 2010, Lubei EID plans to invest 30 billion yuan to build a world-famous EID with advanced technology, intensified knowledge, civilized management, friendly environment, harmonious structure and systematic network. It also plans to set a good example as an EID with great international influence and successful circular economy practice, explore sustainable development of regional economy and contribute to the realization of circular economy in the whole society.

4. **Develop circular economy and build a harmonious and conservation-oriented society**

Harmonious society is a society with harmony among people, resources and environment. The harmonious relationship between human and nature requires us to seek for the best combination among higher productivity, a well-to-do life and sound ecosystem. As a country with large population and relatively scarce per capita resources and weak environment bearing capacity, it is inevitable for China to build a resource-saving and environment-friendly society. Circular economy is the strongest support for building a harmonious society, and also an inevitable choice of resource-saving and environment-friendly society.

Resource-saving society refers to the comprehensive utilization of legal, economic and administrative measures in the areas of production, circulation and consumption to improve the efficiency of resource utilization, obtain the greatest economic and social benefits with minimum resource consumption and guarantee the sustainable development of economy and society. The purpose of building a resource-saving society is to pursue less resource consumption, lower environmental pollution, and greater economic and social benefits, in order to achieve sustainable development.

At present, European developed countries and the United States have explored three mature development patterns for circular economy, shown as follows:

(1) The first pattern of circular economy is within enterprises, and the most famous representative is DuPont Chemical in the United States. Through organizing the material circulation in various processes in the factory, it extends production chains and turns the waste in one workshop into the raw material in another workshop. Through such continuous utilization, waste becomes less and less, and "zero release" can ultimately be realized. By stopping or reducing the use of some harmful chemical substances, and creating new technologies to recover products, the company can reduce the waste plastics generated each year by 25% and the discharge of air pollutant by 70%. Meanwhile, the company has also developed new products like the durable ethylene material by recovering chemical substances from waste plastics.

(2) The second pattern of circular economy is industrial parks, namely, through material, energy and information integration among enterprises, forming an inter-industry symbolism and making the waste gas, waste water, waste residue, waste heat and by-products of one factory the raw materials and energy of another factory. A typical representative is Kalundborg Industrial Park in Denmark. The main bodies of this industrial park are four core enterprises: power plant, oil refinery, pharmaceutical factory and plasterboard manufacturing plant. Through exchanges, each factory utilizes the waste or by-products produced in other factories as the raw materials in their own production, which not only reduces the amount of waste and the expenses on treatment, but also brings very good economic benefits, and contributes to a virtuous circle of economic development and environmental protection.

(3) The third pattern of circular economy is a circular society, which means that following the requirements of circular economy, the whole country and society formulate relevant laws and regulations to achieve cleaner production, clean consumption, resources recycling and clean environment. Japan sets a quite good example. Due to limited resources, Japan pays great attention to resource recycling, particularly focusing on establishing a circular society. Currently, Japan has issued seven laws including "Basic Law for Establishing the Recycle-Based Society." These laws have been implemented since April 2004. It can be assumed that Japan will abandon the original social rules of mass production, mass consumption and mass waste, and gradually move toward a circular society.

With a new round of rapid economic growth, some developed regions in eastern China have successively encountered shortage in energy and raw materials, which provides a great opportunity for China's western regions to develop resource-based industries. At present, we have to confront a serious question: how to promote development while maintaining the coordination among economy, environment and resources, to achieve harmony between human and nature, and to ensure sustainable development by boosting production, improving people's life and protecting the environment? As an important breakthrough of the theory of economic development, circular economy overcomes the disadvantages of traditional economic theory which separates the economic and environmental system. It requires us to utilize natural resources and environmental capacity in an environment-friendly manner, in order to achieve the ecological transformation of economic activities. The tendency of economic globalization shows that the role of tariff barrier is increasingly weakened, while the function of "green barrier" in international trade is getting more prominent. Enterprises are facing more fierce market competition and greater influence exerted by environmental factors. Some developed countries not only require terminal products to confirm to the environmental requirements, but also stipulate that all sectors concerning products, ranging from preparation, development and production to packaging, transportation, utilization and circulation, should conform to the environmental requirements.

Therefore, only by vigorously developing circular economy and promoting cleaner production, we can manufacture products that conform to international standards concerning resources and environmental protection, break the "green barrier," expand foreign trade, alleviate the contradiction among economic development, ecological construction and environmental production, achieve the transformation of economic growth pattern and improve economic efficiency.

We must follow the guidance of the scientific outlook on development, set our objective to build a resource-conserving society, center on optimizing the utilization method of resources and take technological and institutional innovation as the driving force. What is more, we should also make overall planning, adapt to local conditions, highlight the key points, promote cleaner production, advocate green consumption, take practical and effective measures, and encourage the whole society to accelerate the construction of a resource-conserving society with sustainable development. At present, we must focus on the following five aspects:

1. **Accelerate the formulation and actively promote the development of circular economy and eco-industrial strategy**
   We should strengthen research on relevant theories and practice patterns, improve the awareness of governments at all levels and relevant decision-making departments of the importance to develop circular economy, study and formulate the development planning of circular economy, actively strive to become a pilot city

of circular economy and innovate the design philosophy. In the formulation of the "11th Five-year Plan" and the long-term development planning, we should stick to the principle of "reduce, reuse and recycle," give priority to developing major industrial projects with clean production technology and projects that turn terminal treatment into source control, and plan to establish several eco-industrial chains, eco-industrial parks and circular economy circles. In rural areas, we should focus on developing the bases of green food and organic food, promote industrialization layout, standardized production and scale operation, and promote the development of a number of agricultural eco-industrial parks that integrate plantation, cultivation and energy (biogas energy, straw gasification, etc.).

2. **Explore to establish the institutional guarantee system**

   Under the guidance of national macro-industry policies, we should study and formulate policies that are conducive to promoting circular economy, and establish and improve local regulations and normative documents for resource conservation and environmental protection as soon as possible. We should take environmental protection as the objective for economic and social development; formulate and improve the environmental access system, cleaner production system and waste recycling system for projects carried out in industrial parks according to the principles of ecology; and bring the development of leading industries and the cultivation of industry clusters into the eco-industrial chain. In order to prevent a mess condition among enterprises in the industrial parks, projects that do not conform to the development planning of industrial parks must be shut out. We should improve the preferential policy for comprehensive utilization of resources in combination with the reform of investment system, and provide strong support in finance, tax, investment, environmental protection and technology for pilot parks with cleaner production technology and circular economy. We should also resort to both encouragement and restriction, strengthen supervision and law enforcement, strengthen charge for pollution and, in particular, make use of economic leverages like economic incentive and punishment to drive enterprises to develop cleaner production and circular economy.

3. **Make efforts to advance industrial structural adjustment**

   We should put the orderly development and comprehensive utilization of resources in the first place, prohibit the introduction of pollution intensive projects, develop the tertiary industry and high-tech industry with low energy consumption and low emission, and focus on modern service industries like the development of biomass energy, deep processing of minerals and agricultural products, tourist, chain-store operation, logistics distribution and information consultation. Through technology import, joint efforts and reform, we can accelerate the transformation of traditional industries with high and new technology; eliminate outdated process, technology and equipment; strictly restrict industries with high energy consumption, high water consumption and high resource waste; extend the processing chains of products such as minerals, building materials, chemical engineering, cotton and animal products; and promote the resource-based connection of wastes among enterprises. In this way, we can build a number of regional enterprises with resources reproduction and improve the comprehensive utilization of resources.

4. **Prioritize the development of a number of eco-industrial products**

We should carry out a wide range of the pilot work on cleaner production technology and eco-industrial chain in key industries, key fields and industrial parks [24]. By selecting the excellent and supporting the stronger enterprises, we can develop enterprises with a mode of resources recycling, and advance the establishment of an intensive eco-industrial framework by demonstration. Tianjin Taida EID is a national industrial park (see Figure 12.6), also the first domestic economic and technological development zone established and planned as an EID. Now it is becoming a closed industrial chain with product metabolism and waste metabolism. The EID mainly focuses on four pillar industries: electronic and information, bio-pharmacy, automobile, and food and beverage. Through the establishment and improvement of industrial chain, product chain and waste chain, and the reduction of resources and waste, eco-industry can be greatly developed. With symbiotic cooperation among different leading industries, the cross-industry material flow, energy flow, information flow and even capital flow can be realized.

Taida EID adopts the fundamental concept of circular economy. Dedicated to building a symbiotic relationship among enterprises, it promotes that the waste from one enterprise can be used as the raw material of another enterprise. It also aims to build a cooperative mechanism between the government and the community for resource development, cleaner production, ecological design, green consumption and environmental protection, therefore achieving the dream of "zero release" in the entire industrial park. From March 2003 to February 2004, the industrial park had introduced 20 tier one suppliers of Toyota. The constant expanding of automobile industry chain is conducive to the improvement and maturity of this symbiotic network. Apart from automobile industry chain, Taida EID is also going to build and improve the cross-industry material metabolism chain (cross-industry waste exchange and recycling). For instance, it actively carries out internal and inter-industry utilization of steam based on the characteristics of food and beverage industry as with large and stable energy consumption; it also carries out symbiotic cooperation among various industries and sewage treatment plants, new water companies, landscaping companies and municipal companies to build a wastewater metabolism chain. In the meantime, the industrial park is also dedicated to collecting and recycling rainwater, and sorting and recycling waste, in order to form an integral management plan for solid waste in the park. Namely, the hazardous waste accounting for 1% of the

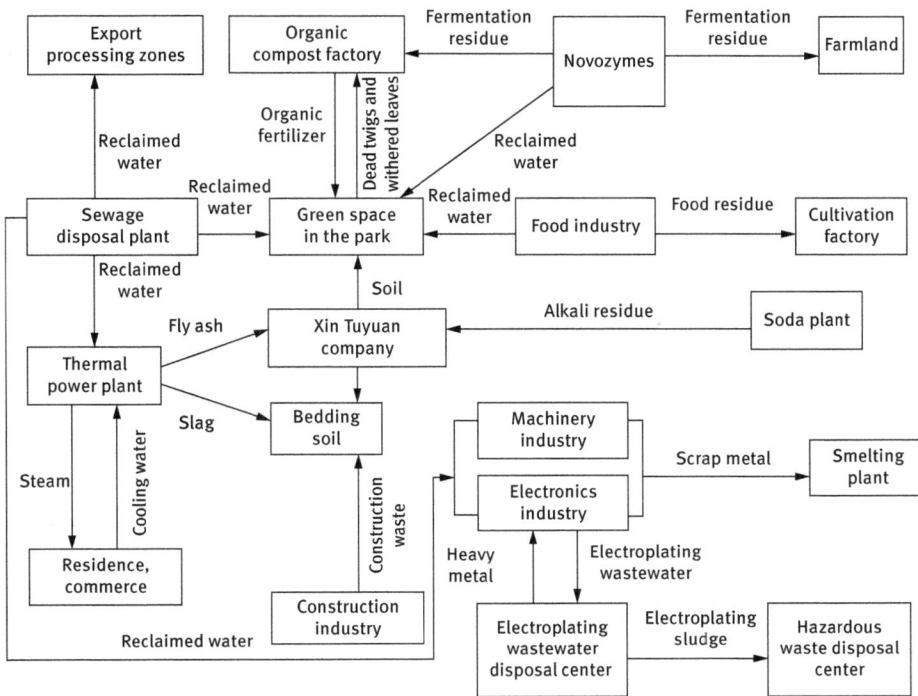

Figure 12.6: Framework of Tianjin Taida EID.

total amount is handed over to qualified units for recycling by classified collection system; the organic waste accounting for 9% of the total amount receives composting treatment; the rest household garbage is incinerated and used to generate electricity after classified recovery and cyclic utilization, and the bottom ashes are sent for landfill disposal.

5. **Strengthen organizational leadership and publicity**

   Governments at all levels should strengthen the organizational leadership for the development of circular economy, build an effective working mechanism, and advance cleaner production and comprehensive resource utilization by steps. They should organize promotion and education activities in various forms, actively advocate the use of environment-friendly products, reduce the use of disposable products, improve all people's consciousness of resource saving and environmental protection, and nurture the ecological concepts of harmony between human and nature. In this way, they can turn energy saving, water saving, material saving, grain saving and classified garbage recovery into conscious behaviors of every citizen, gradually make resource conservation and environmental protection people's lifestyle, and build a good social environment for developing circular economy and a resource-conserving society.

After over 20 years' continuous and rapid economic growth, China's basic national strength and industrial strength have significantly increased. Its technical standards for resource utilization and environmental protection are gradually improving and being in line with the international level, even equal to higher standards of the developed countries. In such circumstances, the resource-consuming technology will become less competitive, and apparently it is no longer a suitable path for industrial development. Therefore, a major task concerning the prospect of China's industrial development is to rely on resource-saving technology to enhance the industrial competitiveness. It is a strategic objective of China's industry in the twenty-first century to achieve strong competitiveness in resource utilization and environmental protection. Over 20 years after the reform and opening up, China's industry has achieved remarkable achievements which have attracted worldwide attention. However, because of the intensive growth mode, the industry developed at the cost of resource consumption and environmental destruction. With the constant economic growth, the society pays growing attention to resources and environment with increasingly higher standards. The path of extensive industrial growth with great resource consumption and environmental destruction can no longer support the continuous development of China's industry. Therefore, China's industry is now in a critical historical period when it is relying more on resource-saving technologies instead of resource consuming technologies. This is an important mutation period for industrial competitiveness. During this period, China should mainly improve its industrial competitiveness by upgrading the industrial structure, advancing industrial technology, and improving national mechanisms and technical standards for resource development and environmental protection until they are close to or reach the level of developed countries. In such a context, significant changes will occur in the growth pattern of industrial economy, the competitive mode among enterprises and the economic management system in China. The basic ideas about economic and social development will also change dramatically. Moreover, it is a correct strategic choice for China in the twenty-first century to "establish the scientific outlook on development," "take a new road to industrialization" and build a "conservation-minded" society.

Questions

1. What is an EID? What is the relationship between the EID and the circular economy?
2. What is the circular economy? What are the basic principles of circular economy?
3. How to understand circular economy in the United States and European developed countries?
4. What is the relationship between circular economy and China?

5. What are the principles followed in the planning of eco-industrial parks?
6. What is the difference between eco-industrial industry and traditional industry?

References

[1] Yang Qing-shan, Xu Xiao-po, Wang Rong-cheng. The theory of industrial ecology and eco-industrial park design: A case study of Jiutai, Jilin Province. Economic Geography, 2002, 22(5): 585–588.
[2] Frosch, R. A., Gallopoulos, N. E. Strategies for manufacturing. Scientific American, 1989, 261(3): 94–102.
[3] Erkman, S. Industrial Ecology. Translated by Xu Xing-yuan. Beijing: The Economic Daily Press, 1999: 31. (Switzerland)
[4] Niu Xue-jie, Li Chang-hong. Evolution research on strategic coordination between circular economy and new energy strategy. China Soft Science, 2013 (12): 146–151.
[5] Lowe, E. (the United States), Geng Yong. Industrial Ecology and Eco-Industrial Park. 1st edition. Beijing: Chemical Industry Press, 2003.
[6] Shi Lei, Zhang Tian-zhu. Chemical industry and circular economy. Modern Chemical Industry, 2004 (7): 1–3, 5.
[7] Jiang Sheng-han. The practice and thinking of circular economy in chemical industry. Chemical Production and Technology, 2005, 12(4): 46–48.
[8] Jiang Shu-hua. Analysis on the value chain of circular economy. Techno-economics & Management Research, 2007 (2): 39–40.
[9] Guan Na-xin, Lv Yong-mei, Zhou Ke-zhong. Analysis on the current status of nitrochlorobenzene production and its market. Chlor-Alkali Industry, 2006 (4): 25–32.
[10] Lv Yong-mei. The development trend of Chlor-Alkali industry in China: Scale, refinement, integration and greening. China Petroleum and Chemical Industries, 2004 (12): 14–16.
[11] Wang Shu-hua. Several thoughts on the development of circular economy by science and technology. Resources Economization & Environment Protection, 2007, 23(01): 28–31.
[12] Li Zhao-qian, QI Jian-guo. A summary on the theory and practice of circular economy. Quantitative & Technical Economics, 2004, (9): 145–154.
[13] Qu Xiang-rong. Cleaner Production and Circular Economy. Beijing: Tsinghua University Press, 2011.
[14] Zhou Hong-chun. Promote circular economy development in new situation. Recyclable Resources and Circular Economy, 2011, 4(11): 4–7.
[15] Wu Ji-song. Circular economy: The Only Way Leading to a Moderately Prosperous Society in All Respects. Bei Jing: Beijing Press, 2003.
[16] Geng Yong, Wu Chun-you. A review of the development of eco-industrial parks. Industry and Environment, 2003 (S): 121–123.
[17] Chen Yan-yan. The latest developing trend of circular economy abroad. Shanghai Business, 2006, (12): 32–33.
[18] Shan Guang-shan. The exploration and practice of developing circular economy in Haihua group. Shandong Chemical Industry, 2005, 34(5): 47–48, 46.
[19] Yang Qin-min, Zhang Qin-ye. Study on the "circular economy" in Haihua group. Chemical Industry Management, 2004 (8): 60–61.
[20] Zhao Rui-xia, Zhang Chang-yuan. Contrast between China and other countries in eco-industrial parks. China Environmental Management, 2003, 22(5): 3–5.

[21] Bai Yi-yan, Ge Cha-zhong, Yang Jin-tian. Analysis on questionnaire investigation of Rizhao eco-industrial park planning. Journal of Industrial Technological Economics, 2005, 23(7): 37–43.

[22] Deng Wei-gen, Chen Lin. The ideas and measures of constructing eco-industrial parks. Journal of Industrial Technological Economics, 2007, 26(1): 31–38.

[23] Xiao Yan-heng, Chen Yan. On the eco-industrial theory and approaches to its realization. China's Population, Resources and Environment, 2001, 12(3): 100–103.

[24] Liu Hong-ci. The development status of eco-industrial parks in China: A case study of typical eco-industrial demonstration parks. Contemporary Economics, 2011 (2): 52–54.

13 Intensification technology and practice in chemical processes

13.1 Overview

13.1.1 The Concept of Chemical Process Intensification

Chemical industry is an industry which produces chemical products by using chemical reactions to change the structure, composition and state of matter. This industry belongs to a knowledge- and capital-intensive industry. With the development of science and technology, it gradually developed into a multi-industry full of rich varieties: from the initial production of a few inorganic products (such as soda ash and sulfate) to a few organic products (such as alizarin dye extracted from plants). There has been a large number of comprehensive utilization of resources and large-scale chemical enterprises emerged, such as inorganic acid, alkali, salt, rare element, synthetic fiber, oil, plastic, synthetic rubber, dyes, paints, fertilizers, pesticides, and so on. After the rapid development in the past 100 years, the chemical industry has become the backbone to fuel economic growth. Today, the human being is facing the most serious environmental crisis. To date, more than 12 billion compounds have been discovered and and create each with their own properties and functions. Agriculture, light industry, heavy industry, food and clothing are closely dependent on chemicals. Chemical makes life more colorful. The chemical industry is closely related to the development of economy and society and the basic human necessities of life. Products such as pharmaceuticals, plastics, rubber, gasoline and so on, are made in the chemical industry. Chemical industries are highly profitable industries and are the mainstay of the national economy. In order to realize the sustainable development of the chemical industry, problems such as high energy consumption, high pollution and others in the process of producing thesechemical products need to be solved. Due to the rapid increase of population, the increase of consumption of resources and the gradual decrease of per capita arable land, freshwater and mineral resources, the contradiction between population and resources is becoming increasingly sharper. As one of the pillar industries of the national economy, the chemical industry plus related industries is a double-edged sword. The advantage is that it makes an important contribution to the material civilization of human beings, and the disadvantage is that it emits a lot of toxic substances during the production, which provides harms to the environment and human health. The pollution caused by the traditional chemical industry to the environment is a very serious issue. At present, the world's annual production of harmful waste amounted to $3 \times 10^8 - 4 \times 10^8$ tons. Especially since the 1930s, the world's eight major public events have also given us a wake-up call.

With the development of modern industrial process, the requirements of constantly updated products, the requirements of increasing environmental protection,

https://doi.org/10.1515/9783110479317-013

the construction of ecological economy and realization of the sustainable development are more in demand. Therefore, people try to apply the principles and methods of green chemical engineering to strengthen the process. Green chemistry focuses on the chemical reaction itself to eliminate environmental pollution, makes full use of resources and reduces energy consumption. But chemical process intensification is emphasized on using new technology and equipment in the process of production under the condition of same production capacity. It can greatly reduce the volume of equipment or improve equipment production capacity, significantly improve the energy efficiency and reduce waste emissions substantially. Chemical process intensification has become an emerging technology to realize the chemical process of efficient, safe, environmental friendly, intensive production and to promote social and economic sustainable development. The United States, Germany and other developed countries have listed chemical process intensification as one of the three areas which is the current priority in the development of chemical engineering. Strengthening the chemical process to accomplish high efficiency, energy saving and no pollution is the effective means to solve the contradiction in the development of the industry pollution and to realize sustainable development.

The characteristics of chemical process intensification are technical innovation and technological process improvement. Under the premise of realizing the set production target, the plant layout should be more compact and reasonable, unit energy consumption should be lower, and the wastage of material and by-product should be less, by greatly reducing the size of the production equipment and the number of devices and other methods. In a broad sense, chemical process intensification includes the development of new devices and new techniques. The development of new devices, such as a new type of reactor, a new type of heat exchanger, high-efficiency packing, a new type of tower plate and so on, is needed to strengthen the production equipment. The development of new techniques – such as the coupling of reaction and separation (e.g., reactive distillation, membrane reaction, reaction extraction, etc.), combination of separation process (e.g., membrane absorption, membrane distillation, membrane extraction, absorption distillation, etc.), field effect (e.g. centrifugal field, ultrasound, radiation, etc.) and others (e.g., supercritical fluid, dynamic response operating system, etc.) – is needed to strengthen the production process. Therefore, the chemical process intensification is a long-term goal of the chemical industry in China and other countries, and it is also one of the main achievements in the research of chemical science and engineering.

The main characteristic of process intensification in chemical engineering is the process of equipment miniaturization and integration, which is the requirement of green chemistry. In 1995, C. Ramshaw first proposed in the *First International Conference on Chemical Process Intensification*: that chemical process intensification is a measure that can significantly reduce the volume of a chemical plant under constant production capacity. He believed that the volume decreasing by more than 100 times can be considered as the process of strengthening. A. I. Stankiewica and

J. A. Molin believed that the reduction in volume of equipment by more than two times, significant reduction in energy consumption per ton of product and a large reduction in waste or by-products can be considered as the process of strengthening.

13.1.2 Origin and Development of Chemical Process Intensification

The history of chemical process intensification can be traced back to the end of the 1970s, when the British chemical industry company first used the concept of the production process to reduce investment. In the mid-1990s, the international appeared intensification technology and practice in chemical processes aim at energy saving, reduction in consumption, environmental protection and intensive production, which is one of the three major areas of chemical engineering. Since 1995 when the first international academic conference was held, *International Conference on Chemical Process Intensification* has been held every three years. In 2001, chemical process intensification was listed as one of the priority areas in the development of chemical engineering in a symposium called *Refocusing Chemical Engineering*, which was held jointly by UEF, NSF and AIChE. In September 2002, UEF held a symposium called *Process Innovation and Process Intensification* in Edinburgh, England. Experts and scholars of various countries have also organized the "Process Intensification Network," and actively carried out academic exchanges and technological cooperation. It can be said that people's understanding of process intensification in chemical engineering has reached a new height, and a large number of research papers have been published in this area. In July 2005, the *Seventh World Conference on Chemical Engineering* was held in the United Kingdom, and the process intensification is one of the most popular research directions. In the twenty-first century, people hope to change the appearance of the chemical industry dramatically by process intensification.

In early years, chemical process intensification was mainly based on hardware. In recent years, high-efficiency trays and innovations in regular and bulk packings have been made one after another, such as new inner parts. China and other countries have made remarkable progress in the use of new tower internals to reform crude oil and atmospheric pressure, ethylene and synthetic ammonia production facilities. This leads to higher efficiency, lower energy consumption, and significant economic benefits. However, the improvement in performance of inner parts of a chemical tower is not much. At present, chemical process intensification emphasizes on the combination of hardware and software – more emphasis on scientific and technological innovation, to pursue higher goals. More and more researchers believe that the goal of chemical process intensification cannot only stay in improving the efficiency of the existing equipment by several percent, and cannot be satisfied with the gradual reform. They should be committed to the volume of equipment, the industrialization of the cycle, energy consumption, material consumption, environmental protection and other aspects to make a breakthrough in the efficiency of factory. People hope to change the appearance of the chemical industry dramatically by process intensification in the twenty-first century. This is a great challenge, but also promotes the process intensification to make some significant progress, such as the successful research and development of the supergravity separator, the high-speed rotating disk reactor, the whole catalyst and the impinging stream reactor. Another developing trend of chemical process intensification is that the chemical science and engineering research has greatly promoted the development of new coupling separation technology, and has been successfully applied to production, such as catalytic distillation, membrane distillation, adsorption distillation, reactive extraction, complexation adsorption, reverse micelle, membrane extraction, fermentation, extraction, chemical absorption and electrophoretic extraction. Coupling technology – of these new types – combining the advantages of various technologies has a unique advantage. The coupling separation technology can also accomplish many tasks, which are difficult to complete by traditional separation technology. Therefore, it has broad application prospects in the fields of biological engineering, pharmacy and new materials. The coupling technology is complicated, and the design amplification is difficult. Therefore, it also promotes the research of mathematical model and design method of chemical engineering.

In recent years, chemical process intensification has attracted more people's attention. In many developed countries such as the United States, chemical process intensification is listed as one of the three areas which is the current priority in the development of chemical engineering. The United Kingdom focuses on basic research. France attaches great importance to the establishment of the theoretical model. Germany focuses on experimental technology and engineering research. Japan has invested a lot in the research of biological engineering and new materials. Canada and Australia take resource utilization as the research focus. Great progress has been made in the research and application of chemical engineering in China. For example, the rise of the petroleum industry has greatly promoted the development of catalysts, reaction engineering and distillation technology, and the development of nuclear fuel reprocessing and wet metallurgy promoted the improvement in the level of the solvent extraction technology. Chemical process intensification technology was listed as the first batch of "11th Five-Year" national "863" plan, in order to save energy and reduce emissions.

Due to space limitations, this chapter introduces only some common chemical process intensification technologies and equipment.

13.2 The Coupling Technology of Reaction Process

The coupling technology of reaction process is used to accomplish the functions by one reactor other than several which were needed traditionally to improve the chemical conversion rate and integration level of the reactor. By using the coupling technique of reaction process, various chemical processes can be integrated into one. It can give full play to the advantages of various chemical processes, to avoid the shortcomings of a single chemical process, which will much more intensify the integration of chemical processes.

13.2.1 Membrane Catalytic Reaction

1. **The concept of membrane catalytic reaction**

   In recent years, membrane catalytic reaction has become a new technology in the field of catalysis, which is to make a catalytic material into a membrane reactor or to operate in a membrane reactor. Integrated catalytic reaction, membrane separation process and then the reactants can selectively penetrate the membrane and react. Or the product can selectively pass through the membrane and leave the reaction zone, thereby regulating the concentration of a reactant (or product) in the reactor to break the balance of the chemical reaction. Or it can strictly control the amount and state of a particular reactant in the reaction, thereby achieving high selectivity.

2. **The membrane catalytic reaction model**

   In the catalytic reaction system, according to different operation modes, the membrane may have different functions. If the membrane itself is catalytic active, it will have the function of catalyst. If the membrane is inert, it can be impregnated with a catalyst active component or buried in the membrane, and the membrane has only the function of selective separation barrier. And some membranes have the dual function of catalytic activity and separation barrier [1]. Membrane catalysis has three modes of operation. First, the catalytic reaction and membrane permeation selectivity coupled together. With the help of membranes implementing catalytic reaction, the products (or one of the products) are removed from the reaction zone by selective membrane, which is particularly advantageous for the reaction controlled by thermodynamic equilibrium. Second, the membrane plays only the role of selective permeation separation. The membrane could only selective separate the target product from the reaction zone, so as to improve the selectivity and the yield of the membrane. Third, through the selective permeation of the membrane, the feed rate of the active reactant is controlled to facilitate the generation of the target product. This is an effective method for the selective oxidation reaction.

3. **The characteristics of membrane catalytic reaction**

   Membrane catalytic reaction will be applied in the catalytic reaction field, and the catalytic reaction and separation process will be carried out simultaneously. There are many outstanding advantages [2].

   (1) Good catalytic activity. The specific surface area of the membrane is relatively large, the atomic (or molecular) share of the unit surface area is high

and there is much active center. Therefore, it can effectively contact with the reaction molecule, which shows a high catalytic activity.

(2) High selectivity. Membrane has a lot of micro pores, whose size distribution has a wide range. Different methods can be used to control the pore size and the distribution of pore size, which is beneficial to the molecular diffusion and to improve the selectivity of the catalyst. Especially, the selectivity of the biological membrane catalyst can reach 100%.

(3) The carrier type of membrane catalyst has the characteristics of high temperature resistance, chemical stability, mechanical strength and the long life of the catalyst.

4. **The membrane materials and classification**

Membrane material is the core of membrane catalytic reaction technology, and the chemical characteristics of membrane materials and the structure of the film play a decisive role in the performance of the membrane. General requirements for membrane materials are good film-forming properties, thermal stability, chemical stability, acid and alkali resistance, microbial erosion and oxidation resistance. The membrane with catalytic function is to fix the catalyst on the membrane surface or inside the membrane. To give the membrane a catalytic reaction, make separation membrane as a reaction part with dual functions of reaction and separation [2].

(1) **Inorganic membrane**

Inorganic membrane is widely used in various kinds of membrane reactors for its good chemical stability, acid and alkali resistance, organic solvent resistance, high-temperature resistance (800–1,000 °C), high-pressure resistance (10 MPa) and strong resistance to biological attack. It consists of metal or alloy membrane, porous metal membrane and porous ceramic membrane. The selective permeable inorganic porous membrane can be used as the support of other membranes, and can also be used as a catalyst or catalyst carrier. At the same time, it can separate the product or residual reactant. Especially in the high-temperature, gas-phase heterogeneous catalytic reaction, the operating temperature has exceeded the thermal stability zone of the organic polymer membrane, so the application of inorganic membrane as a high-temperature resistance of the catalyst and the carrier material is the only choice.

(2) **Polymer membrane**

Polymer membrane material includes polyimide Teflon, polystyrene, polysulfone, silicone polymer and polymer membrane treated by plasma technology. According to the difference of the fixed catalytic material, it can be divided into the polymer–metal complex separation function membrane and the macromolecule catalysis separation membrane. The polymer membrane has good sensitivity, but the main problem is the relationship between the thickness of membrane and the rate of permeability. If the thickness of membrane is between 20 μm and 200 μm, the penetration rate is slow. If the thickness of membrane is between 0.1 μm and 1 μm, the penetration rate is fast. But the membrane is too thin and unstable, so it must have a carrier [3]. In addition, some high polymer membranes cannot be kept stable in some solvents. In China, the main research works in this area include polyvinylidene fluoride microporous membrane, Poly aryl ether ketone and polyether sulphone membrane, which have better heat resistance and oxidation resistance [4].

(3) **Biomembrane**

At present, people have great interest in the immobilized carrier, which uses membrane as a biological catalyst, including soluble biological catalyst membrane system and insoluble biological catalyst membrane system. Enzyme is a highly active biological catalyst. It is immobilized on the membrane surface or membrane pores to form an enzyme membrane reactor. It can be used in research on biochemical engineering, culture of cell, L-amino acid production, continuous production of glutamic acid, continuous fermentation of ethanol and so on [5].

(4) **Composite membrane**

The composite membranes include molecular sieve composite membrane, porous glass composite membrane, metal supported composite membrane and other composite membranes. The composite membrane is characterized by high catalytic activity and good heat resistance. It can permeate high-temperature gas (400–1,000 °C), which can be applied to high temperature reaction. Suzuki makes the mixture of 12.5% n-heptane and 87.5% methyl cyclohexane permeates through the molecular sieve membrane at 1.6 MPa, room temperature, and then he obtained 98.5% heptane. If the ZSM-5 type molecular sieve membrane is used for the dehydrogenation of ethylbenzene, the ethylbenzene conversion rate will increase. Molecular sieve membrane has excellent separation selectivity, but it is very difficult to prepare perfect molecular sieve membrane due to the complexity of molecular sieve crystal growth and the direction of controlling the growth of molecular sieve. Suzuki obtained 98.5% n-heptane from a mixture of 12.5% n-heptane and 87.5% methylcyclohexane through a molecular sieve membrane at 1.6 MPa and room temperature [6]. Recently, there was a kind of copper–palladium composite membrane, combined copper composite membrane with catalytic dehydrogenation function and palladium composite membrane with efficient hydrogen separation function. The outer side of the membrane is provided with a catalytic dehydrogenation activity, and the inner side is provided with a hydrogen separation function.

5. **The membrane catalytic reactor**

The membrane catalytic reactor mainly comprises membranous layer, catalyst and carrier. According to the combination of the membranous layer, catalyst and carrier, there are four kinds of assembling methods for the membrane catalytic reactor [7], as shown in Figure 13.1.

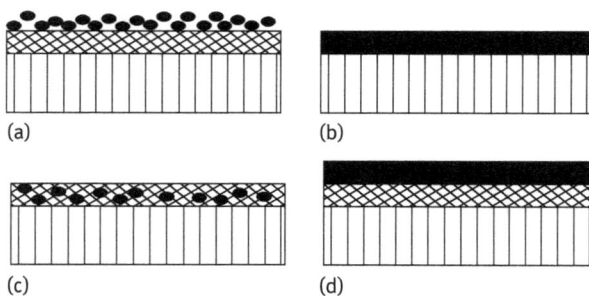

Figure 13.1: Method for assembling membrane catalytic reactor.

Figure 13.1(a) shows that membrane and catalyst are two separate parts. Stick the catalyst particles or small balls on the surface of the film. Then catalyst particles play a catalytic role, which result in the separation of the lower membrane. Figure 13.1(b) shows that membrane material itself has catalytic effect, and can play a role of separation or catalysis. Figure 13.1(c) shows that the catalyst is embedded into the inner layer of the membrane, so that the membrane which has only the separation function also has the catalytic activity. Figure 13.1(d) shows the assembled composite membrane, which is a carrier of catalyst. The upper layer membrane is used as catalytic function, and the lower layer is used for separation.

Using selective permeation properties of porous membranes, the membrane catalytic reactor simultaneously accomplishes catalytic reaction and separation operation in a reactor, which is a typical integration of reaction and separation. According to the different role of the membrane in the membrane catalytic reactor, the membrane catalytic reactor is divided into two types [8], as shown in Figure 13.2.

Figure 13.2(a) shows the inert membrane catalytic reactor. The membrane itself has no catalytic activity, and it only plays a role of separation. The catalyst required for the reaction needs to be loaded separately, such as general membrane bioreactor. Figure 13.2(b) shows the catalytic membrane reactor. The membrane has dual functions of catalysis and separation. According to the specific reaction and separation process, for the preparation of catalytic membrane membrane materials such as inorganic or organic polymer materials can be selected. According to the different types of materials, the catalytic membrane can be divided into inorganic catalytic membrane, polymer catalytic membrane, biological membrane, composite membrane and so on.

The membrane catalytic reactor is an exquisite reaction/separation integration device. In principle, it is a kind of structuring reactor. It combines reaction and separation process by using high selective permeation membrane, and then it can remove a product and improve the equilibrium conversion rate. It can also control the rate of the introduction of a reactant, thereby improving the selectivity.

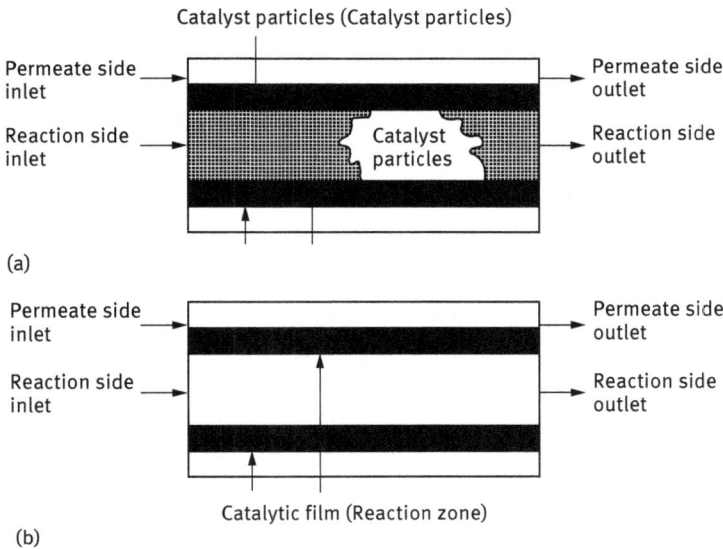

Figure 13.2: Schematic diagram of membrane catalytic reactor [9].

Finally the purpose of high efficiency and energy saving is achieved. The research and application of a membrane catalytic reactor has been gradually extended to all fields of chemical engineering and used in structuring catalytic reactor, which was originally used in separation process and fuel cell. The dehydrogenation reactions of cyclohexane, ethylbenzene and propane are reversible reactions, which have low conversion rate limited by thermodynamic conditions. In the membrane catalytic reactor, removing the reaction product H_2 can break through the limit of thermodynamics, and so increase the conversion rate and the rate of the reaction. Gryaznov et al. [10] found in the hydrogenation reaction experiment that H_2 with high activity in the form of H^+ penetrates through the palladium-based membrane and reacts with hydrocarbons adsorbed in the membrane surface. The diffusion of H_2 through the membrane is not the controlling step of the reaction. In the hydrogenation and dehydrogenation reaction, the yield and selectivity of the reaction can be improved by controlling the amount of H_2. Removing the reaction product H_2 can break through the limit of chemical equilibrium, and the yield is obviously higher than that of the traditional fixed-bed reactor. The catalytic hydrogenation of alkenes and alkynes is a kind of important reaction in the fine chemicals and organic synthesis, and the application research of polymer catalytic membranes and a membrane catalytic reactor has a large part related to this kind of reaction. Cyclopentadiene catalytic selective hydrogenation generally used heterogeneous catalysts to react under pressure and high temperature, but the selectivity of gas-phase catalytic hydrogenation used supported complex catalyst under normal temperature and low pressure. At present, catalytic hydrogenation of alkenes and alkynes, such as different propylene, butadiene, acetylene, propylene, ethylene and so on, by a high-polymer catalytic membrane reactor has been studied. It can obtain both good conversion and selectivity under appropriate reaction conditions. In the chiral catalytic hydrogenation reaction, the transition metal organic complex is generally used as a homogeneous catalyst, which has higher reaction activity and selectivity. But this kind of catalyst is expensive, and the product is not easy to separate after the reaction. The recovery and recycling of the catalyst is difficult, and the final product contains different residual amount of catalyst. Therefore, the purity of the product is not high enough.

The main features of the membrane catalytic reactor are as follows. (1) As for reaction limited by chemical equilibrium, the membrane catalytic reactor can move chemical equilibrium. (2) It is possible to increase the conversion rate of complex reactions. (3) The reaction can react at a lower temperature and pressure. (4) It is possible to operate the chemical reaction, product separation and purification in a membrane catalytic reactor [1].

The membrane catalytic reactor has the function of reaction, catalysis and separation simultaneously. It has the advantages of high reaction efficiency and mild condition, and has incomparable advantages over other reactors. It can be applied to all fields of chemical and biological reactions, and it is especially suitable for the reversible reaction with low equilibrium conversion rate and the biochemical reaction process of product inhibition. It can also carry on two reactions in a reactor at the same time. But in the process of industrialization, it is still in the research and development stage. There are still many problems, such as membrane properties (permeability, mechanical strength, thermal stability, etc.), membrane fouling, design of membrane catalytic reactor, seal up, cost and other issues, which need to be solved in the theory and practice of membrane catalytic reactor. How to solve these problems effectively will be one of the key research topics for us in the future. These problems will be solved with the development of materials science and membrane preparation technology, as well as the application of computer technology in molecular simulation and reactor design. Membrane catalytic reactor is bound to be increasingly more widely used in the industrial fields, such as chemical industry, environmental protection, biological and food industries.

6. **The application of membrane catalysis technology**

Since the 1960s, membrane catalytic reaction technology has been widely used in hydrogenation, dehydrogenation, oxidation, esterification, biochemistry and many other fields. Catalytic hydrogenation is mainly used in the hydrogenation of unsaturated olefins, hydrogenation of olefins, aromatics hydrogenation, selective hydrogenation of C_2, C_3 and so on. Catalytic hydrogenation is mainly used for hydrogenation of unsaturated olefins, cyclic polyenes, aromatics, and selective hydrogenation of C_2 and C_3, etc. Catalytic oxidation of hydrocarbons includes the oxidative coupling of methane in C_1 to olefins, the direct oxidation of methane to methanol, methanol oxidation to formaldehyde, ethanol oxidation to acetaldehyde, propylene oxidation to propylene aldehyde and so on. In the above application example, dehydrogenation of lower alkanes (C_2–C_5) to olefins and C_1 oxidation is of more practical value. In environmental protection, membrane technology has been developed from low-turbidity water treatment to high-turbidity wastewater treatment. The membrane catalytic reaction technology was applied to the biological treatment of organic wastewater. Through the combination of organic membrane and microorganism, efficient and economical treatment of organic wastewater can be realized. Since the 1960s, the membrane bioreactor has been mainly used for the treatment of domestic sewage. Since the 1990s, it has been extended to the high concentration of organic wastewater and refractory industrial wastewater, such as pharmaceutical wastewater, chemical wastewater, food wastewater, tobacco wastewater, papermaking wastewater, printing and dyeing wastewater, and so on. With the rapid development of biological reaction engineering represented by cell culture and enzyme reaction in biotechnology, a membrane bioreactor has been widely used in the fields of fermentation, enzyme catalysis, biological treatment of wastewater, and animal and plant cell culture [11].Kwon et al. [12] produced xylitol by recycling continuous fermentation of biologically active cells, with submerged membrane biological assembly using negative pressure suction. Xylitol production capacity reached 12 g/(L.H.). The production capacity and the total xylitol yield are of 3.4 times and 11.0 times in batch fermentation production, respectively. Shangyong et al. [13] applied the submerged membrane bioreactor to the continuous culture of Cordyceps sinensis. Continuous fermentation after batch fermentation was carried out for 7 days, and then continued for 6 days. The dry weight of mycelium in a fermentation liquid reached 33.2 g/L, the mass concentration of polysaccharide was 5.4 g/L and the yield of polysaccharide was 312 mg/(L.H.), which was ten times of batch fermentation.

13.2.2 Catalytic Distillation/Suspension Catalytic Distillation

1. The concept of catalytic distillation

The catalytic reaction and distillation separation are integrated in a distillation column to complete by catalytic distillation. Generally, the catalytic distillation column is divided into three sections (from top to bottom): distillation section, reaction section and stripping section. The distillation section and the stripping section are not different from the common distillation tower, which can use filler or tower plate. The reaction segment is filled with catalytic activity material, and the reaction function and the separation function are integrated together. In the reaction stage, the reactant is converted into product on the catalyst, and the product is continuously separated from the reaction system. Therefore, the thermodynamic equilibrium of this reaction is broken, so that the product can be obtained more than the amount of the thermodynamic equilibrium conversion. At the same time, the energy efficiency is greatly improved because of the integration of reaction and separation.

2. **The characteristics of catalytic distillation**

   Catalytic distillation has the following characteristics. It uses heat emitted by reaction for distillation separation and saving energy. For a continuous reaction, when the intermediate product is the targeted product, the generated intermediate products can quickly leave the reaction area, to avoid its further reaction to improve the reaction selectivity. For a reversible reaction, the yield of the reaction is limited by the balance, the reaction products is removed from the reaction area timely in catalytic distillation; the equilibrium moves to the right, further improves the yield . The combination of a reactor and a separation tower into one tower simplifies the process and reduces investment in equipment [14].

3. **The catalytic distillation technology application**

   At present, catalytic distillation technology, with its unique advantages, has been widely applied in etherification, ether dissolution , olefin dimerization, alkylation, hydrogenation, isomerization, and dehydration and hydration [14, 15].

   (1) Etherification process. In the presence of large porous acidic cation exchange resin catalyst, isobutylene and methanol can be selectively catalyzed to generate methyl tert-butyl ether (MTBE), a high octane number gasoline addition. Using catalytic distillation,American CDTECH Company developed the process which is widely used in the world, as shown in Figure 13.3.

   (2) Ether solution process. High-purity isobutylene is an important raw material for the production of butyl rubber, using catalytic distillation technology in MTBE as raw materials, and for the production of high-purity isobutene by ether dissolution.

   (3) Olefin dimerization. Olefins in a selective catalytic distillation tower for dimerization can precisely control the reaction temperature and the distribution of reactants,and can generate isomer and trimer, but not copolymer. Using butene mixture to generate octene has been industrialized.

   (4) Alkylation. Catalytic distillation application development is an important field of alkylation (especially phenol) of aromatics. This technology is used to produce various chemical products for drugs, antioxidants,paint, spices and photographic chemicals. For example, ethylene and benzene by catalytic distillation can generate high-purity ethyl benzene. Catalytic distillation technology is also used in the alkylation of butene to produce gasoline blending stock, also has a broad prospect.

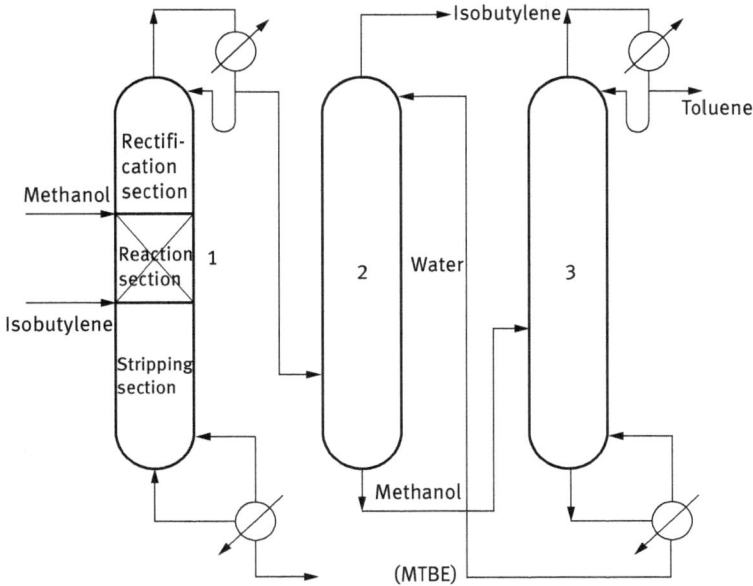

1. Catalyticdistillation column 2. Washing tower 3. Methyl alcohol recovery tower

Figure 13.3: MTBE catalytic distillation separation process.

(5) Hydrogenation and isomerization. With catalytic distillation butene hydrogenation, isomerization, test results were obtained. The equilibrium concentrations
of 1-butene, cis and trans butenes, and isobutene can be adjusted by changing
the operating conditions and the relative feed position of the catalyst bed.
Catalytic distillation technology cannot be applied to all the chemical processes. It
only applies to the reaction process and separation of components using reaction
distillation can be done at the same temperature under the condition of chemical
reaction process. Catalytic distillation does not work if there is a constant boiling
between the components of the reaction, or if the reactants and product have very
similar boiling points. Solid catalyst must be used in the process; it cannot have
miscibility of components and reaction system.

13.2.3 Suspension Catalytic Distillation Technology

1. Putting forward suspension catalytic distillation
Catalytic distillation to heterogeneous catalytic reaction and distillation separation process, coupling simultaneously on the same tower, makes the reaction
and separation, promotes and strengthens each other, and increases the reaction

conversion and selectivity, to reduce the energy consumption, to save investment, and to increase its wide application in the chemical industry. However, the existing studies have shown that in the conventional process of catalytic distillation, the use of "catalyst components" (such as "catalyst bales", structural catalyst components, etc.) in the reaction tower is relatively low. Because the production of "catalyst components" requires larger catalyst particles (usual diameter should be greater than 1 mm), and under the operating conditions of distillation, the influence of diffusion is difficult to overcome, so the efficiency of the catalyst is difficult to get full play. In addition, in the form of conventional fixed-bed catalytic distillation, there is a catalyst component making inconvenience defects such as complicated, loading and unloading and regeneration.

In order to overcome the deficiencies of conventional catalytic distillation, the Research Institute of Petroleum and Chemical Research conducted a research on a novel catalytic distillation process that combines the catalytic reaction of suspended bed and the separation process of distillation [16]. The difference between this new type of catalytic distillation and ordinary catalytic distillation is that the catalyst is not fixed in the reaction tower but suspended and dispersed. Therefore, it is called suspension catalytic distillation (SCD), as shown in Figure 13.4. SCD is characterized by directly using powder catalyst, and no need to make "catalyst components." The efficiency of the catalyst is high, and the catalyst is easy to take out and regenerate, and the presence of the suspended catalyst particles is also conducive to the enhancement of the mass transfer between the gas and liquid phases in the distillation process.

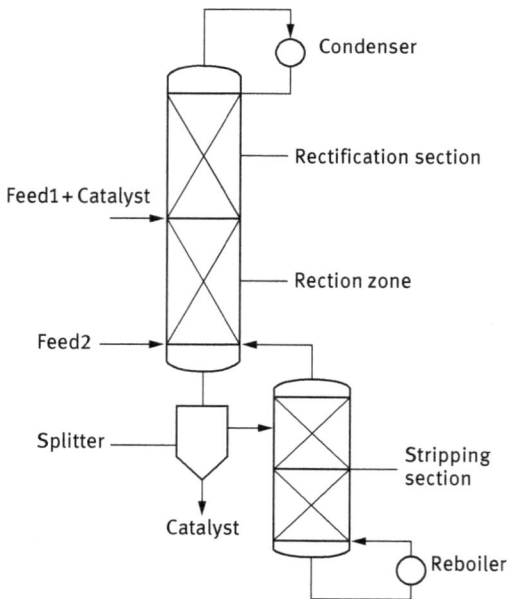

Figure 13.4: Suspension bed catalytic distillation.

2. The suspension catalytic distillation technology application

(1) Suspension catalytic distillation for the synthesis of isopropyl benzene

Isopropyl benzene is one of the important raw materials; preparation of phenol and acetone in the industry at present mainly adopts the fixed-bed process of molecular sieve catalyst. A new SCD process for the synthesis of isopropyl benzene in experiment process is shown in Figure 13.5 [16]. Adopted by the reaction tower of glass towers of 34 mm diameter, the Dixon packing tower is equipped with phi 4 mm, the reaction section is 1 m high and the stripping section is 0.5 m high. Since the alkylation products cumene and polyisopropyl benzene are heavy components, they are produced by the tower reactor and there is no product at the top of the tower. Therefore, the distillation tower does not need a rectifying section, and the tower top adopts full reflux operation. During the test, the catalyst and benzene were made into a suspension by a homogenizer and injected into the upper part of the reaction section by a metering pump. The propylene enters the reaction tower from the lower part of the reaction section after decompression, steady flow and metering. In the reaction zone, the catalyst is held in suspension in a liquid state by the action of rising steam, and catalyzes the alkylation of benzene with propylene while flowing down the surface of the filler. The product cumene, a small amount of polyisopropylbenzene, and the unreacted benzene-carrying catalyst leave the reaction section and enter the stripping section, and pass through the stripping section (most of which is stripped back to the reaction section) into the tower kettle. Tower kettle

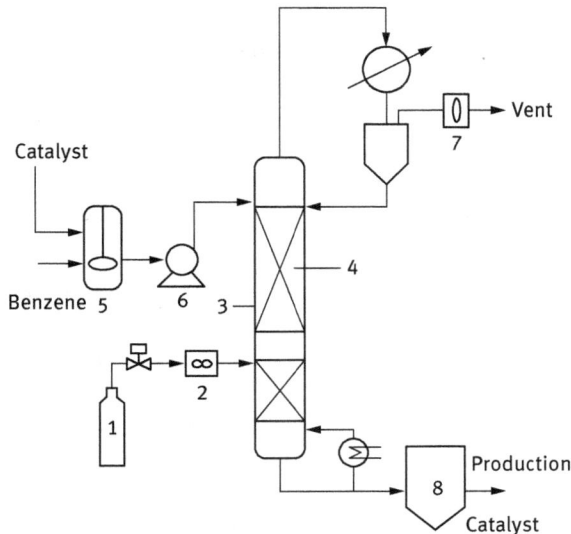

1. Propylene tank 2. Mass flowmeter 3. Reaction tower
4. Distillation packing 5. Homogenizer 6. Metering pump

Figure 13.5: SCD synthesis of isopropyl benzene experimental flow diagram.

produced fluid into the separator for solid–liquid separation; the separation of catalyst made from benzene suspension recycled again.

The experimental results show the synthesis of isopropyl benzene suspension catalytic distillation process. At atmospheric pressure and low temperature (80–100 °C), the conversion rate of propylene is close to 100% and the selectivity of isopropyl benzene is more than 90%. Products prepared by ordinary distillation can be of more than 99.9% purity. Normal propyl benzene impurities and C_8 aromatics content are less than 100 µg/g of isopropyl benzene product.

(2) **Suspension catalytic distillation for the synthesis of linear alkyl benzene**

Linear alkyl benzene (LAB) is an important alkyl benzene product, which is widely used in the manufacture of synthetic detergent. At present, HF is still mainly used as a catalyst for production in the industry, causing serious corrosion and pollution problems. UOP's fixed bed process using solid acid as a catalyst has been industrialized, but the single-pass life of the catalyst is short, and benzene flushing is used for regeneration after 24 hours. Such frequent regeneration is obviously troublesome for fixed bed reactors. In order to research and develop a more advanced LAB production process, based on the study of SCD synthesis of cumene, a new set of experimental equipment for catalytic distillation of suspended beds was established, and the process of SCD synthesis of LAB was explored [17].

SCD synthesis LAB experiment process is shown in Figure 13.6. The reaction tower is a stainless steel sieve tray tower. A catalyst sedimentation separator is connected in series between the reaction section and the stripping section. The 10 to 50 µm supported phosphotungstic acid (PW/SiO 2) catalyst leaves the reaction section and enters the separator for separation and recycling. So as to avoid the occurrence of side reactions caused by the catalyst entering the kettle. The experimental results show that the synthesis of LAB by the suspended bed catalytic distillation process can reach nearly 100% conversion efficiency and selectivity under conditions close to normal pressure (0.06 MPa) and low temperature (100 °C). Further analysis showed that SCD technology preparation LAB product, 1-LAB content, is as high as 35%; fewer impurities such as indan and tetralin can be used as raw materials for the preparation of high-grade detergent.

Results of SCD industrial synthesis of isopropyl benzene and linear alkyl benzene indicate that SCD with fine-particle catalyst, reaction under suspension state, effectively strengthen the mass transfer, improve the efficiency of the catalyst in response to more moderate conditions, effectively inhibit the occurrence of adverse events and reduce feed benzene/ene mole ratio.

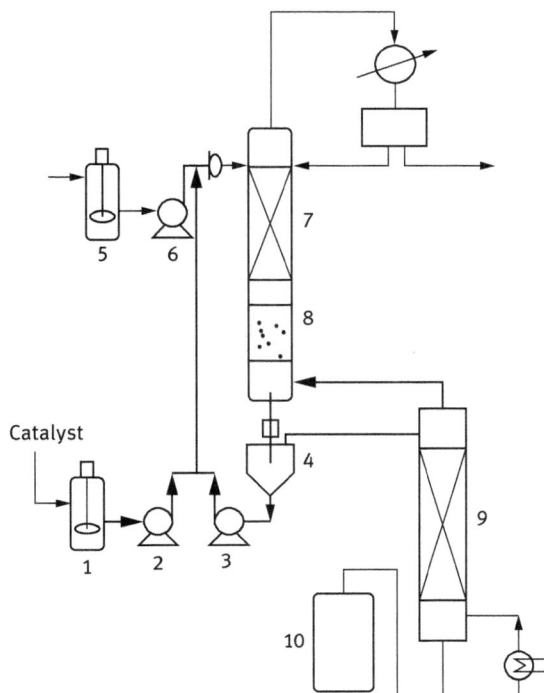

1. Benzene tank 2, 3. Slurry pump 4. Liquid-solid separator
5. Olefin pot 6. Metering pump 7. Distiller 8. Reaction period
9. Stripping section 10. Product tank

Figure 13.6: SCD synthesis LAB experiment flow diagram.

13.2.3 Alternating Flow Reaction

1. Alternating flow reaction concepts

Alternating flow reaction, also known as reflux reaction, controls its direction by controlling the timing of reverse logistics in and out of the reactor, uses heat emitted by reaction to heat cold raw material, makes full use of the heat of reaction, reduces energy consumption and reduces operating costs.

Alternating flow reaction integrates the functions of chemical reaction and heat transfer into one, and its working principle is shown in Figure 13.7.

Alternating flow reaction consists of a reactor, two groups of valve, a heat absorption system, and inert fillers on both ends of the reactor, which are used as heat storage. The reactor gas flow is controlled by two sets of valve device. In the first half of the process cycle (closed valve 1, open valve 2), material mixed gas flows through the lower inert medium at room temperature, heats, and then reaches the catalyst layer; this time the temperature of the mixed gas is enough to make reaction and produce heat. Finally, the upper gas reaches the catalyst layer and heats catalyst to back out. In the first half of the process cycle, the lower part

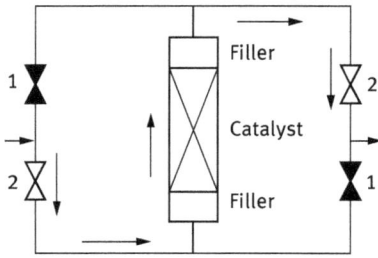

Figure 13.7: Alternating flow reaction working principle diagram.

of the reactor was hot, and then it slowly cooled down; simultaneously, the upper part was cold first, and then gradually it started heating. After a certain period, through a reverse valve (open valve 1, closed valve 2) change the flow of gas; new gas uses upper inert filler during heating. In the second half of the process cycle, the original upward flow of gas changes the flow direction of gas. In this process, the heat generated by the exothermic chemical reaction is absorbed by the heat absorption system in the middle part of the reactor.

2. **The alternating flow reaction characteristics**
   (1) In alternating flow reaction with compact process, the heat recovery is high and the thermal performance has good advantages; it is particularly suited to take advantage of low concentration of reactants in manufacturing chemicals, low-grade fuel gas heating, catalytic combustion removal of organic toxicants in exhaust gas and so on in a series of exothermic reaction.

      In flow reaction technology with compact process, the heat recovery is high and the thermal performance has good advantages.
   (2) Due to the volume heat capacity of gas and solid-phase difference of about three orders of magnitude, wave propagation speed is very slow. Therefore, to transform the cycle of tens of minutes or more, the valve switching frequency is acceptable in engineering.
   (3) For the reaction mixture flow rate and the composition of frequent fluctuations, to transform the steady-state response technology has better stability and maneuverability.

3. **The alternating flow reaction application**
   Industrial applications of alternating flow reactions include the oxidation of volatile organic compounds to purify industrial exhaust gases, the reduction of nitrogen oxides in industrial exhaust gases, and the oxidation of sulfur dioxide to produce sulfuric acid [18]. In the case of fluctuations in the flow rate and concentration of the process gas, a counter-current catalytic combustion reactor is used, and the operating cost can be reduced by 80% compared with the conventional shell-and-tube catalytic combustion reactor. At present, there are dozens of sets of this kind in the world, industrial equipment in operation. For using alternating flow catalytic reactor for ammonia selective reduction of nitrogen oxides, Russia has a built

capacity of about 11,200 m³/h device; the concentration of nitrogen oxide of the outlet of the reactor can be less than 70 mg/m³. For oxidation of sulphur dioxide to produce sulfuric acid, the countercurrent catalytic reactor can reduce the operation cost by 5–20%, and save equipment investment by 20–80%. In China, Henan Province can be used as a belt in the upstream of the catalytic reactor, the reactor diameter is 6.5 m, the capacity is 33,500 m³/h, sulfur dioxide fluctuates within 1–5% and sulfur dioxide conversion rate can be greater than 90%.

13.2.4 Stable Magnetic Field Fluidized Bed

1. The concept of stable magnetic field fluidized bed
Magnetic fluid bed on the basis of normal fluidized bed consists of a force field, a magnetic field and the basic structure; its working principle is shown in Figure 13.8. According to fluidization medium, magnetic field fluidized bed can be divided into liquid–solid fluidized bed magnetic field, gas–solid fluidized bed and gas–liquid–solid three-phase magnetic fluidized bed. According to the magnetic field direction, it can be divided into axial magnetic field of fluidized bed and fluidized bed transverse magnetic field [19]. The most studied magnetic field in fluidized-bed fluids is the uniform magnetic field, which does not change with time, and is usually generated by Holmitz coils or permanent magnets. When the fluidization medium flow rate is higher than the minimum fluidization velocity and speed did not reach out before bed piston with inflation, it is called the stable magnetic fluidized bed. For additional magnetic field, streaming media in the fluidized bed are required to have a magnetic responsiveness. Magnetic particles in the fluidized bed are affected by magnetic field force, but not by gravity, buoyancy and drag effect, and the interaction between the magnetized particles under high magnetic field force. The change in intensity of magnetic field results in a different fluidization phenomenon, as shown in Figure 13.9.

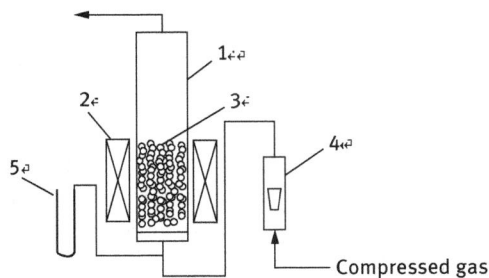

1. Fluidized bed body 2. Magnetic field generating device 3. Magnetic material 4. Rotor flowmeter 5. U-type pressure gauge

Figure 13.8: Fundamental structure and working principle of magnetic field fluidized bed.

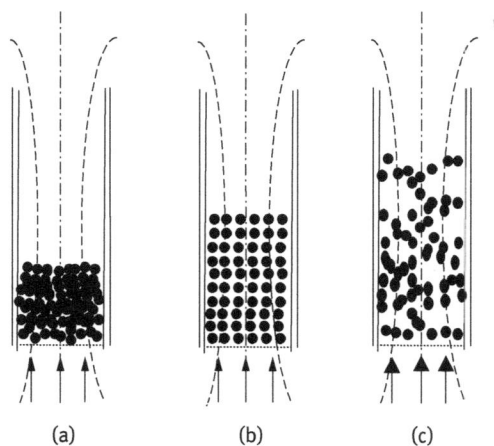

Figure 13.9: Magnetic field fluidized bed state under different magnetic field strength.
(a) Magnetic fixed bed.
(b) Magnetic stabilized bed.
(c) Magnetic bubble bed.

2. **The stable magnetic field characteristics of fluidized bed**

   Stable fluidized bed is a magnetic fluidized bed of special form; it is not in the axial and will change over time even with only weak movement of the stable magnetic field formed under bed. Magnetic field with fixed bed and stable fluidized bed has many advantages; it can use small particles such as fluidized bed solid without high pressure drop. The effect of the external magnetic field effectively controls the back-mixing between phases, and the uniform voidage makes it difficult for the interior of the bed to experience channeling. Liquidity of tiny particles can make loading and unloading of solid very convenient. Stable fluidized bed using magnetic field not only avoids fluidized bed solid particle erosion phenomenon often appearing in the operation, but also avoids the local hotspots which may emerge in the fixed bed. At the same time, magnetic field with stable fluidized bed can have stable operation in a wider scope, can also be broken bubble and improve the mass transfer. Anyway, stable magnetic field fluidized bed is made of different domain knowledge of fluid mechanics, and reaction engineering (magnet) combined to form a model of new ideas is a kind of new and creative forms of bed.

3. **The stable magnetic field fluidized bed applications**

   In the process of caprolactam production, coarse caprolactam solution containing a small amount of property of caprolactam, these substances cannot be removed by conventional separation methods, but the presence of these substances can seriously affect the quality of the finished products, that is, caprolactam; industry generally uses hydrofining method to remove these impurities. In caprolactam hydrogenation in the industry at present we adopt a continuous stirred tank reactor. The industry has the disadvantages of complicated processes, large catalyst consumption, low efficiency, and the need for filtration and separation of the catalyst. Therefore, it is necessary to develop a new type of hydrorefining process.

The Research Institute of Petrochemical Research conducted an exploration of the application of magnetically stabilized bed using the caprolactam hydrorefining process as an example [20]. The effect of various factors on the hydrotreating of caprolactam in a magnetically stable bed was studied on a small test apparatus, and the stability of the SRNA-4 catalyst was investigated. The results show that the continuous use of SRNA-4 catalyst still have higher hydrogenation activity after 1,350 h, compared with the existing kettle type process; hydrogenation effect will be three to five times higher; catalyst consumption will be reduced by more than half. The suitable reaction conditions are as follows: temperature is 60–90 °C, pressure 0.4–0.8 MPa, airspeed 30–50 h^{-1}, hydrogen/liquid feed volume ratio 1.5–3.0 and the magnetic field intensity 15–25 ka/m.

For years a capacity of 7×10^4 tons by caprolactam hydrogenation refining process, using the current industry commonly utilized continuous stirred tank reactor, the reaction volume is 10 m^3; if a magnetic stable bed reactor is used, the reactor volume is only 1.8 m^3. Thus, the stable magnetic bed reactor can make equipment.

4. **Stable magnetic field problem and prospect of fluidized bed**
At present, magnetic field stable fluidized bed applications have some restrictions; further research work is needed in the following areas: 1. research and development of magnetic catalysts. The catalyst remanence should be less when removing the magnetic field; catalyst should have good low-temperature reactivity. 2. Further research is needed on uniform steady magnetic field of magnetic stable bed reactor engineering amplification. 3. Due to the stable magnetic bed particularity, it is necessary to find bed state with the magnetic field, the physical properties of the catalyst and the quantitative relation between the fluid flow. 4. The study of the theory of the stable magnetic bed remains to be further deepened. In the future, more in-depth studies should be conducted on the aspects of local hydrodynamic properties, heat transfer characteristics and heat transfer mechanisms, mass transfer mechanisms, and reactor models.

13.3 The Coupling Technology of Separation Process

In recent years, new types of coupled separation technology, such as catalytic rectification, membrane rectification, adsorptive rectification, reactive extraction, complexation adsorption, membrane extraction, chemical absorption and electrophoretic extraction, which combines the advantages of the two types of separation technology and has a unique feature, have been greatly developed and successfully applied to the production process of chemical industry.

13.3.1 Reaction Separation Coupling

Reaction separation coupling is the integration of chemical reaction and physical separation processes, such as reactive distillation, solvent extraction, reactive absorption and membrane separation, so that the reaction and separation operation is completed in the same equipment. Reaction separation coupling technology can reduce the equipment investment and simplify the process, which has many advantages. For example, catalytic distillation can significantly improve the selectivity of chemical reactions and reduce side effects. For a reversible reaction, the equilibrium of the chemical reaction can be changed significantly, and the yield of the reaction can be improved. For the exothermic reaction, the heat generated by reaction can be effectively utilized, and the heat energy consumption can be reduced. The catalytic distillation operation can change the reaction rate and product distribution by changing the operating pressure, controlling the reaction temperature, changing the vapor partial pressure of the gas phase material, and adjusting the liquid reactant concentration. There are many catalytic distillation applications, such as the process of esterification, alkylation, and hydration and dehydration of etherification. The main products are methyl acetate; ethyl ester and butyl; methyl tert butyl ether; ethylbenzene; cumene; and so on. The commonly used catalysts are ZSM5 and HY zeolite, acidic cation exchange resin and the acidic zeolite [21]. Compared with the traditional production process of methyl acetate, the integrated process put a number of tasks in one device to complete the reaction and separation, which are originally separated. It is a prominent example of chemical process intensification because it not only reduces the cost of investment in equipment and infrastructure, but also reduces the size of the chemical plant significantly.

Research works at Beijing University of Chemical Technology developed a technology that integrated benzene and propylene alkylation, and polyisopropylbenzene trans alkylation reaction in the same equipment, which can reduce the cost of investment in equipment and simplify the process flow. Alkylation of benzene with propylene produces cumene using catalytic reactive distillation technology, and acidic zeolite or acidic ion exchange resin as catalyst. The technological conditions are as follows: temperature 50–300 °C and pressure 0.05–2.0 MPa. The molar ratio of benzene to propylene is (2–10):1, the conversion rate of propylene can be up to 98% and the selectivity of cumene can be up to 90%.

13.3.2 The Coupling of Membrane Separation

Separation of mixture by using membrane as a separation medium is a new separation technology. The membrane is a discontinuous interface between the two phases. Membranes which are usually solid membrane (polymer or inorganic material membrane) and liquid membrane (emulsion liquid membrane or supported liquid membrane) can be divided into gas phase, liquid phase, solid phase or their combination. Membrane separation process is a process of separation of the mixture under different driving forces (pressure, electric field, concentration difference, etc.) by specific selective permeation properties of different membranes. Generally, membrane technology is integrated with other separation techniques. In membrane absorption technology, the function of the membrane is to allow the gas to pass through the membrane without passing through the liquid. Membrane distillation technology is also an integrated technology, which allows liquid to vapor through the membrane, and collects the coacervation on the other side of the membrane. Membrane distillation is a new and efficient membrane separation technology that has been rapidly developed in recent decades.

1. **Membrane distillation**

Membrane distillation is a new type of membrane separation technology, which has been developed rapidly in recent decades. This technique is based on the effect of difference in water vapor pressure on both sides of the membrane. The water vapor on the hot side of the membrane passes through the film hole into the cold side, and then condenses in the cold side. This process is the same as the conventional distillation process, which is evaporation–transfer–condensation. Compared with other membrane separation processes, membrane distillation has unique advantages: it can separate under normal pressure and slightly higher than normal temperature. It can make full use of solar energy, industrial waste heat and low energy, with simple equipment and easy operation. It can be used in seawater and brackish water desalination, ultra pure water preparation, concentrated aqueous solution and medicine, environmental protection and many other aspects [22].

Membrane distillation is a membrane separation process which combined membrane technology and evaporation process. The membrane is a hydrophobic microporous membrane which is not wetted by the solution to be treated. One side (hot side) of the membrane is in direct contact with the hot solution to be treated. And the other side (cold side) is in contact with cold water solution, directly or indirectly. One side of the membrane is in direct contact with the hot solution to be treated (known as the hot side) and the other side is in direct or indirect contact with the cold aqueous solution (known as the cold side). The other specific group is a hydrophobic membrane barrier in the hot side. The volatile components of the hot-side solution vaporize at the membrane surface, enter the cold side through the membrane and are condensed into a liquid phase, and other components are blocked on the hot side by the hydrophobic membrane, thereby achieving the purpose of separation or purification of the mixture. Membrane distillation is a process of simultaneous heat transfer and mass transfer. The driving force for mass transfer is the difference in vapor pressure between the two sides of the membrane. Therefore, the realization of membrane distillation must have two conditions. 1. Membrane distillation must have a hydrophobic microporous membrane. 2. On both sides of the membrane, a certain temperature difference must exist to provide the necessary driving force for mass transfer. The main advantages of membrane distillation technology are as follows. 1. Ions, macromolecules, colloids, cells and other nonvolatile substances cannot completely penetrate through it. 2. The required pressure is not high. 3. Due to the large pore structure, it is less likely to cause fouling on the membrane. 4 It can be used at lower temperatures.

2. **Membrane absorption**

    Membrane absorption is a novel separation technique combining membrane-based gas separation and traditional physical adsorption, chemical absorption and cryogenic distillation. Compared with traditional absorption technology, membrane absorption has attracted much attention because of its characteristics such as large gas–liquid contact area, high mass transfer rate, no entrainment with foam fog and mild operating conditions. Its mass transfer includes absorption, desorption and complexation in membrane pores, and the formation of the solution layer, which are distribution processes of the molecules in a two-phase or multiphase. As a branch of membrane separation technology, its technology has already been known. However, due to the lack of suitable high-efficiency membrane, it does not have any large-scale industrial application in a long time. As the first high-throughput cellulose acetate asymmetric membrane was prepared by Loeb and Sourirajan in 1960 and the first set of membranes for gas separation device was prepared from the United States Monsanto subsidiary Permea in 1979, membrane industry with various functions as the main body has become more complete edge disciplines and emerging industries, and has developed in the direction of the reaction separation coupling and integrated separation technologies. As a representative of this kind of integrated technology, membrane absorption technology has attracted extensive attention and has been applied to the industrial field in the process of making membrane, membrane materials, mass transfer mechanism and model [23].

13.3.3 Adsorptive Distillation

Adsorption is a separation process with the advantages of high separation factor, high purity and low energy consumption. The adsorption process is applicable to the separation of those systems with low relative volatility, which cannot be separated or economically used for ordinary distillation, such as azeotropic or isomeric. But there are shortcomings in the adsorption process. The dosage of adsorbent is large, it is difficult to achieve the operation continuity for the batch operation and the product yield is low. Therefore, the development of composite process combined with adsorption and distillation will be able to offset the respective deficiencies and adverse conditions. Adsorptive distillation was first proposed by the American R. G. Rice. The device consists of a multi-stage continuous flow Kettle-type absorber in series. From the middle of a feed, each level of the tank is filled with adsorbent (the saturation adsorption capacity of the adsorbent at low temperature is generally large) which can selectively adsorb impurities to purify components. R. G. Rice proposed adsorption distillation; in fact, it is adsorption and desorption of a multi-stage fixed bed; each level must be through four steps: feeding, adsorption, drainage and desorption. It is a non-steady-state batch process, which does not reflect the advantages of continuous and large processing capacity of the distillation process. Zhou Ming et al. developed a new separation process called adsorption distillation. The process of adsorption and distillation is carried out in the same adsorptive distillation column to enhance desorption effect, which not only improves the separation factor, but also makes the distillation and desorption operations to carried out in the same distillation desorption tower. Therefore, the adsorption distillation process has the advantages of high separation factor, continuous operation, low energy consumption and high production capacity, and it is suitable for the separation of the constant boiling point system and the similar boiling point system, and the need for high-purity products. At present, the scholars have carried out research on the preparation of absolute ethyl alcohol, and the separation of propane and propylene. The conclusion is that the energy consumption is much lower than that of the conventional distillation [24].

Separation process coupling technology can also solve many questions that were difficult for the traditional separation technology to solve. Therefore, it has broad application prospects in the high technology fields of biological engineering, pharmacy and new materials. For example, a fruitful application of fermentation extraction and electrophoretic extraction in the separation of biological products was successfully achieved. Adsorption resin and organic complexing agent have the characteristics of high separation efficiency and easy analytical regeneration. Highly efficient separation of vitamins can be carried out by electric coupling chromatography. Supercritical extraction of CO_2 and nanofiltration coupling can extract valuable natural products and so on.

13.4 Microchemical Technology

Microchemical technology is the frontier of chemical engineering, which is typically represented by some equipment, such as microreactor, micromixer, microseparator, micro heat exchanger and so on, and focuses on studying the characteristics and rules of "transport and reaction" in micro time scale. It becomes the hotspot of domestic and international academic and industrial research that aims to achieve high efficiency in process, low energy consumption, and safety and control in modern chemical techniques by applying the design ideas of refinement and integration [25, 26]. Microchemical system refers to enhanced reaction and separation process of single-phase or multiphase system with micron-scale dispersion under the role of micro-distructure which is combined by the reaction, mixing, heat transfer and separation device that is precisely made by machining. Compared with the conventional scale system, the microchemical system has the advantages of high heat transfer and mass transfer rate, high intrinsic safety, low energy consumption, high integration, small amplification effect, strong controllability and so on. It can be used to realize the isothermal operation of rapid and strong exothermic/endothermic reaction, the fast mixing of two phases, the synthesis of flammable and explosive compound and the field production of a highly toxic compound, which has a broad application prospect.

13.4.1 Introduction

In recent years, microchemical technology has entered a period of rapid development; researchers of China and other countries have developed a variety of new microchemical equipment. It provides reference and guidance for the understanding of the common law of microchemical process from a new angle of view, rational decoupling of the "three pass and one back" coupling process in microscale and the establishment of microchemical engineering theory system, by the study of the microstructure, the characteristic scale and the effect of the surface/interface. Under the competition of several kinds of fluid force in the microscale, there exist four kinds of dispersion flow patterns, such as extrusion, drop out, jet flow and laminar flow in microchemical equipment. It can form a liquid droplet or bubble with a diameter of 5–1,000 μm and disperse uniformly, which is one to two orders [27–29] of magnitude smaller than that of the traditional chemical equipment. The mass transfer efficiency in a single device can reach more than 90%, and the volumetric heat transfer coefficient can also be increased by 1–2 orders of magnitude. The results show that the volumetric mass transfer coefficients of gas–liquid, liquid–liquid, gas–liquid–liquid and liquid–liquid–solid systems are one to two orders of magnitude higher than those of traditional equipment. The mass transfer efficiency in single equipment can reach 90% while the volume heat transfer coefficient can be increased by one to two orders of magnitude.

13.4.2 The Principle of Microreactor

The microreactor is a kind of microchannel reactor built on the basis of continuous flow, which is used to replace the traditional reactor, such as glass flask, funnel and conventional batch reactors, commonly used in industrial organic synthesis (such as reaction vessels). A large number of microreaction channels are made using precision machining technology in a microreactor, which can provide a large specific surface area and high heat transfer efficiency. In addition, the continuous flow of the microreactor is used to replace the batch operation, which makes it possible to accurately control the residence time of the reactants. These characteristics make the organic synthesis reaction to be controlled precisely on the microscale, which provides the possibility to improve the reaction selectivity and the safety of operation.

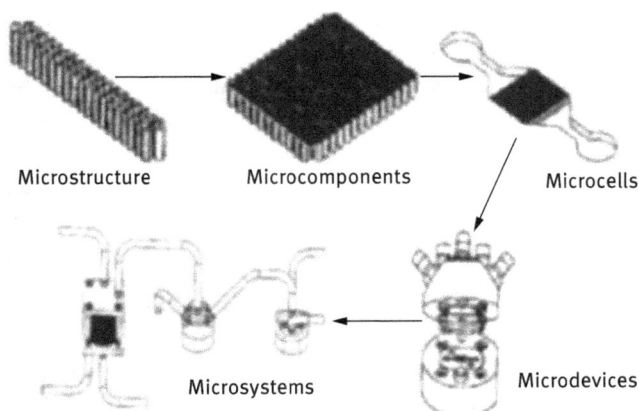

Figure 13.10: Hierarchical structure of micro reaction system.

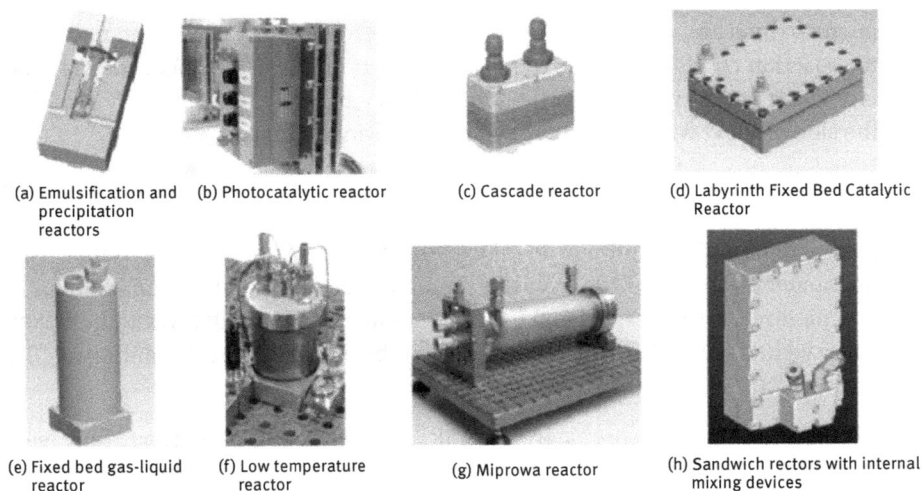

| (a) Emulsification and precipitation reactors | (b) Photocatalytic reactor | (c) Cascade reactor | (d) Labyrinth Fixed Bed Catalytic Reactor |
| (e) Fixed bed gas-liquid reactor | (f) Low temperature reactor | (g) Miprowa reactor | (h) Sandwich rectors with internal mixing devices |

Figure 13.11: Common industrial micro reactors.

Microreactor in the structure often uses a hierarchical structure, first the subunit forms a unit, then the unit forms a larger unit and so on, as shown in Figure 13.10. This kind of characteristic is different from that of the traditional chemical equipment. It is convenient for the microreactor to facilitate the expansion and flexible adjustment on the scale of production by the way of "increasing the number of amplification."

Common microreactors in industry are shown in Figure 13.11.

Compared with the traditional production process, the production process using the microreactor has the following main characteristics.

(1) The precise control of the reaction temperature. The microreactor has a large specific surface area, which determines the great heat transfer efficiency of the microreactor. Even if the reaction moment releases a large amount of heat, the microreactor will also transfer it in a timely manner, to maintain stable reaction temperature. The heat transfer efficiency of a strong exothermic reaction in the conventional reactor is not high enough, and the local overheating phenomenon often occurs. Local overheating often leads to the formation of by-products, which results in a decrease in yield and selectivity. Moreover, if a lot of heat produced in severe reaction during production cannot be exported in a timely manner, it will lead to an explosive accident or even an explosion.

(2) The precise control of the reaction time. The conventional batch reaction, to prevent too severe reaction by gradually adding the reactants, makes a part of the material's residence time too long. In many reactions, it will lead to the generation of by-products whose residence time of reactants, products or intermediate transition states is long under reaction conditions, which reduces the yield of the reaction. The microreactor technology used the continuous flow reaction of the microchannel and can accurately control the residence time of the material under the reaction conditions. Once the optimal reaction time is reached, the material is delivered to the next step or the termination reaction, thus effectively avoiding the side product which is caused by the long reaction time.

(3) Instant evenly mixed material in a precise proportion. In the fast response with fixed ratio of the reaction materials, if the mixture is not good enough, there will be a partial proportion of excess, which resulted in a by-product. This phenomenon is difficult to avoid in the batch reactor. The reaction channel of the microreactor is only a few tens of micrometers. The material can be mixed with accurate and fast mixing ratio, thereby avoiding the formation of the by-product.

(4) The structure to ensure safety. The microreactor adopts continuous flow reaction, which is different from the batch reactor. Therefore, the number of chemicals stuck in the reactor is always very small. And the degree of damage is very limited even in the case of out of control. Moreover, the heat transfer efficiency of the microreactor is very high. Even if the reaction moment releases a large amount of heat, the microreactor will also transfer it in a timely manner, to maintain stable reaction temperature and reduces the possibility of safety and quality accidents. Therefore, the microreactor can easily cope with harsh process requirements, to achieve safe and efficient production.

(5) No amplification effect. The fine chemical industry often uses a batch reactor for production. When the small test process is enlarged, it usually takes a while to explore. The general process is diminutive test–pilot-scale test–mass-produce. When using the microreactor technology for production, the process is not

expanded by increasing the characteristic dimension of the microchannel, but by increasing the number of microchannels. Therefore, the best reaction conditions without any changes can be directly used for production. There is no amplification problem in the conventional batch reactor, thereby greatly reducing the time needed for the product to reach the market from the laboratory.

The microreactor has a discrete three-dimensional structure. It has a number of diameters from several microns to hundreds of microns in the reaction channel. The reaction volume ranges from a few nanoliters to a few microliters, and the total length of the reaction channel is usually a few centimeters. Materials for making microreactors include metals, ceramics, polymers, glass, silicon or a combination of these. There are many ways, such as micromachining, photolithography or electrochemistry, to make a microreactor. It depends on factors such as the amount of material used, the size of the device and the quantity produced. First, take a suitable material for the plate and use the mechanical force, laser beam or etching technology to form a channel. Then these channels are merged into pipes, which can be done by simply bonding or welding a cover plate or by chemical vapor deposition. In this production process, a thin sheet with a microstructure can be used as a cover plate to achieve a number of sheet stacking. At the same time, a large number of different reactor types can be commercialized, ranging from a single mixer and a heat exchanger to a complete integrated system with different residence time units.

The ratio of the wall height to the channel width of the microreactor is larger, which means that the surface area is large and the heat transfer and mass transfer capability is strong. In addition, the fluid in the channel usually has the laminar flow characteristics of Reynolds number 1 to 1000. These two characteristics constitute the main differences between the microreactor and the traditional chemical reactor. Generally speaking, the traditional chemical reactor has a lower specific surface area, and the fluid is in a turbulent state.

13.4.3 The Application and Prospect of Microchemical Technology

Some reactions, such as the direct fluorination of organic compounds by elemental fluorine, which cannot be carried out by the traditional method can be realized by microreaction technology. A kind of falling membrane microreactor was developed in Germany, which was used to fluorination react with toluene. The liquid flows through the microchannel by 35 μm membrane. The specific surface area of the reactor reaches to 20,000 m^2/m^3, which is an order of magnitude higher than that of the traditional contact device. The structure of the reactor is shown in Figure 13.12. In this reactor, the yield of toluene is 20%, which is four times of the bubble column reactor, and the by-product is little.

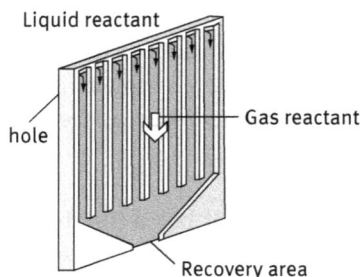

Figure 13.12: Schematic diagram of structure the micro reactor for toluene direct fluoride falling membrane.

Kim applied the fuel cell to a microreactor for the production of hydrogen from sodium borohydride. The reactor has three photosensitive glasses (surface layer, reactor layer and base layer). Nickel is used as catalyst in the reactor, and Co-P-B is applied on the surface of nickel for the hydrolysis reaction of sodium hydroxide. The production rate of hydrogen under 40 °C reached 5.6 mL/min. It provides the maximum output power of 157 mW of the battery, in the case of current of 0.5 A [30].

The application of nanometer calcium carbonate technology, developed by Tsinghua University, in the process of industrialization was realized, and reached the annual production scale of million tons [31]. Industrial-grade microextraction equipment was developed, such as membrane dispersion, microchannel dispersion and micropore array dispersion dispersion. It has been in the process of pilot-scale application, such as deacidification of crude oil, and acid extraction and purification process of phosphoric acid in the preparation of caprolactam. The selectivity of the caprolactam preparation reaction in the pilot-scale toluene process has also been improved. Dalian Institute of Chinese Academy of Sciences has developed a set of mixing, reaction and heat transfer in one of the microchemical systems having an annual processing capacity of 80,000 tons. It has been used for the industrial production of ammonium biphosphate. With small size (micro unit volume less than 6 L), fast response and fast moving speed, the process is easy to control, and the system makes no vibration and noise, emits zero emissions, produces quality product, has stability in producing quality product and so on. So far, more than a year of stable operation, it effectively solves issues such as the safety of the production process, environmental protection, product quality stability and so on. It has successfully developed a set of methanol oxidation reforming, CO selection oxidation, methanol catalytic combustion, feedstock vaporization and micro change thermal subsystems for micro hydrogen source system for kW-level proton exchange membrane fuel cell, with the advantages of small size, fast start, low CO content, high specific power and so on [32, 33].

In recent years, many traditional concepts of chemical industry have changed by the advancement in research and development of microchemical technology. The understanding of the multiphase flow system gradually changed from meter and millimeter to micron and submicron. With the basic laws of multiphase flow, mixing, transfer and reaction in microscale, new chemical equipment is constantly developed, and the process of green, safety and high efficiency is expected to be realized. The successful development and application of microchemical technology will change the performance, volume, energy consumption and material consumption of the existing chemical equipment. It will be a major breakthrough in the existing chemical technology and equipment manufacturing, and will also have a significant impact on the entire chemical industry. As an emerging discipline, there are many problems which need to be further studied: for example, (1) the behavior of complex multiphase flow in microdevice and its control law – including the internal mechanism and the establishment of the physical model of the differential powder, surface interfacial properties, transfer laws and mixing characteristics of multiphase fluids; (2) dynamic interfacial behavior in microscale, development of new testing techniques and methods (contactless measurement technique); (3) development of a new type of microchemical equipment and technology, preparation and characterization of nanocatalysts in microreactor; (4) structural optimization design, parallel amplification and system integration of microreactor; (5) overall performance and structure optimization of micro heat exchanger.

13.5 Intensification Technology Based on Energy Field

Many nontraditional technologies, such as microwave, ultrasonic, radiation and plasma technology, use nonthermal energy to intensify the process.

13.5.1 Microwave Technology

1. **Theory of microwave field**

   Microwave in the electromagnetic spectrum is between the infrared and radio, and the wavelength is in the range of 0.001–1 m (frequency of 0.3–300 GHz) in the region. The microwave wavelength used for heating technology is generally fixed at 0.0122 m (2.45 GHz).

   A microwave generator produces an alternating electric field. The electric field is applied to the object in the microwave field. As the dipole–dipole vibration of the material is similar to the microwave frequency, the polar molecular orientation changes with the change in the direction of electric field. In an electric field with a certain frequency of alternating change, the dipole in the medium can rotate, vibrate or swing rapidly to form a displacement current. Due to the thermal motion of molecules and the interaction of neighboring molecules, the dipole molecules are disturbed and hindered by the change in the direction of the applied electric field. This kind of oscillation often lags behind the change of electromagnetic field, and it has the function of similar friction. The molecules that make the chaotic motion can obtain energy and increase the movement of the molecules, which greatly increase the frequency of collisions between the reactant molecules, and can achieve activation in a very short time. And the number is far greater than the traditional way. Therefore, microwave heating has the characteristic of high heating rate. Microwave heating of matter is from the molecule of the substance, and material molecules absorb electromagnetic energy at a high speed of hundreds of millions of times per second to produce heat, so-called fast internal heating. The energy generated by the intermolecular vibration can be converted into heat energy, which can directly stimulate the reaction between the materials. Compared with conventional heating, microwave has the advantages of fast heating speed, uniform temperature, no temperature gradient, high temperature, small heat loss and so on [34, 35]. In addition, different substances have different dielectric properties and thus have different microwave absorption capabilities, which in turn enables selective heating of microwave radiation. When the material is placed in a microwave field, the electric field can make molecular polarizability, and the magnetic force can move and rotate these charged particles. The diffusion of molecules and the average energy of the molecule are increased. The activation energy of the reaction is reduced, and the speed of chemical reaction can be greatly improved.

2. **Mechanism of microwave effect on chemical reaction process**
   In the process of accelerating chemical reaction, the heat effect of microwave is widely believed to exist. Especially, the nonthermal effect is gradually becoming the focus of controversy for the special effects of microwave in chemical reactions. Many scientists believe that in most cases, the chemical reaction rate in the microwave field is enhanced by the simple thermal/dynamic effects. That is, it is the result of the high-reaction temperature after microwave irradiation on the polar substance, which can promote the reaction. Microwave heating is in the fully closed state. Microwaves penetrate at the speed of light into the interior of an object, which is absorbed by the movement of electrons or ions or by the polarization of the defect dipoles. That is, it is converted into heat to form the effect of "integral" heating inside and outside the material, which greatly reduces the heat loss, reduces the heating time, and can achieve the effect of rapid heating and energy saving. The heat effect of microwave is related to the ability of the material to be converted into heat energy at specific frequency and temperature. At present, it has been reported that the heating rate, the solvent property, the microwave output and the polarity of the microwave energy absorbed in the microwave reaction have been studied. In recent years, people have often used "microwave special effect" or "microwave nonthermal effect" to describe the effect other than the microwave thermal effect, and even a lot of literature works equate special effects and nonthermal effects. In fact, the special effect is the effect of microwave. Nonthermal effects and special effects have the essence of the difference: the difference between the two is

that the special effect does not rule out the correlation with temperature. The specific effects of temperature variation can be explained by thermal effects, which cannot be explained by temperature variation. Microwave on the role of chemical reaction, on the one hand, increases molecular motion after absorbing microwave energy, and energy is rapidly transferred to each other between molecules by collision, resulting in a chaotic movement and an increase of entropy. The other hand is the Lorentz force of the microwave field on the ion or polar molecules. It is forced to move in the way of electromagnetic wave action, which leads to a decrease of entropy. This process is strongly dependent on the working mode and state parameter of electromagnetic wave. Nonthermal effects of microwave supporters believers that the role of microwave in chemical reaction is to change the kinetics, the activation energy and pre-exponential factor of the reaction. And this change is related to the temperature, that is, the influence of selective heating on chemical reaction by microwave (the matching relationship between the molecular structure of the substance and the microwave frequency).

3. **Application types and equipment of microwave in chemistry**
There are two main types of microwave applications in chemistry. The first category is the microwave plasma chemistry: the chemical action of microwave on gaseous substances mainly belongs to this category. It is the use of microwave field to induce the production of plasma, and then to be applied in the chemical reaction. The earliest report on the use of plasma in analytical chemistry appeared in 1952. H. P. Broida et al. used a method for the formation of plasma to determine the content of deuterium isotope in the mixed gas of hydrogen and deuterium by atomic emission spectrometry. Later, they used the technique for nitrogen-stable isotope analysis, and created a new field of microwave plasma-atomic emission spectrometry analysis. Microwave plasma is also used for chemical synthesis. The most successful examples include the preparation of diamonds, polysilicon, ultrafine nanomaterials. The second category is the direct microwave chemistry, that is, the direct effect of the microwave field on the chemical system, so as to promote or change the various types of chemical reactions. Its main object of action is condensed matter. In 1974 J. A. Hesek et al. first heated samples using microwave. The following year, they performed biological sample digestion in the closed vessel of the microwave oven, microwave-induced internal heating and absorption polarization and reached higher temperature and pressure to accelerate the digestion rate, and reduce the amount of oxidant and the loss of trace elements. Microwave digestion technology has been widely used as a standard method for the analysis of sample pretreatment. Microwave is directly used for chemical synthesis [36]. In 1986, the microwave oven was used for esterification, hydrolysis, and oxidation. It has also achieved great success in dozens of synthetic reactions in organic chemistry. The main advantages of this method are that the yield and the reaction time are reduced greatly. In the esterification

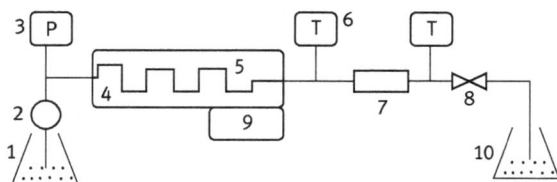

1. Reactant to be pressed 2. Controlled volume pump
3. Pressure transducer 4. Microwave cavity 5. Reaction tube
6. Temperature detector 7. Heat exchanger 8. Pressure governor
9. Microwave program controller 10. Product storage tank

Figure 13.13: Design principle diagram of continuous microwave reactor.

reaction, the reaction rate is increased by 113–1,240 times compared with the conventional heating method. In the inorganic solid-phase synthesis such as the synthesis of zeolite molecular sieve, ceramic material and ultrafine nano powder materials, it has also achieved promising success.

Microwave reactor for the promotion of chemical reactions can be summarized into two parts, microwave devices and reactors. The design principle of the continuous microwave reactor is shown in Figure 13.13. The reactants are introduced into the reaction tube 5 through the pressure pump, and after reaching the required reaction time, they exit the microwave chamber 4 and are cooled down by the heat exchanger 7 and then flow into the product storage tank 10. Continuous microwave reactor can greatly improve the experimental scale, which makes it possible to apply the microwave reaction technology in industrial production. Some continuous reactors can also carry out high pressure reaction. But at present the reactor can only be used in the liquid-phase reaction of a low viscosity system, and it cannot be applied to solid coherent reaction and solid–liquid mixing system. In addition, the temperature measured by the reactor does not reflect the change in temperature gradient of the reaction pipe, so the accurate study of the reaction kinetics cannot be carried out.

In general, as long as the microwave absorption is taking place, the material that microwave can penetrate can be made into a reaction vessel, such as glass, PTFE, polystyrene and so on. Since the microwave heating of the material is internal heating, the temperature rise is very rapid, and the reaction in the closed system is often prone to burst. Therefore, it is required to able to withstand the specific pressure of the closed vessel. There are many pressure reactors; for example, Parr and CEM companies in the United States designed acid digestion systems for acid digestion of samples of ore, biology, which can have, respectively, pressure of 8.1 MPa and 1.4–1.5 MPa. There is also a microwave batch reactor designed by CSIRO that can react at 260 °C and 10.1 MPa. For the reaction in a nonclosed system, the container requirements are not very strict, in general the glass reactor, such as beakers, flasks, conical flasks and so on.

4. **Application examples of microwave technology**
   (1) **Application of microwave technology in chemical synthesis**
       Microwave technology can accelerate the rate of chemical reaction, change the course of the chemical reaction, obtain a new reaction product and realize the reaction which cannot be carried out by some conventional methods. At present, the microwave-assisted synthesis has been successfully applied in many responses, such as alkylation, esterification, saponification, olefin addition, sulfonating, oxidation, cyclization and negative carbon ion condensation. For example, Mallakpour and Dinari [37] have prepared a series of optically active polyamide by microwave radiation. Compared with the conventional solution polymerization, the microwave field can significantly speed up the reaction rate, and the optical activity of polyester is obtained at about 10 min. Zhao et al. [38] synthetized poly isopropyl acrylamide thermosensitive aquogel by microwave radiation. Experimental results show that the microwave radiation synthesis of thermosensitive hydrogel greatly improve the speed of synthesis. The synthesis time (24 h) is shortened by 20 min by ordinary water bath method, and the pore structure of the hydrogel is very uniform.
   (2) **Application of microwave technology in the synthesis of materials**
       Study of microwave technology in the synthesis of inorganic materials, such as hard alloys, high-temperature materials, ceramic materials, nanomaterials, metal compounds, synthetic diamond and so on, is very extensive, and has made good progress in many ways. For example, in microwave sintering of WC–Co cemented carbide, compared with ordinary sintering, the sintering cycle is shortened by 3 h, and energy consumption is reduced to a fraction of ordinary sintering and can improve the performance of the product (such as low porosity, high uniformity of structure, long service life) [39]. Using kaolinite as a raw material by microwave sintered mullite, compared with the traditional method, the synthesis temperature is reduced by 300–400 °C, and the relative density is 98%. Synthesis of silicon nitride bonded silicon carbide bricks using microwave technology and compared with conventional methods synthetic time is reduced by 9/10, and product performance is significantly increased [40]. In the preparation of barium strontium titanate nano ferroelectric ceramics by sol gel method in microwave field, the average grain size was 1 μm and the critical temperature range was widened. In the microwave synthesis of molecular sieves (such as A type, Y type), compared with the traditional method, the method has the advantages of high speed (such as microwave synthesis of Y molecular sieve will take 10 min, and the traditional method 10–50 h), low energy consumption, and small and uniform grain size of the molecular sieve [41].
   (3) **Application of microwave technology in waste disposal**
       Conventional treatment of industrial sludge (emulsions of oil and water containing solid debris) uses a heated demulsification-centrifugal separation-landfill treatment process, which has a large landfill volume and high cost.

However, the use of microwave technology can avoid these adverse factors and improve the processing speed. For example, compared with conventional methods, the processing speed using microwave technology is 30 times faster, the cost is ten times lower and the volume of the treatment system is reduced by 90%. Microwave sterilization has the advantages of low temperature, short time and no secondary pollution [42]. The microwave radiation mold, yeast and other common microorganisms can be heated for ~1 min to ~80 °C and can achieve the purpose of sterilization. At 65–66 °C, microwave irradiation for 2 min can kill the spores of penicillin. A large number of medical wastes are produced every year in the world, which caused serious environmental pollution. If the microwave technology is used, more than 60% of the medical waste can be used as landfill. Compared with the traditional incineration method, it will not produce dioxin with strong toxin and secondary pollutants, and has the characteristics of fast processing speed, good effect, low energy consumption and so on [43].

13.5.2 Ultrasonic Technology

Generally, the frequency of a wave higher than 2×10^4 Hz is called ultrasonic. In 1880 the Cutie found the piezoelectric effect and in 1917 Langevin found the inverse piezoelectric effect; ultrasonic technology and its application won the very extensive and remarkable achievements, corresponding to the formation of various branches such as water acoustics, ultrasound and acoustic. As early as in the 1920s, in the United States, chemical experiments conducted at Princeton University have found that ultrasound has accelerated the role of chemical reactions, but has not attracted the attention of chemists for a long time. Until the mid-1980s, due to the popularity of ultrasound equipment applications, the application of ultrasound in chemistry has developed rapidly, forming a new cross-disciplinary, phonochemistry. Phonochemistry mainly refers to the use of ultrasound to accelerate chemical reaction, which improves the chemical yield of a new cross-discipline. Phonochemistry reaction is not from the direct interaction of sound waves and material molecules, because the liquid commonly used the acoustic wavelength of 0.0015–0.1 m (frequency 15 kHz to 10 MHz), which is far greater than the molecular scale. Phonochemistry reaction is mainly due to acoustic cavitation, formation, turbulence, growth, shrinkage and collapse of vacuoles in liquid, and their physical and chemical changes.

1. **The formation of cavity and its influencing factors**
   Ultrasonic wave is composed of a series of artistically spaced longitudinal wave, which spread around the liquid medium. Like all sound power, the transmission of ultrasonic energy is realized by compressing and expanding the medium. Acoustic cavitation is a very complicated physical process of gathering and

instantaneously releasing the acoustic field energy. It refers to the tiny vesicles in the liquid that are activated by ultrasound, which is activated under the action of ultrasonic wave, which is a series of dynamic processes, such as oscillation, growth, contraction and collapse of the bubble nucleus. When the cavitation bubble collapses, a local high temperature (about 5,000 K), high pressure (about 100 MPa), a temperature change rate of 109 K/s, and a strong shock wave are generated around it. This effect increases and the heterogeneous reaction interface is updated. It enhances the mass transfer and heat transfer processes, increases the reactivity of the reactant molecules, and increases the chance of their colliding with each other. At the same time, this effect forms numerous tiny chemical reactors with extreme physicochemical environment in the liquid, which is favorable for the breakage of chemical bonds, the generation of free radicals and related reactions. The chemical reaction, which is difficult or impossible to realize under general conditions, provides a new and very special physical environment, which opens a new path for chemical reactions.

Sonochemical is affected by three major factors: the frequency of sound field, the energy of sound field and the temperature of a solution.

1. The frequency of sound field: It has a significant effect on sonochemical reaction and frequency of sound field. In general, the effect of pulse sound wave is better than that of continuous sound wave. The modulation mode of the sound source is also very important; when the pulse duty cycle is 1:1–1:1.5, the sonochemical reaction has a higher induction rate.

2. The energy of sound field: It depends on the power of the ultrasonic transducer, and it is the decisive factor of the sonochemical reaction. The acceleration and the opening of the sound chemical reaction originate from the cavitation of ultrasonic. Only when the sound intensity (energy) reaches a certain degree, the bubble can be closed by the frequency of sound field. The higher the sound intensity, the faster the closure and the greater the pressure wave generated. The hotspot temperature and pressure will be higher, and other some cavitation effect is increasingly fiercer, thus triggering and opening a series of sonochemical reaction.

3. The temperature of solution: It is also an important influence factor. In the early years, it was believed that the viscosity of the solution was decreased by the increase in temperature. In this way, the frequency of the cavitation core and the sound field are different, so the number of effective cavitation nuclei can be reduced, and the reaction rate can be decreased. Recent studies have showed that the decline in solution viscosity is not the main reason. But when the temperature of the solution is increased, a large number of gases dissolved in a solution escape, which results in a decrease in the formation of cavitation nucleus and a decrease in the reaction rate.

2. Theoretical explanation of the effect of sonochemical

The main mechanism of the sonochemical effect is acoustic cavitation (including bubble formation, growth and crack). The phenomenon includes two aspects, namely, strong ultrasound produces bubbles in a liquid and the special motion of bubbles under strong ultrasound. Applying an ultrasonic field in a liquid, when the ultrasonic intensity is large enough, a large group of bubbles can be generated in the liquid, which is called "acoustic cavitation bubble." These bubbles are affected by ultrasound. These bubbles are simultaneously subjected to ultrasound, and are subjected to strong ultrasonic waves. When experiencing the sparse and compressive phases of the sound, the bubbles grow, contract, regrow, and recontract. After multiple periodic oscillations, they eventually crack at high speeds. The periodic oscillation or breakup process will produce short local high temperature (about 5,000 K) and high pressure (about 100 MPa), produce heating and cooling rates greater than 10^9 K/s, and generate a strong electric field, causing many mechanical, thermal, chemical and biological effects. The environmental conditions of the reaction system will greatly affect the intensity of the cavitation, which will directly affect the rate and yield of the reaction. These conditions include the reaction temperature, the static pressure of the liquid, the ultrasonic irradiation frequency, the sound power and the ultrasonic intensity. Other factors that have a great influence on cavitation strength include the type and amount of dissolved gases, the choice of solvents, the preparation of samples, and the selection of buffers. In the ultrasonic irradiation system, the acoustic chemical reaction can occur in three areas, namely, the gas-phase zone, the gas–liquid transition zone and the bulk liquid-phase region of the cavitation bubble. 1. The gas-phase area of the cavitation bubble is composed of a mixture of the cavitation gas, water vapor and volatile solute vapor, which is in the extreme conditions of cavitation. If the cavitation bubble bursts in a short period, the bubbles of the water vapor broken down by thermal decomposition reaction, which results in the production of OH and H radicals, and nonpolar, easy volatile solute vapor will also direct thermal decomposition. 2. The gas–liquid transition zone is a thin layer of super-hot liquid around the gas phase, containing volatile components and surfactants (if any). In the middle of the cavitation condition, there is a high concentration of OH radicals, and water will become supercritical. 3. The bulk liquid-phase region is basically in the environmental condition. The oxidizing agent, such as OH free radicals, which is not consumed in the first two areas, will continue to react with the solute in the region, but the amount of the reaction is very small. The reaction of nonvolatile solutes is mainly in the boundary region (gas–liquid transition zone) or in the bulk solution. The cavitation at the interface between liquid and solid is very different from that in the pure liquid. Because the liquid in the field is uniform, the bubble will remain spherical in the process of cracking. The cavitation bubble near a solid surface bursts for nonspherical; the bubble burst will produce high-speed micro jets and shock waves. The impact of jet impingement can cause the erosion of the solid surface, and the surface of the oxide layer can be removed. On the solid surface, the cavitation bubble bursts because of high temperature and high pressure, which can greatly promote the reaction.

3. **Application of ultrasonic wave**
   (1) **Application of ultrasonic wave in chemical synthesis**
   Starting from the principle of acoustics, it can be considered that the ultrasonic wave plays a key role in the sounding reaction. It is a unique form of interaction between the substance and the substance – cavitation. Under the effect of ultrasonic wave, the series of dynamic behavior of microbubbles in the fluid is oscillatory, expansion, contraction and burst or collapse. It is this cavitation effect of ultrasonic waves that forms high-temperature, high-pressure, strong shock waves, microjets, and local high-energy environments such as charge and discharge, and luminescence in the interior of liquids. It is obviously enough to be the center of initiating or accelerating the reaction to cause the molecules to be thermally dissociated, ionized, produce free radicals, and so on, resulting in a series of chemical reactions. Li Dezhan et al. [44] studied to promote the synthesis of benzene dibromide by ultrasound. Experimental results show that without ultrasonic effect, only in the higher temperature and longer time, it can be synthesized into paradibromobenzene, and the yield is low. MC Nulty [45] reported, in response to traditional, with electronic substituent aromatic aldehydes and nitro-paraffin in acetic acid and ammonium acetate system, that the yield is relatively low in a reaction for a few hours at 100 °C (for example, the yield of 2,3-dimethoxy benzaldehyde and nitrocarbol reaction is only 35%). And the formation of noncrystalline, resin-like material often leads to the pollution of mother liquor. In the same system, the yield of the product can reach 99–89%, and the pollution problem will be solved by using the ultrasonic wave at 22 °C for 3 h. At room temperature, without ultrasound radiation, the reaction would not have occurred at all. The experimental results show that the effects of ultrasonic wave on the aromatic aldehydes with the electron-donating groups and nitro-paraffin reaction are promoted.

   (2) **Application of ultrasonic wave in polymerization reaction**
   Ai et al. [46] studied the emulsion polymerization of styrene and butyl acrylate under ultrasonic irradiation. They dissolved the PS waste in butyl acrylate and added water and initiator, and then the graft copolymer was prepared under the irradiation and stirring of the ultrasonic wave. The higher the ultrasonic power, the longer the irradiation time; the higher the reaction temperature, the lower the coagulation rate of the grafted products. The type and amount of emulsifier, the total concentration of the emulsifier and other factors also affect the condensation rate of the product. M. A. Bahattab et al. studied the emulsion polymerization of vinyl acetate under ultrasonic irradiation. When there is no initiator and emulsifier, only the effect of ultrasonic on the ambient

temperature can lead to the emulsion polymerization of vinyl acetate. The conversion of polymerization and the polymer yield were improved after the use of the redox initiator system and the ultrasonic irradiation. Ultrasound plays an important role in the initiation of reaction and control of the polymer structure.

(3) Application of ultrasonic in environmental protection
Ultrasonic technology in environmental protection can be used for sewage treatment, solid waste treatment, gas purification and oil cleaning. In addition, it can be used in conjunction with other treatment technologies, including ultrasonic-ozone combination, ultrasonic-hydrogen peroxide combination, ultrasound-ultraviolet light combination and ultrasonic-UV/TiO_2 combination. Xiao-Hui Bai [47] conducted a lot of experiments on the sludge treatment by ultrasonic, and obtained the best effect of ultrasonic frequency in decomposition of biological solid which is 41 kHz. The final objective of the ultrasonic decomposition of sludge is to improve the reaction efficiency and the degree of degradation of anaerobic process, and the pilot experiment was carried out. The results showed that the digestion time of the sludge treatment could be reduced from 22 days to 8 days, the average degree of ultrasonic cell decomposition was 12% and the anaerobic degradation process was accelerated by the addition of ultrasonic treatment. Lin made the oxidation of 2-chlorophenol in water by ultrasonic/H_2O_2, and Trablsi made degradation of phenol in water by ultrasonic electrochemical method; both have achieved good results.

13.5.3 Radiation Technology

Radiation technology is a subject which is closely related to polymer materials science, environmental science, biotechnology and medicine. At present, in the field of polymer material, radiation technology has been used in the radiation cross-linking of polyolefin, radiation curing of unsaturated polyester resin, radiation vulcanization of rubber, radiation degradation of polymer and radiation grafting modification; there are many products which realize industrialization production [48, 49].

1. Radiation cross-linking
Radiation cross-linking, as an industrial technology, has been widely used in the electric wire and cable for lighting and electronic equipment circuit for automotive, home appliances, aircraft and spacecraft. In the United States, all aircraft cables used radiation cross-linking products, and flame retardant wire and cable has also been widely used in an offshore oil platform. Because of the cross-linking reaction of the polymer chain free radical after radiation of wire and cable with polyethylene and polyvinyl chloride as the base material, the heat resistance, insulation resistance, chemical corrosion resistance, anti-atmospheric aging and mechanical strength of the material have been greatly improved. For example, the use of radiation cross-linked polyethylene temperature limit can be increased to 200–300 °C.

Another type of product produced by radiation cross-linking technology is a kind of thermal shrinkage material with special "memory effect." Radiation cross-linking technology uses crystalline polymer materials (such as polyethylene after the expansion of heating, and then cooling forming). It uses a crystalline polymer material such as polyethylene to expand after heating, and then cools it. When it is reheated, the material returns to its pre-expanded state. It uses its shrinkable properties to make insulation materials or anticorrosive coatings at wire and cable joints. The radiation cross-linked heat-shrink materials produced by Raychem and BCS in the United States have brought considerable economic benefits.

2. **Radiation curing**

Compared with chemo setting, radiation curing has the advantages of high cure rate, low energy consumption, good product quality and so on. It is generally welcomed because it avoids the use of solvents and does not cause pollution. Now it has been more mature for the radiation curing coating, such as metal, tape, ceramic, paper and other products of the surface processing. In addition, due to the high penetration of electron beam radiation, it has unique advantages in the development of lightweight, such as high strength, high modulus, corrosion resistance, abrasion resistance, impact resistance and damage resistance of advanced composite materials. These reinforced composites can be widely used in transportation, sports equipment, infrastructure, aerospace and military industries. Today, Canada has the radiation curing technology of Airbus aircraft fuselage and cowl repair test, and plans to further develop the electron beam curing repair of aircraft composite parts.

3. **Radiation vulcanization**

In the rubber industry, the natural rubber latex or rubber molecules can be cross-linked under irradiation. It is similar to the process of rubber vulcanization, called radiation vulcanization. But this kind of radiation vulcanization cannot add additives such as curing agent and accelerant. Avoid the cross-linking of traditional chemical heat vulcanization due to the use of the cross-linking agent in the uneven distribution of the substrate, and the effect of temperature gradient caused material performance degradation. Recently, Beijing Ray Application Center has developed radiation vulcanized rubber, with excellent ozone resistance, aging resistance, wear resistance and fatigue resistance. It is very suitable to be used as truck tyre, seal gasket and the long-term outdoor use rubber products, such as plastic steel window sealing strip and auto wipers.

4. **Radiation grafting modification**

Radiation grafting technique is one of the effective methods for the development of new materials with excellent properties or for the modification of the original materials. Since radiation grafting is initiated by radiation, it is not necessary to add an initiator to the system, and a very pure graft polymer can be obtained, which is an effective method for synthesizing a medical polymer material. With the continuous development of technology in the medical field, and the emergence of artificial organs, a large number of polymer materials began to be applied in the medical field. In order to improve the anti-clotting properties of polymers and reduce the formation of thrombus, the polymer is usually modified by bulk or surface, such as graft hydrophilic monomer. The radiation-induced grafting modification technology was used to introduce different functional groups into the surface of polypropylene film by South Korea Isotope Radiation and Application Group, which can improve the blood compatibility of the membrane and can be used for artificial lung. The radiation grafting modification of polymer materials (polyethylene, polypropylene) with excellent properties and low price has been of very concern, and some valuable new materials have been obtained, such as ion exchange resin and blending. Radiation grafting of polar molecules to the surface of polyethylene can improve the surface affinity, advantageous for secondary bonding of materials, printing and coating. In addition, grafting acrylamide or acrylic monomer in cotton or silk can improve the fabric surface properties, such as improving silk crease resistance and so on. Now, it has also made the stage progress that study on using radiation technology for the preparation of powdered rubber by graft modification for natural rubber hole. The modified powdered rubber can be made of rubber products, which can be used as a toughening agent and as a plasticizer for toughening of engineering plastics. In addition, radiation degradation of butyl rubber can also be used as lubricating oil additives.

5. **Radiation degradation**

Under the irradiation of radiation, the polymer not only produced cross-linking, but also broken down at main chain, namely, radiation degradation. Radiation degradation also has an industrial application value, such as the treatment of waste plastics and the recycling of rubber. Waste PTFE and the processed leftover material after irradiation can be used as lubricant and wear-resistance improver. In the 1990s, in China, research and development of radiation regenerated butyl rubber test, compared with other rubber regeneration method, has the advantages of low energy consumption, simple process and not producing "three wastes." Instead of some imported butyl rubber, butyl rubber regenerated by radiation method can be used to produce rubber products. And the incorporation of reclaimed rubber radiation can improve the processing technology of butyl rubber, such as large intensity of semi-finished rubber, small shrinkage and mouth size easy to grasp.

6. **Radiation processing technology**

Radiation processing technology is an important part of the peaceful use of atomic energy. Radioactive cobalt 60 (^{60}Co) or cesium 137 (^{137}Cs) release high-energy gamma rays and electron accelerators produce energy of the 0.2–10 Mev electron beam, referred to as the electron radiation. The principle of radiation processing technology is to use electronic radiation to the irradiated material, and produce ionization and excitation, so as to release the track electron, forming a free radical. Thus, the physical properties and chemical composition of the irradiated material can change which will make it become a new material needed by people, or make it the living organisms (microorganisms, etc.) under the nonrecoverable loss and damage. This new processing technology is called radiation processing technology. This technology is different from the traditional mechanical processing and thermal processing technology, which is known as the third revolution of human processing technology. The characteristic is that the radiation source releases the ray which has the very strong penetration ability, can go deep into the material interior to carry on "the processing" and carries on under the room temperature condition. Therefore, it is a clean processing technology, and its reaction is easy to control and the processing flow is simple. Therefore, it has advantages of low energy consumption, no residue and no environmental issues.

7. **Irradiation technique**

This technique uses the interaction between radiation and matter, ionization and excitation of the activation of atoms and molecules, to ionize and motivate generated activated atoms and molecules. A series of physical, chemical, and biochemical changes occur with the substance, resulting in degradation, polymerization, cross-linking, and modification of the substance. As a result, it provides a new way for the removal of some pollutants which are difficult to be removed by conventional treatment methods. For example, treat domestic sewage and industrial wastewater by irradiation, and use gamma ray irradiation to treat solid waste. Irradiation technology can effectively deal with harmful organic substances, such as detergent, organ mercury pesticides, plastic agents, nitrite amines and phenols. The combination of irradiation technology and common wastewater treatment technology (such as coagulation, activated carbon adsorption, ozone-activated sludge) has a synergistic effect, which can improve the treatment effect. When combined with the activated carbon, after the adsorption of organics for charcoal, the activated carbon can be regenerated by gamma ray irradiation, which is beneficial to the continuous use of the activated carbon. Since the late 1980s, China has carried out the further research works, such as radiation disinfection of drinking water, radiation treatment of organic dye wastewater and coking plant wastewater, and has obtained the good effect.

13.5.4 Plasma Technology

The plasma, the ionized gas, is a collection of electrons, ions, atoms, molecules or free radicals, which are usually generated by an applied electric field that causes the dissociation or ionization of the gas molecules. Regardless of partially ionized or fully ionized gas, the total number of positive charge and negative charge on the numerical is always equal; in the macro the charge is neutral, called plasma. The plasma can be divided into two categories based on the relative height of the charged particle energy in the plasma (usually expressed as electron temperature). One is the high-temperature plasma, that is, the electron temperature in the tens of electron volts (1 eV = 11,600 K) above the plasma. The other is the low-temperature plasma, that is, the electron temperature in the plasma of tens of electron volts. Low-temperature plasma has been widely used in many fields, such as materials, information, energy, chemical, metallurgy, machinery, military and aerospace [50, 51].

1. **Application of plasma technology in chemical synthesis**
    Almost all kinds of plasma-rich particles are active chemical active substances. The various particles in the plasma state have strong chemical activity and can obtain relatively complete chemical reactions under certain conditions. The high temperature of the plasma or the active particles and radiation therein promote certain chemical reactions to obtain new substances. For example, boron nitride ultrafine powder is prepared by arc plasma, and titanium dioxide (titanium dioxide) powder and the like are prepared by high-frequency plasma.

2. **Application of plasma technology in machining**
    A high-temperature and high-speed jet produced by plasma spray gun can be used for many machining processes, such as welding, spraying, cutting, heating, etc. Micro plasma arc welding produced in 1965, the torch size being only 2–3 mm, can be used to process very small work piece. Wear-resistant, corrosion-resistant and high-temperature-resistant alloys can be deposited on the parts by plasma arc surfacing and can be used for processing all kinds of special valves, drills, cutting tools, mold and crankshaft. The high temperature and strong jet force of arc plasma can spray metal or nonmetal on the surface of the work piece, which can improve the wear resistance, corrosion resistance, high-temperature oxidation resistance and seismic resistance of the work piece. It is the cut metal rapid local heating to melting state by arc plasma, at the same time with the high-speed airflow blown molten metal and the formation of narrow incision, which is called plasma cutting. Plasma heating cutting is appropriate to set a plasma arc in front of the tool, so that the metal is heated before cutting, which changes the mechanical properties of the processing material, making it easy to cut. This method improves the work efficiency by 5–20 times compared with the conventional cutting method.

3. **Application of plasma technology in metallurgy**
    Since the 1960s, people started to use hot plasma to melt and refine metals; now the plasma arc melting furnace has been widely used in melting heat-resistant alloy and refining high-grade alloy steel. It can also be used to promote the chemical reaction and to extract the desired product from the mineral.

4. **Application of plasma technology in surface treatment**
    There is a significant effect that the surface of metal or nonmetal solid is treated by cold plasma. A 10 μm organic silicon monomer film deposited on the surface of the optical lens can improve the scratch resistance and reflection index of the lens. The surface wettability of polyester fabric can be changed by cold plasma treatment. This technique is often used for the cleaning and etching of metal solid surfaces.

In addition, the combustion-generated plasma is used for the electric power generation of magnetic fluids. Since the 1970s, people have used the interaction force of current and magnetic field in ionized gas to make the thrust generated by high-speed jetting of gas to create magnetic plasma thrusters and pulsed plasma thrusters. Their specific impulse (the ratio of rocket exhaust velocity to acceleration of gravity) is much higher than that of a chemical fuel propeller, and has become an ideal propulsion method in aerospace technology.

5. **Problems and prospects of plasma technology**
In order to solve the bottleneck problem that restricts the further development of a plasma-enhanced chemical process, the following plasma-related basic research needs to be strengthened: (1) the relationship of physical and chemical properties between plasma and plasma components and (2) plasma-related multi-scale structures and their transfer and reaction characteristics. With the development of a plasma-related multi-scale structure and its transfer and reaction characteristics, the traditional definition and methods of plasma cannot meet the requirements of development. But to achieve a substantial breakthrough in theory, there are still considerable difficulties in the two aspects of thermodynamics and kinetics. 3. It is necessary to realize that the control is the future development of plasma subjects, such as electron temperature (energy), electron density, excited state energy parameter, free radical and its density.

Plasma technology intensification uses the chemical process as an interdisciplinary science technology; it provides a new method and a new idea for chemical workers to solve the problems related to energy, resources and environment in chemical industry. There are a large number of innovative development opportunities for the two aspects of theoretical and practical applications; potential economic benefits and social benefits are very significant.

13.6 Other Intensification Techniques

13.6.1 Hydrodynamic Cavitation Technology

Hydraulic cavitation technology is a new water treatment technology. When the fluid flows through the section of the pipeline (such as the geometry of orifice plates or Venturi tubes), due to the limitation of the contraction fitting, the flow rate of the current limited area increases with decrease in pressure. When the pressure is lower than the saturated vapor pressure at the corresponding temperature of the fluid, the gas dissolved in the fluid is released. At the same time, the fluid itself also produces a large number of cavitation bubbles. With the rapid recovery of the surrounding fluid pressure, cavitation bubbles instantly collapse accompanied by a variety of physical and chemical effects, which result in the phenomenon of hydraulic cavitation.

1. **Mechanism of hydrodynamic cavitation**

   The hydrodynamic cavitation reactor is developed by changing the geometry of the fluid system (such as the fluid passing through the Venturi tube and the perforated plate); the flow velocity of the liquid is increased and the pressure is decreased. The liquid flow rate is accelerated instantly, resulting in a great drop in pressure. When the pressure is reduced to the saturated vapor pressure (or air separation pressure) of the liquid at the working temperature, the liquid begins to "boiling", quickly "vaporizes", a large number of cavitation bubbles are generated inside, and the continuity of the liquid is destroyed. With the decrease of pressure, the bubble expands continuously. When the pressure recovers, the bubble collapses instantaneously to produce high temperature (1,000–5,000 K) and instantaneous high pressure (1–5 × 107 Pa), which forms the so-called hot spot [52]. This is the equivalent of a large discrete energy input, which can make the water vapor at high temperature and high pressure to split and chain reaction to occur; the reaction is as follows:

$$H_2O \rightarrow \bullet OH + \bullet H \tag{1}$$
$$\bullet O\,H + \bullet OH \rightarrow H_2O_2 \tag{2}$$
$$2\bullet H \rightarrow H_2 \tag{3}$$

   The cavitation bubble collapse generated shock wave and jet, and produced free radical with high chemical activity $\bullet OH$ and strong oxidizing agent H_2O_2. Subsequently, with the organic pollutants in the solution oxidation reaction, most of the organic pollutants in water can be degraded into harmless substances, so as to achieve the purpose of purifying water quality. In addition, the strong hydraulic shearing force can make the carbon bond in the main chain rupture and destroy the microbial cell wall, so as to achieve the role of the polymer degradation and microbial inactivation.

2. **Hydrodynamic cavitation reactor**

   The commonly used hydraulic cavitation reactor consists of a high-speed homogenization device, a high-pressure homogenization device, a micro-fluidized device, a porous plate and so on. High-speed homogenization (HSH) is generally composed of a rotor and a stator, cavitation is generated by adjusting the speed of the rotor, and the energy consumption of the device is very high. High-pressure homogenization (HPH) is a constant volume type pump with throttling device. The typical high-pressure homogenization device is composed of a water tank, two throttle valves and a high-pressure pump, and the fluid in the water tank can be increased by the throttle valve under the action of the pump. A typical high-pressure homogenization device is composed of a water tank, two throttle valves and a high-pressure pump, and the fluid pressure in the water tank can be increased by the throttle valve under the action of the pump. HPH is mainly used for the emulsification process, but its cavitation intensity is small and difficult to control. Perforated plate equipment has the greatest flexibility and adaptability, can be applied to a variety of occasions, such as degradation of various organic pollutants, disinfection, and easy to industrialization, commercialization and engineering.

3. **Application of hydraulic cavitation technology**

   (1) **Sewage disposal**

   With the rapid development of industry, the harmful synthetic chemicals and organic compounds in the water are increasing year by year, and the traditional water treatment methods cannot meet the new requirements of environmental protection. The potential application of hydraulic cavitation technology in the treatment of toxic and difficult to degrade organic polluted water bodies has prompted many researchers to study the application of hydraulic cavitation in wastewater treatment. Sivakumar and Pandit [53, 54] used porous plates of six different geometries in a circulating hydraulic system to generate cavitation with different intensities to degrade 50 L of dye wastewater containing 5–6 µg/L of Rhodamine B. After 1 h it was found that the biodegradable aromatic amine was degraded, and the solution was decolorized, and the energy efficiency was 1.5 times that of ultrasonic cavitation.

   (2) **Hydrolysis reaction**

   Pandit and Joshi [55] did an experiment on the hydrolysis of fat oil. In the experiment, the valve is used as a throttling device to produce cavitation, the oil–water mixture is used for circulation flow in the equipment, the treatment capacity of the hydraulic cavitation device is 200 L and the treatment quantity of ultrasonic cavitation is 220 mL. Experiments show that the energy (1,080 J/mL) of hydrodynamic cavitation is 30% less than that of ultrasonic cavitation (1,384 J/mL) at the same degree of hydrolysis. The equipment is simple, is efficient and has the advantages of high treatment capacity, so that the hydraulic cavitation has the absolute advantage of the industry.

   (3) **Disinfection**

   Jyoti and Pandit [56, 57] conducted comparison experiment for disinfection of well water with a variety of equipment. In the experiment, the porous plate in a hydraulic device is used to generate cavitation and to treat 75 L well water at 0.17, 0.35 and 0.52 MPa. After 1 h, it was found that the number of microorganisms in water decreased with the increase in pressure. In 0.52 MPa, the total *E. coli* and *Streptococcus* were reduced by 85%. The total *E. coli* was 28% after the same H_2O_2 treatment and the ultrasonic cavitation was 75%. Compared with other physical treatment methods, such as ultrasonic cavitation equipment, high-speed homogeneous reactor and high-pressure homogeneous reactor, the energy consumption of the hydrodynamic cavitation is the least. However, compared with the chemical treatment method, the hydraulic cavitation energy consumption is higher than that of several orders of magnitude, such as chlorination, but the process of the treatment of the hydraulic cavitation avoids the generation of toxic substances. Therefore, the hydrodynamic cavitation is an effective method for disinfection of drinking water and bath water. In subsequent experiments, they combined hydraulic cavitation with other methods, such as ultrasonic cavitation, ozone technology and chlorine, and found that the effect of disinfection was better than that of a single process; the eliminated quantity of *E. coli* reached 97%.

(4) Other aspects of the application

Hydraulic cavitation has been applied in many other aspects. For example, Shirgaonkar et al. studied the cell division split matrix in high-pressure and high-speed equipment. Research shows that the pressure pulsation and collapse of a cavitation bubble is the main reason of cell division. Chaivate and Pandit [58] conducted experiments to study the solution of the polymer solution by cavitation. In the experiment, the solution of critical micelle concentration is 0.5%, which is produced by the porous plate. The viscosity of the solution decreased from 30 mPa·s to 5 mPa·s after 2 h. Dandoth [59] did bleaching of sulfite pulp by cavitation. The experimental results show that the rate of bleaching is seven to nine times the rate of hypochlorite bleaching in 20–60 °C, and the pH value is increased from 7.0 to 8.0.

Hydrodynamic cavitation can also be used in combination with other treatment methods, such as ozone technology, hydrogen peroxide oxidation and ultraviolet radiation, which can improve the oxidation efficiency and energy efficiency. Buckley [60] combined hydraulic cavitation and UV radiation technology; the results show that the degradation rate is greatly improved. At present, a kind of CAV-OX has been applied in the commercial field. It is an integrated technique in which hydraulic cavitation, UV irradiation and hydrogen peroxide are combined. It can largely degrade pentachlorophenol, benzene, toluene, benzene ethyl ester, xylene, cyanide, phenol, atrazine and other refractory organics in water.

In the hydraulic cavitation equipment, produce cavitation only needs the installation of the mouth (such as the throttle valve, the tube and the porous plate). The cavitation intensity can be controlled by adjusting the size and form of the shrinking mouth, so that the hydraulic cavitation has the advantages of simple operation, low energy consumption, less pollution and large amount of treatment. And the technology can be used alone or can also be combined with other technologies (such as UV radiation and H_2O_2). Although various kinds of application of a hydraulic cavitation reactor are confined to small scale, its advantages show that it has a strong vitality and broad prospects for development, and is a promising water treatment technology.

13.6.2 Supercritical Fluid Technology

1. **Introduction**

    The supercritical fluid is the state when the temperature and pressure of the material are at critical point. It has many unique properties which are different from those of the traditional solvent. Supercritical fluid has the characteristics of small gas viscosity, large diffusion coefficient, high liquid density and good solubility. And the properties (density, viscosity, diffusion coefficient, dielectric constant, interfacial tension) of fluid near the critical point are mutable and adjustable. It is easy to control the phase equilibrium characteristics, transfer characteristics and reaction characteristics of the system by adjusting the temperature and pressure, so that separation, reaction and other chemical processes are more controllable [61].

    Supercritical fluid technology, as a "green" strengthening method, not only greatly reduces the pollution of chemical process in the environment – the diffusion coefficient is much larger than that in ordinary solvents – but also significantly improves the efficiency of mass transfer, so as to improve the efficiency of separation and reaction of chemical process.

2. **Application of supercritical fluid technology**

    As early as in 1822, Cagniard discovered the existence of critical phenomena. The critical parameters of CO_2 were measured by Andrews in 1869. In 1879, Hanny and Hogarth found that the supercritical fluid has the ability to dissolve the solid, which provides the basis for the application of supercritical fluid technology. Although there has been more than 100 years of history since the discovery of critical phenomena, its rapid development is only a matter of more than 30 years. With the development of deep research in theory and application in recent years, supercritical fluid has been widely used in the process of extraction, reaction, granulation, chromatography and cleaning technology, and it shows a broad application prospect in the field of chemical, pharmaceutical, food, environmental protection, and materials [62, 63].

    (1) **Supercritical fluid extraction**

    Supercritical fluid extraction technology is one of the most studied technologies. Early studies mainly focused on the theoretical aspects, including the supercritical fluid density and viscosity measurement and correlation, the supercritical phase equilibrium data of determination and thermodynamic model establishment, and kinetics of mass transfer in the extraction process in the supercritical state. In recent years, many researchers have studied the molecular interactions in the supercritical state from the microscopic view, and try to explain the mechanism of selective extraction from the molecular level. In an application, the supercritical fluid extraction technology is mainly used for the extraction of effective components from natural products, and can also be used for the removal of trace elements such as metal ions and pesticides. Since the establishment of the first caffeine removal plant using supercritical fluid extraction technology in Germany by HAG company, in 1978, its industrialization has achieved rapid development. The United States, Japan, Canada, Italy and China have also established the production equipment, and it is used in the extraction of beer floral essence, natural flavors, colors and oils [62, 63]. Although the research on supercritical fluid extraction technology is becoming increasingly more mature, the dissolution of CO_2 is still a bottleneck to restrict the development of the technology. To this end, it was used to improve the protein and other biological macromolecules or metal ions in the supercritical CO_2 solubility that the formation of CO_2 microemulsion is supercritical or adds a chelating agent [63, 64]. These intensification methods have greatly expanded the application range of supercritical fluid extraction.

(2) Supercritical fluid chemical reaction

Supercritical fluid chemical reaction is the reaction of supercritical fluid as reaction medium or as reactant. Compared with the traditional method, the unique properties of supercritical fluids have improved the reaction rate, yield and conversion rate, catalyst activity, life and product separation. The chemical reactions in supercritical CO_2 include oxidation, hydrogenation, alkylation, carbonyl, polymerization and enzyme catalytic reactions [65, 66]. The reaction system phase behavior and the influence of the interaction between molecules have been studied extensively not only from the theory to the reaction mechanism and reaction kinetics, but have been carried on the industrialization to explore. For example, DuPont Co's annual output of 1,100 tons supercritical reaction device has been officially put into operation. The supercritical water oxidation reaction can be used in the treatment of toxic wastewater and organic waste. At present, it is a leading environmental protection technology, which realized industrialization. In addition, due to the current energy crisis, the conversion of biomass in supercritical water has aroused people's attention, but the research in this area is still in the primary stage.

(3) Other applications of supercritical fluids

Supercritical fluid crystallization technology can be used for the preparation of ultrafine particles of drugs, polymers and catalysts. Supercritical fluid chromatography is particularly suitable for the separation of high value-added substances such as chiral drugs or natural products. In addition, the supercritical fluid technology can also be used in the cleaning of semiconductor, textile printing, dyeing and other fields.

3. The problem and prospect of supercritical fluid

With the rapid development of supercritical fluid technology, the research of theory and application is gradually deepening. In recent years, as the application of supercritical fluid technology continues to expand, a variety of supercritical fluid technologies have been realized in the industrialization. However, due to the nonideality of supercritical fluids and the scarcity of research methods under high pressure, the study of the theory of the supercritical fluid is more concentrated on the macrolevel of thermodynamics and dynamics. Little research is done on the molecular level , and the research on the multisystem in the supercritical state is still lacking. Therefore, the theoretical basis of the supercritical fluid technology still needs to be strengthened. Because the molecular simulation is not essential to carry out the experiment of high pressure, the useful information such as thermodynamics, transmission characteristics and spectral properties can be obtained. The technology of supercritical fluid extraction and reaction has been industrialized, but it is restricted by the limited solubility of the supercritical fluid, resulting in a large volume of equipment and high investment, which is only applicable to the high added value of the material. Adding modifier can improve the solubility of supercritical fluid to a certain extent, but it also reduces the "green," which increases the difficulty of post treatment. If we can make a breakthrough in the scale and automation of the new medium and high-pressure equipment, it will greatly improve the economic performance of the supercritical fluid technology. In addition, the supercritical fluid technology and other technologies (such as distillation, adsorption, membrane separation, etc.) help to improve the efficiency of chemical processes, reduce costs and will become a new trend in supercritical fluid technology research.

13.6.3 Pulse Combustion Drying Technology

Pulse combustion drying technology is a new and promising high-efficient drying technology, which uses high-temperature, high-speed oscillation air produced by pulsating combustion to dry materials. The use of pulsating burning tail gas as a source of dry heat has greatly improved the transmission characteristics of the drying process and improved the drying rate and energy utilization, effectively reducing the pollution emissions.

1. **Pulse combustion drying system**
   The main equipment of the pulse combustion drying system is a pulse combustor. The pulse combustor is the drying medium generator. It is three types: Schmidt, Helmholtz and Rijke [67]. Pulse combustion is carried out periodically, and it also produces very strong acoustic resonance phenomena. The temperature of the exhaust gas stream produced by combustion is usually more than 800 °C, the pulse frequency is 50–300 Hz and the oscillating air flow rate can reach 100 m/s. Such characteristics of the flow field and the strong sound wave can greatly enhance the heat transfer and mass transfer processes. In addition, the fuel source has a wide range, the fuel combustion is full, the exhaust pollution is low and the structure is simple, the basic, with no moving parts. Thus, it is a very ideal drying medium generator. Two types of the combustor has been applied in the production of pulse combustion dryer with a tail tube as the drying chamber and the end of

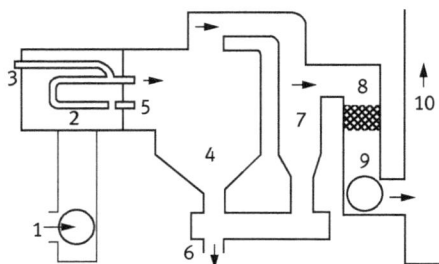

1. Induced draft fan 2. Pulse combustor 3. Inlet pipe
4. First separator 5. Return pipe 6. Discharged gate
7. Second separator 8. Filter 9. Air outlet 10. Air-vent

Figure 13.14: Pulse combustion drying system.

the tube connection drying room. Figure 13.14 shows as a drying chamber a pulse combustion drying system developed by American Sonodyne Corporation. The air is blown in by draught fan 1, and the liquid material is fed into the tail pipe of the pulse combustor by feeding tube 1. In the jet of the pulsating flow in the stern tube, the material is evaporated by atomization, and the material is dried and separated in first-stage separator 4. The tail gas flow and some fine particles are separated in two-stage separator 7. The dry material is discharged from discharge port 7 by the feeder. The exhaust gas is discharged through exhaust port 10 by filter 8 in exhaust blower 9. The backflow of the pulse combustion is introduced into drying chamber 4 by the pulse combustor of the drying device adopting a U-type nonvalve simple structure, and the backflow of the pulse combustion is introduced into the drying chamber to improve the heat utilization rate of drying chamber 5 and to improve the heat utilization rate of the drying chamber. The tail pipe is used as the pulse combustion dryer of the drying chamber, the heat energy and the sound energy are fully utilized, the sound wave can be used for crushing the material in the tail pipe, the atomization effect is good and the structure is simple. However, the material interferes with the pulsating airflow in the tail pipe and affects the combustion process. It is not appropriate to process the material in large quantities.

2. **Characteristics of pulse combustion drying technology**
   Compared with the conventional drying technology, pulse combustion drying technology has the following advantages [67].
   (1) Adapt to a wide range of materials. Pulse combustion drying can be used in food, agricultural products, chemical products and industrial products, and other materials drying; compared with the traditional drying, material adaptability is more extensive.
   (2) Strong evaporation capacity and low energy consumption. Pulse combustion drying with atomization effects makes the material atomized into fine droplets. The area of evaporation surface and the drying rate are greatly increased, but the energy consumption is reduced. The traditional dryer unit

heat consumption is generally 500–9,000 kJ/kg, and the pulse combustion dryer unit heat consumption is only 2,900–3,500 kJ/ kg.

(3) High dry product quality. Although the drying medium temperature is up to 900 °C, the material has a very short drying time, which is usually not more than 1 s. Therefore, the material temperature is generally not more than 50 °C, and this feature is especially suitable for the treatment of heat-sensitive materials and to protect the quality of dried. But the traditional flash dryer material retention time is about 10–60 min. Therefore, the traditional high-temperature dryer should not be used for efficient and rapid drying of thermal sensitive material.

3. **Application of pulse combustion drying technology**
Pulse combustion technology uses high-temperature exhaust gas flow produced by pulsating combustion with strong oscillation characteristics to dry material. The experimental study of pulsation combustion drying lignite adopts 205 kW valveless pulse burner with propane as fuel. The exhaust temperature is 370–790 °C and the operating frequency is 15 Hz. The water content of lignite can be reduced from 35% to below 10%. The productivity is 2×104 kg/h, also has drying and conveying function [68]. The power of the pulse combustor in an experimental study on drying corn was 88 kW. Because the temperature of the drying medium has certain requirements in grain drying, in the drying system, the secondary air flow rate is increased in order to reduce the temperature of the drying medium. In addition, a pulse combustor with a mixture of propane, residual oil and mixture of 3% residual oil and lignite as fuel is used for lignite drying [69]. In the comparative experiment of pulsating combustion drying and hot-air drying of fruits and vegetables, the dried material was Huqib, the air temperature was 200–300 °C, and the material was dried for 10 minutes. Using pulse combustion drying technology on the material for 10 min, the weight loss rate is 63%, while the weight loss rate of hot air drying material is only 15%, down to the same weight loss rate needs 35 min [70]. In 1989, an American gas company developed pulse combustion drying equipment, which was mainly used for drying paste and liquid material; material stay in the heating zone is about 0.01 s short. In the mixing chamber, the pulsed hot gas flow disperses the liquid and slurry materials and atomizes them into fine droplets, thereby increasing the evaporation surface area of the material and increasing the drying rate. The dryer exhaust temperature is 93–104 °C and the product temperature is only 38–49 °C, especially suitable for drying thermal sensitive materials. The dehydrating rate of a drying system is 1,330 kg/h, the unit heat consumption is only 2,800 kJ/kg and dry materials include yeast, caseinate, coffee mate and so on [71]. In 1992, the American production of a Unison pulse combustion dryer, used for animal feed additives, achieved success. The machine does not require a nozzle and a high-pressure pump, and high-speed air flow atomization is produced by a pulse combustor. It not only saves maintenance costs, but also improves the quality of dry products.

Pulse combustion drying technology can make short the residence time of the material in the dryer, the temperature is low and it is conducive for the protection of the quality of products. But the combustion mechanism is complex. The high-temperature oscillation airflow generated by the pulsating combustion can accelerate the drying process, and the sound wave generated by the pulse can be crushed and dried; there is no quantitative description of the mathematical model, which is only partly qualitative. Therefore, it is very difficult to guarantee the best working condition of the drying system and the working stability of the drier, and it is necessary to strengthen the research on the optimum condition of the drying system and the working stability of the dryer.

13.6.4 Supergravity Intensification Technology

1. Introduction

Supergravity is the force that material is subjected to in an environment of gravitational acceleration much larger than the earth. The supergravity separation technology, which originated from NASA's space experiments in 1979, is a typical chemical process intensification technology. In the earth, the simple way to realize the supergravity environment is to simulate the realization by rotating the centrifugal force field. Such a rotating device is called hige device or rotating packed bed (RPB).

According to the theory of the supergravity technology, the molecular diffusion and interphase mass transfer processes of different size molecules are much faster than those of the conventional gravity field under the supergravity environment. In the supergravity environment, different materials flow contact in complex flow channels. The strong shearing force will break the liquid phase material into tiny film, filaments and drop, and will produce huge and fast update phase interface. The interphase mass transfer rate is one to three orders of magnitude higher than that of the traditional tower, and the molecular mixing and mass transfer processes are highly enhanced. At the same time, the gas line speed can also be greatly improved, which increased the unit volume of equipment production efficiency by one to two orders of magnitude; the volume of equipment can be significantly reduced [72].

2. Characteristics of supergravity engineering
(1) Greatly intensify the transfer process (the height of the mass transfer unit is only 1–3 cm).
(2) Greatly reduce the size and mass of equipment (it not only reduces the investment, but also improves the environment).
(3) The residence time of the material in the equipment is very short (10–100 ms).

(4) The pressure drop of gas through equipment is similar to that in the traditional equipment.
(5) Easy to operate, easy to start and stop – from the start to enter the steady-state operation needs a very short time (within 1 min).
(6) The degree of convenience for operation and maintenance can be compared with the centrifuge or centrifugal fan.
(7) Can be installed in any direction, not afraid of vibration and bumps. It can be installed in the moving objects (such as ship and aircraft) and offshore platform.
(8) Fast and uniform micromixing.

3. **Application of supergravity technology**
China has the leading position in the world in the research on the supergravity technology. In 1994, Chen Jianfeng [73], Beijing University of Chemical Technology, found that the phenomenon of micromolecular mixing enhanced one hundred times under high-gravity environment. Accordingly, break through the limitations of high-gravity technology in the field of international limited separation, the new idea and the new technology of the supergravity-enhanced molecular mixing and reactive crystallization process are presented. Subsequently, the successful industrialization was developed, and eight industrial production lines for preparing nanoparticles using the high-gravity method. The nanometer calcium carbonate (average particle 30 nm in diameter) production line capacity reached 10,000 tons/year; the products are exported to Europe, the United States and other countries. This progress has been mentioned in the international review as "an important milestone in the history of the development of solid synthesis." In 1998, Zheng Chong, Chen Jianfeng and Guo Kai, Beijing University of Chemical Technology, were the first to realize the commercial application of the super deoxidize technology by gravity water. The high-gravity machine having water capacity of 250 t/h was successfully used in No. 2 offshore platform in the Shengli oil field in Chengdao. In 1999, the Dow Chemical Company of the United States, under the technical cooperation of Beijing University of Chemical Technology, applied supergravity technology to the industrial production process of hypochlorite. The coupling of the operation of a plurality of units, such as coupling absorption, reaction and separation, is carried out in a high-gravity machine. The yield of chlorine acid was increased by 10% or more, the chlorine cycle was decreased by 50%, the volume of equipment was reduced by more than 70% and the remarkable intensification effect was achieved [72]. Chen Jianfeng group also made supergravity technology successfully applied in the production process of MDI. After technical transformation, the production capacity increased from 1.6×10^5 tons/year to 3×10^5 tons/year, the process increased the energy saving by 30%, product impurities decreased significantly and MDI total capacity reached 9×10^5 tons/year after the application of technology promotion. High-gravity technology has also been used for product preparation or production of nanomedicine (5,000 tons/year),

nanodispersions, butyl rubber, and oxidation process in industrial exhaust gas SO_2 removal, wastewater in the volatile organic matter removal, refinery acid gas removal of H_2S, boiler feed water deoxygenation, fermentation, biodegradable polymer and other industrial process.

4. **Prospect of supergravity technology**

   After decades of development, it has been proved that the supergravity technology is a very promising and competitive process intensification technology. It has the characteristics of miniaturization, high efficiency, energy saving, high quality of products and easy to enlarge. It is in line with the developing trend of modern process industry to the resource-saving and environment-friendly model transformation. In the process of mass transfer and molecular mixing limitation and some special requirements of the industrial process (such as high viscosity, heat sensitive or expensive materials processing), the supergravity strengthening technology has outstanding advantages. It can be widely used in absorption, desorption, distillation, polymer, emulsion and other elements of the operation process and in the preparation of nanoparticles, sulfonated, polymerization and other reaction processes and reaction crystallization process [74, 75].

13.6.5 Mechanochemical Process

Mechanochemical process is a cross-discipline that specialty physical and chemical changes induced by high-energy mechanical forces involved: mechanical mechanics, inorganic chemistry, organic chemistry, surface chemistry, solid chemistry, structural chemistry, synthetic chemistry, mineral processing, materials science and other disciplines. At present, the research on mechanochemical process has become a hotspot in metallurgy, chemical industry, material, mineral processing, environmental protection and other high-tech fields. It is widely used in the preparation of micro- and nanopowders, nanocomposite materials, surface modification of powder, metal refining, mineral and waste treatment, and synthetic new phase fields.

1. **Introduction**

   The concept of mechanochemical was proposed by Peter in the early 1960s, which is defined as follows: "the effect of the mechanical forces of the material and the chemical changes or physical and chemical changes of the phenomenon." From the point of view of energy conversion, it is understood that the energy is converted into chemical energy. Mechanochemical technology uses mechanical energy to induce chemical reactions and changes in the structure and properties of materials, to prepare new materials or to modify the material.

2. **Mechanism of mechanochemical process**

   The effect of mechanical force on the properties of substances is a complex process; its energy supply and dissipation mechanism are not clear. There is no theory which can completely, quantitatively and reasonably explain the mechanical chemistry effect generated by numerous phenomena. Following theories are available on the mechanism of mechanochemical [76].

(1) Plasma model

The mechanically acting plasma model states that the mechanical forces cause lattice relaxation and structural cracking, stimulating high energy electrons and plasma regions. In general, when the temperature of a thermochemical reaction is higher than 1,000 °C, the electron energy is not more than 4 eV. Even the photochemical UV energy of the electron does not exceed 6 eV. But under the action of mechanical force, the electron energy induced by the plasma produced by the high excited state can be more than 10 eV. Therefore, it is possible to carry out the reaction which cannot be carried out by the thermochemical reaction, so that the reaction temperature of the solid substance is reduced and the reaction rate is accelerated.

(2) Solid-state synthesis reaction model

According to the diffusion theory, Xi Shengqi et al. analyzed the diffusion characteristics of a high-energy ball milling process and put forward the solid-state reaction model. The results show that the solid-state reaction can occur, depending on the degree of energy increase during high-energy ball milling. And whether the reaction can be performed is controlled by the diffusion process in the system, namely, subject to the grain refinement degree and the powder collision temperature. On the one hand, due to the strong plastic deformation of the particles in the ultrafine grinding process, it generated the stress and strain. Lattice defects and crystal transformation and noncrystallization within the particles significantly reduced the element diffusion activation energy; the group element at room temperature can significantly perform atom or ion diffusion. Particles continuously performed cold welding, fracture, microstructure refinement and the formation of numerous diffusion reaction pairs. On the other hand, due to the chemical bond breaking of the particle surface, the unsaturated bonds, free ions and electrons can be produced, which leads to the increase in the internal energy of the crystal, and the rapid development of the crack causes the increase in top temperature and pressure. The equilibrium constant and the reaction rate constant of the substance are obviously increased. Stress, strain, defects, and a large number of nanograin boundaries and phase boundaries create high energy storage in the system and increase the activity of the powder, which may cause solid-state reactions at nanometer sizes and sometimes even induce heterogeneous chemical reactions.

(3) Hotspot theory

The elastic stress caused by the mechanical force on the solid particles is an important factor of the mechanical force chemical effect. Elastic stress causes the stress concentration at the atomic level, which generally changes the binding constant between atoms. Thus, it not only changes their original frequency, but also changes the atomic spacing and valence bond angle. The result changed the chemical binding energy, which increased the reaction ability. The elastic stress can also cause relaxation, which can lead to the occurrence of chemical reactions. This energy in the stress point appears in "hotspots" form. Although the macro temperature is generally not more than 60 °C, the local collision point temperature is much higher than 60 °C. This temperature will cause the chemical reaction of the nano size, at the collision point, to produce high impact force (up to 3.30–6.18 GPa). Such high impact forces contribute to the diffusion of crystal defects and aberrations as well as the rearrangement of atoms, so the heating of the local collision points may be a contributing factor to the mechanical chemical reaction.

3. **Application of mechanochemical process**
   (1) **Synthesis of inorganic materials**

   (1) Preparation of nanocomposite materials. Mechanochemical method is a new method for preparing nanocomposite materials. Iwase et al. ball milled the Ti and Si_3N_4 system and prepared TiN-$TiSi_2$ composite nanopowder, and the nanoceramic composite materials were prepared, which have super-plastic behavior at high temperature. UzukiTS ball milled Al–Ti alloy with heptane as grinding aids, $TiAl$–Ti_4Al_2C nanocomposite powder was prepared by chemical reaction and the composite material was prepared by hot pressing and forming [77].

   (2) Preparation of nanoceramic materials. Calcium titanate ceramics are widely used in the world. But the sintering temperature is high, the variation in the sintering temperature is narrow and the grains grow up fast, so that they cannot be used in production. Therefore, it is difficult to prepare ultrafine $CaTiO_3$ by traditional solid-phase sintering method. Wu Qisheng et al. [78] mixed CaO and TiO_2 in a planetary mill under certain conditions. The results show that CaO and TiO_2 can be mechanically and chemically synthesized into $CaTiO_3$ nanocrystals with a particle size of 20–30 nm.

   (3) Preparation of alloy materials. An alloy material not only has the properties of its components, but also has mechanical and physical properties which its components do not possess. In recent years, the method of mechanical alloying (MA) by high-energy ball milling has attracted the attention of many researchers. The effect of alloying elements on the magnetic properties of α-Fe Nd-Fe-B/α-Fe dual-phase nanocrystalline alloys was studied by adding alloying elements [79]. Results show that the alloy made by the addition of Zr can significantly improve the material's thermal stability. In the range of 20–140 °C, temperature coefficient of remanence and temperature coefficient of coercive force are much lower than that of the conventionally sintered Nd-Fe-B.

(2) Synthesis of organic materials

The application of mechanochemical in the synthesis of organic polymers has three main aspects: polymer ploymerization, polymer condensation and surface polymer grafting of inorganic material [80].

(1) Polymer polymerization. Mechanochemical plays an important role in polymer polymerization. It not only can be used as a substitute for the initiator, but also, more importantly, can be used to modify the polymer. The structure and properties of the synthetic polymer have greater advantages compared with other methods, and the reaction can be carried out at room temperature without the need of warming up. Clepatravasiliua Oprea et al., without any initiator or catalyst in the vibration mill, used the single system of acrylonitrile to obtain the polymer. The results show that the mechanochemical synthesis of Pan shows good thermal stability, solvent stability and chemical stability.

(2) Polymer condensation. Under the mechanical force of the polymer, the bond can be broken to generate macromolecular free radicals. In this case, the macromolecule polycondensation can occur if suitable small molecules are encountered. Christofor Simionescu et al. made the mechanical chemical condensation polymerization of poly(ethylene terephthalate) and ethanediamine to obtain polyester polyamide fragment, and then obtained a complex of vanadium as the center with the action of V^{3+}. Studies have indicated that this composite has many excellent properties.

(3) Polymer grafting. The crystal lattice distortion and defects on the surface of inorganic materials are produced by mechanochemical action. An increase in surface free energy, caused by chemical bond breakage and restructuring, can be in the fresh fracture surface with unsaturated bond and positive and negative areas. The graft can be made of inorganic and polymer, and the purpose of modification can be achieved. In addition, the mechanical and chemical technology in metallurgy, environmental protection, food, medicine, feed and other industries and fields has a wide range of applications.

13.7 Chemical Process Intensification Equipment

With the development of chemical process integration, a lot of new and high-efficient chemical unit equipment has been developed in recent years. Due to the adoption of new fluid mixing technology, new equipment has the characteristics of special catalyst, special structure of the reactor and alternative energy technologies. The use of a suitable chemical process can greatly reduce the volume of the plant, save investment, simplify operation, reduce energy consumption, reduce environmental pollution, increase production capacity and strengthen the production process.

13.7.1 Static Mixing Reactor

1. Structure and performance of a static mixing reactor

Static mixer was a new type of equipment developed in the beginning of the 1970s. It is a kind of mixing mechanism with no moving parts, but with special properties. It relies on the special structure of the equipment and the movement of the fluid to separate the immiscible liquids and mix them with each other to achieve a good mixing effect. When the fluid passes through the static mixer, the mixing element causes the material flow to be on the left, sometimes to be on the right, and the direction of the flow is changed continuously. The liquid flow in the center is pushed to the periphery, and the surrounding fluid is pushed to the center. To obtain a good radial mixing effect, there is no dead zone in the tube, and no short circuit phenomenon occurs. At the same time, the fluid spinning will occur at the interface of the adjacent elements. Compared with traditional equipment, a static hybrid reactor has the advantages of simple process, compact structure, less investment cost, less energy consumption, large production capacity, low operation cost and easy to realize continuous mixing process. It is ideal equipment to solve the liquid–liquid, liquid–solid, liquid–gas, gas–air mixtures and emulsification, absorption, extraction, reaction and heat transfer enhancement. Figure 13.15 shows a schematic diagram of the structure of JSSK-type static mixer. The fluid is repeatedly turbulent and sheared during movement, and the mixing efficiency is very high.

Static mixer in the application mainly plays the following roles. 1. Direct application of the dispersion performance of static mixers, e.g., for mixing, emulsification, dissolution and other processes. 2. Disperses the passed fluid well, the specific area of the phase boundary increases, and the turbulence is intense, thereby strengthening the mass transfer and mass transfer controlled reaction processes, such as for extraction, absorption, liquid–liquid reaction, liquid–gas reaction process, and provides reaction conditions for the control of tubular reactors. Enhance the heat

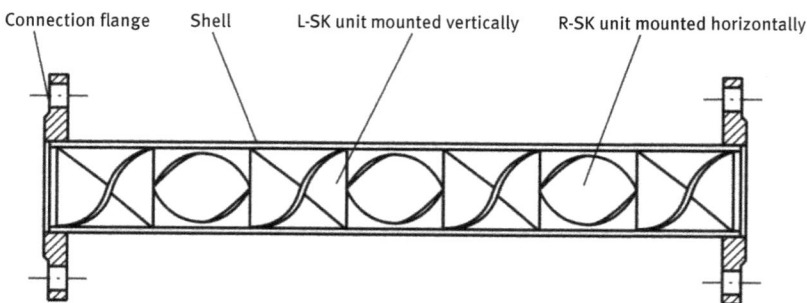

Figure 13.15: Schematic diagram of the structure of JSSK type static mixer.

transfer process, such as cooling or heating plastics, paints and viscous materials, guiding the plug flow in the polymerization reaction to control the reaction temperature and the like.

A major flaw in the static mixer is it is easy to plug, so it cannot be used in the case that needs to use slurry catalyst. Sulzer company overcomes this difficulty by using special filler which has a very good mixing performance and can be used as a catalyst carrier. Sulzer-type static mixing reactor with its mixing unit is composed of a heat transfer pipe. It can be used for mixing; a reaction needs to provide a large amount of heat, to produce a large amount of heat and to remove the processes, such as nitration reaction and neutralization reaction [81].

2. **Application of static mixing reactor**

(1) **Nitration of organic compounds**

TNT explosive is widely used in national defense and mining, which is produced by nitration of an organic compound [82]. The nitration reaction rate of the organic matter is very fast, it can be completed instantly and a lot of heat generated during the reaction is released at the same time. If the heat generated during the reaction cannot be removed from the system in time, it will cause an explosion. The traditional nitration reaction is generally carried out in a stirred tank reactor with cooling jacket. Because of the small heat exchange area, the heat transfer efficiency is very limited and had to control the reactant feed rate to avoid the runaway reaction caused by heat accumulation. Therefore, the volume of the reactor is large, and the reaction time is long. With an annual output of 15 tons nitro compounds as an example, the reactor volume is 13 m^3; each nitration reaction time is up to 18 h or more. Sulzer has successfully developed a reactor that uses heat exchange tubes as a static mixing component to enhance the mixing of materials. It can rapidly remove the reaction heat from the system while achieving efficient mixing of the materials. It can greatly reduce the volume of equipment and increase production capacity, and is especially suitable for a strong exothermic reaction process. The static mixing reactor technology is used in the production of TNT; the volume of the reactor is reduced to 200 ml, only 1/6,500 of conventional jacketed reactor volume. Nitration reaction time is only 0.25 s, which is 1/259,200 of original; the annual production capacity has increased by 2.2 times [82]. Simultaneously, due to the very short time of nitration reaction, the formation of by-products is basically eliminated. And the environmental pollution is reduced [82].

(2) **Ketone reduction reaction**

Ketone reduction reaction is carried out by Grignard reagent. The reaction is a part of the process of producing a fine chemical in German Merck company, which can be done in a few seconds. In August 1998, Merck established a fully automated continuous production pilot plant with five small mixers operating in parallel and successfully operated. The yield of pilot production was 92%, which was significantly higher than that in the actual batch production of 72%. In addition, the reaction time is shortened from the previous 5 h to less than 10 s. It is worth mentioning that the use of a small mixer can achieve the reaction at a higher temperature, thereby effectively reducing the technical investment of cooling equipment, and can save energy.

13.7.2 Monolithic Reactor

1. The structure and characteristics of a monolithic reactor

In recent years, more and more people have realized that the design of a new type of catalyst material is no longer the synthesis of a new compound, but the perfect structure design is made based on the active material. The active component is distributed on the structure of multiscale optimal design, which emphasizes the concept of multi-scale optimization design and the idea of multi-scale optimization design of reactor. Figure 13.16 shows some channel structures of a typical integral reactor metal carrier, which is applied to the reduction reaction of nitric oxide in the presence of oxygen. The load of the noble metal component is only in a narrow range of temperature, and the emission of exhaust gas has a wide range of temperature. In order to improve the performance of the catalyst, a structured catalyst with different activity and temperature range was designed [83, 84].

Integral fixed-bed reactor is a new device to replace the conventional particle packed bed reactor. Compared with the traditional packed bed, the overall reactor pressure drop is very small, usually one to two orders of magnitude lower than the traditional method. The geometrical area of the unit reactor volume is high, which is usually 1.5–4 times higher than that of the catalyst bed reactor. Because the coating is very thin, the diffusion path is very short; thus, it has high catalytic efficiency, the actual amount of up to 100%. It can be installed in the pipeline, such as the static mixer unit, which has the advantages of compact structure, low cost, easy maintenance and good safety.

Due to the isolation of the channel of the monolithic reactor and the lack of radial distribution, the only heat transfer path of the system is through the thermal conductivity of the monolithic reactor material; thus, one of its shortcomings is

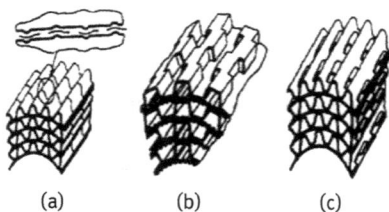

(a)   (b)   (c)

Figure 13.16: Some channel structures of the metal carrier in the catalytic reactor.

poor heat transfer effect. Modification of the heat exchanger by coating or introducing a catalytic active component can overcome the shortcomings mentioned above.

2. **Monolithic catalyst**

The monolithic catalyst is a catalyst, which is generally composed of a carrier, a coating and an active component. The carrier has the function of carrying the coating and the active component, and provides a suitable fluid channel for the catalytic reaction. Early overall carrier usually processed into straight or curved parallel regular channels, often honeycomb sintered oxide, the so-called honeycomb ceramics. There are two types of monolithic catalysts that have been studied so far, namely honeycomb ceramic carriers and honeycomb metal carriers monolith catalysts. For the automobile exhaust gas purifier, the common carrier has a uniform parallel channel, and the number of open holes in 1 cm^2 is 31–62. The geometric characteristic of a parallel, straight channel honeycomb catalyst is determined by the shape, size and wall thickness of the channel. These factors determine the catalyst bulk density, porosity, surface area and dynamic diameter. There is a limited radial mixing in the straight path, and there is almost no mass transfer between the adjacent channels. The active component of the catalyst is supported on the wall of the passage. The monolithic catalysts can be designed for all kinds of channels, depending on the synthesis requirements of the reaction, the cost of processing and the cost of operation. For example, foamed carrier and fiber braided carrier can be used under specific conditions [85].

The carrier materials that monolithic catalysts commonly used are various kinds of ceramics, metals and alloys. Cordierite ($2MgO \bullet 2Al_2O_3 \bullet 5SiO_2$) is the most widely used in the ceramic materials. For example, an automobile exhaust purification converter mostly uses a carrier made from this material. This carrier basically contains no pores, so there is no gas radial mixing and diffusion, and no flow of radial heat transfer. The radial thermal conductivity of the ceramic wall is small, the reactor is close to the adiabatic operation and it is advantageous to speed up the reaction. But it is not suitable for high reaction heat, and the higher volume ratio and the production cost of the ceramic carrier have limited its further application. Compared with the ceramic carrier, the metal carrier has the advantages of better thermal conductivity, easy processing and high mechanical strength, thus showing a more potential application prospect. Stainless steel or aluminum ferritic alloy is commonly used as metal carrier; especially high-temperature FeCrAl alloy is the most widely used. The hole density of the metal honeycomb carrier can reach 600 cells/in^2, in the same volume, which is 40% larger than that of the honeycomb ceramic carrier, and the quality is 45%.

3. **Application of a monolithic reactor**

Monolithic reactor is mainly used in waste gas treatment, catalytic combustion and catalytic distillation. Compared with the traditional pellet packed bed reactor, it can improve the mass transfer and heat transfer, it can increase the utilization of catalytic materials or it is more convenient to operate. In the early 1970s, honeycomb ceramic monolith catalyst has been successfully applied in the automobile exhaust gas catalytic converter; three-way catalyst is the core part of the catalytic converter. The honeycomb-shaped cordierite or metal is used as carrier, attaching a layer of high specific surface area of the thin Al_2O_3 coating on the surface, and loading the activity of PD, Pt or Rh in the precious metal components. Mei Hong et al. studied the transfer performance of the metal-based monolithic catalytic reactor, catalytic combustion and steam reforming of methane by numerical simulation. Subsequently, the coupling process of methane catalytic combustion and heat absorption of methane steam reforming in a self-made metal-based monolithic catalytic reactor was simulated. The simulation results show that monolithic catalytic reactor-based metal applied to the coupling of the absorption/exothermic reaction has a great potential for research and development. Many operating parameters, such as the ratio of reforming-side inlet gas velocity to combustion-side inlet gas velocity, gas inlet temperature, and the volume flow rate of methane in the combustion and reforming parts, are the important factors that affect the performance of the reactor. In addition, the performance of the segmented tubular metal base monolithic catalytic reactor is studied by numerical simulation, and some valuable results are obtained [83, 84].

13.7.3 Rotating Disk Reactor

The rotating disk reactor is developed for the extremely fast exothermic liquid reaction. The turntable rotates at a high speed of about 1,000 rev/min. The liquid on the turntable surface forms a film which is about 100 μm thick. The two phases have a high rate of heat transfer, and the heat transfer rate is very high. It has special advantages for some nitration, sulfonated and polymerization reaction. Traditional polymerization reactor structure is shown in Figure 13.17. The reactor has great limitation in heat transfer and mass transfer. The structure and working principle of the rotating disk reactor are shown in Figure 13.18. The rotating disk reactor is used to enhance the mixing and transfer properties, which is suitable for the dynamic control of the reaction.

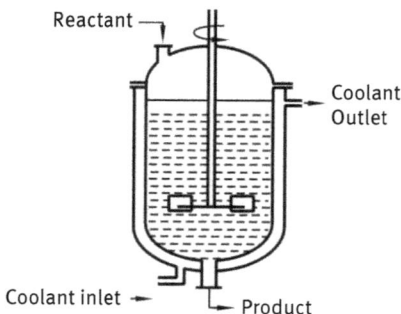

Figure 13.17: Traditional polymerization reactor structure.

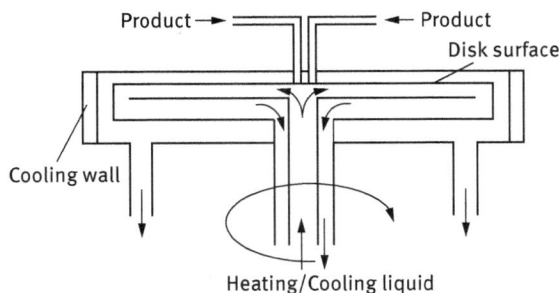

Figure 13.18: Structure and working principle of rotating disk reactor.

Irina et al. [86] found that the rotating disk reactor can absorb the incident light more effectively than the conventional ring reactor, and the average volume rate is also an order larger than that of the conventional ring reactor. When the mass transfer rate was increased obviously, the maximum surface reaction rate of the rotating disk reactor was two times of that of the traditional reactor, so the photocatalytic rotating disk reactor was a very promising process intensification technology. The research group of Chen Jianfeng, Zou Haikui and other subjects of Beijing University of Chemical Technology has systematically studied three aspects, after more than 20 years of basic theory, new technology and engineering application. The new idea of strengthening the control reaction process of the rotating packed bed reactor is presented, and the large-scale industrial applications are realized in the fields of bulk chemicals, carbon fiber and gas separation. A new method to enhance the control of the rotating packed bed reactor is proposed, which is "subject to the micromixing or transfer restriction." A new process technology is developed, in which a rotating packed bed reactor is used in the process of condensation, sulfonation, polymerization and oxidation. It has been successfully applied to the industrial production of three products with a total capacity of 1 million tons/year of polyurethane monomer MDI and 110,000 tons/year of caprolactam, resulting in remarkable energy saving, consumption reducing and yield increasing effects. For example, when used in MDI manufacturing, compared with the original reactor process, the acceleration of the condensation reaction process is 100%, the production capacity increased by 75%, the product impurity content decreased by 30% and the technology after the system integration unit product energy consumption decreased by more than 30%.

13.7.4 Oscillating Flow Reactor

Oscillating flow reactor (OFR) is a new type of process intensification equipment, which is used in liquid phase or multiphase flow reaction with liquid phase. It has good mixing and good heat transfer and mass transfer performance, and the transfer process is easy to control. Under the suitable condition of continuous operation, it has an ideal residence time distribution, which is close to the plug flow, and can provide better performance than the conventional tubular reactor and stirred tank reactor.

As a new type of chemical production equipment, the oscillating flow reactor is composed of an oscillating mechanism, an inlet section, a reaction section and an outlet section which are connected in turn. A feeding port is arranged at the inlet section, a discharging port is arranged at the outlet section and the reaction section is composed of a hollow cylinder which is vertically installed. The inner space of the cylinder is divided into a plurality of chamber structures by equidistant baffle plate. The outer surface of the cylinder can also be provided with a heat exchange jacket, and the heat transfer medium in the jacket sleeve is provided with a heating, cooling or heat preservation condition. Oscillation mechanism (piston, oscillating film or pulse oscillator) makes the liquid move back and forth. The vortex is formed by liquid through a series of baffles in the reactor. The formation and disappearance of the vortex intensified the mixing effect in the reactor and improve the transfer rate. At the same time, the flow state is maintained closed to the plug flow. The periodic motion can be controlled by the physical structure and operation parameters of the reactor, so that the residence time distribution of the reactor can be independent of the feed flow rate, such as baffle diameter, baffle spacing, oscillation frequency and amplitude. The structure and working principle of the tube oscillating flow reactor are shown in Figure 13.19.

The whole structure is divided into the starting vibration part and the reaction part. The oscillation device of the starting vibration part has a different structure, which is composed of an eccentric mechanism to generate a regular vibration which can be adjusted, and the vibration is transferred to the main body of the flow to generate oscillation. A variable-frequency motor, a driving plate, an eccentric plate, a top rod and a rubber diaphragm can be formed as an oscillation device. The vibration principle is that the frequency converter starts the motor, drives the driving plate and the eccentric plate to rotate, and the eccentric plate drives the top rod to move up and down. In this way, the push rod can push and pull the rubber diaphragm constantly, so as to cause the oscillation of the fluid. Reis et al. [87] proposed a novel, continuous oscillation screening for a meso reactor consisting of

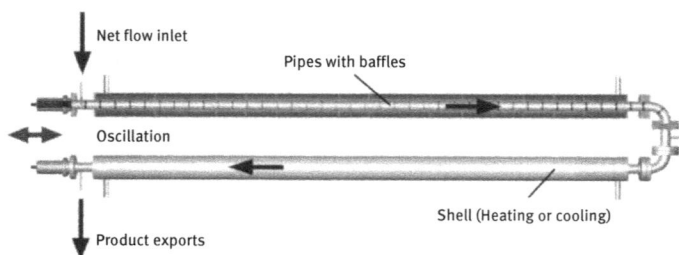

Figure 13.19: Schematic diagram of the basic structure of oscillating flow reactor.

tubes with smooth periodic constraint. It can provide a very good material mixing, as well as a series of different settling velocity of suspended catalyst. Therefore, it is particularly suitable for those involved in the process of solid catalyst screening. At the same time, this kind of reactor can be directly amplified and applied to industrial production.

Compared with the conventional tubular reactor, the main characteristics of the oscillating flow reactor are as follows. 1. The residence time distribution of the material in the reactor is very narrow, which is very close to the ideal plunger flow. 2. The flow state in the reactor is mainly determined by oscillatory flow, and is basically independent of net flow rate. Therefore, it can obtain good mixing, and good heat transfer and mass transfer performance at low speed under the lower net flow rate.

It is found that the oscillating flow reactor is characterized by a certain condition; the residence time of the fluid in the oscillating flow reactor is similar to that of the flat push flow, and has a longer residence time distribution. Therefore, if the oscillating flow reactor is applied to the chemical reaction, the conversion rate of the chemical reaction is very high, and the reaction component is mixed well, so that the reaction can reach the equilibrium quickly. The British Harvey [88] used oscillating flow reactor for the continuous production process of preparation of biodiesel from rapeseed oil and vinegar exchange. The experimental results show that the oscillatory flow reactor can enhance the reaction process, and it can achieve a better reaction result with shorter residence time on the basis of the realization of the continuous reaction.

Oscillating flow reactor is a kind of high-efficiency process intensification equipment. Although its application in industry is not much, targeted research is still in the stage of application basis and application research in laboratory. However, due to its wide applicability and convenient operation, it has broad application prospects in many chemical units' operation.

13.7.5 Impinging Stream Reactor

Impinging stream reactor is a new type of high-efficiency chemical reactor. Because the material inside the reactor is hitting itself, the fluid material is broken and the contact area of the interphase is enhanced, so that the heat transfer rate and the mass transfer rate increase accordingly. The concept of impinging stream was first proposed by Elperin, and then a series of basic and applied research works was carried out on the impinging stream by Elperin and Tamir et al. In the impact, the surface of the material is constantly updated because there is a decrease in friction, the mass transfer layer thickness and mass transfer resistance. Thus, the transfer coefficient increases and the heat transfer and mass transfer rate are enhanced. The chemical reaction which is influenced by the transfer process is promoted, and the production capacity of the chemical device is improved. Figure 13.20 shows a common submerged impinging stream reactor in industry.

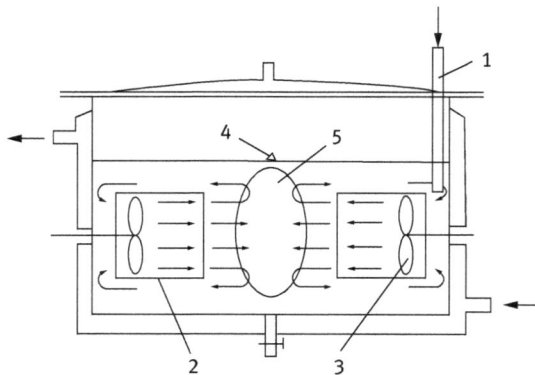

1. Feed pipe 2. Guide tube 3. Screw propeller 4. Flow-off 5. Impact zone

Figure 13.20: Schematic diagram of the structure of the submerged circulation impinging stream reactor.

1. **Basic principles and characteristics of impinging streams**

   After two equal amounts of gases fully accelerating the solid particles, gas–solid two-phase flow is formed. The two equal amounts of gas accelerate the solid particles to form a gas-solid two-phase flow. The two-phase flow flows coaxially and at high speed and collides with each other at the impact surface in the middle of the two accelerating tubes to form a high-speed turbulent flow zone with the highest particle concentration. This provides excellent conditions for enhanced heat and mass transfer. In a two-phase system such as gas–solid suspension with great density difference, the particles can penetrate into the reverse fluid from one stream of fluid because of the inertia, and the relative velocity reaches the maximum at the beginning of the reverse flow. After infiltration of the reverse flow, the particles slow down due to the friction resistance of the reverse flow. After reaching the zero speed, the air flow is accelerated to the impact surface, and then the air is infiltrated. Therefore, several times after the reciprocating motion of damped oscillation, the axial velocity of particles gradually disappeared, and finally brought out of the impact zone by the air radial flow after impact.

   The main characteristics of the transmission mode of an impinging stream are as follows. 1. The phase transfer can be greatly increased by the relative velocity between the particles and the reverse flow. 2. The penetration of particles in the opposite reciprocating air extended their residence time in the active zone. To some extent, the conditions of enhanced transmission can be extended to a certain extent. In the impinging stream of a gas–liquid system, the high phase velocity and particle collision promote the liquid phase surface renewal, reduce the liquid film resistance and increase the total mass transfer coefficient. Continuous opposite impact of the liquid–liquid phase and reciprocating oscillating motion of particles lead to a strong mixing of the impact zone, resulting in homogenization of temperature and composition. This is conducive to improving the average driving force to promote the transfer process [89].

2. **Application of impinging stream reactor**

   Due to its small size and simple experimental process, the submerged impinging stream reactor has been used in the preparation of nanomaterials and has achieved good results. Zhou Yuxin et al. [90] used the submerged impinging stream reactor as a reaction settling device to prepare "ultrafine" carbon-white by one-step method. In the submerged circulative impinging stream reactor, ultrafine carbon-white of 2.1 μm particle size was obtained, the particle size of the reaction product was stable and no change occurred in the subsequent treatment. Bao Chuanping et al. [91] used SCISR for semi-works production of ultrafine carbon-white, the results showed that the average particle size of the product is 2–3 μm and the specific surface area is as high as 322 m^2/g. Therefore, the reactor has excellent performance and can be produced in large amount. Yuan Jun et al. [92] studied the reaction–precipitation method of $NH_4H_2PO_4$ and $Ca(NO_3)_2$ in an ammonia solution to obtain nanohydroxyapatite. Under suitable operating conditions, the diameter is about 15 nm, the length is 50–70 nm and the shape is very regular. There are a lot of similar studies in the field of nanomaterials, such as strontium carbonate, titanium dioxide, barium titanate, zinc phosphate, all of which have achieved the ideal size of the product. Therefore, in the industry, its superior micromixing conditions make the saturation very uniform, and the circulation flow can inhibit the growth of the crystal and the passivation of the surface. This results in nanoscale products that are small in diameter and do not gel. Therefore, the impinging stream has a great potential in the practical production of nanometer grade products.

   In the field of chemical industry, the process of mass transfer and heat transfer cannot be found, and any kind of process in the chemical field can be realized by impinging stream. Compared with the traditional heat transfer and mass transfer method, the impinging stream is more efficient and consumes less energy. Impinging streams have been increasingly more used in practical production. However, the application of impinging stream in industrial practice is still relatively lacking. At present, more and more scholars and enterprises begin to pay attention to this problem.

13.7.6 Supergravity Reactor

The high-gravity reactor uses the centrifugal force generated by the rotation to accelerate the reaction/separation rate.

1. **The structure and principle of the supergravity reactor**

   The basic structure of the supergravity reactor is shown in Figure 13.21.

   The core component of the supergravity reactor is a high-speed rotating annular rotor. In the rotor, the surface channel of the gas–liquid phase is formed

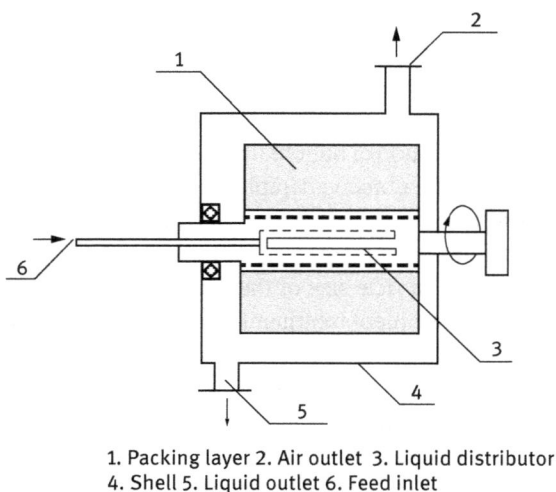

1. Packing layer 2. Air outlet 3. Liquid distributor
4. Shell 5. Liquid outlet 6. Feed inlet

Figure 13.21: Schematic diagram of the basic structure of the super gravity reactor.

by the tower plate or the packing material. The liquid from the stationary liquid distributor is introduced into the center of the rotor. After the predistribution by distributor, it sprayed to the inner rotor and thrown out under the action of centrifugal force. The gas arrived into the rotor from the rim, depending on the gas pressure, from the outside in contact with the liquid.

2. **Application of a supergravity reactor**

The supergravity reactor can be used in the preparation of various nanomaterials, which can make the reactants in the reaction system to realize the instant uniform micromixing. When the nanomaterials are prepared by the supergravity reactor, the reaction nucleation zone and the crystal growth zone are separated, so that the growth zone of the crystal is placed in a completely macroscopic mixing zone, the product component distribution is uniform and the range of grain size distribution is narrow. The preparation of the catalyst by using the supergravity coprecipitation method can significantly improve the dispersion of the active component of the catalyst. Huang Weili et al. [93] used a supergravity reactor for the production of a gas–liquid–solid phase catalyst. By discussing the synthesis mechanism of the $CuO/ZnO/Al_2O_3$ catalyst, the feasibility and existing problems of the material of the gas–liquid–solid three phase by using the supergravity rotating bed were studied, and the solution was put forward. Under the guidance of the transformation of the supergravity rotating bed, a large-scale supergravity machine with 50 t $CuO/ZnO/Al_2O_3$ annual output catalyst was constructed and used for the preparation of the catalyst. $CuO/ZnO/Al_2O_3$ catalysts are widely used in the process of methanol synthesis, low-temperature conversion of CO, methanol steam reforming and hydrogen production. Zhao Chenxi et al. [94] carried out an experimental study on the absorption of CO_2 in the mixed gas of sodium glycine in a supergravity reactor. The effects of rotating speed, temperature of absorption liquid, concentration of absorption liquid and gas/liquid ratio on CO_2 absorption rate were investigated. The results show that the absorption rate of CO_2 increases with the increase in rotating speed. When the speed reaches 1,000 rev/min, the absorption rate is basically stable. The absorption rate of CO_2 increases with the increase in temperature of the absorption liquid, which can reach more than 83% at 90 °C. In the experimental range, the absorption rate of CO_2 increases slightly with the increase in concentration of the absorption liquid, and decreases with the increase in the gas/liquid ratio and tends to be stable.

Questions

1. What is chemical process intensification? List an example to illustrate the method of chemical process intensification.
2. What is the reaction process coupling technology? Briefly introduce several common reaction coupling technologies, and their main characteristics and applications.
3. What is the separation coupling technology? Briefly introduce some common separation coupling techniques, and their main characteristics and applications.
4. What is microchemical technology? Briefly introduce the main application of microchemical technology.
5. Compared with the traditional production process, what are the main features of the use of microreactor in the chemical production process?
6. Briefly introduce several common applications of the use of energy fields in chemical process intensification.
7. Briefly introduce the mechanism and main applications of hydraulic cavitation.
8. Briefly introduce supercritical fluid technology and its application.
9. Briefly introduce the characteristics and main application of pulse combustion drying technology.
10. Briefly introduce the supergravity technology, and its main features and applications.
11. Briefly introduce the mechanism and main application of mechanical force chemical technology.
12. Briefly introduce the structure, main properties and application of static mixer.
13. Briefly introduce the integral catalyst composition and the application of the integral reactor.
14. Briefly introduce the main application of rotating disk reactor.
15. Briefly introduce the main application of oscillating flow reactor.
16. Briefly introduce the main application of impinging stream reactor.

References

[1]   Chen Longxiang, You Tao, Zhang Qingwen. Progress in membrane reactors and its application. Modern Chemical Industry, 2009, 29(4): 87–90.

[2]   Wang Fang, Cui Bo, Zheng Shiqing. Application and development of membrane catalytic technology. Journal of Qingdao Institute of Chemical Technology, 2002, 23(2): 18–22.

[3]   Roberts, S. L., Koval, C. A., Nob le, R. D., et al. Strategy for selection of composite membrane materials. Industrial & Engineering Chemistry Research, 2000, 39(6): 1673.

[4]   Tu Xueren. Development center of water treatment technology. Technology of Water Treatment, 1999, 25(1): 1–3.

[5]   Shi Yuehong, et al. Progress in preparation technology of inorganic membrane. Shanxi Chemical Industry, 1999, 19(4): 14.

[6]   Suzuki, H. Composite membrane having a surface layer of an ultrathin film of cage-shaped zeolite and processes for production thereof. US Pat: 4699892, 1987.

[7]   Xue Junbin. Preparation and performance of TS-1 zeolite catalytic film. Dalian: Dalian University of Technology, 2007.

[8]   Gu Xiujun. Study on Fe Al(PO4)2 catalysts for 1,2-dichloropropane to form propylene oxide in membrane reactor. Tianjin: Tianjin University, 2002.

[9]   Wang Maogong. Study on catalysts and membrane reactor of isobutane partial oxidation with carbon dioxide couple 'catalysis-separation' reaction. Tianjin: Tianjin University, 2007.

[10]   Gryaznov, V. M., Ermilova, M. M., Orekhova, N. V. Membrane catalyst systems for selectivity improvement in dehydrogenation and hydrogenation reactions. Catalysis Today, 2001, 67(1/2/3): 185–188.

[11]   Wang Longyao, Wang Lan. Applications of membrane technology in biotransformation. Chemical Industry and Engineering Progress, 2008, 27(6): 804–808.

[12]   Kwon, S. G., Park, S. W., Oh, D. K. Increase of xylitol productivity by cell recycle fermentation of Candida tropicalis using submerged membrane bioreactor. Journal of Bioscience and Bioengineering, 2006, 101(1): 13–18.

[13]   Ning Shangyong, Liu Miaomiao, Xu Zhiqiang, Li Shizhong. Continuous culture of Cordyceps sinensis in submerged membrane bioreactor. Chemical Industry and Engineering Progress, 2008, 27(8): 1269–1271.

[14]   Feng Dong-mei, Liu Xue-nuan. Research actualities and application of catalytic distillation technology. Chemical Industry and Engineering, 2004, 21(5): 388–391.

[15]   Qian Bozhang. Catalytic distillation technology and its application. Speciality Petrochemicals, 1990, 18(4): 3–10.

[16]   Wen Langyou, Min Enze, Pang Guici. Synthesis of cumene by suspension catalytic distillation process. Journal of Chemical Industry and Engineering (China), 2000, 51(1): 115–118.

[17]   Min Enze, Meng Xiangkun, Wen Langyou. New catalytic materials and chemical process intensification. Petroleum Processing And Petrochemicals, 2001, 32(9): 1–6.

[18]   Zhang Yongqiang, Min Enze, Yang Keyong. Impacts of process intensification on the future of chemical industry. Petroleum Processing and Petrochemicals, 2001, 32(6): 1–6.

[19]   Zeng Ping, Zhou Tao, Chen Guanqun. Research and application of magnetic fluidized bed. Chemical Industry and Engineering Progress, 2006, 25(4): 371–376.

[20]   Meng Xiangkun, Zong Baoning, Mu Xuhong, et al. Study on process of purification of caprolactam in magnetically stabilized bed reactor. Chemical Reaction Engineering and Technology, 2002, 18(1): 26–30.

[21]   Xu Xien, Mng Xiangkun. Advances in catalytic distillation processes. Advances in Chemical Engineering, 1998, 17(1): 7–12.

[22] Shen Long, Gao Ruichang. Research and application progress of membrane distillation. Chemical Industry and Engineering Progress, 2013, 33(2): 289–295.

[23] Song Weijie, Hang Xiaofeng, Wan Yinhua. Application of membrane technology in chemical wastewater treatment. Engineering Sciences, 2014, 16(12): 59–75.

[24] Zhou Ming, Xuchunjian, Yuguozong. New separation process of adsorption and distillation. Progress in Natural Science, 1995, 5(2): 147–152.

[25] Service, R. F. Miniaturization puts chemical plant where you want them. Science, 1998, 282: 400.

[26] Chen Guangwen, and Yuan Quan. Micro-chemical technology. CIESC Journal, 2003, 54(4): 427–439.

[27] Utada, A. S., Fernandez-Nieves, A., Stone, H. A., et al. Dripping to jetting transitions in coflowing liquid streams. Physical Review Letters, 2007, 99: 094502.

[28] Xu, J. H., Li, S. W., Tan, J., et al. Preparation of highly monodisperse droplet in a T-junction microfluidic device. AIChE Journal, 2006, 52: 3005–3010.

[29] Tan, J., Li, S. W., Wang, K., et al. Gas-liquid flow in T-junction microfluidic devices with a new perpendicular rupturing flow route. Chemical Engineering Journal, 2009, 146: 428–433.

[30] Kim, T. Hydrogen generation from sodium borohydride using microreactor for micro fuel cells. International Journal Hydrogen Energy, 2011, 36: 1404–1440.

[31] Luo Guangsheng, Lan Wenjie, Li Shaowei, et al. Research progress on preparation of functional materials by microfluidic devices. Petrochemical Technology, 2010, 39(1):1–6.

[32] Chen, G. W., Yuan, Q., Li, H. Q. et al. CO selective oxidation in a microchannel reactor for pem fuel cell. Chemical Engineering Journal, 2004, 101(1–3): 101–106.

[33] Shen, J. N., Zhao, Y. C., Chen, G. W., et al. Investigation and improvement of nitration processes of alcohols with mixed acid in a microreactor. Chinese Journal of Chemical Engineering, 2009, 17(3): 412–418.

[34] Guo, D. J. Novel synthesis of PtRu/multi-walled carbon nanotube catalyst via a microwave-assisted imidazolium ionic liquid method for methanol oxidation. Journal of Power Sources, 2010, 195(21): 7234–7237.

[35] Liu Xinhua, Liu Jinxin, Bai Linshan, et al. Novel dihydropyrazole derivatives linked with 4H-chromen: Microwave-promoted synthesis and antibacterial activity. Letters in Organic Chemistry, 2010, 7(6): 487–490.

[36] Gedye, R., Smith, F., Westaway K., et al. The use of microwave ovens for rapid organic synthesis. Tetrahedron Letters, 1986, 27(3): 279–282.

[37] Mallakpour, S., Dinari, M. A green route for synthesis of different polyureas based on phenylurazole: Rapid solid-state, microwave-assisted technique. High Performance Polymers, 2010, 22(3): 314–327.

[38] Zhao, Z. X., Li, Z., Xia, Q., et al. Fast synthesis of temperature-sensitive PNIPAAm hydrogels by microwave irradiation. European Polymer Journal, 2008, 44(4): 1217–1224.

[39] Li Xuefang. New application of microwave technique in hard tool materials industry. Cemented Carbide, 2000, 17(2): 196–199.

[40] Peng hu, Li Jun. Advances in microwave high temperature heating technology. Materials Review, 2005, 19(10): 100–103.

[41] Li Wei, Tao Keyi and Li Hexuan. Application of microwave technology in molecular sieve catalysts. Petrochemical Technology, 1999, 27(2):691–694.

[42] Jacqueline, M. R., Bélanger, J. R., Jocelyn Paré. Applications of microwave-assisted processes to environmental analysis. Analytical & Bioanalytical Chemistry, 2006, 386: 1049–1058.

[43] Werner Curt. Hospitals using new medical waste disposal methods to save money. Hospital Materials Management, 1993, 18(7): 10–13.

[44] Li Dezhan, Li Yinkui, Long Yongfu. Ultrasonic synthesis of two bromobenzene. Chemical Reagents, 2000, 22(2): 184–185.

[45] Mc Nulty, J., Steere, J. A., Wolf, S. The ultrasound promoted knoevenagel condensation of aromatic aldehydes. Tetrahedron Letters, 1998, 39: 8013–8016.

[46] Ai Z Q, Zhou Q L, Rong G, et al. Preparation and properties of polystyrene-g-poly(butyl acrylate) copolymer emulsions with ultrasonic radiation. I. Preparation technology and coagulum ratio[J]. Journal of Applied Polymer Science, 2010, 96(4):1405–1409.

[47] Bai Xiao-hui. Application of ultrasonic technique to sewage sludge and refractory wastewater treatment. Industrial Water Treatment, 2000, 20(12): 8–14.

[48] Luo Yanling, Zhao Zhenxing. Radiation cross-linking of polymers & its progress. PolymerBulletin, 1999, 12 (4): 88–99.

[49] Wu Minghong, Liu Ning, Xu Gang. Application of radiation technology in environmental protection. Progress in Chemistry, 2011, 23(7): 1547–1557.

[50] Jacobson, R. E., Rowell, R. M., Caufield, D. F., et al. United States based agricultural " waste products" as fillers in polypropylene homopolymer. F. S. Denes, S. Manolache/ Progress in Polymer Science, 2004, 29: 815–885, 878. In: Proceedings of Second Biomass Conference of the Americas: Energy, Environment, Agriculture, and Industry, August 21–24, 1995; Portland, OR, Golden, CO: National Renewable Energy Laboratory, 1995:1219.

[51] Denes, F., Young, R. Surface modification of polysaccharides under cold plasma conditions. In: Ed. Dumitriu S, editor. Polysaccharides, structural diversity and functional versatility. New York: Marcel Dekker, 1998: 1087

[52] Gogate, P. R., Pandit, A. B. Hydrodynamic cavitations reactors: A state of the art review. Divisions of Chemical Engineering, 2001, 17 (1): 1–85.

[53] Sivakumar, M., Pandit, A. B. Hydrodynamic cavitation assisted rhodamine B degradation: A technologically viable waste water treatment technique. International Conference on Science and Technology, New Delhi, India, 2001, 10: 12–13.

[54] Sivakumar, M., Pandit, A. B. Wastewater treatment: A novel energy efficient hydrodynamic cavitational technique. Ultrasonics Sonochemistry, 2002, 9(3): 123–131.

[55] Joshi, J. B., Pandit, A. B. Hydrolysis of fatty oils: Effect of cavitation. Chemical Engineering Science, 1993, 48(19): 3440–3442.

[56] Jyoti, K. K., Pandit, A. B. Hybrid cavitation methods for water disinfection: simultaneous use of chemicals with cavitation. Ultrasonics Sonochemistry, 2003, 10(4–5): 255–264.

[57] Jyoti, K. K., Pandit, A. B. Water disinfection by acoustic and hydrodynamic cavitation. Biochemical Engineering Journal, 2001, 7(3): 201–212.

[58] Chivate, M. M., Pandit, A. B. Effect of hydrodynamic and sonic cavitation on aqueous polymeric solutions. Indian Chemical Engineer, 1993, 35(l–2): 52.

[59] Danfoth, D. N. Effect of refining parameters on paper properties. Process International Conference New Technologies in Refining, 1986, 2(11): 9–11.

[60] Botha, C. J., Buckley, C. A. Disinfection of potable water: The role of hydrodynamic cavitation. Water Supply, 1995, 13(2): 219–229.

[61] Han buxing. Supercritical Fluid Science and Technology. Beijing: China Petrochemical Press, 2005.

[62] Brunner G. Applications of supercritical fluids. Annual Review of Chemical and Biomolecular Engineering, 2010, 1: 321–342.

[63] Herrero, M., Mendiola, J. A., Cifuentes, A., et al. Supercritical fluid extraction: Recent advances and applications. Journal of Chromatography A, 2010, 1217(16): 2495–2511.

[64] Zhang, J. L., Han, B. X. Supercritical CO2-continuous microemulsions and compressed CO2-expanded reverse microemulsions. The Journal of Supercritical Fluids, 2009, 47(3):531–536.

[65] Ramsey, E., Sun, Q., Zhang, Z., et al. Mini-review: Green sustainable processes using supercritical fluid carbon dioxide. Journal of Environmental Sciences, 2009, 21: 720–726.

[66] Keil, F. J. Modeling of Process Intensification. Weinheim: Wiley-VCH, 2007.

[67] Li Baoguo, Cao Chongwen, Liu Xiangdong. Pulsating combustion technology and its Application. Journal of China Agricultural University, 1998, 3(2): 36–40.

[68] Ellman, R. C., Belter, J. W., Dockter, L. Adapting a pulse-jet combustion system to entrained drying of lignite. Fifth International Coal Preparation Congress, Pittsburgh, PA. 1966: 463–476.

[69] Muller, J. L. The development of a resonant combustion heater for drying application. South African Mechanical Engineer, 1967, 16(7): 137–147.

[70] Tamburello, N. M., Hill, G. A. The development o f a vegetable and fruit dehydration with a pulse combustion Chamber. Symposium on Pulse-Combustion Applications Proceedings, Chicago, IL. 1982, 1: 1–7.

[71] Swientek R J. Pulse combustion burner dries food in 0.01 sec[J]. Food Processing, 1989, 7: 9–10.

[72] Chen Jianfeng. High-Gravity Technology and Its Application – A New Generation of Reaction and Separation Technology. Beijing: Chemical Industry Press, 2003.

[73] Chen Jianfeng, Zhou Xumei, Zheng Chong. Preparation of ultrafine particles. CN: 95105344.2. 1995.

[74] Zhao, H., Shao, L., Chen, J. F. High-gravity process intensification technology and application. Chemical Engineering Journal, 2010, 156: 588–593.

[75] Trent, D. Chemical processing in high-gravity fields. In: Stankiewicz, A., Moulijn, J. A., editors.. Re-engineering the Chemical Processing Plant: Process Intensification. New York: Marcel Deckker, 2004: 33–67.

[76] Shuai Ying, Zhang Shaoming, Lu Chengjie. Progress and prospect of mechanochemistry. New Technology & New Process, 2006, 28(11): 21–24.

[77] Chen Ding, Yan Hongge, Huang Peiyun. Development of mechanochemical process. Chinese Journal of Rare Metals, 2003, 27(2): 293–296.

[78] Wu Qisheng, Zhang Shaoming. Study on nano-crystalline $CaTiO_3$ synthesized by mechanochemistry. Journal of the Chinese Ceramic Society, 2001, 29(5): 479–483.

[79] Jurcay, K., Cook, M., Collocott, J. S. Application of high energy ball milling to the production of magnetic powders from NdFeB-type alloys. Alloys and Compounds, 1995, 217(1): 65–68.

[80] Wu Wei, Shao Lei, Lu Shou-ci. The main application of mechanochemistry in macromolecule composition. New Chemical Materials, 2000, 28(2): 10–13.

[81] Li, H. Z., Fasol, Ch, Choplin, L. Hydrodynamics and heat transfer of rheologically complex fluids in a sulzer SMX static mixer. Chemical Engineering Science, 1996, 51(10): 1947–1955.

[82] Zhang Shuang, Zhao Xun, Gao Huajing. The applications of static mixing reactor. Shandong Chemical Industry, 2015, 44(21): 79–81.

[83] Li Yongdan, Zhang Haijuan. Advances and perspectives in development of structured catalysts. Petrochemical Technology, 2005, 34(1): 1–5.

[84] Zhao Yang, Zheng Yafeng, Xin Feng. Properties and applications of monolithic catalysts. Chemical Reaction Engineering and Technology, 2004, 20(4): 357–361.

[85] Cybulski, A., Moulijn, A. Structured Catalysts and Reactors. New York: Marcel Dekker Inc., 1998.

[86] Irina, B., Stuart, N., Darrell, A. P. The case for the photocatalytic spinning disc reactor as a process intensification technology: Comparison to an annular reactor for the degradation of methylene blue. Chemical Engineering Journal, 2013, 225: 752–765.

[87] Reis, N., Harvey, A. P., Mackley, M. R. Fluid mechanics and design aspects of a novel oscillatory flow screening mesoreactor. Transl ChemE, Part A, Chemical Engineering Research & Design, 2005, 83(A4):357–371.

[88] Harvey, A. P., Mackley, M. R., Seliger, T. Process intensification of biodiesel production using a continuous oscillatory flow reactor. Journal of Chemical Technology and Biotechnology, 2003, 78: 338–341.

[89] Tamir, A. Principle and Application of Impinging Stream Reactor.Wu Yuan, translated. Beijing: Chemical Industry Press, 1996.

[90] Zhou Yuxin, Zhu Xuping, Wu Yuan. Preparation of ultrafine silica by impinging stream reaction-precipitation method. Chemical Engineering and Equipment, 2007, 36(4):1–4.

[91] Bao Chuanping, Liu Haizhou, Guo Jia, et al. Pilot experiment for preparation of Ultrafine white carbon black by precipitation in impinging stream reactor. Chemical Engineering, 2010, 38(12): 53–55, 92.

[92] Yuan J, Wu Y, Zheng Q. Synthesis of hydroxyapatite nano-rods/whiskers by precipitation in impinging streams. Orthopaedic Biomechanics Materials and Clinical Study, 2006, 3(5): 1–3.

[93] Huang Weili, Hou Jin, Wang Cilin, Zhang Bangliang. Preparation of catalyst with gas-li-quid-solid three-phase products using high gravity reactor. Chemical Industry and Engineering Progress, 2010, 29(5): 807–811.

[94] Zhao Chenxi, Zou Hai-kui, Chu Guang-wen. Study of carbon dioxide absorption by aqueous sodium glycine in a rotating packed bed. Journal of Chemical Engineering of Chinese Universities, 2015, 29(2): 438–441.

14 Green chemistry assessment and practice

Green chemistry is the frontier domains in the international chemistry, the development of the chemical industry and the research of science and technology in chemical engineering. Green chemistry is the technology aimed at supporting sustainable development of research by integrating social, economic and ecological factors in order to eliminate pollution and maximize utilization of natural resources while protecting the ecological environment. Therefore, it is a significant theoretical subject in the assessment of green properties of a chemical process to develop an efficient green technology in chemical industries. Although the professional organization of green chemistry has paid attention to the assessment of chemistry and chemical process since the twenty-first century in some countries, has not been formed standard specification for the assessment of green chemistry and chemical process. The reason is that the assessment of green chemistry is not only involved in chemistry, chemical engineering, environmental science and other subjects but also closely connected with biology, medicine, physics, materials science and information science. Therefore, according to the basic principles of green chemistry, the assessment methods of chemistry and chemical process are introduced and discussed in this chapter.

14.1 Basic Principles of Green Chemistry Assessment

14.1.1 Twelve Well-known Principles of Green Chemistry

The objective of green chemistry is to prevent waste formation rather than to treat it after it is formed by using chemical principles and new chemical technologies. In this regard, the famous 12 well-known principles of green chemistry have been proposed by P.T. Anastas and J.C. Warner, with the aim of providing guidance for the development of green chemistry and chemical processes. The 12 well-known principles involve the various aspects of the synthetic chemistry and the chemical processes, such as the greenization of raw material, process and products; the production cost; the consumption of energy; and the safety technique [1].

14.1.2 Twelve Additional Principles of Green Chemistry

In order to improve on the principles of green chemistry proposed by P.T. Anastas and J.C. Warner, N. Winterton put forward the 12 additional principles of green chemistry for developing and improving research methods in the laboratory and assessing the relatively greenization of the chemical process [2].

https://doi.org/10.1515/9783110479317-014

14.1.3 Twelve Principles of Green Chemical Engineering

The 12 well-known principles and 12 additional principles of green chemistry play an important role in providing guidance to researchers for the development of chemical reaction process. It is important to note the role played by chemical engineering technology in the practice of green chemistry. P. T. Anastas further proposed the 12 principles of green chemical engineering technology to guide the design of chemical engineering, and to develop new environmentally friendly green chemical engineering technology [3, 4]:

(1) As far as possible, energy inputs and outputs as well as all raw materials should be nontoxic and harmless.
(2) It is better to prevent waste formation than to treat it after it is formed.
(3) The design of separation and purification operations should ensure less consumption of energy and materials.
(4) Products, processes and systems should be designed to maximize quality, and energy, space and time efficiency.
(5) Product, process and the whole system should be the traction of "the output materials" rather than the propulsion of "the input materials and energy".
(6) It is beneficial to have a full understanding of the inherent complexity of the chemical process during the design of regeneration, recycling and other useful process.
(7) Chemical product durability rather than permanence should be emphasized.
(8) Design projects should focus on actual need to minimize the excess.
(9) Material diversity in complex products should be reduced, and the value of raw material should be preserved.
(10) It should overall consider the relevant conditions of the raw material and energy, and enhance the integration of material flow and energy flow.
(11) The design of the products, processes and the whole system should overall consider the treatment and recycling after their using functions ended.
(12) Materials and energies in the design should be renewable rather than disposable.

In view of the practical considerations involved in the science of chemical engineering in implementing green chemical technology, nine additional green chemical engineering principles, namely Sandestin principles, were put forward in the green chemical engineering conference in Florida, Sandestin [5].

(1) System analysis should be adopted in the design of product and engineering methods, in addition to integrating environmental impact assessment tools in the project.
(2) Protection and improvement of ecological systems should be considered when products and engineering methods are designed to protect the human health and social well-being.

(3) Life cycle thinking should be introduced in all chemical engineering activities.
(4) All energy and material inputs and outputs should be safe and environmentally friendly.
(5) The consumption of natural resources should be reduced as far as possible.
(6) Production of waste should be avoided.
(7) The chemical engineering developed and implemented should be accordance with the local actual situation and requirements. In addition it also should be approved by the local geographic and cultural circumstance.
(8) Improvements, innovation and invention of technology should be in accordance with the principle of sustainable development.
(9) The society and stakeholders should take an active part in the design and development of engineering solutions.

The above principles of green project are no longer confined in the green chemical engineering, but are expanded in the whole engineering field. Furthermore, the above principles also pay more attention to the harmony between man and nature as well as the security and sustainable development of society [6, 7].

14.2 Life Cycle Assessment

14.2.1 Meaning of Life Cycle Assessment

Life cycle assessment (LCA), which was developed in the 1970s, involves the use of methods that can evaluate the ecological environment impact associated with a product or an industrial process during its whole life cycle and can reduce these influences [8–10] from raw material acquisition through processing, product manufacture and use until disposal and finally waste treatment. In other words, life cycle is the whole process from cradle to grave. According to the International Standard Organization, LCA is a comprehensive assessment of a product's inputs, outputs and its potential environmental impacts, including resource utilization, human health and ecological consequences, throughout its life cycle. For example, the basic information of the chemical production LCA is shown in Figure 14.1, which can be regarded as the design of environmental consciousness.

LCA has been generally accepted worldwide as a preventative environmental protection and a new method of environmental management [11–12]. The applications of LCA include the assessment of the utilization of materials and energy, assessment of environmental impacts of a product or an industrial process, assessment of the emissions resulting from treatment options and assessment of alternatives for environmental improvement. Because LCA can give a whole quantitative assessment in a relatively wider time-scale, LCA is an important method for assessing the green chemistry and a tool of the industrial ecological design.

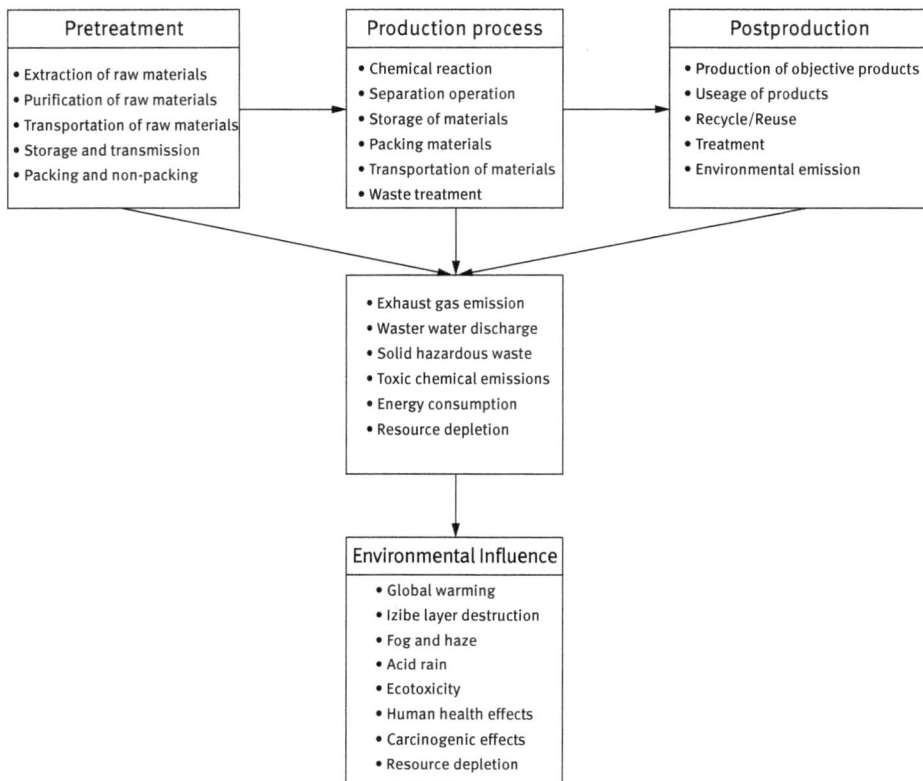

Pretreatment	Production process	Postproduction
• Extraction of raw materials • Purification of raw materials • Transportation of raw materials • Storage and transmission • Packing and non-packing	• Chemical reaction • Separation operation • Storage of materials • Packing materials • Transportation of materials • Waste treatment	• Production of objective products • Useage of products • Recycle/Reuse • Treatment • Environmental emission

• Exhaust gas emission
• Waster water discharge
• Solid hazardous waste
• Toxic chemical emissions
• Energy consumption
• Resource depletion

Environmental Influence

• Global warming
• Izibe layer destruction
• Fog and haze
• Acid rain
• Ecotoxicity
• Human health effects
• Carcinogenic effects
• Resource depletion

Figure 14.1: Life cycle assessment of chemical production.

14.2.2 Steps of LCA

LCA is an objective assessment process of the environmental impact caused by the products, production process and production activities. The main steps involved in the LCA process are as follows. First, the relevant information and data are collected; second, scientific calculation methods are established; third, the qualitative and quantitative assessment can be obtained in the aspects of the consumption of resources, the health of human and the ecological environmental impact; finally, the chance and the opportunity of improvement environmental performance should be pursued.

According to the standard of ISO14040, the implementation of LCA can be divided into four steps: (1) identification of the target and scope; (2) analysis of checklist; (3) assessment of the influence; and (4) explanation of the results. Figure 14.2 illustrates the relationship between the above steps.

1. **Identification of the target and scope**

   The target and scope of LCA should be clearly identified to comply with the expected application.

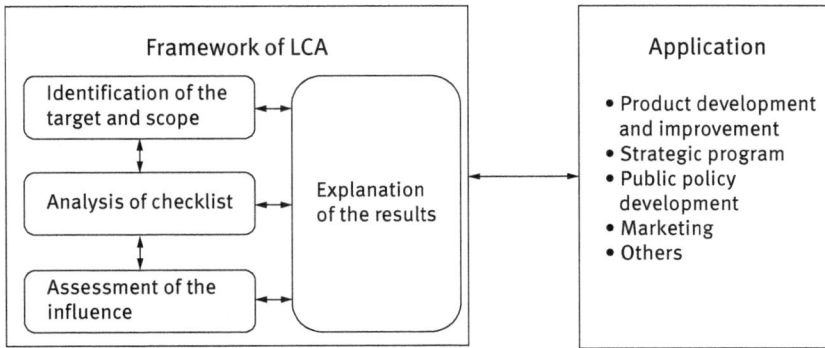

Figure 14.2: Framework of life cycle assessment.

2. **Analysis of checklist**

   The detailed lists of the input and output related with the production system should be compiled. This list includes the utilization of resources and the emissions of air, water and pollution. In addition, the list also contains the collection of data and calculation method used to obtain the relationship between the input and output.

3. **Assessment of the influence**

   The results of the analysis should be assessed for all potential significant environmental impacts of the product during its life cycle. Firstly, the results of the detail list analysis can be attributed to the various types of the environmental influence. Secondly, the characteristic of the different environmental influence type should be identification and quantification. Finally, the results of the analysis and judgment can be obtained. Generally speaking, it should be obtained the relationship between the detail list data and the specific environmental impact and realized the essence of the influences during assessment of the influence. Moreover, during the assessment process, the impacts on human health, ecological system and other aspects should be considered.

4. **Explanation of the results**

   A conclusion should be drawn from the comprehensive consideration of the results of the analysis and assessment of its impacts should reflect the research objectives. Moreover, improvement measures should be put forward to reduce the environmental impacts, which is the ultimate goal of LCA. As a result, the explanation of results can be regarded as the foundation that assists in the decision-making of the client.

14.2.3 Purposes of LCA

The aim of LCA is to look for appropriate pollution prevention technology in order to reduce environmental pollution and protect the ecological system as far as possible, while at the same time developing and utilizing resources efficiently, by saving nonrenewable resources and energy, and ensuring maximum recycling of raw materials and waste in order to promote the sustainable development of economy and society. Hence, LCA are mainly used in the following aspects:

(1) The assessment and quantization of the environmental problems associated with the use of the chemical product and the production process from cradle to grave are often regarded as management tools for evaluating the greenization of the chemical products and industrial process.

(2) The environmental influence assessment indexes should involve the environment evaluation index of the product and engineering, the evaluation index and the identification index of the environment mark of products.

(3) LCA can be regarded as one of the marketing economy strategies. And there are various LCA measures, such as the environmental statement, the environmental protection propaganda, the environmental label, and so on. Among the different LCA measures, the environmental label is a proved trademark of the product, which means not only the quality of the product is up to standard, but also the product meet the requirements of environmental protection in its life cycle.

14.3 Assessment of Green Chemistry and Chemical Process

Green chemistry is a sustainable development chemistry, meanwhile, green chemistry assessment is a research field of multidisciplinary cross area. How to determine whether the chemical process is green or not? The answer is as follows: first, considering the human health and safety, the assessment should examine whether toxic and harmful materials or products are produced or used in the process; second, with regard to environmental protection, it should examine whether any pollutants are being discharged into the surrounding environment; third, from the perspective of economic development, it should assess the quality of the product, energy and raw material utilization efficiency and the overall economic benefits. Therefore, assessment of green chemistry and chemical process is a very complicated system engineering [13–15].

14.3.1 Greenization of Chemical Reaction Process

Efficiently using chemical principles and methods are the core of the green chemistry. The green chemistry is the improvement production and technology process of reducing and eliminating the use of harmful substances and of producing environment-friendly chemicals. In other words, green chemistry is the prevention of pollution from the sources in order to implement the greenization of the chemical

Greenization of raw materials	Greenization of chemical reaction	Greenization of products
• Non-toxic harmless raw materials • Renewable raw materials	• Atom economy • High selectivity • High conversion • Low energy consumption	• Environmentally friendly products

Greenization of catalysts	Greenization of solvents
• Non-toxic harmless catalysts	• Non-toxic harmless solvents

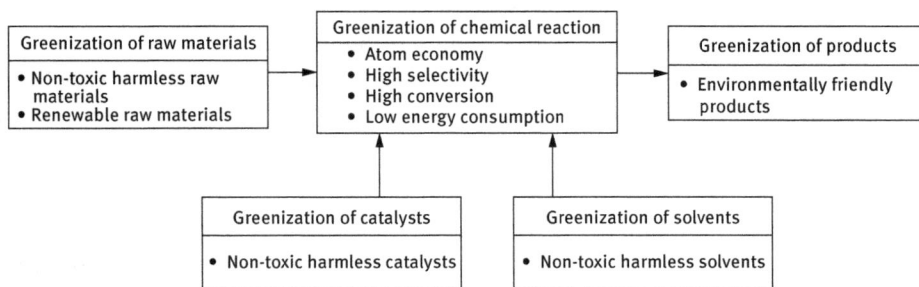

Figure 14.3: Diagram of green chemistry process [16–18].

industry. Hence, to realize green chemistry, the process should include the greenization of the raw materials, chemical reaction, synthesis technology, engineering technology and the products (Figure 14.3).

14.3.1.1 Greenization of Raw Materials
This involves:
(1) Using nontoxic and harmless raw materials as much as possible
(2) Using renewable resources as raw materials as much as possible
(3) Recycling utilization of materials and economy utilization of atomic as much as possible

14.3.1.2 Greenization of Chemical Reaction and Synthesis Technology
(1) It is better to use environmentally friendly solvents and additives as much as possible.
(2) It is better to develop high-selectivity and efficient new catalysts and catalytic technologies in order to simplify the process operation and improve the atom economy (AE) of the reaction.
(3) It is better to optimize the reaction pathways, develop green synthesis methods and enhance the coupling of green technology in order to improve the utilization of resources and energy.
(4) The key actions of greenization in the fine chemical enterprises are reforming the process (the operation) and implementing the clean production process, which can result a better economic benefits and social benefits.

14.3.1.3 Greenization of Engineering Technology
This involves:
(1) Developing new reaction engineering technologies, such as the biological engineering technology and membrane technology.

(2) Developing microchemical technology.
(3) Strengthening the coupling of the chemical technology.
(4) Strengthening the optimization integration of the material flow, the energy flow and the information flow

14.3.1.4 Greenization of Products

The greenization of products refers to using safe chemicals, also called green chemicals. Though there is no standard definition for green chemicals, from the point of LCA, green chemicals are the initial raw materials and should be renewable resources as far as possible, the products of which should be safe and healthy without causing harm to environment and health, and the used products should be recyclable or degradable to nontoxic material. As a result, green chemical products can be obtained from the advanced production technology and can be used in a safety and security condition in its whole life cycle. Furthermore, the green chemical products also meets the contemporary international recognized environmental protection standard.

The assessment system of green chemicals is composed of the fundamental property, the environment property, the resources property, the energy property, the economic property and other properties.

14.3.2 Measures of Greenization Chemistry and Chemical Process

For a long time, selectivity or the yield was used as the evaluation standard for measuring whether the chemical reaction process or the synthesis process was green or not. In fact, the concept of selectivity or yield is established on the basis of the maximum economic benefits without regard to the amount of waste involved; as a result, some processes with a high yield cause serious damage to the ecological environment. Then, it is obviously that yield cannot be regarded as the only evaluation index along with the development of the modern chemical industry. Green chemistry not only strives for maximum efficiency of chemical process but also aims to prevent pollution in the whole chemical process so as to be environment friendly. Therefore, the primary problem in the development and evaluation of chemical processes is to establish measures or evaluation indexes of greenization in chemical process.

14.3.2.1 Atom Economy

The concept of AE was proposed by organic chemistry professor B.M. Trost in Stanford University in 1991 [19]. The AE of a reaction refers to the maximal use of each atom of materials in organic synthesis, leading to zero emissions. AE can be represented as follows:

$$\text{Atom economy} = \frac{\text{molecular weight of target product}}{\text{sum of molecular weight of reaction materials}} \times 100\% \quad (14.1)$$

For the general synthesis reaction: $A + B \rightarrow C$

$$\text{Atom economy} = \frac{\text{molecular weight of } C}{\text{molecular weight of } (A + B)} \times 100\% \quad (14.2)$$

AE indicates how many reactants are converted into the final products [20]. If the conversion of the reactants is 100%, it can be said that the AE of the reaction is 100%. An ideal AE reaction is the reaction that does not produce protection groups and by-products. Therefore, the addition reaction, the molecular rearrangement reaction and other high-efficiency reactions are always green reactions. On the contrary, the AEs of the elimination reaction and substitution reaction are poor.

AE is one of the important principles of green chemistry, one of the major measures of the work guidance of the chemists and chemical engineers [21]. The AE of a reaction can be improved through reasonable designing of the organic synthesis methods and through quantitative analysis of the chemical process. In addition, a high AE reaction consumes less resource, less energy and is highly efficient. Hence, AE is regarded as a useful assessment index by chemical and chemical engineering industries. In practice, without considering the yield of products, excess reactants, usage of agents, loss of solvents and the consumption of energies, adopting AE as the index of the chemical reaction process is too simplified. Therefore, the AE should be combined with the other assessment indexes to make scientific conclusions.

14.3.2.2 Environmental Factor and Environmental Coefficient

Environmental factor (E-factor) is a standard metric proposed by organic chemistry professor R. A. Sheldon of Holland in 1992. According to the definition, E-factor can be illustrated as follows:

$$\text{E-factor} = \frac{\text{total amount of wastes (kg)}}{\text{products (kg)}} \quad (14.3)$$

In eq. (14.3), the wastes refer to the by-products. Moreover, greater E-factor implies more wastes, which has a negative impact on the environment. Therefore, the value of zero is an optimal value for E-factor.

Due to the complicated chemical reactions and process operations, E-factor is not only related to the reaction but also associated with other unit operations. Hence, E-factor must be calculated from the data of actual production process. In fact, E_{actual} can be expressed as follows:

$$E_{actual} = E_{theory} + E_1 + E_2 + E_3 + E_4 + E_5 + E_6 + E_7 \quad (14.4)$$

In eq. (14.4), E-factor must be calculated from the data of actual production process, at the same time, the chemical reaction and process operation is complicated, then the E-factor is not only related to the reaction but also associated with the other unit operation. As for E_1, most of the reaction is the chemical equilibrium reaction, which caused a less than 100% of the actual yield, as a result, E_1 can be expressed the contribution of the waste emissions. As for E_2, the excess amounts of cheap material is always been used in the reaction in order to make full use of the expensive reactant, as a result, E_2 can be expressed the contribution of the excessive part of the cheap reactant. As for E_3, the acid and alkali is added in the separation process to produce the inorganic waste, as a result, E_3 can be expressed the contribution of the inorganic waste. As for E_4, the group protection reagent is always added in the reaction and eliminated from the reaction due to the more reaction steps, as a result, E_4 can be expressed the contribution of the addition and the elimination of the group protection reagent. As for E_5, because of the different optical isomers, the useless and harmful isomer must be separated and abandoned even for one product reaction which is common in the pharmaceutical industry, as a result, E_5 can be expressed the contribution of the useless and harmful isomer. As for E_6, due to the limitation of the separation engineering technology, the separation process always is not a completely separation, as a result, E_6 can be expressed the contribution of the by-product in the environment from the incomplete separation process. As for E_7, solvent is usually used in the separation process, as a result, E_7 can be expressed the contribution of the incomplete recovered solvents.

The E_{theory} can be calculated from the AE or quality strength in the absence of the experimental data of E_1-E_7.

Strictly speaking, E-factor only considers the amount of waste rather than the quality of the waste; hence, E-factor is not a reasonable index for the assessment of environmental impacts. For example, the influence of 1 kg sodium chloride on the environment is not similar to that of 1 kg chromium salt. Therefore, a parameter, known as environmental quotient, was proposed by R.A. Sheldon, which equals the product of E-factor and the environmental-unfriendly factor Q listed as follows:

$$\text{Environmental Quotient} = E - \text{factor} \times Q \qquad (14.5)$$

In eq. (14.5), a less toxic inorganic intermediate, such as NaCl, can be assigned $Q = 1$. Accordingly organic intermediates and some fluoride compounds can be assigned $Q = 100-1,000$ depending on its toxicity and LD_{50} value. Sheldon believed Environmental Quotient and the related metrics to be important assessment indexes in green chemistry assessment.

14.3.2.3 Mass Intensity
In order to comprehensively assess the efficiency of organic synthesis reactions, mass intensity (MI) is proposed and can be expressed by eq. (14.6). In eq. (14.6), the total mass consumed in the processes or reactions includes all the raw or auxiliary

materials such as reactants, agents, solvents, catalysts and the consumption of acid, alkali, salt and organic solvents in the process of extraction, crystallization and washing. Because of the harmless essence of water to the environment, of special note is that the total mass does not contain the mass of water:

$$\text{Mass Intensity (MI)} = \frac{\text{total mass of consumed in the process or reaction (kg)}}{\text{mass of the product (kg)}} \quad (14.6)$$

The concept of MI considers the yield, stoichiometry, solvents and agents used in the reaction mixtures, and also the excess problem of reactants. In the ideal condition, MI should be close to 1. Generally, the lower the MI the better, since it results in a lower production cost, lower energy consumption and the reaction has less negative influence on the environment. Hence, for synthetic chemists and business leaders and managers, MI is a useful assessment index for evaluating the synthesis processes or the chemical production processes.

According to the definition of MI, the relation between the E-factor and MI can be expressed as follows:

$$E\text{-factor} = \text{MI--1} \quad (14.7)$$

Some useful metrics of green chemistry can be derived from the MI concept:
(1) Mass productivity (MP):

$$\text{Mass Productivity} = \frac{1}{\text{MI}} \times 100\% = \frac{\text{mass of the product (kg)}}{\text{total mass of consumed in the process or reaction (kg)}}$$

$$(14.8)$$

(2) Reaction mass efficiency (RME):

$$\text{RME} = \frac{\text{mass of product (kg)}}{\text{mass of reactant (kg)}} \times 100\% \quad (14.9)$$

(3) Carbon efficiency:
Reaction efficiency can be indicated by the conversion of carbon atoms because the organic compounds always contain carbon atoms, which is also called carbon efficiency (CE). CE can be expressed as follows:

$$\text{CE} = \frac{\text{moles of product} \times \text{number or atom/atoms in product}}{\text{moles of reactant} \times \text{number of atom/atoms in reactant}} \times 100\% \quad (14.10)$$

Example: sulfonic acid ester (23.6 g, 0.09 mol, FW 262.29) can be obtained from the reaction of benzyl alcohol (10.81 g, 0.1 mol, FW 108.1) and p-toluenesulfonyl chloride (21.9 g, 0.115 mol, FW 190.65) in the solvent mixture containing toluene (500 g) and triethylamine (15 g). The yield of sulfonic acid ester is 90%.

Then:

$$AE = \frac{262.29}{108.1 + 190.65} \times 100\% = 87.8\%$$

$$CE = \frac{0.09 \times 14}{0.1 \times 7 + 0.115 \times 7} \times 100\% = 83.7\%$$

$$RME = \frac{23.6}{10.81 + 21.9} \times 100\% = 70.9\%$$

$$MI = \frac{10.81 + 21.9 + 500 + 15}{23.6} = 23.2 \, g/g = 23.2 \, kg/kg$$

$$MP = \frac{1}{MI} \times 100\% = 4.3\%$$

From the results, AE is less than 100% due to the formation of the by-product HCl, CE is less than 100%, RME is 70.9% due to the excess of reactants, and the yield of product is 90.0%.

14.3.3 Assessment of Green Chemistry and Chemical Process

14.3.3.1 Assessment System of Green Chemistry and Chemical Process
According to the requirements of sustainable development, the basic principles of green chemistry and engineering technology proposed by P.T. Anastas and J.C. Warner have provided guidance and basic standards for the greening of chemistry and chemical process. From the previous research, the evaluation index should be a complete evaluation system, which involves not only the green chemical technology and green chemical engineering technology but also the production cost relationship and environmental factors, and so on [22–25]. Part of the assessment indexes of greening are shown in table 14.1.
(1) Mass assessment index
    Including AE, %; MI, total mass /product mass, kg/kg; additional solvent intensity, solvent mass/product mass, kg/kg; wastewater intensity, wastewater mass/ product mass, kg/kg; RME, %; quality of product, %purity.
(2) Energy assessment index
    Consumption energy of heating, MJ/kg (product); consumption energy of cooling, MJ/kg (product); electrical energy of process, MJ/kg (product); circular dissipation of cooling, MJ/kg (product).
(3) Pollutant assessment index
    For example, persistent toxic materials and biological cumulative toxic materials; kg/kg(product); greenhouse gas, MJ/kg (product).
(4) Safety factor
    This includes generation of thermal pollution, dangerous chemicals, pressure damage and harmful by-product.

Table 14.1: Part of the assessment indexes of greening.

Category	Unit
Mass	
Total mass/mass of product	kg/kg
Total mass/product mass	kg/kg
Mass of total solvents/mass of product	%
Mass of single product /mass of total reactants × 100%	%
Mole mass of product/mole mass of total reactants	%
Energy	
Total energy/mass of product	MJ/kg
Energy of solvent recovery/mass of product	MJ/kg
Poison and pollutant emission	
Total mass of long-time existence and biological accumulation/mass of product	kg/kg
Toxicity	
Total mass of long-time existence and biological accumulation/EC50 in raw material	kg
Human health	
Total mass of raw materials/ACGIH	Kg/(μg/g)
POCP	
Mass of solvent × POCP × vapor pressure/(mass of product × vapor pressure of methylbenzene × POCP of methylbenzene)	kg/kg
Greenhouse gas emissions	
Total mass of greenhouse gas emission from energy consumption (kg equivalent of CO_2)/mass of product	kg/kg (according to CO_2)
Security	
Thermal damage	Remarkable
Reagent damage	Remarkable
Pressure (high/low)	Remarkable
Toxic by-product	Remarkable
Solvent	
Number of solvent	Number
Overall recovery efficiency estimation	%
Required energy form solvent recovery	MJ/kg
Net mass intensity	kg/kg

Note: EC50, half fatality rate; ACGIH, American Conference of Government Industrial Hygienists; POCP, Photochemical Ozone Creation Potential.

14.3.3.2 Cost Relationship

It is certainly not a comprehensive assessment criterion if only considering the problem based on mass relationship, and also take into account the influence of raw material cost when discuss the assessment criteria in chemistry and chemical reaction process. For example, a low AE assessment index of the reaction process implies first that all the reactant molecules are not converted into the target product, which means the raw materials and energy have not been efficiently utilized; second, it means that the synthesis technology is complicated, that the technology steps are much more and the technology process is longer; third, it means the by-products, unconverted reactants, agents and solvents should be removed by purification and separation processes. The above three problems certainly increase the amount of work and cost involved in the management of raw materials, the treatment of wastes and treating the environment. D.J.C. Constable and A.D. Curzons discussed the relationship between the AE and the production costs through four kinds of pharmaceutical synthesis reactions [26].

Cost minimization model 1: minimum stoichiometry of process + standard production yield, stoichiometric reactant and solvent. In this model, all reactants and agents should be added in the reaction according to the stoichiometric ratio. The other production costs follow the standard measurement obtained from the data of practical application and calculation.

Cost minimization model 2: 100% AE of reaction + standard production yield, solvent and stoichiometry of the process. In this model, all of the reactants should be converted into the products, and the AE is 100%. The other production costs follow the standard measurement obtained from the data of practical application and calculation.

Cost minimization model 3: production yield of 100% + standard solvent and stoichiometry of the process. In this model, the production cost is based on the reactants, but, at the same time, the amount of chemicals and solvent added is the standard amount.

Cost minimization model 4: 100% recycling of solvent + standard yield and stoichiometry of the process. In this model, all various solvents should be recycled. The other production costs follow the standard measurement obtained from the data of practical application and calculation.

Cost minimization model 5: 100% AE of reaction, stoichiometry of the process and recycling of solvent. In this model, all of the reactants should be converted into the products, and the AE is 100%. At the same time, chemicals and solvents should be added in the reaction according to the stoichiometric ratio, and all various solvents should be recycled.

Cost minimization model 6: production yield of 100%, recycling of solvent and standard stoichiometry of the process. In this model, the production cost is based on the production yield of 100% and the recycling of solvents of 100%. The other production costs follow the standard measurement obtained from the data of practical application and calculation.

Cost minimization model 7: production yield of 100%, recycling of solvent, stoichiometry of the reactants and process. In this model, it should be a theoretical minimization of production cost model. Furthermore, all reactants and processes

Table 14.2: Production cost models of four kinds of medicines.

Cost minimization model	Total production cost (%)			
	Medicine 1	Medicine 2	Medicine 3	Medicine 4
1	86	99	92	97
2	87	40	84	69
3	71	32	56	57
4	63	84	64	55
5	36	22	40	21
6	34	16	20	11
7	20	15	12	8

should be in a stoichiometric relationship, all various solvents should be recycled and the production yield of various steps should be 100%.

The total costs in Table 14.2 include the various materials applied in pharmaceutical synthesis reaction.

From Table 14.2, the yield and stoichiometry are the important driving force for determining the production cost. In theory, minimization of cost can be achieved under the conditions of no excess reactant materials according to the stoichiometric ratio, cyclic utilization of solvent and catalyst and the yield of 100%. For the pharmaceutical synthesis reaction, using high-yield synthesis reaction, reducing excess reactants and improving the circulation and cyclic utilization of solvent are the effective ways to reduce the production cost and improve economic benefits.

14.3.3.3 Technical Factor

An ideal green chemical process should be environment friendly during its whole life cycle, and includes the green reactant material, the green chemical reaction and synthesis, the green engineering technology and the green products. Therefore, the green chemical process and green chemical engineering technology should be developed by the cooperation of synthetic chemists and chemical engineers for the green design of product, the simulation of processes, analysis of system, optimization of synthesis reaction, multifunction and miniaturization of equipment, system integration of high selectivity, high efficiency and advanced technology.

Generally, the synthetic chemists always pay attention to the conditions of synthesis, the mechanism of the reaction and the application of reagents. Sometimes, they neglect the related chemical engineering aspects of the reaction. When the synthesis reaction cannot be carried out normally, they consider changing the reaction conditions rather than researching the chemical reaction place (reaction equipment). The problems related to the transmission process of mass and energy, the process of mixing, phase transition and the design of reactor are normally the focus of chemical engineers rather than the synthetic chemists. In fact, many researches rely heavily on the cooperation between synthetic chemists and chemical engineers. Hence, the the green chemical engineering will be truly put into effect with the joint development of the green chemical process and the green chemical reaction engineering technology.

14.3.3.4 Case Analysis

Based on the above researches and discussions, the model of green technology guide proposed by A.D. Curzons and J.C. Constable can be an expert system in the assessment of green chemistry and chemical process, especially in the research and development of fine chemical engineering [27, 28].

In the synthesis of fine chemicals, the reaction of carbonyl compounds and organometallic reagents is a frequently encountered type of reaction. For example, consider the synthesis of fine chemical I:

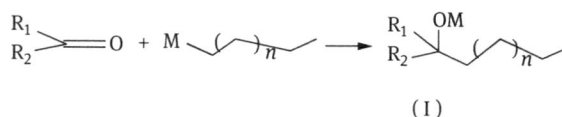

$$(\mathrm{I})$$

This reaction is carried out in the liquid phase. Meanwhile, the reaction is an exothermic reaction and the standard heat of reaction (ΔH^{θ}) is -300 kJ/mol. In addition, the main reaction and side reaction are carried out quickly, and the reaction residence time is less than 10 s. In the system, all of the materials are sensitive to the reaction temperature.

1. **Assessment of mass intensity**

   The reactors used in this reaction are the micro-reactor, lab-scale reactor, laboratory batch reactor (0.5 L flask) and batch reactor of industrial production (6,000-L reactor with stirring impeller). The characteristics of the above reactors are listed in Table 14.3, and the experiment results of MI are listed in Table 14.4.

   From Tables 14.3 and 14.4, in the microreactor and lab-scale reactor, due to high concentration gradient and temperature gradient, the transport velocities of mass and energy can be accelerated, which can result in a more uniform reaction condition, less side reactions and fewer by-products. Therefore, the conversion rate, selectivity and yield can be greatly improved by using micro-reactor technology.

Table 14.3: Characteristics of four kinds of reactors.

Reactor	Temperature (°C)	Residence time	Yield (%)	Area/volume (m²/m³)	Size of reactor
Bottle	−40	0.5 h	88	80	0.5 L
BR	−20	5 h	72	4	6,000 L
Microreactor	−10	<10 s	95	10,000	2 × 16 channels of $w \times h = 40 \times 220$ μm
Lab-scale reactor	−10	<10 s	92	4,000	capacity of 3×10^{-5} m³/s (30 mL/s)

Table 14.4: Comparisons of mass intensity.

Intensity	Bottle	BR	Microreactor	Lab-scale reactor	Theory value
Mass intensity (without solvent)	2.27	2.78	2.10	2.17	2.00
Additional solvent intensity	–	–	⌐	–	–
Additional water intensity	0	0	0	0	0
Residue intensity (without solvent)	0.27	0.78	0.10	0.17	0
Yield (%)	88	72	95	92	100

Table 14.5: Energy consumption of four different reactors.

Item	Bottle	BR	Microreactor	Lab-scale reactor
Cooling energy (MJ/mol product)	0.42	0.42	0.42	0.42
Electric energy (MJ/mol product)	0.167	0.107	0.080	0.080

Table 14.6: Amount of the pollutants (unit: g/mol product).

Pollutant	Microreactor	Lab-scale reactor	Bottle	BR
CO_2	13.4	13.4	17.9	27.8
CO	3.66×10^{-3}	3.66×10^{-3}	4.88×10^{-3}	7.58×10^{-3}
CH	4.72×10^{-2}	4.72×10^{-2}	6.28×10^{-2}	9.76×10^{-2}
VOC	3.09×10^{-3}	3.09×10^{-3}	4.12×10^{-3}	6.40×10^{-3}
NO_x	2.85×10^{-2}	2.85×10^{-2}	3.79×10^{-2}	5.89×10^{-2}
SO_x	3.99×10^{-2}	3.99×10^{-2}	5.31×10^{-2}	5.25×10^{-2}
COD	4.07×10^{-2}	4.07×10^{-2}	0	0
BOD5	1.63×10^{-3}	1.63×10^{-3}	0	0
TDS	5.43×10^{-2}	5.43×10^{-2}	5.42×10^{-2}	8.42×10^{-2}
Solid waste	6.01×10^{-1}	6.01×10^{-1}	2.17×10^{-3}	3.37×10^{-3}

2. **Assessment of energy consumption**

Because the reaction is exothermic, the system should be cooled during the reaction. As a result, the electrical energy required for cooling is the main energy consumed. The consumption of energy in different reactors is listed in Table 14.5. From Table 14.5, the energy needed to cool the system is basically the same due to the standard heat of reaction, while the consumption of electric energy is different. A smaller reactor involves less consumption of electric energy.

3. **Assessment of pollutant**

Table 14.6 lists the emissions of the pollutants during the whole life cycle of the process.

From Table 14.6, the amount of the main pollutants, such as the volatile organic compounds (VOC), hydrocarbons (CH), nitrogen oxide (NO_x), sulfur oxides (SO$_x$) and CO, discharged from the microreactor technology is less than that from the batch reactor technology. That is to say, when the microreactor is used in this synthetic reaction, the resources can be effectively utilized and the pollution caused can be reduced and eliminated from the source of the reaction process, which is beneficial to the environment.

4. **Safety assessment**

It has been found that a relatively small volume of reaction system and effective control of the reaction heat can improve the safety of the reaction process in a microreactor. In the application, though harmful materials may be produced, the risk involved can be minimized through reasonable design and control of the downstream conditions of the process.

5. **Micro chemical engineering technology**

With the development of fine chemical engineering technology, the micro chemical engineering technology is a small size of few microns (or slightly larger) of reaction equipment (production process) and a high volume production capacity. Using the microchemical engineering technology is beneficial in terms of monitoring the process, improving the reaction residence time, increasing the selectivity, yield and quality of the products, shortening the period of research and development, and quickening the development of new products and new process.

In conclusion, the comprehensive assessment results are shown in Table 14.7 using the green technology guide. The colors green, yellow and red stand for excellent, medium and inferior, respectively. Among them, green can meet the demand of sustainable development and is environment friendly.

Green technology guide regarded as the assessment system can well explain and evaluate the greening of the chemical reaction process and technology. In addition, green technology guide is very simple and clear and is easily mastered by users. However, the theoretical model of the green technology guide is too simple, moreover, the assessment of green chemical engineering process is mostly confined to the qualitative aspects rather than quantitative analysis, and, as a result, the green technology guide needs to be further complemented and developed.

In addition, based on the analysis of thermodynamics, energy efficiency can be regarded as the basic measure of quantitative sustainability. For more on quantitative sustainability one may refer to the literature [28].

Table 14.7: Comparison of green technologies in various reactors.

Technology	Environment	Safety	Efficiency	Energy
Microreactor	Green	Green	Yellow	Green
Lab-scale reactor	Green	Green	Green	Green
BR	Yellow	Yellow	Red	Yellow

14.4 Building Green Chemical Industry and Promoting Green Development

The so-called green chemical industry refers to the modern chemical industry aimed at developing and promoting concepts of recycling and green chemical technology based on the basic principles of green chemistry and ecological, industrial development patterns. In a modern chemical industry, firstly, its traditional model should transit to the optimization industrial structure; secondly, it should insist the development direction of "refinement, specialty, high-end oriented and greening"; thirdly, it should implement the clean production technology. In other words, we should build a resource-conserving, technologically innovative and environment-friendly modern chemical industry.

1. **Develop green chemical engineering with the concept of circular economy**

   Circulation economy, having high efficiency and recycling and reuse of resources as its central idea, and the concept of "reduction, reuse and recycle" as its principle, and the concept of "low consumption, low emission and high efficiency" as its basic feature, is a suitable model of economic development under the sustainable development ideology. In the technology combining green chemical engineering with the concept of circulation economy, resources can be effectively utilized in the most environmentally friendly way and utmost protection to the environment, and economic, social and ecological benefit can be obtained with minimum cost.

2. **Develop the chemical industry with green chemical technology**

   Green chemical process is a kind of technology with the characteristics of harmless to human health, safety to social and friendly to environment. Moreover, it is also a kind of sustainable development and the mainstream of industry technology.

3. **Implement clean production technology**

   The production process of the existing enterprises is assessed with green chemistry principles and green chemical engineering technology. In order to develop its green chemical technology, implementation of clean production technology can strengthen the technical innovation, improve the production process, save the whole energy and reduce the pollutant emissions.

4. **Strengthen the coupling symbiosis of chemical industry and related industries**

   Coupling symbiosis of chemical industry and the related industries have the advantages of the optimum distribution of resources, the comprehensive utilization of resources, and the promotion of industry intensification and industry clustering development.

5. **Ecoindustrial park**

   Ecoindustrial park is an industrial park set up for the development of green chemical engineering practices and the establishment of ecological industry.

Ecoindustrial park was proposed based on the idea of circular economy development, the basic principle of industrial ecology, to meet the demand of clean production technology. And ecoindustrial park is a kind of construction mode of a positive cycle of production, utilization of resources and environmental protection.

Questions

1. Describe the significance of green chemistry assessment briefly.
2. What are the basic principles of green chemistry assessment?
3. What are the major methods of green chemistry assessment?
4. What is LCA? What are the important applications of LCA?
5. Why do you say green chemistry assessment is a complicated system?
6. What contents do the greening chemical reaction processes have?
7. What is mass intensity? What are the important applications of the mass intensity index?
8. What principles and aspects should be considered in the comprehensive assessment of a green chemical engineering process?

References

[1] Anastas, P. T., Bartlett, L. B., Kirchoff, M. M., et al. The role of Catalysis in the design, development and implementation of green chemistry. Catalysis Today, 2000, 55: 11–22.
[2] Winterton, N. Twelve more green chemistry principles. Green Chemistry, 2001, 3: G73–G75.
[3] Anastas, P. T., Zimmerman, J. B. Design through the 12 principles of green engineering. Environmental Science & Technology 2003, 37(5): 94A–101A.
[4] Anastas, P. T., Heine, L., Williamson, T. C. Green engineering. Washington DC. American Chemical Society, 2000.
[5] Ritter, S. K. A green agenda for engineering. C&EN, 2003, 81(29): 30–32.
[6] Sue Hail. The greening of engineering. Green Chemistry, 1999, 1: G31–G33
[7] Bashkin, J., Rains, R., Stern, M. Taking green chemistry from laboratory to chemical plant. Green Chemistry, 1999, 1: G41–G43.
[8] Chen Hui, Wen Ya, Waters, M. D., et al. Design guidance for chemical processes using environmental and economic assessments. Industrial & Engineering Chemistry Research, 2002, 41: 4503–4513.
[9] Lankey, R. L., Anastas, P. T. Life-cycle approaches for assessing green chemistry technologies. Industrial & Engineering Chemistry Research, 2002, 41: 4498–4502.
[10] Herrchen, M., Klein, W. Use of the life-cycle assessment (LCA) toolbox for an environmental evaluation of production processes. Pure Applied Chemistry, 2000, 72(7): 1247–1252.
[11] Anastas, P. T., Lankey, R. L. Life cycle assessment and green chemistry. Green Chemistry, 2000, 2(6): 289–295.
[12] Domenech, X., Ayllon, J. A., Peral, J., et al. How green is a chemical reaction? Application of LCA to green chemistry. Environmental Science & Technology, 2002, 36: 5517–5520.
[13] Sheldon, R. A. Catalysis: The key to waste minimization. Journal of Chemical Technology and Biotechnology, 1997, 68: 381–388.

[14] Curzons, A. D., Constable, D. J. C., Mortimer, D. N., et al. So you think process is green, how do you know? –Using principles of sustainability to determine what is green – a corporate perspective. Green Chemistry, 2001, 3: 1–6.

[15] Constable, D. J. C., Curzons, A. D., Freitas, L. M., et al. Green chemistry measures for process research and development. Green Chemistry, 2001, 3: 7–9.

[16] Graedel, T. Green chemistry in an industrial ecology context. Green Chemistry, 1999, 1: G126–G128.

[17] Tsoka, C., Johns, W. R., Linke, P., et al., Towards sustainable and green chemical engineering: tools and technology requirements. Green Chemistry, 2004, 6: 401–406.

[18] Gong, C. S. Assessment of green chemical engineering process. Modern Chemical Engineering (China), 2005, 25(2): 67–69.

[19] Trost, B. M. The atom economy – a search for synthetic efficiency. Science, 1991, 254: 1471–1477.

[20] Rouhi, A. M. Atom economical reactions help chemists eliminate waste. C&EN. 1995 (19 Jun): 32–35.

[21] Sheldon, R. A. Organic synthesis–past, present and future. Chemistry & Industry, 1992, (7 Dec): 903–906.

[22] Greadel, T. E. Green chemistry as systems science. Pure Applied Chemistry 2001, 73(8): 1243–1246.

[23] Clark, J. H. Green chemistry: Challenges and opportunities. Green Chemistry, 1999, 1: 1–11.

[24] Dewulf, J., Van Langenhove, H., Mulder, J., et al. Illustrations towards quantifying the sustainability of technology. Green Chemistry, 2000, 2: 108–114.

[25] Gong, C. S., Shan, Z. X. Introduction of Green Fine Chemical Engineering. Beijing: Chemical Industry Press, 2005.

[26] Constable, D. J. C., Curzons, A. D., Cunningham, V. L. Metrics to 'green' chemistry – which are the best? Green Chemistry, 2002, 4: 521–527.

[27] Grimaldi, S., Couturier, J.-L. A new green tool for selective fine chemical oxidation reactions. Speciality Chemicals Magazine, 2006, 26(10): 32–33.

[28] Haswell, S. J., Watts, P. Green chemistry: Synthesis in microreactors. Green Chemistry, 2003, 5: 240–249.

Index

https://doi.org/10.1515/9783110479317-015

www.ingramcontent.com/pod-product-compliance
Lightning Source LLC
Chambersburg PA
CBHW060954210326
41598CB00031B/4825